ENCYCLOPEDIA OF ELECTROCHEMISTRY OF THE ELEMENTS

VOLUME I

ORGANIZATIONAL CHART

	Symbol	Element No.	Volume	Chapter
Actinium	Ac	24		
Aluminum	Al	21		
Americium	Am	24		
Antimony	Sb	11		
Argon	Ar	1	I	8
Arsenic	As	9		
Astatine	At	6	I	3
Barium	Ba	18	I	7
Berkelium	Bk	24		
Beryllium	Be	16		
Bismuth	Bi	12		
Boron	B	14		
Bromine	Br	5	I	2
Cadmium	Cd	20	I	4
Calcium	Ca	18	I	7
Californium	Cf	24		
Carbon	C			
Cerium	Ce	22		
Cesium	Cs	15		
Chlorine	Cl	5	I	1
Chromium	Cr	31		
Cobalt	Co	36		
Copper	Cu	38		
Curium	Cm	24		
Dysprosium	Dy	22		
Einsteinium	Es	24		
Erbium	Er	22		
Europium	Eu	22		
Fermium	Fm	24		
Fluorine	F	4		
Francium	Fr	15		
Gadolinium	Gd	22		
Gallium	Ga	14		
Germanium	Ge	13		
Gold	Au	40		
Hafnium	Hf	26		
Helium	He	1	I	8
Holmium	Ho	22		
Hydrogen	H	2		
Indium	In	14		
Iodine	I	6	I	3
Iridium	Ir	41		
Iron	Fe	37		
Krypton	Kr	1	I	8
Lanthanum	La	22		
Lawrencium	Lw	24		
Lead	Pb	28	I	5
Lithium	Li	15		
Lutetium	Lu	22		
Magnesium	Mg	17		
Manganese	Mn	34	I	6
Mendelevium	Md	24		

ORGANIZATIONAL CHART

	Symbol	Element No.	Volume	Chapter
Mercury	Hg	10		
Molybdenum	Mo	32		
Neodymium	Nd	22		
Neon	Ne	1	I	8
Neptunium	Np	24		
Nickel	Ni	35		
Niobium	Nb	30		
Nitrogen	N	8		
Nobelium	No	24		
Osmium	Os	41		
Oxygen	O	3		
Palladium	Pd	41		
Phosphorus	P	9		
Platinum	Pt	41		
Plutonium	Pu	24		
Polonium	Po	7		
Potassium	K	15		
Praseodymium	Pr	22		
Promethium	Pm	22		
Protactinium	Pa	24		
Radium	Ra	18	I	7
Radon	Rn	1	I	8
Rhenium	Re	42		
Rhodium	Rh	41		
Rubidium	Rb	15		
Ruthenium	Ru	41		
Samarium	Sm	22		
Scandium	Sc	22		
Selenium	Se	7		
Silicon	Si	13		
Silver	Ag	39		
Sodium	Na	15		
Strontium	Sr	18	I	7
Sulfur	S	7		
Tantalum	Ta	30		
Technetium	Tc	42		
Tellurium	Te	7		
Terbium	Tb	22		
Thallium	Tl	14		
Thorium	Th	24		
Thulium	Tm	22		
Tin	Sn	27		
Titanium	Ti	25		
Tungsten	W	33		
Uranium	U	24		
Vanadium	V	29		
Xenon	Xe	1	I	8
Ytterbium	Yb	22		
Yttrium	Y	22		
Zinc	Zn	19		
Zirconium	Zr	26		

ADVISORY BOARD

RALPH N. ADAMS

Department of Chemistry, The University of Kansas, Lawrence, Kansas

N. A. BALASHOVA

Institute of Electrochemistry, Academy of Sciences (USSR), Moscow

ANDRE DEBETHUNE

Department of Chemistry, Boston College, Chestnut Hill, Massachusetts

K. S. G. DOSS

Department of Chemical Engineering, Indian Institute of Technology, Madras

NORMAN HACKERMAN

Rice University, Houston, Texas

HENNING LUND

Department of Organic Chemistry, University of Aarhus, Denmark

G. MILAZZO

Laboratorio di Chimica, Istituto Superiore di Sanità, Rome

ROGER PARSONS

School of Chemistry, University of Bristol, England

M. POURBAIX

Centre Belge d'Etude de la Corrosion, Brussels

K. VETTER

Freie Universität Berlin, Institüt für Physikalische Chemie, Berlin

PETR ZUMAN

Department of Chemistry, Clarkson College of Technology, Potsdam, New York

NOBUYUKI TANAKA

Department of Chemistry, Tohoku University, Sendai, Japan

ENCYCLOPEDIA OF ELECTROCHEMISTRY OF THE ELEMENTS

EDITOR

Allen J. Bard

Department of Chemistry
University of Texas
Austin, Texas

VOLUME I

Ar	Cl	Pb
At	He	Ra
Ba	I	Rn
Br	Kr	Sr
Ca	Mn	Xe
Cd	Ne	

MARCEL DEKKER, INC. New York

MARCEL DEKKER, INC.
95 Madison Avenue, New York, New York 10016

LIBRARY OF CONGRESS CATALOG CARD NUMBER 73-88796

ISBN 0-8247-6093-X

PRINTED IN THE UNITED STATES OF AMERICA

CONTENTS

I–5. LEAD

Thomas F. Sharpe

I–6. MANGANESE

C. C. Liang

I–7. CALCIUM, STRONTIUM, BARIUM AND RADIUM

Shinobu Toshima

I–8. INERT GASES

Bruno Jaselskis and R. H. Krueger

LIST OF CONTRIBUTORS

PIER GIORGIO DESIDERI, Institute of Analytical Chemistry, University of Florence, Italy

G. FAITA, Laboratory of Electrochemistry and Metallurgy, University of Milan, Italy

NOEL A. HAMPSON, Department of Chemistry, Loughborough University of Technology, Leicestershire, England

DANIELA HEIMLER, Institute of Analytical Chemistry, University of Florence, Italy

BRUNO JASELKIS, Department of Chemistry, Loyola University, Chicago, Illinois

R. H. KRUEGER*, Department of Chemistry, Loyola University, Chicago, Illinois

ROGER J. LATHAM, School of Chemistry, City of Leicester Polytechnic, Leicestershire, England

LUCIANO LEPRI, Institute of Analytical Chemistry, University of Florence, Italy

C. C. LIANG, Laboratory for Physical Science, P. R. Mallory & Co., Inc., Burlington, Massachusetts

T. MUSSINI, Laboratory of Electrochemistry and Metallurgy, University of Milan, Italy

THOMAS F. SHARPE, Electrochemistry Department, Research Laboratories, General Motors Corporation, Warren, Michigan

SHINOBU TOSHIMA, Department of Applied Chemistry, Faculty of Engineering, Tohoku University, Sendai, Miyagi, Japan

*Present address: Borg-Warner Corporation, Des Plaines, Illinois

INTRODUCTION

The aim of this series is to provide a critical, systematic, and comprehensive review of the electrochemical behavior of the elements and their compounds. Part I deals with inorganic electrochemistry and Part II with organic electrochemistry. The field of electrochemistry has undergone extensive growth in recent years and electrochemical techniques and concepts have been applied to many areas of basic and applied research and technology. While the fundamentals of electrochemistry and the recent advances in this field have been described in a number of textbooks, monographs, and review series, the vast literature concerning the descriptive electrochemistry of inorganic and organic compounds has been relatively neglected. This series is designed to provide the natural starting point for new electrochemical investigations and to suggest areas where further research is needed. Each chapter, written by experts on that subject, contains the best available information on the electrochemical behavior and applications of the element.

The classification of elements generally follows the scheme used in Gmelin, with some grouping and rearrangement. Each element is assigned a number (n). This number determines in what chapter a compound or alloy will be considered; a chapter on element n will discuss compounds of element n with elements of number lower than n. For example, zinc-amalgam is treated in the zinc chapter (element number 19), rather than in the mercury chapter (element number 10). A listing of the elements and the element numbers is given inside the rear cover. To minimize delays in publication of the chapters, these are published in the order received. The volume and chapter number for each element are also located in the table on the inside back cover.

The chapters are generally organized into five sections:

1. Introduction and Standard Potentials
    1.1. Aqueous Solutions
    1.2. Nonaqueous Solvents
    1.3. Fused Salts
    1.4. Other Data
2. Voltammetric Characteristics
    Presentation of Polarographic and Other Voltammetric Results

3. Kinetic Parameters and Double-Layer Properties

   Presentation of available rate constants, exchange current densities, transfer coefficients, potential of zero charge, double-layer capacities, etc.

4. Electrochemical Studies

   A survey and critical review of the electrochemical reactions of the element and its compounds. Description of known mechanisms of the electrode reactions, oxidation and reduction products, current efficiencies, reaction orders, etc. Discussion of passivation phenomena, oxide films, and anodization.

5. Applied Electrochemistry

   The use of electrochemistry in the isolation or purification of the element, the electrochemical production of compounds, and the use and behavior of the element and its compounds in electrochemical devices.

   5.1. Electrowinning and Electrorefining
   5.2. Electrodeposition, Electroplating, and Electropolishing
   5.3. Electrosynthesis
   5.4. Corrosion
   5.5. Batteries and Cells
   5.6. Other

Conventions generally follow the recommendations of IUPAC [1]. Half reactions corresponding to a standard electrode potential are uniformly written as reductions and the sign of the electrode potential follows the Gibbs-Stockholm convention [2]. Notation generally follows that in the "Manual of Symbols and Terminology for Physicochemical Quantities and Units" [3]. Modified and frequently-used symbols are given below.

1. POTENTIALS

E: potential vs NHE, SCE, or other reference electrode. When corresponding cell is to be indicated, it may be written

$$E(Zn^{2+}/Zn) \text{ or } E(1/2)Zn^{2+} + e = (1/2)Zn$$

E°: standard electrode potential. It is the EMF of the cell in which the reaction is the reduction of the oxidized species by hydrogen, all reactants being in their standard state, e.g.,

$$(1/2)Zn^{2+} + (1/2)H_2 = (1/2)Zn + H^+ \qquad E^\circ(Zn^{2+}/Zn)$$

$E^{\circ\prime}$: formal electrode potential or conditional electrode potential corresponding to unit concentration quantities.

η: overpotential, the deviation of the potential of an electrode from its equilibrium value. Negative values are associated with reductions and positive values are associated with oxidations.

Polarographic and Voltammetric Potentials

$E_{1/2}$: half-wave potential in dc polarography, i.e., potential at which $i = (1/2)i_d$.

$E_{1/4}$, $E_{3/4}$: 1/4 and 3/4 wave potentials in dc polarography.

E_p, $E_{p/2}$: peak and half-peak potential, e.g., in linear scan or cyclic voltammetry. Cathodic and anodic waves are indicated by E_{pc} and E_{pa}.

$E_{\tau/4}$: quarter-wave potential in chronopotentiometry.

E_Z: point of zero charge.

E_{mix}: mixed potential.

2. CURRENT

i: current, instantaneous current, total current.

i_a: anodic current.

i_c: cathodic current.

i_o: exchange current.

i_d: diffusion current.

i_l: limiting current.

i_p: peak current.

Anodic or cathodic waves are indicated by i_{pa} and i_{pc}, i_{da} and i_{dc}, etc. Average value of current is indicated by a bar over the appropriate symbol, e.g., $\bar{i}_d$: average diffusion current.

j: current density, subscripted as above to indicate j_a, anodic; j_c, cathodic; j_o, exchange, etc.

I, $\bar{I}$: diffusion current constant in dc polarography

$$I = i_d/m^{2/3}t_d^{1/6} \qquad \bar{I} = \bar{i}_d/m^{2/3}t_d^{1/6}$$

with i_d in μA, m in mg/s and t in s.

3. CONCENTRATION

C, C_O, C_R: concentration of O, R, ..., at electrode surface. Unless stated otherwise, [O], [R], ..., may be used in discussing equilibria in solution.

C°, C_R°: bulk concentrations, bulk concentration of species R.

$C_R(x, t)$: concentration of species R at distance x from electrode at time t.

4. TIME

t: time.

t_d: drop time of DME.

τ: transition time in chronopotentiometry.

5. RATE CONSTANTS

For the reaction $O + ne = R$:

$$j = i/A = nF(k_a C_R - k_c C_O)$$

n: charge number of elementary step.

k_a: formal or conditional rate constant of anodic (oxidation) reaction at potential E vs reference electrode, in cm/sec (k_a° refers to value of k_a at E = 0 vs reference electrode).

k_c: formal or conditional rate constant of cathodic (reduction) reaction at potential E vs reference electrode, in cm/sec (k_c° refers to value of k_c at E = 0 vs reference electrode).

The current convention corresponds to an anodic current positive.

$$k_c = k^\circ \exp\{-\alpha z F(E - E^{\circ\prime})/RT\}$$

$$k_a = k^\circ \exp\{(1 - \alpha) z F(E - E^{\circ\prime})/RT\}$$

k°: formal or conditional standard rate constant for the electrode reaction.

α: the electrochemical charge transfer coefficient for the cathodic reaction.

$1 - \alpha$: the electrochemical charge transfer coefficient for the anodic reaction.

$$j_0 = i_0/A = nFk^\circ C_O^{1-\alpha} C_R^{\alpha}$$

i_0: formal or conditional exchange current (A).

j_0: formal or conditional exchange current density (A/cm^2).

If the rate constants are determined by extrapolation to zero ionic strength so that they are written in terms of activities (a_O, a_R) in place of concentrations (C_O, C_R), the symbols are primed: k_c', k_a', $k^{\circ\prime}$, j_0', i_0'.

For rate constants corrected for double layer effects, the subscript t is used; k_t°, j_t° where $k_t^\circ = k^\circ \exp -\{(\alpha n - z)F\phi_2/RT\}$.

6. OTHER

A = electrode area (cm^2).

F = Faraday constant (C/mol).

u_i: mobility of species i (cm^2/V/sec).

z_i: charge number of species i, positive for cations and negative for anions.

t_i: transport number of species i.

D_O, D_R: diffusion coefficient of species O, R.

δ: thickness of diffusion layer (cm).

m_O, m_R: mass transport constant of species O, R (cm/sec)

$$m_R = D_R/\delta_R$$

Q = quantity of electricity, charge.

C = capacitance (μF/cm^2).

C_{dl} = double-layer capacitance.

m = rate of mercury flow in DME (mg/sec).

v = potential scan rate (V/sec).

ω = angular frequency of rotation (sec^{-1}).

Abbreviations for words and journals follow the American Chemical Society Handbook for Authors and the Chemical Abstracts Service Guide for Abbreviating Periodical Titles. Some special abbreviations used in the tables in this compilation are given below.

1. TECHNIQUES

E swp: potential sweep.
i swp: current sweep.
CV: cyclic voltammetry.
E stp: potentiostatic or potential step.
V stp: voltage step.
i stp: chronopotentiometry, galvanostatic, or current step.
ac pol: ac polarography.
ac harm: ac harmonic method.
farad. imp: faradaic impedance (ac bridge) method.
chramp: chronoamperometry.
chrcoul: chronocoulometry.
coul: coulostatic method.
farad. rect: faradaic rectification method.
dc pol: dc polarography.

2. ELECTRODES

rot: rotating.
hng: hanging.
vib: vibrating.
stat: stationary.
stir: stirred.
bub: bubbling.
soln: solution.
drp: drop.
dsk: disk.
wr: wire.
pl: pool.
pwd: powder.
pst: paste.
fl: foil.
rd: rod.
DME: dropping mercury electrode.
SCE: saturated calomel electrode.
NHE: normal hydrogen electrode.

Formula of element or compound is used to complete description of electrode, e.g., rot Pt wr (rotating platinum wire electrode), C pst - stir soln (carbon paste electrode in stirred solution), etc.

3. DESCRIPTIONS OF REACTIONS AND WAVES

rev: reversible.
irr: irreversible.

sl: slightly.

q: quasi.

w: well-defined.

i: ill-defined.

fw: fairly well-defined.

fi: fairly ill-defined.

do: drawn out.

mb: merges with background.

The general philosophy of this compilation was to let each author choose the extent and scope of his chapter, subject only to limitations of format and notation, since it would clearly be impossible to prescribe too closely the type of treatment appropriate for the different elements. Supplementary volumes and, eventually, new editions are planned which will contain new results and corrections. The authors and the editor would welcome comments, suggestions, and corrections for inclusion in the supplements. The editor is indebted to the members of the advisory board, his colleagues, and his students for suggestions and assistance in preparing this volume. Special thanks are due to Mrs. Gaynel Klingemann for her secretarial and administrative efforts.

REFERENCES

1. J. Electroanal. Chem., 7, 417 (1964).
2. A. J. deBethune, J. Electrochem. Soc., 102, 288C (1955); T. S. Licht and A. J. deBethune, J. Chem. Educ., 34, 433 (1957).
3. Pure Appl. Chem., 21, 3 (1970).

Allen J. Bard
Austin, Texas

Chapter I-1

CHLORINE

T. MUSSINI and G. FAITA

Laboratory of Electrochemistry and Metallurgy
University of Milan
Milan, Italy

1. OXIDATION STATES AND STANDARD POTENTIALS

1.1. OXIDATION STATES

The more reliable oxidation numbers for which the existence of chlorine compounds was stated are: -1 in hydrochloric acid and in the chlorides; +1 in the oxide Cl_2O, in the corresponding hypochlorous acid HClO and in the hypochlorites; +2 in the oxide ClO; +3 in chlorous acid and in the chlorites; +4 in the oxide ClO_2; +5 in chloric acid and in the chlorates; +6 in the oxide Cl_2O_7, in the corresponding perchloric acid $HClO_4$, and in the perchlorates.

The values of the fundamental thermodynamic functions for the more significant species based on chlorine in different oxidation states are collected in Table 1.1.1. Only recently was it possible to obtain these thermodynamic data for the trichloride ion Cl_3^- [1]. A comprehensive compilation of thermodynamic data for the basic chlorine compounds is available in a recent publication by the National Bureau of Standards [2].

1.2. STANDARD POTENTIALS

1.2.1. Aqueous Solutions

1.2.1.1. The Chlorine/Chloride Electrode. The potential of the reversible chlorine/chloride electrode can be expressed in terms of either the electrode

$$Cl_2(g) + 2e^- = 2Cl^-(aq) \tag{1.2.1}$$

or

$$Cl_2(aq) + 2e^- = 2Cl^-(aq) \tag{1.2.2}$$

A third form of electrode reaction must be considered by analogy with the bromine/bromide and iodine/iodide electrodes, i.e.,

$$Cl_3^-(aq) + 2e^- = 3Cl^-(aq) \tag{1.2.3}$$

since chlorine does react with chloride to form the trichloride ion Cl_3^- according to either

$$Cl_2(g) + Cl^-(aq) = Cl_3^-(aq) \tag{1.2.4}$$

or

$$Cl_2(aq) + Cl^-(aq) = Cl_3^-(aq) \tag{1.2.5}$$

Extensive and accurate work for the determination of these standard potentials with related thermodynamic functions was carried out recently [1, 3, 4]; the temperature range covered was 25 to 80°C.

The first approach [4] was based on measurements of the reversible emf of the cell

$$Pt|Ag|AgCl|1.75\ m\ HCl|N_2\text{-}Cl_2\ 10\%|\ Pt\text{-}Ir\ 45\%,\ Ta|Pt \tag{1.2.6}$$

This emf is independent of chloride ion concentration but is a function of the chlorine pressure p_{Cl_2}, and is given in terms of the standard molal potentials of the relevant half cells by

$$E_6 = E^\circ_m(Cl_2, g + 2e^- = 2Cl^-) - E^\circ_m(AgCl + e^- = Ag + Cl^-) + (k/2)\log p_{Cl_2} \tag{1.2.7}$$

where $k = 2.303RT/F$. Taking into account the atmospheric pressure and the equilibrium partial pressures of H_2O and HCl, the reversible emf's E_8 of the cell

$$Pt|Ag|AgCl|1.75\ m\ HCl|Cl_2, 1\ atm|\ Pt\text{-}Ir\ 45\%,\ Ta|Pt \tag{1.2.8}$$

were obtained from the corresponding E_6's. Because

$$E_8 = E^\circ_m(Cl_2, g + 2e^- = 2Cl^-) - E^\circ_m(AgCl + e^- = Ag + Cl^-) \tag{1.2.9}$$

and $E^\circ_m(AgCl + e^- = Ag + Cl^-)$, the standard molal potential of the silver/silver-chloride electrode, is known [6], the required standard molal potentials of the chlorine/chloride electrode can be readily determined from the E_8's. These values at temperatures from 25 to 80°C are reported in Table 1.2.1 together with the relevant limits of error.

TABLE 1.1.1. Basic Thermodynamic Functions for Chlorine at Different Oxidation States at 25° C[a]

Formula and description	State	ΔG_f° (kcal/mole)	ΔH_f° (kcal/mole)	S° (cal/deg mole)	Refs.
Cl	g	25.262	29.082	39.457	[2]
Cl_2	g	0	0	53.288	[2]
Cl_2, std state, m = 1	aq	1.72	-5.04	30.5	[1]
	aq	1.65	-5.6	29.	[2]
Cl^-	g		-58.8		[2]
Cl^-, std state, m = 1	aq	-31.325	-39.85	-2.0[a]	[1]
	aq	-31.331	-39.895	-2.079[a]	[3]
	aq	-31.372	-39.952	-2.1[a]	[2]
HCl	g	-22.777	-22.062	44.646	[2]
HCl, std state, m = 1	aq	-31.325	-39.85	13.6	[1]
	aq	-31.331	-39.895	13.525	[3]
	aq	-31.372	-39.952	13.5	[2]
Cl_3^-, std state, m = 1	aq	-28.70	-47.26	17.6[a]	[1]
Cl_2O	g	23.4	19.2	63.60	[2]
ClO^-, std state, m = 1	aq	-8.8	-25.6	-5[a]	[2]
HClO	g			56.54	[2]
HClO, undiss; std state, m = 1	aq	-19.1	-28.9	34	[2]
ClO	g	23.45	24.34	54.14	[2]
ClO_2^-, std state, m = 1	aq	4.1	-15.9	8.6[a]	[2]
$HClO_2$, undiss; std state, m = 1	aq	1.4	-12.4	45.0	[2]
ClO_2	g	28.8	24.5	61.36	[2]
ClO_2, std state, m = 1	aq	28.1	17.9	41.4	[2]
ClO_3^-, std state, m = 1	aq	-0.8	-23.7	23.2[a]	[2]
$HClO_3$, std state, m = 1	aq	-0.8	-23.7	38.8	[2]
ClO_3	g		37		[2]
ClO_4^-, std state, m = 1	aq	-2.06	-30.91	27.9[a]	[2]
$HClO_4$, std state, m = 1	aq	-2.06	-30.91	43.5	[2]
H_3OClO_4	c		-91.35		[2]

[a]The entropy values of ions are quoted on the electrochemical scale, i.e., they are referred to $S^\circ(H^+, aq) = \frac{1}{2} S^\circ(H_2, g) = 15.604$ cal/deg mole which follows from the electrochemical convention $E^\circ(H^+, aq + e^- = \frac{1}{2}H_2, g) = 0$ at all temperatures [4, 5]. To convert these entropy values to the convention $S^\circ(H^+, aq) = 0.000$, they must be added to the quantity $15.604\, z_i$ cal/deg mole, where z_i is the ionic valency <u>with sign.</u>

TABLE 1.2.1. Standard Molal Potentials, $E^\circ_m(Cl_2, g/Cl^-, aq)$, of the Chlorine Electrode in Aqueous Solution, in Absolute Volts, at Various Temperatures, with Reference to $E^\circ_m(H^+, aq/H_2, g) = 0$ at all temperatures

T (°C)	From Ref. 4	From Ref. 3	From Ref. 1	Recommended values
25	1.35830 ± 0.00005	1.35852 ± 0.00011	1.35827 ± 0.00002	1.35828 ± 0.00003
30	1.35213 ± 0.00006	1.35224 ± 0.00012	-	1.35215 ± 0.00004
40	1.33919 ± 0.00007	1.33919 ± 0.00013	1.33910 ± 0.00002	1.33911 ± 0.00003
50	1.32559 ± 0.00007	1.32574 ± 0.00013	-	1.32562 ± 0.00006
60	1.31114 ± 0.00007	1.31144 ± 0.00014	1.31154 ± 0.00001	1.31153 ± 0.00004
70	1.29651 ± 0.00008	1.29679 ± 0.00016	-	1.29657 ± 0.00012
80	1.28104 ± 0.00014	1.28153 ± 0.00019	1.28172 ± 0.00001	1.28172 ± 0.00004

The experimental conditions represented in the Cell (1.2.6), correspond to an optimum compromise ensuring minimization of errors, and lead to a cumulative uncertainty of ±0.04 mV for the E_8 values uniformly over the entire temperature range of the experiments. In particular:

1. The tantalum-supported platinum-45% iridium substrate ensured freedom from corrosive attack by the Cl_2/Cl^- couple and absence of spurious mixed potentials
2. With the HCl concentration chosen (1.75 m), the exchange currents for the Cl_2/Cl^- couple were as high as 16 mA/cm^2 (apparent geometrical surface) at 25°C, 20 mA/cm^2 at 45°C, and even higher at 80°C. At the same time, the AgCl solubility in HCl was still small
3. The Cl_2 pressure used (0.1 atm) was low enough to minimize the amount of Cl^- ions converted to Cl_3^- by Reaction (1.2.4) but, at the same time, was compatible with adequate accuracy in the analysis of the chlorine-nitrogen mixture.

A value of 0.01 atm^{-1} [7] for the equilibrium constant of Reaction (1.2.4) was used for these calculations of limits of uncertainty over the entire range 25 to 80°C; the actual values were, however, determined accurately in a subsequent work [1].

The E°_m data thus obtained (Table 1.2.1) as well as the E_8 values were least-squared, giving the following polynomials for E°_m and E_8 as functions of absolute temperature T:

$$E^\circ_m(Cl_2, g + 2e^- = 2Cl^-) = 1.47252 + (4.82271 \times 10^{-4})T - (2.90055 \times 10^{-6})T^2 \quad \text{abs V} \quad (1.2.10)$$

$$E_8 = 1.28958 - (4.31562 \times 10^{-4})T - (2.79220 \times 10^{-7})T^2 \quad \text{abs V} \quad (1.2.11)$$

Both equations fit the relevant data within ±0.15 mV.

From $E^\circ_m(Cl_2, g + 2e^- = 2Cl^-)$ and the corresponding first isothermal temperature coefficient $(dE^\circ_m/dT)_{isoth}$ (taken from the first derivative of Eq. (1.2.10), the standard thermodynamic functions ΔG°, ΔH°, and ΔS° at 25°C and 1 atm for Reaction (1.2.1) were obtained (cf. Table 1.2.3 for comparison purposes). Similarly, from E_8 and the corresponding temperature coefficient $(dE_8/dT)_{isoth}$ (taken from the first derivative of Eq. (1.2.11), the ΔG°,

ΔH°, and ΔS° values at 25°C and 1 atm were obtained for the overall reaction of the Cell (1.2.8):

$$Ag + \frac{1}{2}Cl_2, g = AgCl \quad (1.2.12)$$

and can be seen in the same Table 1.2.3. As a result, for the standard entropies of aqueous Cl^- ion and of solid AgCl the following values were obtained: $S^\circ(Cl^-, aq) = -2.125 \pm 0.006$ cal deg^{-1} $mole^{-1}$ ("electrochemical scale," cf. footnote to Table 1.1.1) and $S^\circ(AgCl) = 23.056 \pm 0.004$ cal deg^{-1} $mole^{-1}$, both values being more accurate than the corresponding ones then available in the literature.

Not only are Eqs. (1.2.10) and (1.2.11) useful for calculations of the standard entropy changes ΔS° for the relevant cell reactions through the first isothermal temperature coefficient ($\{dE^\circ/dT\}_{isoth} = \Delta S^\circ/nF$), but they are also suitable for evaluations of standard heat capacity changes ΔC_p° through the second isothermal temperature coefficient ($\{d^2E^\circ/dT^2\}_{isoth} = \Delta C_p^\circ/nFT$). These temperature coefficients are reported in Table 1.2.2.

A second approach [3] was based on measurements of the reversible emf of the cell

$$Pt|H_2, 1\ atm|HCl(m)|N_2\text{-}Cl_2\ 10\%|Pt\text{-}Ir\ 45\%, Ta|Pt \quad (1.2.13)$$

in which the chlorine/chloride electrode had the same structure and characteristics as in the Cell (1.2.6) previously discussed, and the hydrogen electrodes were of the capillary-imbibition type recently described by Bianchi et al. [8-10]. Cell (1.2.13) was mainly devised for determinations of mean molal activity coefficients $\gamma_\pm$ of aqueous hydrochloric acid at high concentrations (1 to 11 mole/kg) and temperatures (25 to 80°C). In fact, using the familiar alternative cell

$$Pt|H_2, 1\ atm|HCl(m)|AgCl|Ag|Pt \quad (1.2.14)$$

would have caused AgCl to become solubilized in HCl, an effect which increases strongly with increasing HCl concentration. Of course, the consequence of using Cell (1.2.13) was Reaction (1.2.4) which forms Cl_3^- ions, thus removing some Cl^- ions, but this effect was likely to be small compared with the AgCl solubility in HCl, considering the low value of the equilibrium constant for Reaction (1.2.4) (see Table 1.2.6, and Ref. 7).

The emf of Cell (1.2.13), neglecting Cl_3^- formation, is given by

$$E_{13} = E_m^\circ(Cl_2, g/Cl^-) + (k/2)\log(p_{Cl_2}p_{H_2}) - 2k\log(m\gamma_\pm)_{HCl} \quad (1.2.15)$$

Let E''_{13} and E'_{13} be the emf's corresponding to two different molalities m" and m', the respective mean molal activity coefficients being γ'' and γ'. Working with p_{Cl_2} and p_{H_2} constants ($p_{Cl_2} = 0.1$ atm, for the reasons explained above, and $p_{H_2} = 1$ atm), from Eq. (1.2.15) one obtains

$$E''_{13} - E'_{13} - 2k\log(m'\gamma'_\pm/m''\gamma''_\pm)_{HCl} \quad (1.2.16)$$

Evidently $\gamma''_\pm$ can be calculated from the measured E_{13}'s if, at the reference molality m' (which was m' = 1.000 for convenience), the activity coefficient $\gamma'_\pm$ is known from independent measurements (this was obtained from emf measurements of Cell (1.2.14). Of course, independent knowledge of $\gamma'_\pm$ at fixed m' = 1.000 also allows $E_m^\circ(Cl_2, g/Cl^-)$ to be obtained from Eq. (1.2.15). The data of $E_m^\circ(Cl_2, g/Cl^-)$ so obtained are quoted in Table

1.2.1 with the relevant limits of error. These $E^\circ_m(Cl_2, g/Cl^-)$ values fit the least-squares polynomial

$$E^\circ_m(Cl_2, g/Cl^-) = 1.479874 + (4.315277 \times 10^{-4})T - (2.812217 \times 10^{-6})T^2 \quad \text{abs V} \qquad (1.2.17)$$

the mean deviation of fit being 0.02 mV.

The third approach [1], and the most accurate one, is based again on measurements of the reversible emf of a cell of type (1.2.13) but with the partial pressure p_{Cl_2} of chlorine varying from 0.05 to 1 atm. Let this cell be indicated by the following simplified notation

$$H_2, 1\ \text{atm} | HCl, 1m | Cl_2, p\ \text{atm} \qquad (1.2.18)$$

Unlike the work in Ref. 3, in this case the influence of Reaction (1.2.4) was not neglected but the corresponding equilibrium constant K_4 was determined with an original procedure leading to the simultaneous determination of $E^\circ_m(Cl_2, g/Cl^-)$ and K_4. Measurements of chlorine solubility were run in parallel to the emf measurements of Cell (1.2.18) to also obtain K_5, the equilibrium constant for Reaction (1.2.5).

Due to the occurrence of Reaction (1.2.4) in the compartment of the chlorine electrode, the molality of the Cl^- ion, $m^*_{Cl^-}$ (the asterisk is used throughout this chapter to denote m_{Cl^-} in the chlorine electrode compartment), differs from m_{HCl} as some Cl^- ions are converted into Cl_3^- ions; in the hydrogen electrode compartment, $m_{Cl^-} = m_{H^+} = m_{HCl}$. Reaction (1.2.4) does not cause changes in the ionic strength of the HCl solution, thus the activity coefficient for the Cl^- ion, γ_{Cl^-}, remains unchanged in the two half-cell compartments. The equilibrium constant for Reaction (1.2.4) is

$$K_4 = m_{Cl_3^-}\gamma_{Cl_3^-}/(p_{Cl_2} m^*_{Cl^-}\gamma_{Cl^-}) \qquad (1.2.19)$$

where $m^*_{Cl^-}$ is defined by

$$m_{HCl} = m^*_{Cl^-} + m_{Cl_3^-} \qquad (1.2.20)$$

Inserting the value of $m_{Cl_3^-}$ obtained from Eq. (1.2.19) into Eq. (1.2.20), one obtains

$$m^*_{Cl^-} = m_{HCl}/(1 + K_4 p_{Cl_2}\gamma_{Cl^-}/\gamma_{Cl_3^-}) \qquad (1.2.21)$$

Cl^- and Cl_3^- are influenced by the same ionic strength and should not be much different in solvated-ion size, and therefore it is reasonable to assume $\gamma_{Cl^-} = \gamma_{Cl_3^-}$. This assumption probably introduces an uncertainty of the order of 2-3% in the calculated K_4 values, working with unit HCl molality. From Eq. (1.2.21) and the above assumption, the expression of the emf of the Cell (1.2.18) with $p_{H_2} = 1$ atm becomes

$$\begin{aligned} E_{18} &= E^\circ_m(Cl_2, g + 2e^- = 2Cl^-) + (k/2)\log p_{Cl_2} - k\log(m_{H^+}\gamma_{H^+}m^*_{Cl^-}\gamma_{Cl^-}) \\ &= E^\circ_m(Cl_2, g + 2e^- = 2Cl^-) + (k/2)\log p_{Cl_2} - 2k\log(m\gamma_\pm)_{HCl} \\ &\quad + (RT/F)\ln(1 + K_4 p_{Cl_2}) \end{aligned} \qquad (1.2.22)$$

K_4 is of the order of 1.0×10^{-2} and $p_{Cl_2} \leq 1$ atm, therefore the last logarithmic term in Eq. (1.2.22) can be linearized according to $\ln(1 + K_4 p_{Cl_2}) = K_4 p_{Cl_2}$ without introducing any appreciable error. Thus rearranging Eq. (1.2.22) yields

$$\begin{aligned} &E_{18} + 2k\log(m\gamma_\pm)_{HCl} - (k/2)\log p_{Cl_2} \\ &\quad = E^\circ_m(Cl_2, g + 2e^- = 2Cl^-) + K_4 p_{Cl_2} k/2.303 = E^{\circ\prime} \end{aligned} \qquad (1.2.23)$$

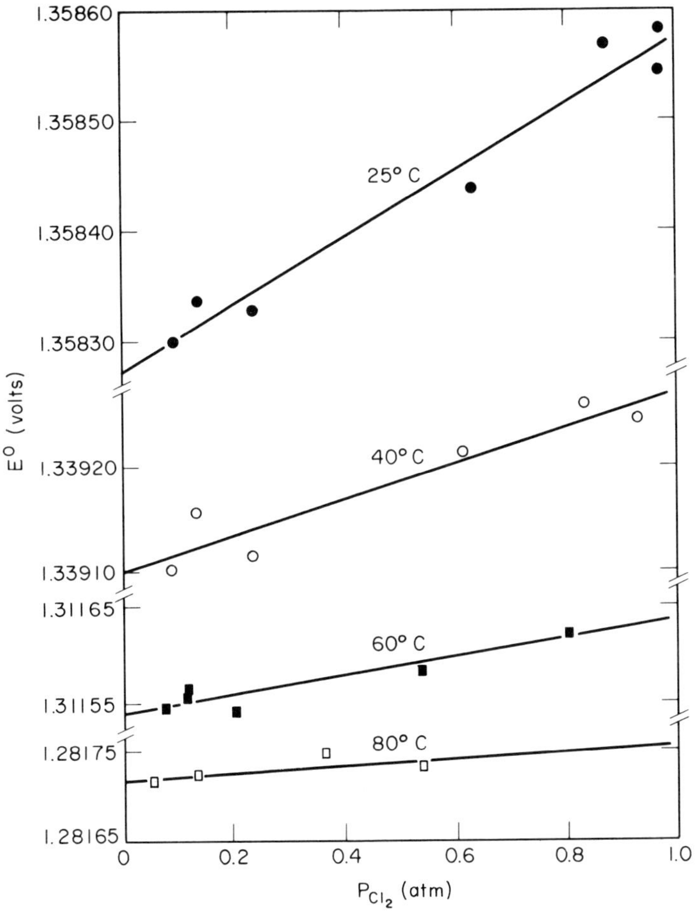

Fig. 1.2.1. Simultaneous determination of $E_m^\circ(Cl_2, g + 2e^- = 2Cl^-, aq)$ and K_4 from the intercept at p_{Cl_2} and the slopes of the $E^{\circ\prime}$ vs p_{Cl_2} straight lines predicted by Eq. (1.2.23).

Equation (1.2.23) predicts that a plot of $E^{\circ\prime}$ vs p_{Cl_2} should be a straight line whose slope yields $K_4k/2.303$ and whose intercept gives $E_m^\circ(Cl_2, g + 2e^- = 2Cl^-)$, that is, K_4 and E_m° are determined simultaneously. The results obtained conform to the expected behavior (see Fig. 1.2.1). HCl (1 m) was used throughout and the relevant $\gamma_\pm$ values over the range 25 to 80°C were obtained from emf measurements of Cell (1.2.14) and are the same used previously [3].

The $E_m^\circ(Cl_2, g + 2e^- = 2Cl^-)$ values so obtained are collected in Table 1.2.1, fourth column, with the corresponding uncertainties and can be reproduced with a maximum deviation of 4 μV — in practice, with all the significant figures — by the least-squares polynomial

$$E_m^\circ(Cl_2, g + 2e^- = 2Cl^-) = 1.3873947 - (1.0939812 \times 10^{-3})T - (2.8370370 \times 10^{-6})T^2 \quad \text{abs V} \qquad (1.2.24)$$

where T is the absolute temperature.

The agreement between the three sets of data in Table 1.2.1 is very good, considering the effect of AgCl solubility in HCl and the limits of uncertainty for the values of standard potentials of the silver/silver-chloride electrode [11]. The "recommended values" in Table 1.2.1 have been obtained as weighted means of the above entries at each temperature by giving each datum a weight inversely proportional to the square of the relevant error. These recommended values for the standard potential of the chlorine/chlorine electrode fit the polynomial

$$E^\circ_m(Cl_2, g + 2e^- = 2Cl^-) = 1.484867 + (3.958492 \times 10^{-4})T - (2.750639 \quad 10^{-6})T^2 \quad \text{abs V} \qquad (1.2.25)$$

with a mean deviation of 0.10 mV.

Once K_4 is known, K_5 for Reaction (1.2.25) can be determined in turn if the total chlorine in solution, viz.,

$$m_{totCl_2} = m_{Cl_3^-} + m_{Cl_2} \qquad (1.2.26)$$

is known at each p_{Cl_2} used.

K_5 can be expressed as

$$K_5 = m_{Cl_3^-}\gamma_{Cl_3^-}/(m_{Cl_2}\gamma_{Cl_2}m^*_{Cl^-}\gamma_{Cl^-}) \qquad (1.2.27)$$

If the value for $m_{Cl_3^-}$ obtained from Eq. (1.2.27) is substituted in Eq. (1.2.26), remembering the assumption $\gamma_{Cl_3^-} = \gamma_{Cl^-}$, one obtains

$$m_{Cl_2} = m_{totCl_2}/(1 + K_5 m^*_{Cl^-}\gamma_{Cl_2}) \qquad (1.2.28)$$

where m_{Cl_2} is the molality of free chlorine in solution and γ_{Cl_2} is the corresponding activity coefficient. Equations (1.2.20), (1.2.26), and (1.2.28) lead to

$$m_{Cl_2} = m_{totCl_2} + m^*_{Cl^-} - m_{HCl} = m_{totCl_2}/(1 + K_5 m^*_{Cl^-}\gamma_{Cl_2}) \qquad (1.2.29)$$

Inserting Eq. (1.2.21) into Eq. (1.2.29) and solving for K_5, one obtains

$$K_5 = (1 + K_4 p_{Cl_2})/\gamma_{Cl_2}\{(m_{totCl_2} + m_{totCl_2}/(K_4 p_{Cl_2}) - m_{HCl})\} \qquad (1.2.30)$$

The values of K_5 in Table 1.2.6 are the averages of the values calculated using all of the values of m_{totCl_2} at each temperature (K_4 being known from the slope of the $E^{\circ\prime}$ vs p_{Cl_2} plot). As for the value to be assigned to γ_{Cl_2} when $m_{HCl} = 1$ (in Eq. 1.2.30), Sherrill and Izard showed [7] that such a γ_{Cl_2} is very close to unity at 25°C. A value of $\gamma_{Cl_2} = 1.00$ at unit m_{HCl} was then used over the entire range, 25 to 80°C. This could probably have introduced an uncertainty of the order of 3% for K_5.

Now that $E^\circ_m(Cl_2, g/Cl^-)$, K_4, and K_5 are known as functions of temperature, all the basic standard thermodynamic functions of the species relevant to the $Cl_2/Cl^-/Cl_3^-$ system are determinable (cf. data in Table 1.1.1).

Moreover, it is possible to determine the standard potentials of the chlorine/chloride electrode in terms of Reaction (1.2.2). In fact, the actual potential of the chlorine/chloride electrode can be expressed in terms of either (Cl_2, g) or (Cl_2, aq):

$$E(Cl_2/Cl^-) = E^\circ_m(Cl_2, g + 2e^- = 2Cl^-) + (k/2)\log p_{Cl_2} - k\log(m^*_{Cl^-}\gamma_{Cl^-})$$

$$= E^\circ_m(Cl_2, aq + 2e^- = 2Cl^-) + (k/2)\log(m_{Cl_2}\gamma_{Cl_2}) - k\log(m^*_{Cl^-}\gamma_{Cl^-}) \qquad (1.2.31)$$

so that, taking into account Eqs. (1.2.19) and (1.2.27) one obtains

$$E_m^\circ(Cl_2, aq + 2e^- = 2Cl^-) = E_m^\circ(Cl_2, g + 2e^- = 2Cl^-) + (k/2)\log(K_5/K_4) \quad (1.2.32)$$

The values of $E_m^\circ(Cl_2, aq + 2e^- = 2Cl^-)$ obtained from Eq. (1.2.32) are given in Table 1.2.2 and can be reproduced by the least-squares polynomial

$$E_m^\circ(Cl_2, aq + 2e^- = 2Cl^-) = 1.40922 - (4.30504 \times 10^{-4})T - (4.87922 \times 10^{-6})T^2 \quad \text{abs V} \quad (1.2.33)$$

with a maximum deviation of 0.6 mV.

Finally, the standard molal potentials for the chlorine/chloride electrode in terms of Reaction (1.2.3) are readily determined as

$$E_m^\circ(Cl_3^- + 2e^- = 3Cl^-) = E_m^\circ(Cl_2, g + 2e^- = 2Cl^-) - (k/2)\log K_4 \quad (1.2.34)$$

and can be reproduced with a mean deviation of fit of 0.12 mV by the polynomial

$$E_m^\circ(Cl_3^- + 2e^- = 3Cl^-) = 1.359357 + (6.407284 \times 10^{-4})T - (1.969988 \times 10^{-6})T^2 \quad \text{abs V} \quad (1.2.35)$$

Table 1.2.2 collects the values of the standard potentials of the chlorine/chloride electrode in terms of the various electrode reactions with the corresponding values of the isothermal temperature coefficients.

1.2.1.2. Cl(ox)/Cl(red) Couples other than Chlorine/Chloride. The standard potentials for these redox couples are generally not amenable to determination by direct emf methods. (For instance, attempts to evaluate the standard potentials for the couple $ClO_2, aq/ClO_2^-$ by direct emf measurements over a range of temperature and pH and with different electrode substrates led to somewhat discrepant and uncertain results [17-19]. E° values of 0.92 and 0.80 V, respectively, for the reactions $ClO_2, aq + e^- = ClO_2^-$ and $ClO_2, aq + H^+ + e^- = HClO_2$ — both not quoted in Table 1.2.4 — should be accepted with some caution. From the thermodynamic data one would correspondingly obtain 1.16 and 1.04 V, respectively.)

Instead, the relevant E° values are best calculated from thermodynamic data for which there is good agreement among the literature data [12-16]. Table 1.2.4 provides a selection of these data and compares them with the values calculated on the basis of the more recent thermodynamic data in Table 1.1.1. Most of the redox reactions whose standard potentials are quoted in Table 1.2.4 were considered in Pourbaix's potential-pH equilibrium diagram for chlorine [20].

1.2.2. Nonaqueous Solutions

As is true for any kind of redox couple, there is still a severe shortage of data of standard potentials for chlorine-based redox couples in nonaqueous solutions in spite of the growing interest of electrochemists for the latter area of research.

The available data essentially concern the chlorine/chloride couple in terms of the electrode reactions

TABLE 1.2.2. Standard Molal Potentials E_m° over a Range of Temperatures for Different Electrode Reactions Concerning the System $Cl_2/Cl^-/Cl_3^-$ in Aqueous Solution, Together with the Isothermal Temperature Coefficient $(dE_m^\circ/dT)_{isoth}$ and the Second Isothermal Temperature Coefficient $(d^2E_m^\circ/dT^2)_{isoth}$. The Thermal Temperature Coefficient at 25°C is Obtainable from $(dE_m^\circ/dT)_{therm} = (dE_m^\circ/dT)_{isoth} + 0.871$ mV/deg, the Constant 0.871 mV/deg Being the Thermal Temperature Coefficient for the SHE at 25°C

Reaction	T (°C)	E_m°(V)	$\left(\frac{dE_m^\circ}{dT}\right)_{isoth}$ (mV/deg)	$\left(\frac{d^2E_m^\circ}{dT^2}\right)_{isoth}$ (μV/deg^2)	Remarks
$Cl_2(g) + 2e^- = 2Cl^-(aq)$	25	1.35828	-1.2444	-5.502	Based on the "recommended values" of E_m° in Table 1.2.1. Second isothermal temperature coefficient taken from the second derivative of Eq. (1.2.25); compare with the value -5.454 μV/deg^2 in Ref. 13.
	30	1.35215	-1.2719	-5.502	
	40	1.33911	-1.3269	-5.502	
	50	1.32562	-1.3819	-5.502	
	60	1.31153	-1.4369	-5.502	
	70	1.29657	-1.4919	-5.502	
	80	1.28172	-1.5469	-5.502	
$Cl_2(aq) + 2e^- = 2Cl^-(aq)$	25	1.396	-0.716	-8.2	From Ref. 1. Early value at 25°C, E_m° = 1.39 V, in Ref. 12. Second isothermal temperature coefficient taken as second derivative of Eq. (1.2.33).
	40	1.384	-0.840	-8.2	
	60	1.366	-1.004	-8.2	
	80	1.344	-1.169	-8.2	
$Cl_3^-(aq) + 2e^- = 3Cl^-(aq)$	25	1.4152	-0.534	-3.94	From the recommended values of E_m° in Table 1.2.1 and the equilibrium constants for the Reaction (1.2.4) in Ref. 1. Second isothermal temperature coefficient taken as second derivative of Eq. (1.2.34).
	40	1.407	-0.593	-3.94	
	60	1.394	-0.672	-3.94	
	80	1.38	-0.751	-3.94	

Table 1.2.3. Standard Thermodynamic Functions for Some Important Reactions Referring to the Chlorine/Chloride Electrode in Aqueous Solution at 25°C and 1 atm. Reference: $E_m^\circ(H^+ + e^- = 1/2H_2) = 0$ at all Temperatures

Reaction	ΔG° (kcal/mole)	ΔH° (kcal/mole)	ΔS° (cal/deg mole)	Refs.
$\frac{1}{2}Cl_2(g) + e^- = Cl^-$	-31.326	-39.903	-28.768	[4]
	-31.325	-39.85	-28.6	[1]
$\frac{1}{2}Cl_2(g) + \frac{1}{2}H_2(g) = H^+ + Cl^-$	-31.331	-39.895	-28.713	[3]
$Cl_2(g) + Cl^- = Cl_3^-$	+2.63	-7.44	-33.8	[1]
$Cl_2(aq) + Cl^- = Cl_3^-$	+0.91	-2.40	-11.1	[1]
$Ag(s) + \frac{1}{2}Cl_2(g) = AgCl(s)$	-26.198	-30.310	-13.793	[4]

TABLE 1.2.4. Standard Molal Potentials E_m° and Corresponding Isothermal Temperature Coefficients at 25°C for Aqueous Redox Couples other than Chlorine/Chloride. Underlined Values Calculated from Thermodynamic Data in Table 1.1.1.

Reaction	E_m° (V)	$\left(\frac{dE_m^\circ}{dT}\right)_{isoth}$ (mV/deg)	Refs.	Remarks
$HClO + H^+ + e^- = \frac{1}{2}Cl_2(g) + H_2O$	1.63 (1.63)	-0.14 (-0.27)	[12-15]	
$HClO + H^+ + 2e^- = Cl^- + H_2O$	1.50 (1.49)	- (-0.76)	[14], [16]	
$ClO^- + H_2O + 2e^- = Cl^- + 2OH^-$	0.89 (0.890)	-1.079 (-1.96)	[13-15]	Basic solution
$HClO_2 + 2H^+ + 2e^- = HClO + H_2O$	1.645 (1.67)	-0.55 (-0.55)	[12-15]	
$ClO_2^- + H_2O + 2e^- = ClO^- + 2OH^-$	0.66 (0.681)	-1.454 (-1.45)	[13-15]	Basic solution
$ClO_2(g) + H^+ + e^- = HClO_2$	1.275 (1.19)	-1.44 (-1.39)	[12-15]	
$ClO_2(g) + e^- = ClO_2^-$	1.16 (1.07)	-2.22 (-2.29)	[13-15]	Basic solution
$ClO_3^- + 6H^+ + 6e^- = Cl^- + 3H_2O$	1.45 (1.450)	- (-0.496)	[16]	
$ClO_3^- + 3H^+ + 2e^- = HClO_2 + H_2O$	1.21 (1.18)	-0.25 (-0.180)	[13-15]	
$ClO_3^- + H_2O + 2e^- = ClO_2^- + 2OH^-$	0.33 (0.295)	-1.47 (-1.47)	[13-15]	Basic solution
$ClO_3^- + 2H^+ + e^- = ClO_2(g) + H_2O$	1.15 (1.17)	- (1.04)	[12], [14]	
$ClO_3^- + H_2O + e^- = ClO_2(g) + 2OH^-$	-0.50 (-0.481)	- (-0.634)	[14]	Basic solution
$ClO_4^- + 8H^+ + 8e^- = Cl^- + 4H_2O$	1.35 (1.3877)	- (-0.476)	[16]	
$ClO_4^- + 2H^+ + 2e^- = ClO_3^- + H_2O$	1.19 (1.20)	-0.41 (-0.416)	[12-15]	
$ClO_4^- + H_2O + 2e^- = ClO_3^- + 2OH^-$	0.36 (0.374)	-1.24 (-1.25)	[13-15]	Basic solution

TABLE 1.2.5. Standard Molar Potentials E_c° for the Chlorine/Chloride Couple in Various Nonaqueous or Mixed Solvents and at Different Temperatures

Reaction	Solvent	E_c° (V)	T (°C)	Refs.	Remarks
$Cl_2 + 2e^- = 2Cl^-$	Methanol	1.116	25	[21-23]	Referred to $E_c^\circ(H^+/H_2) = 0$ in same solvent and T
$Cl_2 + 2e^- = 2Cl^-$	Methanol	4.16	25	[24]	Referred to $E_c^\circ(Rb^+/Rb) = 0$ in same solvent and T
$Cl_2 + 2e^- = 2Cl^-$	Ethanol	1.048	25	[22], [23], [25]	Referred to $E_c^\circ(H^+/H_2) = 0$ in same solvent and T
$Cl_2 + 2e^- = 2Cl^-$	Acetonitrile	0.58	25	[22]	Referred to $E_c^\circ(H^+/H_2) = 0$ in same solvent and T
$Cl_2 + 2e^- = 2Cl^-$	Acetonitrile	3.75	25	[24]	Referred to $E_c^\circ(Rb^+/Rb) = 0$ in same solvent and T
$Cl_2 + 2e^- = 2Cl^-$	Liq ammonia	3.96	-50	[23], [24]	Referred to $E_c^\circ(Rb^+/Rb) = 0$ in same solvent and T
$Cl_2 + 2e^- = 2Cl^-$	HCOOH	4.22	25	[24]	Referred to $E_c^\circ(Rb^+/Rb) = 0$ in same solvent and T
$Cl_2 + 2e^- = 2Cl^-$	Nitromethane	1.165	20	[26]	Referred to $E_c^\circ(Ag/AgCl/Cl^-) = 0$ in same solvent and T
$3Cl_2 + 2e^- = 2Cl_3^-$	Nitromethane	0.50	20	[27]	Referred to E_c°(ferricinium/ferrocene) = 0 in same solvent and T
$3Cl_2 + 2e^- = 2Cl_3^-$	Sulfolane	0.48	22	[28]	Referred to E_c°(ferricinium/ferrocene) = 0 in same solvent and T
$Cl_3^- + 2e^- = 3Cl^-$	Sulfolane	0.21	22	[28]	Referred to E_c°(ferricinium/ferrocene) = 0 in same solvent and T
$Cl_3^- + 2e^- = 3Cl^-$	Acetonitrile	0.41	20	[27]	Referred to E_c°(ferricinium/ferrocene) = 0 in same solvent and T

$Cl_2 + 2e^- = 2Cl^-$

$Cl_3^- + 2e^- = 3Cl^-$

and (1.2.36)

$3Cl_2 + 2e^- = 2Cl_3^-$

Table 1.2.5 contains a selection of values of standard potentials E_c° (molar scale) for the chlorine/chloride couple in various solvents and at different temperatures, and quoted vs different reference potentials.

1.3. EQUILIBRIUM DATA

1.3.1. Aqueous Solutions

1.3.1.1. The Formation Constant for the Trichloride Ion. For the equilibrium constants K_4 and K_5 for the reactions

$$Cl_2,g + Cl^-,aq = Cl_3^-,aq \tag{1.2.4}$$

and

$$Cl_2,aq + Cl^-,aq = Cl_3^-,aq \tag{1.2.5}$$

accurate and extensive sets of data were obtained recently by Mussini et al. by the emf method [1] which was described in Section 1.2.1.1 in conjunction with the determination of the standard potentials for the redox couple Cl_2/Cl^-. The temperature range covered was 25 to 80° C.

Earlier values of K_4 and K_5 determined at 25°C by Sherrill and Izard [7] from chlorine solubility measurements, and of K_5 at 25°C determined by Zimmermann and Strong [29] from spectrophotometric measurements, are in satisfactory agreement with the emf-based data [1], as shown in Table 1.3.1.

1.3.1.2. Mean Molal Activity Coefficients, and Partial Molal Thermodynamic Functions for Hydrochloric Acid at High Concentrations and Temperatures, from emf Measurements of Hydrogen/Chlorine Cells. The reversible emf's E_{13} of Cell (1.2.13) (cf. Section 1.2.1.1) have been measured working with $p_{H_2} = 1$ atm and $p_{Cl_2} = 0.1$ atm. If $E_{13}{}^1$ denotes the corresponding emf's at $p_{H_2} = 1$ atm and $p_{Cl_2} = 1$ atm, the $E_{13}{}^1$'s can be obtained from the E_{13}'s by simply adding the term (k/2). The corresponding cell reaction is

$$\tfrac{1}{2}H_2(g, 1\text{ atm}) + \tfrac{1}{2}Cl_2(g, 1\text{ atm}) = HCl(aq, m) \tag{1.3.1}$$

Thus from $E_{13}{}^1$ and $dE_{13}{}^1/dT$ the values can be obtained for the partial molal free energies, enthalpies, and entropies of HCl at various molalities, the required $dE_{13}{}^1/dT$ values at each HCl molality being taken from the first derivatives of least-squares polynomials of the type

$$E_{13}{}^1 = a + bT + cT^2 \tag{1.3.2}$$

TABLE 1.3.1. Equilibrium Constants for Reactions (1.2.4) and (1.2.5) at Various Temperatures in Aqueous Solutions

T (°C)	25	40	60	80	Refs.
K_4 (atm^{-1})	0.0119 ± 0.0009	0.0064 ± 0.0009	0.0032 ± 0.0007	0.0012 ± 0.0010	[1]
	0.010	-	-	-	[7]
K_5 (kg/mole)	0.215 ± 0.022	0.18 ± 0.03	0.14 ± 0.03	0.07 ± 0.06	[1]
	0.176	-	-	-	[7]
	0.191	-	-	-	[29]

The partial molal functions $\bar{G}$, $\bar{H}$, and $\bar{S}$ for HCl at 25°C at various molalities are collected in Table 1.3.2. The limits of error are ±2 cal mole^{-1}, ±5 cal mole^{-1}, and ±0.009 cal deg^{-1} mole^{-1} for $\bar{G}$, $\bar{H}$, and $\bar{S}$, respectively. Table 1.3.2 also includes the values of the apparent molal enthalpies ϕ_H obtained by means of the relationship

$$\phi_H = \bar{H} - (n_1/n_2)\int (x_2/x_1)\, d\bar{L} \tag{1.3.3}$$

where mole fractions x_1 and x_2, and numbers of moles n_1 and n_2, refer to water and HCl, respectively, and $\bar{L}$ is the relative partial molal enthalpy of HCl, defined by

$$\bar{L} = -2RT^2(\partial \ln \gamma_\pm/\partial T) \tag{1.3.4}$$

Finally, Table 1.3.2 reports the mean molal activity coefficients for HCl at the various molalities at different temperatures, determined as described in Section 1.2.1.1.

Aqueous hydrochloric acid is among those strong electrolytes whose thermodynamic and electrochemical parameters have been extensively determined by many physically independent methods (see the classical books by Harned and Owen [30] and by Robinson and Stokes [31]). Its activity coefficients were determined with the greatest accuracy, and this is the reason for selecting and proposing the value of the mean molal activity coefficient of 0.01 molal aqueous HCl as a standard value for indirect checks of standard potentials of chloride-reversible electrodes of different kinds and constructions [11].

Determinations of $(\gamma_\pm)_{HCl}$ at high temperatures (up to 275°C) were carried out by Greeley et al. [32].

The transference numbers of aqueous HCl, previously determined by Harned and Dreby [33] by the emf method over the temperature range 5 to 50°C at HCl concentrations from 0.005 to 3 mole/kg, were extended over a molality range up to 15 mole/kg by Lengyel [34], again by the emf method. In both the above works [33, 34] the key formula was the thermodynamic equation

$$t_{H^+} = (dE_t/dm_{HCl})/(dE/dm_{HCl}) = dE_t/dE \tag{1.3.5}$$

i.e., a ratio of differential quotients [35] of cell emf's, where E_t is the emf of a "transference cell" of the type

TABLE 1.3.2. Thermodynamic Functions for Aqueous Hydrochloric Acid at Various Molalities (from Ref. 3). Temperature 25°C Unless Otherwise Indicated. $\bar{G}$, $\bar{H}$, and ϕ_H are in cal/mole; $\bar{S}$ in cal/deg mole. Underlined Data Are Taken from Ref. 2 for Comparison.

m_{HCl}	n_{HCl}/n_{H_2O}	$\bar{G}$	$\bar{H}$	ϕ_H	$\bar{S}$	$(\gamma_\pm)_{25°}$	$(\gamma_\pm)_{30°}$	$(\gamma_\pm)_{50°}$	$(\gamma_\pm)_{70°}$	$(\gamma_\pm)_{80°}$
Std state (hypothetical m = 1)	-	-31331	-39895	-39895 (-39952)	13.526	(1.000)	(1.000)	(1.000)	(1.000)	(1.000)
1.000	1/55.51	-31586	-39315	-39543 -	16.342	0.8063	0.8004	0.7709	0.7413	0.7287
3.084	1/18	-29634	-38424	-39135 -	12.786	1.358	1.307	1.217	1.101	1.051
3.701	1/15	-29228	-38067	-38927 (-38964)	12.621	1.594	1.535	1.401	1.255	1.193
4.626	1/12	-28643	-37539	-38590 (-38762)	12.430	2.088	2.011	1.783	1.569	1.481
5.551	1/10	-28087	-37136	-38371 (-38556)	11.915	2.784	2.676	2.306	1.986	1.856
6.938	1/8	-27305	-36473	-38051 (-38242)	11.517	4.308	4.120	3.406	2.825	2.586
7.930	1/7	-26785	-35992	-37837 -	11.385	5.844	5.561	4.466	3.589	3.225
9.251	1/6	-26143	-35425	-37582 (-37687)	11.136	8.616	8.121	6.281	4.814	4.197
10.092	1/5.5	-25764	-35125	-37447 -	10.868	10.87	10.17	7.69	5.70	4.86
11.102	1/5	-25340	-34758	-37318 (-37231)	10.679	14.14	13.10	9.63	6.83	5.65

$$Ag|AgCl|HCl(m_2, \text{varied})||HCl(m_1, \text{fixed})|AgCl|Ag \qquad (1.3.6)$$

the double vertical bar denoting the junction between two HCl solutions, and E is the emf of the corresponding "cell without transference"

$$Ag|AgCl|HCl(m_2, \text{varied})|H_2(1\ \text{atm})|Pt|H_2(1\ \text{atm})|HCl(m_1, \text{fixed})|AgCl|Ag \qquad (1.3.7)$$

An interesting discussion of the influence of the diffusion processes on the accuracy of the t_{H^+} determined through Eq. (1.3.5) is given in Lengyel's work [34].

1.3.1.3. Other Equilibrium Data (Aqueous Solutions). The equilibrium constants K_8, K_9, and K_{10} for the reactions of dissociation of hypochlorous and chlorous acids

$$HClO = H^+ + ClO^- \qquad (1.3.8)$$

$$HClO_2 = H^+ + ClO_2^- \qquad (1.3.9)$$

and of chlorine hydrolysis (disproportionation)

$$Cl_2,aq + H_2O = HClO + H^+ + Cl^- \qquad (1.3.10)$$

are key data for the calculations of the values of the standard potentials of redox couples involving the hypochlorite and the chlorite ions.

The more recent data for K_8 and K_9 are critically presented in the exhaustive review by Perrin [36] on dissociation constants ($pK_8 = 7.30$ to 7.537; $pK_9 = 1.94$ at 25°C). K_{10} was accurately redetermined by Zimmermann and Strong [29] by a spectrophotometric method ($pK_{10} = 3.47$ at 25°C).

1.3.2. Nonaqueous Solutions

1.3.2.1. The Formation Constant for the Trichloride Ion in Nonaqueous Solvents. Very few data are available for the formation constant, $K_{Cl_3^-} = a_{Cl_3^-}/(a_{Cl_2}a_{Cl^-})$, of the trichloride ion in nonaqueous solvents. Following a general trend similar to that in aqueous solutions, the formation constant for Cl_3^- is lower than that for Br_3^- and even lower than that for I_3^- according to the order $K_{I_3^-} > K_{Br_3^-} \gg K_{Cl_3^-}$ [26-28, 37]: in sulfolane, for example, the relative values are $10^{7.4}$, $10^{6.8}$, and $10^{3.1}$, respectively, at 22°C. However, there was a claim [38] that in aprotic solvents the above order was reversed, say, $K_{Cl_3^-} \gg K_{Br_3^-} > K_{I_3^-}$, and the following enormous values were quoted for $K_{Cl_3^-}$: $>10^{13}$, $>10^{12}$, and 10^{10} in nitromethane, acetone, and acetonitrile, respectively [38]. In contrast, arguments and experimental proofs were put forward leading to the conclusion that in such aprotic solvents the Cl_3^- ion was unstable [26-28, 37]. It is evident that this matter requires resolution and supplementary data by independent methods are awaited.

1.3.2.2. Activity Coefficients and Thermodynamic Functions for HCl in Nonaqueous or Mixed Solvents. There is a substantial body of activity coefficient data, essentially based on reversible emf measurements, for HCl at various concentrations and temperatures in different nonaqueous or mixed solvents. The greatest part of this work was carried out by Harned's school [39] using the cell

$$Pt\,|H_2, 1\ atm\,|HCl(m)\ in\ H_2O\ (x) + Solvent\ S\ (100 - x)\,|AgCl\,|Ag\,|Pt \qquad (1.3.11)$$

and by Schwabe's school [40-46] using the alternative cell

$$Pt\,|H_2, 1\ atm\,|HCl(m)\ in\ H_2O\ (x) + Solvent\ S\ (100 - x)\,|Hg_2Cl_2\,|Hg\,|Pt \qquad (1.3.12)$$

Both cells have an emf expressed by

$$E = E^\circ - 2k \log\left({}^{s}_{s}\gamma_{\pm} m\right)_{HCl} \qquad (1.3.13)$$

where superscript "s" indicates working in the S solvent (nonaqueous or mixed) and subscript "s" indicates that the activity coefficient is measured relative to unity at infinite dilution in the said solvent S.

Equation (1.3.13) requires the classical procedure implying the extrapolation of the plot of a suitably arranged function of E vs m_{HCl} to obtain E° as ordinate at $m_{HCl} = 0$, and the subsequent calculation of the required ${}^{s}_{s}\gamma_{\pm}$ from Eq. (1.3.13) directly. Partial molal quantities and medium effects can also be derived.

The nonaqueous solvents used (alone or in mixture with water) were [39-46] methanol, ethanol, n-propanol, isopropanol, ethylene glycol, glycerol, sucrose, glucose, fructose, acetone and 1,4-dioxane. (In the very extensive investigation by Harned's school on HCl in H_2O-dioxane mixtures, the transference numbers for HCl were also determined over a wide range of HCl concentrations, dioxane percentages, and temperatures [33] by the transference cell method mentioned in Section 1.3.1.2.)

Valuable $\left({}^{s}_{s}\gamma_{\pm}\right)_{HCl}$ data were recently obtained by working with solvents other than those considered above, e.g., acetonitrile [47], 1,2-dimethoxyethane [48], tetrahydrofuran [49], methylcellosolve [50], propylene glycol [51], triethylene glycol [52], methyl ethyl ketone [52], and formamide [53].

2. KINETIC STUDIES

2.1. KINETIC PARAMETERS AND MECHANISMS

2.1.1. Aqueous Solutions

2.1.1.1. The Cl_2/Cl^- Couple. The electrochemical kinetics of the Cl_2/Cl^- couple has been widely studied in the last decade and, at present, a large amount of information is available on the reaction mechanism and, particularly, on the rate-determining steps.

Tables 2.1.1 to 2.1.5 collect the existing data on diffusion coefficients, transfer coefficients (or "symmetry factors"), anodic and cathodic Tafel's coefficients, exchange currents,

TABLE 2.1.1. Diffusion Coefficients D for the Aqueous Cl^- Ion at 25°C

$D \times 10^5$ (cm^2/sec)	Method	Electrolyte	Refs.
0.82	Chronopotentiometry	0.1 M HCl + 1 M H_2SO_4	[54]
1.77	Rot Pt dsk electrode	0.0095 M HCl + 1.88 M H_2SO_4	[57]

charge-transfer resistances, standard rate constants, activation energies, stoichiometric numbers, and reaction numbers. Data on the degree of coverage of adsorbed Cl^- ions are shown in Table 2.1.6.

2.1.1.1.1. Platinum Electrode-Cathodic Reduction of Molecular Chlorine. Although the greatest part of the efforts has been devoted to the study of the anodic oxidation of Cl^-, owing to the industrial interest in this reaction, considerable attention has also been given to the cathodic reaction $Cl_2 + 2e^- = 2Cl^-$, aq.

The mechanisms considered are

$$Cl_2, aq = 2Cl_{ads} \quad (2.1.1a)$$

$$Cl_{ads} + e^- = Cl^-, aq \quad (2.1.1b)$$

$$Cl_2, aq + e^- = Cl_{ads} + Cl^-, aq \quad (2.1.2a)$$

$$Cl_{ads} + e^- = Cl^-, aq \quad (2.1.2b)$$

and/or

$$2Cl_{ads} = Cl_2, aq \quad (2.1.2c)$$

and finally

$$Cl_2, aq = Cl_{2, ads} \quad (2.1.3a)$$

$$Cl_{2, ads} + e^- = Cl_{ads} + Cl^-, aq \quad (2.1.3b)$$

$$Cl_{ads} + e^- = Cl^-, aq \quad (2.1.3c)$$

and/or

$$2Cl_{ads} = Cl_2, aq \quad (2.1.3d)$$

where the subscript "ads" denotes a species adsorbed at the electrode surface. The first important contribution on this subject was provided by Frumkin and Tedoradse [56]. On the basis of experimental kinetic parameters, i.e., ν, n_{Cl_2}, and n_{Cl^-} (cf. Table 2.1.5), these authors suggested Mechanism (2.1.2) as the most probable one, with Step 2.1.2a) being rate-determining; the lack of appropriate anodic measurements, however, prevented the authors from clearing up completely the nature of the second step, which could be Eq. (2.1.2b) as well as Eq. (2.1.2c). It should be pointed out that, in the first case, a stoichiometric number of 2 would be obtained if the exchange currents j_0 of Steps (2.1.2a) and (2.1.2b) are only slightly different [56, 89]. Frumkin and Tedoradse's mechanism has recently been confirmed by Faita et al. [86] who studied electrolytically platinum-coated titanium electrodes. They found a stoichiometric

TABLE 2.1.2. Cathodic Transfer Coefficients α and Tafel Slopes b for the Reaction $Cl_2 + 2e^- = 2Cl^-$ at 25°C (Unless Otherwise Stated)

α	b (V/dec.)	Method	Solution	Refs.	Comments
0.69	0.085	Rot Pt dsk electrode	2.2 M $HClO_4$ + 0.012 to 0.063 M HCl	[56]	Electrochemically activated bright Pt electrodes
0.27	-	Quasi-steady E/j curves	1 M H_2SO_4 + 0.05 M HCl	[61]	Bright Pt electrodes
0.27	-	Quasi-steady E/j curves	1 M H_2SO_4 + 0.05 M HCl	[63]	Pt-coated Ti and Ta electrodes
0.28	-	Quasi-steady E/j curves	1 M H_2SO_4 + 0.05 M HCl	[79]	Ag/AgCl electrodes
-	0.18	Rot dsk electrode	1 M $NaClO_4$ + 1% Cl_2 + 0.02 to 0.14 M NaCl	[86]	T = 30°C; electrochemically activated Pt-Ir 30% electrodes (Ti substrate); E > 1.15 V vs SHE
-	-	Rot dsk electrode	1 M $NaClO_4$ + 1% Cl_2 + 0.02 to 0.14 M NaCl	[86]	T = 30°C; electrochemically activated Pt-Ir 30% electrodes (Ti substrate); E = 0.95 to 1.1 V SHE
-	0.4 to 0.5	Rot dsk electrode	1 M $NaClO_4$ + 1% Cl_2 + 0.02 to 0.14 M NaCl	[86]	T = 30°C; electrochemically activated Pt-Ir 30% electrodes (Ti substrate); E < 0.85 V vs SHE
-	0.2	Rot dsk electrode	1 M $NaClO_4$ + 1% Cl_2 + 0.02 to 0.14 M NaCl	[86]	T = 30°C; Pt-coated, electrochemically activated, Ti electrodes; E > 1.15 V vs SHE
-	0.4 to 0.5	Rot dsk electrode	1 M $NaClO_4$ + 1% Cl_2 + 0.02 to 0.14 M NaCl	[86]	T = 30°C; Pt-coated, electrochemically activated, Ti electrodes; E = 0.85 to 1.1 V vs SHE

TABLE 2.1.3. Anodic Transfer Coefficients (1 - α) and Tafel Slopes b (V/dec) for the Reaction $Cl_2 + 2e^- = 2Cl^-$ at 25°C (Unless Otherwise Stated)

1 - α	b	Solution	Refs.	Remarks
0.41	0.143	1 M H_2SO_4 + 0.1 to 0.2 M HCl	[54]	Chronopotentiometry; bright Pt electrodes
0.205	-	0.5 M NaCl at pH = 5.2	[55]	Quasi-steady E/j curves; chemically and electrochemically activated bright Pt electrodes. E < 1.6 V vs SHE
0.235	-	0.5 M NaCl at pH = 0.75	[55]	Quasi-steady E/j curves; chemically and electrochemically activated bright Pt electrodes. E < 1.6 V vs SHE
0.400	-	2.0 M NaCl at pH = 5.2	[55]	Quasi-steady E/j curves; chemically and electrochemically activated bright Pt electrodes. E < 1.6 V vs SHE
0.143	-	0.5 M NaCl at pH = 5.2	[55]	Quasi-steady E/j curves; chemically and electrochemically activated bright Pt electrodes. E > 2.0 V vs SHE
0.130	-	0.5 M NaCl at pH = 0.75	[55]	Quasi-steady E/j curves; chemically and electrochemically activated bright Pt electrodes. E > 2.0 V vs SHE
0.162	-	2.0 M NaCl; pH = 5.2	[55]	Quasi-steady E/j curves; chemically and electrochemically activated bright Pt electrodes. E > 2.0 V vs SHE
0.48	-	1.0 M HCl	[55]	Quasi-steady E/j curves; bright Pt electrodes
-	0.040	0.1 M KCl	[58]	Quasi-steady E/j curves; bright Pt electrodes; with electrochemical pretreatment
-	0.450	0.8 to 3.3 M KCl	[59]	Quasi-steady E/j curves; bright Pt electrodes; E range 1.6 to 2.5 V vs SHE
0.72	-	1 M H_2SO_4 + 0.05 M HCl	[61]	Bright Pt electrodes
0.72 to 0.74	-	1 M H_2SO_4 + 0.05 M HCl	[63]	Pt-coated Ti and Ta electrodes
-	0.120		[65]	Bright Pt electrodes in "active" state. Cyclic voltammetry
-	0.270		[65]	Bright Pt electrodes in "passive" state. Cyclic voltammetry
-	0.12	1 M $NaClO_4$ + 1% Cl_2 + 0.02 to 0.14 M NaCl	[86]	Pt-coated Ti electrodes; rot dsk method; E < 1.55 V vs SHE

-	0.60	1 M $NaClO_4$ + 1% Cl_2 + 0.02 to 0.14 M NaCl	[86]	Pt-coated Ti electrodes; rot dsk method; 1.55 < E < 1.75 V vs SHE
-	0.030	1 M to satd NaCl + 1% of Cl_2	[87]	Rot dsk, Pt-coated Ti electrodes; up to j = 0.14 A/cm^2 in 1 M NaCl; up to j = 0.30 A/cm^2 in satd NaCl; Pt-Ir (0.5 to 4%) in all j-ranges. T = 30°C
-	0.15 to 0.20	1 M $NaClO_4$ + 1% Cl_2 + 0.02 to 0.14 M NaCl	[86]	Pt-Ir 30%-coated Ti electrodes; E < 1.55 V (SHE); T = 30°C
-	0.60	1M $NaClO_4$ + 1% Cl_2 + 0.02 to 0.14 M NaCl	[86]	Pt-Ir 30%-coated Ti electrodes; 1.55 < E < 1.75 V
1.0	0.060	1.5 M HCl + 2.9 M NaCl	[72]	Steady E/j curves; electrodes: graphite, plastic-treated graphite, pyrographite. j < 0.01 A/cm^2
0.5	0.120	1.5 M HCl + 2.9 M NaCl	[72]	Steady E/j curves; electrodes: graphite, plastic-treated graphite, pyrographite. j > 0.01 A/cm^2
0.38	0.151	1 M HCl + 4 M NaCl	[74]	Nonaged graphite electrodes; 0.0005 < j < 0.1 A/cm^2. T = 14.5°C
0.46	0.127	1 M HCl + 4 M NaCl	[74]	Nonaged graphite electrodes; 0.0005 < j < 0.1 A/cm^2. T = 25°C
0.49	0.123	1 M HCl + 4 M NaCl	[74]	Nonaged graphite electrodes; 0.0005 < j < 0.1 A/cm^2. T = 30°C
0.61	0.102	1 M HCl + 4 M NaCl	[74]	Nonaged graphite electrodes; 0.0005 < j < 0.1 A/cm^2. T = 40°C
0.63	0.101	1 M HCl + 4 M NaCl	[74]	Nonaged graphite electrodes; 0.0005 < j < 0.1 A/cm^2. T = 50°C
0.372	-	Satd NaCl	[78]	Graphite electrodes; E/j curves; ohmic-drop errors accounted for. T = 20°C
0.462	-	Satd NaCl	[78]	Graphite electrodes; E/j curves; ohmic-drop errors accounted for. T = 40°C
0.659	-	Satd NaCl	[78]	Graphite electrodes; E/j curves; ohmic-drop errors accounted for. T = 60°C
0.712	-	Satd NaCl	[78]	Graphite electrodes; E/j curves; ohmic-drop errors accounted for. T = 80°C
-	0.040	1 M NaCl + 1% Cl_2	[88]	RuO_2-coated Ti electrodes; steady E/j curves. T = 30°C
-	0.032	4 M NaCl + 1% Cl_2	[88]	RuO_2-coated Ti electrodes; steady E/j curves. T = 30°C
0.174	-	0.5 M NaCl	[75]	Steady E/j curves; PbO_2 anodes
0.27	-	2.0 M NaCl	[75]	Steady E/j curves; PbO_2 anodes
0.69	-	1 M H_2SO_4 + 0.05 M HCl	[79]	Steady E/j curves; Ag/AgCl electrodes

TABLE 2.1.4. Exchange Currents, j_0 (A/cm^2), Charge-Transfer Resistances, $R_t = (\partial\eta/\partial j)_{j=0}$ (ohm/cm^2), Standard Rate Constants, k° (cm/sec), and Activation Energies, $\Delta G^\ddagger$ (kcal/mole of Cl_2), for the Reaction $Cl_2 + 2e^- = 2Cl^-$ at 25°C

$10^3 j_0$	R_t	$10^3 k^\circ$	$\Delta G^\ddagger$	Refs.	Remarks
5.7	4.0	-	-	[56]	Bright Pt electrodes; 2.2 M $HClO_4$ + 0.012 M HCl; cathodic
8.0	3.0	-	-	[56]	Bright Pt electrodes; 2.2 M $HClO_4$ + 0.021 M HCl
13	2.0	-	-	[56]	Bright Pt electrodes; 2.2 M $HClO_4$ + 0.037 M HCl
16	1.5	-	-	[56]	Bright Pt electrodes; 2.2 M $HClO_4$ + 0.063 M HCl
5	-	-	-	[55]	Bright Pt electrodes; 1.0 M HCl
-	-	1.73	-	[54]	Chronopotentiometry; bright Pt electrodes; 1 M H_2SO_4 + 0.1 to 0.2 M HCl
-	-	1.13	-	[61]	Bright Pt electrodes; 1 M H_2SO_4 + 0.05 M HCl
-	-	-	2.26	[62]	Bright Pt electrodes; 1 M H_2SO_4 + 0.05 M HCl
-	-	1.1 to 1.28	-	[63]	Bright Pt electrodes; 1 M H_2SO_4 + 0.05 M HCl; Pt-coated Ti and Ta electrodes
1.85	-	-	-	[55]	Bright Pt electrodes; 0.5 M NaCl, pH = 5.2; extrapolation from E/j curves for E < 1.6 V vs SHE
1.70	-	-	-	[55]	Bright Pt electrodes; 0.5 M NaCl, pH = 0.75; extrapolation from E/j curves for E < 1.6 V vs SHE

4.45	-	-	-	[55]	Bright Pt electrodes; 2 M NaCl, pH = 5.2; extrapolation from E/j curves for E < 1.6 V vs SHE
0.17	-	-	-	[55]	Bright Pt electrodes; 0.5 M NaCl, pH = 5.2; extrapolation from E/j curves for E > 2.0 V vs SHE
0.49	-	-	-	[55]	Bright Pt electrodes; 0.5 M NaCl; pH = 0.75; extrapolation from E/j curves for E > 2.0 V vs SHE
0.37	-	-	-	[55]	Bright Pt electrodes; 2.0 M NaCl, pH = 5.2; extrapolation from E/j curves for E > 2.0 V vs SHE
0.27 ± 0.03	-	-	-	[74]	Apparent exchange current; aged graphite electrodes; roughness factor = 30; 4 M NaCl + 1 M HCl
0.25	-	-	-	[74]	Apparent exchange current; nonaged electrodes; roughness factor = 2; 4 M NaCl + 1 M HCl
-	-	-	8.4	[74]	Apparent exchange current; nonaged electrodes; roughness factor = 2; 4 M NaCl + 1 M HCl at E (1 atm Cl_2)
0.008	-	-	-	[75]	PbO_2 electrodes; 0.5 M NaCl, pH = 5.2
0.0004	-	-	-	[75]	PbO_2 electrodes; 2 M NaCl, pH = 5.2
-	-	1.26	-	[79]	Ag/AgCl electrodes; 1 M H_2SO_4 + 0.05 M HCl
-	-	-	9.2	[58]	Bright Pt electrodes; 0.1 M KCl, E = 1.39 V (SHE)

TABLE 2.1.5. Stoichiometric Numbers ν and Reaction Orders n for the Reaction $Cl_2 + 2e^- = 2Cl^-$ at 25° C (Unless Otherwise Stated)

ν	Anodic n_{Cl_2}	Anodic n_{Cl^-}	Cathodic n_{Cl_2}	Cathodic n_{Cl^-}	Refs.	Remarks
-	-	1.1	-	-	[59]	Calcd from E/j curves; smooth Pt electrodes; 0.8 to 3.3 M (neutral) KCl
1	0	2	0.9	0	[68]	Smooth Pt electrodes; 1 M H_2SO_4 + 0.006 to 0.2 M HCl; p_{Cl_2} = 0.0891 to 1 atm; E = -0.15 V vs E(Cl_2/Cl^-) in 1 M H_2SO_4 + 0.2 M HCl + Cl_2 (1 atm)
1	0	2	0.8	0	[68]	Smooth Pt electrodes; 1 M H_2SO_4 + 0.006 to 0.2 M HCl; p_{Cl_2} = 0.0891 to 1 atm; E = -0.1 V vs E (Cl_2/Cl^-) in 1 M H_2SO_4 + 0.2 M HCl + Cl_2 (1 atm)
1.25	0	1.7	0.6	0	[68]	Smooth Pt electrodes; 1 M H_2SO_4 + 0.006 to 0.2 M HCl; p_{Cl_2} = 0.0891 to 1 atm; E = -0.05 V vs E (Cl_2/Cl^-) in 1 M H_2SO_4 + 0.2 M HCl + Cl_2 (1 atm)
2	0	1	0.5	0	[68]	Smooth Pt electrodes; 1 M H_2SO_4 + 0.006 to 0.2 M HCl; p_{Cl_2} = 0.0891 to 1 atm; E = 0.0 V vs E (Cl_2/Cl^-) in 1 M H_2SO_4 + 0.2 M HCl + Cl_2 (1 atm)
2	0	1.1	0.5	0	[68]	Smooth Pt electrodes; 1 M H_2SO_4 + 0.006 to 0.2 M HCl; p_{Cl_2} = 0.0891 to 1 atm; E = 0.05 to 0.25 V vs E (Cl_2/Cl^-) in 1 M H_2SO_4 + 0.2 M HCl+Cl_2 (1 atm)
1	0	2	1	0	[68]	Smooth Ir electrodes; 1 M H_2SO_4 + 0.006 to 0.2 M HCl; p_{Cl_2} = 0.0891 to 1 atm; E = -0.2 to -0.1 V vs E (Cl_2/Cl^-) in 1 M H_2SO_4 + 0.2 M HCl+Cl_2 (1 atm)
1	0	1.9	1	-0.1	[68]	Smooth Ir electrodes; 1 M H_2SO_4 + 0.006 to 0.2 M HCl; p_{Cl_2} = 0.0891 to 1 atm; E = -0.05 V vs E (Cl_2/Cl^-) in 1 M H_2SO_4 + 0.2 M HCl + Cl_2 (1 atm)
1	0	1.6	1	-0.4	[68]	Smooth Ir electrodes; 1 M H_2SO_4 + 0.006 to 0.2 M HCl; p_{Cl_2} = 0.0891 to 1 atm; E = 0.0 V vs E (Cl_2/Cl^-) in 1 M H_2SO_4 + 0.2 M HCl + Cl_2 (1 atm)
1	0	1.4	1	-0.6	[68]	Smooth Ir electrodes; 1 M H_2SO_4 + 0.006 to 0.2 M HCl; p_{Cl_2} = 0.0891 to 1 atm; E = 0.05 V vs E (Cl_2/Cl^-) in 1 M H_2SO_4 + 0.2 M HCl + Cl_2 (1 atm)

1	0	1	1	-1	[68]	Smooth Ir electrodes; 1 M H_2SO_4 + 0.006 to 0.2 M HCl; p_{Cl_2} = 0.0891 to 1 atm; E = 0.1 to 0.2 V vs E (Cl_2/Cl^-) in 1 M H_2SO_4 + 0.2 M HCl + Cl_2 (1 atm)
1	-0.1	2	0.9	0	[68]	Smooth Rh electrodes; 1 M H_2SO_4 + 0.006 to 0.2 M HCl; p_{Cl_2} = 0.0891 to 1 atm; E = -0.2 V vs E (Cl_2/Cl^-) in 1 M H_2SO_4 + 0.2 M HCl + Cl_2 (1 atm)
1	-0.11	1.8	0.8	-0.1	[68]	Smooth Rh electrodes; 1 M H_2SO_4 + 0.006 to 0.2 M HCl; p_{Cl_2} = 0.0891 to 1 atm; E = -0.1 V vs E (Cl_2/Cl^-) in 1 M H_2SO_4 + 0.2 M HCl + Cl_2 (1 atm)
1.25	0	1.5	0.8	-0.2	[68]	Smooth Rh electrodes; 1 M H_2SO_4 + 0.006 to 0.2 M HCl; p_{Cl_2} = 0.0891 to 1 atm; E = -0.05 V vs E (Cl_2/Cl^-) in 1 M H_2SO_4 + 0.2 M HCl + Cl_2 (1 atm)
1.50	0	1.1	0.7	-0.4	[68]	Smooth Rh electrodes; 1 M H_2SO_4 + 0.006 to 0.2 M HCl; p_{Cl_2} = 0.0891 to 1 atm; E = 0.0 V vs E (Cl_2/Cl^-) in 1 M H_2SO_4 + 0.2 M HCl + Cl_2 (1 atm)
1.60	0	1	0.7	-0.4	[68]	Smooth Rh electrodes; 1 M H_2SO_4 + 0.006 to 0.2 M HCl; p_{Cl_2} = 0.0891 to 1 atm; E = 0.05 V vs E (Cl_2/Cl^-) in 1 M H_2SO_4 + 0.2 M HCl + Cl_2 (1 atm)
-	0	1	-	-	[68]	Smooth Rh electrodes; 1 M H_2SO_4 + 0.006 to 0.2 M HCl; p_{Cl_2} = 0.0891 to 1 atm; E = 0.1 V vs E (Cl_2/Cl^-) in 1 M H_2SO_4 + 0.2 M HCl + Cl_2 (1 atm)
-	0	1	-	-	[68]	Smooth Rh electrodes; 1 M H_2SO_4 + 0.006 to 0.2 M HCl; p_{Cl_2} = 0.0891 to 1 atm; E = 0.2 V vs E (Cl_2/Cl^-) in 1 M H_2SO_4 + 0.2 M HCl + Cl_2 (1 atm)
1.9 to 2.0	0	-	-	-	[86]	Pt- and Pt-Ir 30%-coated Ti electrodes; 1 M $NaClO_4$ + 0.02 to 0.14 M NaCl + Cl_2 (0.01 atm); rot dsk, anodic data, E = 1.35 to 1.50 V vs SHE. T = 30°C
2.3	-	-	1	0	[86]	Pt-coated Ti cathode; 1 M $NaClO_4$ + 0.02 to 0.14 M NaCl + Cl_2 (0.01 atm); rot dsk, anodic data, E = 1.25 to 1.35 V vs SHE. T = 30°C
2.0	-	-	0.5	0	[86]	Pt-Ir 30%-coated Ti cathode; 1 M $NaClO_4$ + 0.02 to 0.14 M NaCl + Cl_2 (0.01 atm); rot dsk, anodic data, E = 1.25 to 1.35 V vs SHE. T = 30°C
-	-	-	1	0	[86]	Pt-coated Ti cathode; 1 M $NaClO_4$ + 0.02 to 0.14 M NaCl + Cl_2 (0.01 atm); rot dsk, anodic data; E = 1.15 V vs SHE. T = 30°C
2	-	-	1	0	[56]	Smooth Pt electrodes; 2.2 M $HClO_4$ + 0.012 to 0.063 M HCl. Cathodic data

TABLE 2.1.6. Degree of Coverage θ for Adsorbed Cl^- Ions on Pt Electrodes at 25°C as Determined with Different Methods

E^a	θ (radiochemical)	θ (electrochemical)	θ (ellipsometric)	Refs.
0	-	0.07	-	[70]
0.177	0	-	-	[69]
0.200	-	0.28	-	[70]
0.377	0.018	0.27	0	[69]
0.400	-	0.46	-	[70]
0.577	0.07	0.35	0.01	[69]
0.600	-	0.48	-	[70]
0.777	0.101	0.50	0.05	[69]
0.800	-	0.50	-	[70]
0.977	0.12	-	0.11	[69]

[a]Referred to SHE.

number $\nu = 2$ and reaction orders $n_{Cl_2} = 1$ and $n_{Cl^-} = 0$. More recently, Frumkin and Tedoradse's work has been reviewed critically [60] and the reaction order with respect to Cl_2 has been shown to be 0.6 to 0.7 on the basis of a more sophisticated theory. In addition, it has been pointed out that Frumkin's data allowing calculation of the reaction orders were obtained within an overpotential range of about 40 mV. This fact casts some doubt on the reliability of a mechanism where the reverse rate of Step (2.1.2a) has been completely neglected. Moreover, the reaction order $n_{Cl_2} = 1$ was confirmed on the grounds of the observed linearity of the j^{-1} vs $\omega^{-1/2}$ plot (ω = angular velocity of the rotating disk electrode). This kind of test might be insensitive, as has been recently shown by Dickinson et al. [64]. In fact they observed that — within some range of chlorine concentration — either equation

$$1/j = 1/kC_{Cl_2} + 1/(BC_{Cl_2}\ \omega) \qquad (2.1.4)$$

(valid for a reaction order $n_{Cl_2} = 1$) and

$$j^{3/2} = k^{3/2}C_{Cl_2} - jk^{3/2}/(B\sqrt{\omega}) \qquad (2.1.5)$$

(valid for $n_{Cl_2} = 2/3$) fit the experimental data equally well. (As a matter of fact, both equations are particular cases of the general equation

$$j^{-1/n} - j_L^{-1/n} = 1/\left[j^{(1/n)-1}\omega^{\frac{1}{2}} B j_L^{(1/n_K) - (1/n)}\right] \qquad (2.1.6)$$

described in an earlier work [66].) As for the several symbols in Eqs. (2.1.4), (2.1.5), and (2.1.6), $j_L = kC^n$ is the kinetic limiting current, k is the rate constant of the electrochemical reaction, and B is the product of the characteristic constants in Levich's equation for the rotating-disk electrode. This interesting finding is probably due to the range of ω explored (400 to 2400 rpm in Dickinson's work) which does not allow a sufficient change of the surface concentration of molecular chlorine to occur, so that the method is characterized by an intrinsically low sensitivity. As a matter of fact, the ratio between the Cl_2 concentrations at 400 and at 2400 rpm is given by

$$\frac{(C_{Cl_2})_{400\ rpm}}{(C_{Cl_2})_{2400\ rpm}} = \frac{\{(j_D - j)/j_D\}_{400\ rpm}}{\{(j_D' - j)/j_D\}_{2400\ rpm}} \approx 1.7 \tag{2.1.7}$$

j_D denoting the limiting diffusional current at each rotation speed considered [64]. Consequently, in Dickinson's work the reaction order n_{Cl_2} has been calculated using the relationship

$$\log(j) = \log(k) + n_{Cl_2}\log\{(C_{Cl_2})(j_D - j)/j_D\} \tag{2.1.8}$$

which proved to be more reliable; n_{Cl_2} turns out to be a function of the chlorine concentration and is 1 at low C_{Cl_2} and 2/3 at high C_{Cl_2}. This fact, the observed dependence of the reaction rate on the Cl^- concentration, and the nonlinearity of the graph $\log\{j_D j/(j_D - j)\}$ vs the electrode potential suggest that the most probable mechanism should be Mechanism (2.1.2), but now the rate-determining step is (2.1.2b). In addition, the reverse rate of Step (2.1.2a) seems to be of the same order of magnitude of the forward rate of Step (2.1.2b). The latter conclusion is not surprising because the reverse rate of Step (2.1.2a) is a function of the Cl^- concentration which in Dickinson's work was considerably higher than in Frumkin and Tedoradse's. The Mechanism (2.1.3) — the rate-determining step being (2.1.3c) — is equally apt to account for the experimental data. However, at high cathodic overpotentials, Step (2.1.3a) should become rate-determining and the current (now a kinetic current) should be independent of the electrode potential. This behavior is not met in practice, and the adsorption step has been discarded. It is worthwhile to remember that an adsorption stage for molecular chlorine prior to the charge-transfer reaction was tentatively suggested [56] to explain the lack of linearity in the plots log(j) vs potential obtained by Chang and Wick [67].

The most recent and thorough work [68] on the cathodic kinetics of the Cl_2/Cl^- couple is based on short-time galvanostatic pulses (of about 10 msec) — the previous works were based on quasi-steady methods — whereby stoichiometric numbers and reaction orders were shown to be functions of the electrode potentials (cf. Table 2.1.5). This interesting finding points out that in a relatively narrow potential range a change in the reaction mechanism takes place.

In particular, at electrode potentials above 1.42 V vs SHE (standard hydrogen electrode), the reaction follows Path (2.1.1) with Step (2.1.1b) rate-determining. For the sake of comparison, the present values of potential have been transposed onto the SHE scale by setting the ionic activity coefficient $(\gamma_-)_{Cl^-}$, which appears in the equation for the reference electrode potential in Ref. 68, equal to the mean ionic activity coefficient appropriate to the ionic strength of the supporting electrolyte.

At electrode potentials lower than 1.3 V vs SHE, Path (2.1.1) holds again, but Step (2.1.1a) becomes rate-determining as shown by a unit stoichiometric number and a reaction order of 0.9 with respect to Cl_2. The kinetics is independent of Cl^- concentration in the potential range investigated.

A change of the rate-determining step is quite possible because Step (2.1.1a) is, in principle, independent of electrode potential whereas Step (2.1.1b) is a typical electrochemical reaction whose rate is a function of the quantity $\exp(-\alpha \Delta EF/RT)$. At 1.3 to 1.42 V vs SHE potentials, an intermediate situation emerges and both Steps (2.1.1a) and (2.1.1b) can be considered as equally rate-determining ($1 < \nu < 2$; $0.5 < n_{Cl_2} < 0.9$, cf. Table 2.1.5).

A basic difficulty which was treated differently in the various studies is caused by the formation of adsorbed oxygen; this exerts an inhibiting influence, as is well known. This influence can be put in terms of a decrease of the available surface (the oxide-covered sites being obviously lost) and an increase of free energy of adsorption of the intermediates. Inhibition of the cathodic reduction of chlorine is demonstrated quantitatively by the change of the transfer coefficients α_1 and α_2 for Steps (2.1.2a) and (2.1.2b), according to Dickinson et al. [64]. These values as functions of electrode pretreatment (anodization) are:

Pretreatment vs SHE (V)	Time (sec)	α_1	α_2
1.1	60	0.32	0.13
1.8	10	0.23	0.13
1.9	10	0.19	0.12
2.0	60	0.18	0.13

The decrease in the α values is probably caused by the reduction of molecular chlorine taking place at a surface becoming more and more covered by adsorbed oxygen.

In the above work, acid solutions were almost always used for lowering the degree of oxygen coverage as much as possible. It should, however, be remembered that in acid solutions and in the presence of chlorides platinum corrodes [71] so that some doubt exists about the true nature and the reproducibility of the electrode surface. In view of this fact some authors operated in less acidic conditions [86]. A common feature of the published data is that the amount of adsorbed oxygen changes during experiments, more or less depending on the electrode pretreatment. On the other hand, Enyo and Yokoyama [68], who used short-time galvanostatic pulses instead of quasi-steady techniques, were unable to observe any particular effect connected with a change of degree of oxygen coverage. This fact might be related both to the short time required for recording the electrode potential and to the known slow rate of formation and reduction of the oxygen layer in chloride-containing media [73].

In conclusion, it can be said that the surface conditions of platinum electrodes are reasonably reproducible within each of the cited works; these conditions may, however, be quite different from author to author according to electrode pretreatment and recording procedure. In this light, the observed disagreements among the proposed mechanisms and rate-determining steps are easily understood.

2.1.1.1.2. Platinum Electrode-Anodic Oxidation of Cl^- Ions. From a purely thermodynamic point of view, oxygen discharge should be energetically favored instead of that of chlorine, and the equations for the reversible potentials are

$$E_{O_2/H_2O} = 1.23 + k \log\left(a_H + p_{O_2}^{\frac{1}{4}} a_{H_2O}^{-\frac{1}{2}}\right) \qquad (2.1.9)$$

$$E_{Cl_2/Cl^-} = 1.35828 + k \log\left(p_{Cl_2}^{\frac{1}{2}}/a_{Cl^-}\right) \qquad (2.1.10)$$

However, the chlorine discharge is actually favored kinetically, and chlorine is obtained electrolytically with excellent current efficiencies even at high current densities (cf. Refs. 78 and 88, for instance).

The kinetics of anodic oxidation of Cl^- is strongly dependent on the anode potential range, as was clearly shown by Littauer and Shreir [55] who pointed out that the log(j) vs E_{an} relationship is linear over two distinct potential ranges, i.e., at $E_{an} < 1.6$ V (SHE) and at $E_{an} > 2.0$ V (SHE). Considering Tafel's b coefficients and the exchange currents, the reaction mechanisms appear to change mainly because of the strongly varying coverages by adsorbed oxygen. As a matter of fact, at $E_{an} > 2.0$ V the reaction is quite likely to take place at a completely oxidized electrode surface whose catalytic properties are quite different from those of a bare metal. The available kinetic data mainly refer to the potential region below 1.6 V (SHE).

An almost general agreement exists on the reaction mechanism, which can be schematized as

$$Cl^-, aq - e^- = Cl_{ads} \qquad (2.1.11a)$$

$$2Cl_{ads} = Cl_2 \qquad (2.1.11b)$$

On the contrary, conflicting opinions can be found in the literature about the most probable rate-determining step. Tedoradse [57] observed that Mechanism (2.1.11) fits the experimental data best with Step (2.1.11a) the rate-determining one (such a conclusion was previously reached by Ershler [76]).

Quite recently Faita et al. [86] established that for an electrochemically-Pt-coated Ti electrode behaving cathodically according to Path (2.1.2) an anodic mechanism symmetrical to the cathodic one should be disregarded because the experimental data (charge-transfer resistances and exchange currents) did not fit Vetter's equation [77] for the case of a symmetrical two-stage charge-transfer mechanism:

$$\left(\frac{\partial n}{\partial j}\right)_{j \to 0} = \frac{RT}{4F}\left(\frac{1}{j_{0,\ anode}} + \frac{1}{j_{0,\ cathode}}\right) \qquad (2.1.12)$$

Earlier researchers [58, 80], on the contrary, suggested the possibility of Step (2.1.11b) being rate-determining; more recently [87] Mechanism (2.1.11) with Step (2.1.11b) as rate-determining has been chosen as the only way to explain the low Tafel b coefficients (0.030 V/decade at 30°C, cf. Table 2.1.3) obtained working with thermally-Pt-coated Ti anodes in 1 M to saturated NaCl solutions. This disagreement can be settled by taking into account that a change of mechanism (or rather a change of rate-determining step) can take place depending on both the Cl^- concentration and the electrode potentials affecting the rates of either Step (2.1.11a) or Step (2.1.11b). In fact, the rate for Step (2.1.11a) is expressed by

$$v = \left(\frac{kT}{h}\right) a_{Cl^-}(1 - \theta)\exp(-\Delta G_1^{\ddagger}/RT + 1 - \alpha \; F\Delta E/RT) \qquad (2.1.13)$$

where k is Boltzmann's constant, h is Planck's constant, $(1 - \theta)$ is the available surface, and $\Delta G_1^{\ddagger}$ is the chemical activation energy for the forward rate; the other terms have their usual significance. The rate for Step (2.1.11b) — Step (2.1.11a) is supposed to be at equilibrium — is given by

$$v = \left(\frac{kT}{h}\right) K a_{Cl^-}^{\;2} \exp(2F\Delta E/RT) \qquad (2.1.14)$$

where the constant K refers to the equilibrium for Step (2.1.11a). It has recently been demonstrated [68] that at electrode potentials above 1.42 V (SHE) Step (2.1.11a) is rate-determining (cf. Table 2.1.5), whereas at potentials below 1.3 V (SHE) the rate-determining step becomes (2.1.11b). On careful inspection of the experimental conditions adopted by the various authors, we conclude that the earlier works are in satisfactory agreement with the kinetic picture given in Ref. 68. For instance, the data by Faita, Fiori, and Augustynski [86] refer to a potential region above 1.36 V (SHE), whereas those by Faita, Fiori, and Nidola [87] correspond to potentials lower than 1.35 to 1.37 V. Analogous considerations can be made about Ref. 58, although less agreement is found in this case.

A factor which could explain some of the discrepancies found between the existing data is the degree of oxidation of the electrode surface: here the degree of oxidation is, of course, much higher than in the cathodic potential ranges and (as in the case of the cathodic reaction) it varies from author to author according to pretreatment and to measurement techniques. The influence of the adsorbed oxygen layer on electrochemical parameters is adequately pointed out in the literature [55, 57, 59, 65, 80-82, 86, 87]. A theoretical analysis of the effect of adsorbed oxygen has been tentatively given by Tedoradse [57], whose starting points were the kinetic Eq. (2.1.13), the experimental finding that the reaction rate decreases exponentially as a function of the oxygen coverage, and the hypothesis that the adsorbed oxygen is at equilibrium with respect to (H^+, aq) and H_2O. Then the reaction rate becomes

$$v = \frac{kT}{h} a_{Cl^-}(1 - \theta)\exp(-\Delta G_1^{\ddagger}/RT + (1 - \alpha)F\Delta E/RT) \quad \exp(-K a_{H^+}^{-2} \exp 2F\Delta E/RT) \qquad (2.1.15)$$

Equation (2.1.15) predicts an exponential increase of current with increasing electrode potential up to a maximum and a subsequent decrease, as is actually observed, more or less markedly depending on the potential scan rate. From Eq. (2.1.15) one can easily obtain the expression for the maximum current j_{max}:

$$j_{max} \propto a_{Cl^-} a_{H^+}^{\frac{1}{2}} \qquad (2.1.16)$$

an expression which fits the experimental data fairly well. The described procedure does not give a physicochemically significant picture of the reaction occurring at an oxidized surface, and can be criticized especially as regards the hypothesis of adsorption of oxygen in equilibrium with H^+, aq and H_2O; this is not verified in practice, as is clearly shown by the dependence of j_{max} on the potential scan rate. In conclusion, Eq. (2.1.15) can be considered only as an operational expression.

The influence of adsorbed oxygen is qualitatively explained by a general equation which was developed [83] by taking into consideration the decrease of available area for Cl^- ions and reaction intermediates as well as the increase of free energy of adsorption for those species which are brought about by the adsorption of foreign species (oxygen, in the present case). Such a general equation relies upon correct physicochemical hypotheses and was shown to be able to explain the observed j vs E relationships found in the oxidation of Cl^- ions.

The adsorption of chlorides is competitive [85] with that of oxygen. Through the cyclic voltammetry method (scanning rate, 30 mV/sec; lower potential limit, hydrogen discharge; upper potential limit, chlorine discharge) it was shown that oxygen adsorption is more and more hindered by increasing concentration of Cl^-. At Cl^- concentrations above 0.01 M the characteristic peak corresponding to the reduction of the oxidized surface disappears from the voltammograms; of course, this fact holds within the range of time which is typical of the cited method (50 sec). It seems more likely that oxygen adsorption slows down with increasing Cl^- concentration. However, oxygen adsorption at platinum is typically irreversible [84], i.e., adsorbed oxygen cannot be displaced by Cl^-; hence the buildup of the adsorption layer increasingly hinders the electrode, thus affecting the available surface and the kinetic parameters for the Cl^- oxidation.

Finally, it should be borne in mind that the coverage by oxygen is also a function of the pH of the solution, temperature, and electrode potential. It was particularly pointed out that the surface oxidation of a platinum electrode in a chloride-containing solution at pH = 3 is strongly affected by species contained in the bulk solution. According to a tentative explanation [86], ClO^- ions formed by hydrolysis of molecular chlorine (cf. Eq. 1.3.10) have a role in the formation of adsorbed oxygen. In addition, at potentials above 2.0 V (SHE) the oxidation of the platinum surface is very rapid even in the presence of high Cl^- concentrations [55]. Finally, it should be remembered that in the literature data are available concerning the coverages by Cl^- and Cl_{ads} [69, 70] as well as the theoretical elaboration of the Cl^- adsorption in terms of Temkin's isotherm [70] and of adsorption and desorption kinetics [70]. However, a satisfactory treatment of these topics is rather lacking at present so that a detailed picture of the kinetic mechanism concerning the Cl_2/Cl^- couple is not yet possible.

<u>2.1.1.1.3. Iridium Electrode-Anodic and Cathodic Processes for the Chlorine/Chloride Couple.</u> Comparatively few data are available on the kinetics of the Cl_2/Cl^- couple at iridium electrodes.

Recently it has been shown [68] that the anodic and cathodic reaction orders (with respect to molecular chlorine and Cl^- ions) are functions of the electrode potential; the stoichiometric number is, however, unity in all the ranges investigated.

The proposed mechanism is

$$Cl^-, aq - e^- = Cl_{ads} \qquad (2.1.17a)$$

$$Cl_{ads} + Cl^-, aq - e^- = Cl_2, aq \qquad (2.1.17b)$$

both in the cathodic and the anodic direction.

At electrode potentials above 1.57 V (SHE), Step (2.1.17a) is rate-determining, with Step (2.1.17b) in equilibrium; within the potential region below 1.3 V (SHE), Step (2.1.17b) becomes rate-determining. Once more, this change of mechanism has been ascribed to a different dependence of the reaction rates on the electrode potential. E.g., the rate for Step (2.1.17a) — anodic direction — is given by the previously cited Eq. (2.1.13) with a Tafel coefficient $b = RT/(1 - \alpha_1)F$; the Tafel coefficient for Step (2.1.17b) is $RT/(2 - \alpha_2)F$.

In a recent work [87] a thermally-Ir-coated Ti anode has been tested in 0.1 M NaCl solution (pH = 3, T = 30°C). The observed low slopes of the log(j) vs E straight lines (0.030 V/decade, cf. Table 2.1.3) led to the selection of Mechanism (2.1.11) as the only one able to explain the experimental data. This disagreement could be attributed to the catalytic properties of the iridium surface used, which were certainly different in Refs. 68 and 87. Actually, the slopes of the log(j) vs E lines in the potential region below 1.42 V (SHE) would be $RT/(2-\alpha_2)F$ for Mechanism (2.1.17) — with Step (2.1.17b) rate-determining — and $RT/2F$ for Mechanism (2.1.11) with Step (2.1.11b) rate-determining, and almost coincide if α_2 is $\ll 0.5$.

2.1.1.1.4. Platinum-Iridium Electrodes. The only cathodic data available at present refer to electrodes consisting of platinum-iridium 30 wt% alloy thermally deposited on a titanium substrate [86]. The proposed reaction mechanism once more follows Path (2.1.11) at electrode potentials above 0.86 V (SHE):

$$Cl_2, aq = 2Cl_{ads} \qquad (2.1.11b)$$

$$Cl_{ads} + e^- = Cl^-, aq \qquad (2.1.11a)$$

More precisely, at potentials above 1.16 V (SHE), Step (2.1.11a) is rate-determining as is shown by the reaction order 1/2 with respect to Cl_2 (cf. Table 2.1.5), whereas in the region between 1.16 and 0.86 V (SHE) a kinetic limiting current is reached. At potentials lower than 0.86 V (SHE), Path (2.1.17) is obeyed, namely, the rate of the electrochemical reaction $Cl_2, aq + e^- = Cl^-, aq + Cl_{ads}$ at high cathodic overpotentials overcomes the rate of the chemical reaction $Cl_2, aq = 2Cl_{ads}$.

The anodic mechanism is exactly symmetrical to the cathodic one, and Step (2.1.11a) is rate-determining (cf. b-coefficients in Table 2.1.3). More recently, the anodic behavior of a number of platinum-iridium alloys as a function of iridium content (0.5 to 30 wt%) has been investigated [87]; these alloys were obtained by thermal deposition methods. In this case the anodic discharge of Cl^- should again follow Path (2.1.11) but Step (2.1.11b) is now rate-determining. The behavior of these anodes is essentially independent of iridium content, and the conflict between the conclusions in Refs. 86 and 87 may be solved by taking into account the comparatively much higher Cl^- concentrations (0.1 M to saturated NaCl) used in Ref. 87 which substantially increased the rate of Step (2.1.11a) with respect to Step (2.1.11b).

One of the more interesting features in the behavior of the platinum-iridium electrodes is linked to the special properties of chemisorbed oxygen. Contrary to the case of platinum, iridium shows a nearly reversible behavior with respect to the buildup and the reduction of the oxygen layer; the breaking of the iridium-oxygen bond is characterized by the absence of significant energy barriers, and thus an irreversible deactivation of the surface cannot occur. Experiments within the range 1 M to saturated NaCl have provided evidence that very low percentages of iridium (0.5%-wt in the Pt-Ir alloys) are effective in decreasing the irreversibility of the oxygen coverage.

2.1.1.1.5. Rhodium Electrodes. Rhodium electrodes are characterized by significant hysteresis, even in acid solutions, so that the observed stoichiometric numbers and reaction orders are not completely reliable (cf. Table 2.1.5).

A tentative reaction scheme has been suggested [68]:

$$Cl^-, aq = Cl_{ads} + e^- \quad (2.1.18a)$$

$$Cl_{ads} + Cl^-, aq = Cl_2, aq + e^- \quad (2.1.18b)$$

$$2Cl_{ads} = Cl_2 \quad (2.1.18c)$$

At low potentials, $E < 1.2$ V (SHE), the mechanism involves Steps (2.1.18a) and (2.1.18b), the latter being rate-determining. At $E > 1.2$ V, Step (2.1.18c) becomes increasingly faster, and a mechanism involving Steps (2.1.18a) —rate-determining — and (2.1.18c) — at equilibrium — is likely to be obeyed.

2.1.1.1.6. Graphite Electrodes. The kinetic behavior of the Cl_2/Cl^- couple on graphite electrodes was thoroughly investigated by Krishtalik et al. [155-158] who primarily considered the anodic process. First [155] the authors examined the influence of pH on the chlorine evolution overvoltage, η_{an}, and at each NaCl concentration they found a threshold pH value at which an abrupt increase of the η_{an} values takes place (with 4.4 M NaCl at 25°C, threshold pH = 3.5). This fact is linked with the formation of adsorbed oxygen which affects the Cl_2/Cl^- kinetics through both double-layer effects and increases in the free energy of adsorption of atomic chlorine.

Contrasting with the behavior of platinum, the overvoltage on graphite becomes constant and practically independent of pH as the threshold pH value is overcome: this happens because almost all of the formed oxygen is eventually evolved as carbon oxides. With regard to the kinetics of Cl^- oxidation, Krishtalik and Rothenberg [156] found that the electrode potential vs log(j) plot exhibits two distinct linear sections with different slopes: 0.060 V/decade at $j < 0.01$ A/cm^2 and 0.110 to 0.120 V/decade at $j > 0.01$ A/cm^2. This change in the Tafel b-coefficient has been justified by assuming the electrochemical reaction $Cl^-, aq - e^- = Cl_{ads}$ occurs as a slow barrierless discharge (b = 0.06 V/decade, and $|1 - \alpha| = 1$) at low j's and as a normal slow discharge (b ≅ 0.120 V/decade, and $|1 - \alpha| = 0.5$) at higher j's. This mechanism finds further support in the fact that the overpotentials are insensitive to additions of SO_4^{2-} and Fe^{3+} ions which are known to influence the anodic kinetics through double-layer effects. This insensitivity takes place because in a barrierless electrochemical reaction the overpotential is independent of the double-layer structure and composition.

There have been recent objections [74] that in Krishtalik's treatment the graphite electrodes were assumed to behave only as porous electrodes whose apparent b-coefficients are known to be higher than those of smooth electrodes by a factor of two [158]. (As a matter of fact, Krishtalik's experimental b-values were 0.12 V/decade at low j's and 0.22 to 0.24 V/decade at j's greater than 0.01 A/cm^2, respectively.) On the contrary, in Ref. 74 the electrochemically active area for a graphite electrode has been shown to be practically coincident with the apparent geometrical area because of the slow diffusion of the Cl^- ions into the pores. The above objection undoubtedly holds for porous graphite, but it must also be pointed out that Ehrenburg and Krishtalik [157] obtained quite similar results for pyrographite and for polyethylene-$PbCl_2$-AgCl-filled graphite, the porosity of which amounts to only several per cent. Moreover, the slow barrierless discharge followed by the usual electrochemical desorption step is confirmed by cathodic polarization experiments and adequately fits differential capacitance data and stoichiometric numbers.

From activation energies (7.2 to 7.3 kcal/mole of molecular chlorine evolved at $PbCl_2$-impregnated electrodes) and heats of formation of Cl atoms (29 kcal/mole), Ehrenburg and Krishtalik were able to determine the adsorption heat for Cl atoms as 22 kcal/mole.

The anodic behavior of the Cl_2/Cl^- couple at graphite has been thoroughly investigated by Janssen and Hoogland, too [74], who found $b = 2.3RT/(1 - \alpha)F$ at $j > 0.01$ A/cm^2 whereas at lower j's the observed b's were substantially lower (e.g., 0.042 V/decade at 25°C).

With <u>aged</u> electrodes linear E vs log(j) plots were never observed [74, 160]; it was, however, shown [159] that a newly prepared graphite electrode <u>is</u> covered by a stable layer of adsorbed oxygen which makes the surface resistant to corrosion, whereas an aged one is not. The collapse of the protective layer (for instance, by aging the graphite) is followed by intense formation of carbon oxides with resulting roughening of the surface.

The following reaction schemes were suggested [74]:

$$Cl^-, aq - e^- = Cl_{ads} \qquad (2.1.19a)$$

$$2Cl_{ads} = Cl_2, aq \qquad (2.1.19b)$$

or

$$Cl^-, aq - e^- = Cl_{ads} \qquad (2.1.20a)$$

$$Cl_{ads} + Cl^-, aq - e^- = Cl_2, aq \qquad (2.1.20b)$$

Mechanism (2.1.19) — often referred to as the Volmer-Tafel mechanism — has been discarded because it does not fit the experimental data either at high or low j's.

Mechanism (2.1.20) — often referred to as the Volmer-Heyrovsky mechanism — is appropriate to explain the experimental data for <u>aged</u> electrodes provided that the exchange-current ratio $j_{0(2.1.20a)}/j_{0(2.1.20b)}$ is higher than 10. In such a case a theoretical

$b = 2.3RT/|1 - \alpha|_{(2.1.20b)}F$ must be obtained at $j > 0.01$ A/cm^2 — and the observed value is satisfactorily close to it; at $j < 0.0025$ A/cm^2 a theoretical slope $b = 2.3RT/|2 - \alpha|_{(2.1.20b)}F$ is obtained — and the observed value is again in satisfactory agreement. In the case of newly prepared electrodes, Mechanism (2.1.20) as in Krishtalik's work applies once more if the exchange-current ratio $j_{0(2.1.20a)}/j_{0(2.1.20b)}$ ranges from 1 to 10; this condition implies Steps (2.1.20a) and (2.1.20b) to be equally rate-controlling. Consequently the disagreement between Ref. 74 and Refs. 156 and 157 is essentially confined to the nature of the rate-determining step within the same reaction mechanism. The quantity of adsorbed atomic chlorine has been titrated [161] following the reaction

$$4Cl_{ads} + 4OH^-aq = 4Cl^-,aq + 2H_2O + O_2,g \qquad (2.1.21)$$

which allows the atomic chlorine to be distinguished from molecular chlorine. Indeed, the latter would react with alkaline solutions giving hypochlorites instead of molecular oxygen. The degree of coverage of adsorbed chlorine amounts to about 0.05 at the reversible potential of the Cl_2/Cl^- couple in 4 M NaCl + 1 M HCl at 25°C and increases up to 0.5 at high anodic overpotentials.

2.1.1.1.7. Electrode Materials other than Platinum-Group Metals and Graphite. At present little data are available for oxide electrodes in connection with the Cl_2/Cl^- couple. PbO_2 [75] is characterized by overpotentials higher than platinum. Mixed IrO_2-RuO_2 anodes have proven to be very interesting from the standpoint of electrocatalytic properties. They exhibit outstanding mechanical and chemical stability and low slopes in the E vs log(j) plots, thus suggesting an electrochemical-chemical reaction mechanism. The anodic and cathodic transfer coefficients and the standard rate constant have been determined for Ag/AgCl anodes [79].

2.1.1.2. Oxidation State Cl(I) in Aqueous Solution. The literature on the electrochemical behavior of hypochlorous acid and hypochlorites is relatively poor. The cathodic reduction of these on bright platinum and on graphite (or glassy carbon) has been most extensively investigated. The reduction of HClO was studied in a paper by Schwarzer and Landsberg [89]; they found that the rate as well as the key characteristics of the process strongly depend on the nature of the surface which in turn is a function of electrode potential and time. In fact, hypochlorous acid is catalytically decomposed [89] by the oxygen layer adsorbed on the platinum surface according to

$$HClO \overset{(Pt\text{-}Ox)}{=} Cl_{ads} + OH_{ads} \qquad (2.1.22)$$

$$Cl_{ads} + e^- \overset{(Pt\text{-}Ox)}{=} Cl^- \qquad (2.1.23)$$

$$OH_{ads} + e^- \overset{(Pt\text{-}Ox)}{=} OH^- \qquad (2.1.24)$$

whereas hypochlorous acid is directly reduced only at that fraction of surface which is free of oxide according to

$$H^+ + HClO + e^- = Cl_{ads} + H_2O \tag{2.1.25}$$

$$Cl_{ads} + e^- = Cl^- \tag{2.1.26}$$

It is noteworthy that the observed reaction order, n_{H^+}, is less than unity owing to the occurrence of the pH-independent Step (2.1.22).

The reduction of the ClO^- ions at platinum electrodes in alkaline solutions was also investigated [89-91]. According to Müller's work [91], hypochlorite gives rise to two waves, the first of which is linked once more with the catalytic activity of the oxygen adsorbed on the electrode surface. ClO^- is adsorbed on the oxidized surface with consequent breaking or loosening of the Cl—O bond and the adsorbed intermediates react faster than the original ClO^- molecule. The second wave can, on the other hand, be ascribed to the direct reduction of ClO^- on the reduced-platinum surface. The reaction order, n_{ClO^-} is unity in both cases, and both waves give well-defined Tafel slopes which suggest a charge-transfer step as rate-controlling. A detailed analysis of the influence of the oxygen coverage on the reaction kinetics is available [91].

Similar conclusions [90] have been arrived at in the case of glassy carbon, although in this case more experimental difficulties are encountered because of the sluggishness of the surface oxide reduction. This point has been confirmed recently for paraffin-impregnated graphite electrodes [92].

As regards the anodic oxidation of both ClO^- and HClO, the following reaction scheme [93-95] has been suggested for graphite and platinum electrodes in NaCl solutions within a wide range of temperatures:

$$6ClO^- + 3H_2O - 6e^- = 2ClO_3^- + 4Cl^- + 6H^+ + \frac{3}{2}O_2 \tag{2.1.27}$$

$$6HClO + 3H_2O - 6e^- = 2ClO_3^- + 4Cl^- + 12H^+ + \frac{3}{2}O_2 \tag{2.1.28}$$

This electrochemical process plays an important role in the industrial preparation of chlorates (cf. Section 3.2). As was pointed out by Hammar and Wranglén [96], Reaction Schemes (2.1.27) and (2.1.28) are rather ambiguous in that hypochlorite undergoes no net oxidation. These schemes could be obtained as an overlapping of the processes

$$6ClO^- = 2ClO_3^- + 4Cl^- \tag{2.1.29}$$

$$3H_2O - 6e^- = \frac{3}{2}O_2 + 6H^+ \tag{2.1.30}$$

However, it was demonstrated [96] that the anodic evolution of oxygen is directly linked with the hypochlorite concentration and involves formation of Cl^- ions.

So far, no mechanism has been suggested for either Reaction (2.1.27) or (2.1.28); it is, of course, clear that processes with such high reaction orders are very unlikely as elementary steps and the reaction must certainly proceed in several steps. A key for solving the problem could be found in the above cited work concerning the cathodic reduction of ClO^- and HClO. The dissociative adsorption of these species at an oxidized platinum or graphite surface leads to the formation of active adsorbed oxygen which possibly causes oxidation of Cl^- and ClO^- to ClO_3^- according to

$$ClO^- = Cl_{ads} + O_{ads}^- \text{ (at oxidized surface)} \qquad (2.1.31)$$

$$O_{ads}^- - e^- = O_{ads} \qquad (2.1.32)$$

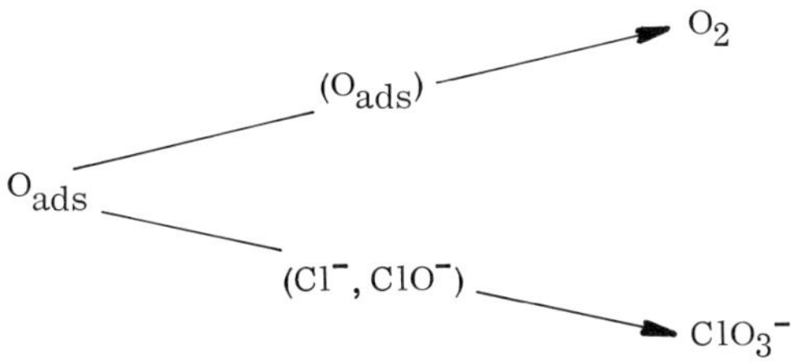

(2.1.33)

2.1.1.3. The Cl(III) and Cl(IV) Oxidation States in Aqueous Solutions. In addition to earlier works [97, 98], two recently published papers make a substantial contribution to the understanding of the kinetics of the electrochemical processes for ClO_2^- and ClO_2. Schwarzer and Landsberg [99] pointed out that the ClO_2/ClO_2^- couple behaves reversibly (cf. also Refs. 17, 18, and 19 and Section (1.2.1.2) in the pH range 4 to 7; this finding would provide the basis for an analytical determination of both species in the same solution. However, suitable polarization curves are only observed at specially prepared graphite-azobenzene (or -diphenyl) electrodes. Use of the rotating disk electrode technique allowed the diffusion coefficients $D_{ClO_2^-} = 0.8 \times 10^{-5}$ cm^2/sec and $D_{ClO_2} = 1.1 \times 10^{-5}$ cm^2/sec, the anodic and cathodic transfer coefficients (both essentially 0.5), and the reaction orders $n_{ClO_2} = 1$ (cathodic) and $n_{ClO_2^-} = 1$ (anodic) to be determined. On the basis of these results, the charge-transfer step should be

$$ClO_2^- - e^- = ClO_2 \qquad (2.1.34)$$

with the relevant standard potential, $E^\circ = 0.945$ V vs SHE at 20°C (but see Section 1.2.1.2).

More recently, Raspi and Pergola [100] confirmed the irreversibility of Reaction (2.1.34) at bright platinum, but found a reversible behavior at platinized platinum. At pH's of 5 to 9 only one electron is involved in the anodic process as is also shown by coulometric measurements; the relevant half-wave potential $E_{1/2} = 0.949$ V (SHE) at 25°C agrees rather well with Schwarzer and Landsberg's figure.

At pH's higher than 10.5 the electrochemical process remains unchanged but the chemical reaction (disproportionation)

$$2ClO_2 + 2OH^- = ClO_2^- + ClO_3^- + H_2O \qquad (2.1.35)$$

causes the limiting anodic current to be twice as great as that at lower pH and yields chlorate as the final reaction product.

A similar behavior is observed for the cathodic wave at a pH lower than 2. Here the reaction

$$5HClO_2 = 4ClO_2 + Cl^- + H^+ + 2H_2O \qquad (2.1.36)$$

regenerates the electrochemically reacting ClO_2 and the final product is Cl^-; this is confirmed by coulometric experiments showing that five electrons are involved in the reduction of one molecule of ClO_2.

2.1.1.4. The Cl(V) Oxidation State in Aqueous Solution. The better known reactions for the formation of chlorates involve ClO^- and/or HClO as intermediates:

$$6ClO^- + 3H_2O - 6e^- = 2ClO_3^- + 4Cl^- + 6H^+ + \frac{3}{2}O_2 \tag{2.1.37}$$

(electrochemical reaction, cf. Section 2.1.1.2)

$$ClO^- + 2HClO = ClO_3^- + 2Cl^- + 2H^+ \tag{2.1.38}$$

(chemical reaction, cf. Section 3.2).

However, ClO_3^- ions can be obtained by anodic oxidation of Cl^- in acidic solutions, as was recognized by earlier workers [101, 102] and more recently re-examined by Flisskii [103-106].

In strongly acidic solutions the hydrolytic equilibrium

$$Cl_2 + H_2O = H^+ + Cl^- + HClO \tag{1.3.10}$$

cannot take place and a reaction mechanism other than (2.1.37) and (2.1.38) must be assumed. The E vs log(j) plots obtained for the anodic oxidation of Cl^- in 0.5 M H_2SO_4 solutions at platinum electrodes show two linear sections separated by an abrupt potential jump: Tafel's b-coefficients are 0.110 V/decade at potentials up to 2.2 V (SHE) and 0.400 V/decade at potentials higher than 2.4 V (SHE). The current yield of $HClO_3$ increases steeply after this potential jump and attains a maximum value of 30% in 0.04 M HCl + 0.5 M H_2SO_4 solutions at 25°C. According to the suggested mechanism, Cl^- ions are converted to $HClO_3$ by the active oxygen atoms adsorbed onto the oxidized surface of platinum at potentials higher than 2.4 V (SHE):

$$PtO + mH_2O - 2me^- = PtO(O)_m + 2mH^+ \tag{2.1.39}$$

$$PtO(O)_m + Cl^- = PtO(O)_{m-3} + ClO_3^- \text{ (rate-controlling)} \tag{2.1.40}$$

Similar behavior was observed at graphite electrodes [106].

The chlorate anion is difficultly reducible in neutral medium; only in strongly acidic solutions (say, 8 M H_2SO_4) was a polarographic wave observed at mercury electrodes [107]. This wave was, however, ascribed to ClO_2 formed by decomposition of $HClO_3$. Nevertheless, the ClO_3^- reduction can occur through a heterogeneous or homogeneous catalytic mechanism. For instance, it has long been known that a cause of the loss in current yields during the preparation of chlorates lies in the cathodic reduction of ClO_3^- at the mild steel cathodes, which are known to play a catalytic role. A detailed treatment is available in the literature [108].

As for homogeneous catalysis, an earlier paper [109] showed that ClO_3^- can be reduced at platinum cathodes in the presence of VO^{2+}; the latter is the actual electrochemical reactant, and the product, V^{3+}, reduces ClO_3^- to ClO_2 in turn. Recently [110, 111] it has been

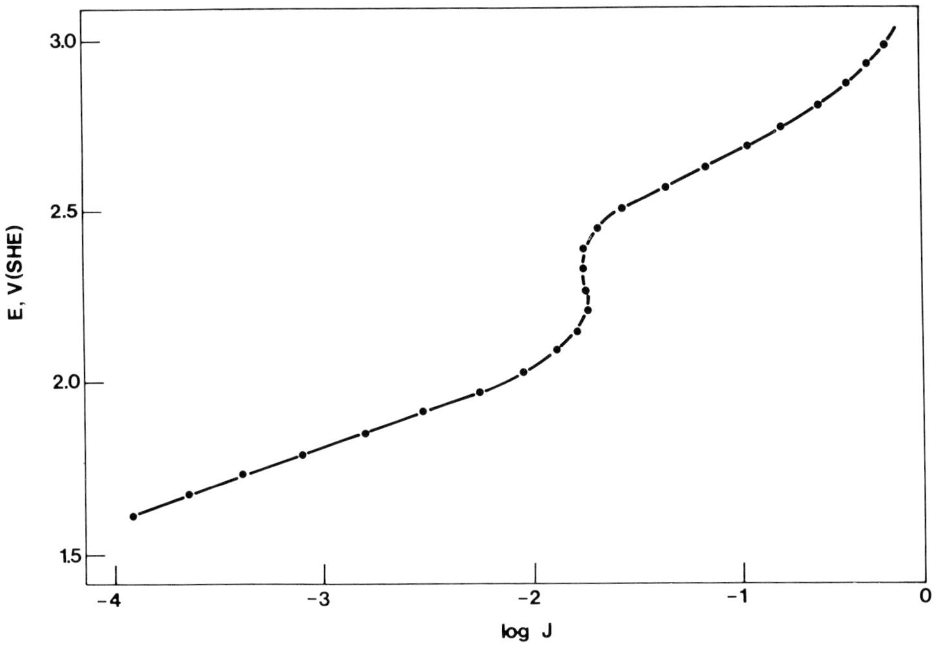

FIG. 2.1.1. Anodic E vs log(j) plot for 1 M $NaClO_3$ solution at 25°C. Bright platinum electrode; pH = 5. SHE = Standard hydrogen electrode.

shown that the reduction of chlorates is catalyzed by Mo(VI) and by mixtures of Mo(VI) and W(VI). ClO_3^- ions react with Mo(V) and Mo(III) which are formed electrochemically at mercury electrodes; this reaction could be of analytical value.

Chlorates are easily oxidized at platinum and lead dioxide anodes, respectively, at electrode potentials higher than 2.4 and 2.1 V (SHE) as seen from Fiori's [112] and Osuga's [113] work. The E vs log(j) plots are characterized by two distinct linear sections separated by a sharp potential increase (cf. Figure 2.1.1); it is noteworthy that these plots are quite similar to those obtained in the high-potential oxidation of Cl^- in strongly acidic solutions [103]. The current yields of ClO_4^- increase steeply in the region above the potential jump, attaining a value of practically 100% for platinum anodes [112] at 25°C and with $NaClO_3$ concentrations higher than 0.5 M, whereas a maximum 80% yield has been reported for PbO_2 anodes [113] in 5 M $NaClO_3$ at 50°C. Almost nothing is known about the relevant reaction mechanism for which a number of possible paths has been suggested.

Sugino [114] and Fiori [112]:

$$ClO_3^- - e^- = (ClO_3)_{ads} \qquad (2.1.41a)$$

$$2(ClO_3)_{ads} = Cl_2O_6 \qquad (2.1.41b)$$

or better

$$(ClO_3)_{ads} + ClO_3^- - e^- = Cl_2O_6$$

$$Cl_2O_6 + H_2O = ClO_3^- + ClO_4^- + 2H^+ \qquad (2.1.41c)$$

Sugino [114]:

$$ClO_3^- + e^- = (ClO_3)_{ads} \qquad (2.1.42a)$$

$$OH^- - e^- = OH_{ads} \qquad (2.1.42b)$$

$$(ClO_3)_{ads} + OH_{ads} = ClO_4^- + H^+ \qquad (2.1.42c)$$

De Nora, Gallone et al. [115] and Narasimham et al. [116]:

$$ClO_3^- - e^- = (ClO_3)_{ads} \qquad (2.1.43a)$$

$$(ClO_3)_{ads} + H_2O - e^- = ClO_4^- + 2H^+ \qquad (2.1.43b)$$

Grootheer and Cook [117]:

$$H_2O - e^- = OH_{ads} + H^+ \qquad (2.1.44a)$$

$$OH_{ads} - e^- = O_{ads} + H^+ \qquad (2.1.44b)$$

$$ClO_3^- + O_{ads} = ClO_4^- \qquad (2.1.44c)$$

The data on Tafel's b-coefficients available so far range from 0.12 V/decade (in 0.47 to 4.7 M $NaClO_3$ at 50°C, Ref. 115) to 0.60 V/decade (in 5.6 M $NaClO_3$ at 17°C, Ref. 114). The reaction order, $n_{ClO_3^-}$, varies from 0.5 (at 50°C in 0.47 to 4.7 M $NaClO_3$, Ref. 115) to 1 (at 25°C, in 0.1 to 1.0 M $NaClO_3$, Ref. 118).

Looking over the observed b-coefficients, reaction orders, and results of radiochemical measurements [119], a charge-transfer step involving the discharge of ClO_3^- ions has been suggested as rate-controlling [112, 114, 115]. The chemical reaction Step (2.1.44c) has generally been discarded on the basis of the experimental observation that ClO_4^- begins to form just beyond the potential jump observed in the polarization curves; perchlorates do not form at lower potentials where adsorbed oxygen undoubtedly exists at the electrode surface. This reasoning is, however, contradicted by the data about the high-potential oxidation of Cl^- in strongly acidic solution [103-105]. As a matter of fact, it was shown that the potential jump in the E vs log(j) plots corresponds to the appearance of a new kind of active (i.e., much more reactive) oxygen adsorbed on the already oxidized surface. Hence, Mechanism (2.1.44) cannot be discarded a priori; it remains to be seen whether such a mechanism is able to fit the observed b-values and reaction orders.

2.1.1.5. The Cl(VII) Oxidation State in Aqueous Solution. The ClO_4^- anion is commonly regarded as extremely stable in aqueous solution. Yet, it was shown [110, 111, 120] that its reduction is effectively catalyzed by Mo(VI) [or better by the reduction product Mo(III)].

Recent work [121, 122] has also shown that the perchlorate anion loses its chemical inertness when adsorbed at the electrode surface under intense potential gradients, and thus it participates in the O_2 evolution reaction.

2.1.2. Nonaqueous Solvents

2.1.2.1. The Cl_2/Cl^- Couple in Acetonitrile Solutions. The pioneer contribution to this topic is that by Kolthoff and Coetzee [123] who found that the anodic oxidation of chlorides at a rotating platinum disk electrode in acetonitrile with 0.1 M Et_4NClO_4 + 0.0005 M LiCl proceeds via two waves. The first wave is exactly twice the height of the second one, and the half-wave potentials are respectively 1.1 V (SCE, satd calomel electrode) and 1.7 V (SCE) at 25°C. The following overall reactions were postulated:

$$6Cl^- - 4e^- = 2Cl_3^- \quad \text{(first wave)} \tag{2.1.45}$$

$$2Cl_3^- - 2e^- = 3Cl_2 \quad \text{(second wave)} \tag{2.1.46}$$

The formation of Cl_3^- as an intermediate and the high electrode potentials required for the Cl_3^- oxidation to Cl_2 were explained on the basis of a claimed outstanding stability of the Cl_3^- ion in acetonitrile [38], but more convincing evidence has recently been presented to show that the Cl_3^- stability is not so great [26-28, 37].

As an alternative, the second wave could be ascribed to the oxidation of Cl^- to a higher-valency compound (e.g., Cl^+); the 2:1 ratio observed in the heights of the two waves and assumed as the proof of Reactions (2.1.45) and (2.1.46) could be questioned because of side reactions between molecular chlorine and the solvent [124]. In fact, it is well known that solutions of chlorine and bromine are unstable in a number of solvents such as acetonitrile, dimethylsulfoxide, dimethylformamide, and nitromethane due to the occurrence of a reaction such as

$$Cl_2 + CH_3CN = ClCH_2CN + HCl \tag{2.1.47}$$

and the regeneration of chlorides is likely to affect the height of the first wave observed by Kolthoff and Coetzee [123].

A paper has recently been published on the electrochemical kinetics of the Cl_2/Cl^- couple on the platinum electrode [125]. Tafel b-coefficients, cathodic and anodic transfer coefficients, stoichiometric numbers, and reaction orders with respect to Cl^- were determined; a selection of these is given in Table 2.1.7. The following elementary steps were considered:

$$Cl_2 + e^- = Cl_{ads} + Cl^- \tag{2.1.48a}$$

$$Cl_{ads} + e^- = Cl^- \tag{2.1.48b}$$

$$2Cl_{ads} = Cl_2 \tag{2.1.48c}$$

On the basis of the calculated kinetic parameters, a reaction path made up of Steps (2.1.48a) (as rate-determining) and (2.1.48c) has been suggested for the whole cathodic region as well as at anodic potentials higher than 1 V (SCE); however, a combination of Steps (2.1.48a) and (2.1.48b) (as rate-determining) has been found to be applicable at anodic potentials lower than 1 V (SCE). Inspection of the log(j) vs E plots in Ref. 125 shows the Tafel b-coefficients to be as low at 0.07 V/decade within the anodic potential range below 1 V (SCE). Moreover, the anodic reaction order, n_{Cl^-}, approaches 1.5 at Cl^- concentrations from

TABLE 2.1.7. Reaction Orders n_{Cl^-}; Tafel's b-Coefficients V/decade; Transfer Coefficients α (cathodic) and $\{1 - \alpha\}$ (anodic); and Stoichiometric Numbers ν for the Reaction $Cl_2 + 2e^- = 2Cl^-$ in 0.1 M Et_4NClO_4 + 0.001 to 0.1 M Et_4NCl in Acetonitrile at 25°C. Platinum Wire Electrode [125].

n_{Cl^-}	α	$1 - \alpha$	b	ν	Remarks
0 (cathodic)	0.40	-	0.150	2	Within the whole range of potentials investigated
1 (anodic)	-	0.52	-	1	At electrode potentials lower than 1 V (SCE-satd calomel electrode)
1 (anodic)	-	0.41	0.100	2	At electrode potentials higher than 1 V (SCE)

0.01 to 0.1 M. Consequently, an alternative mechanism consisting of Steps (2.1.48b) and (2.1.48c) could possibly describe the actual reaction path over the whole investigated range of potentials. The observed decrease in n_{Cl^-} down to 1 could be explained as due to a change from Langmuir-type to Temkin-type kinetics. More precisely, for the anodic region Step (2.1.48c) would be rate-controlling at $E < 1$ V (SCE); Step (2.1.48b) would be rate-controlling in the remaining anodic region and over the whole cathodic region. It is worthwhile to note that the alternative mechanism closely matches the kinetic situation at platinum electrodes in aqueous solutions (cf. Section 2.1.1.1.2). In Ref. 125 the effect of various amounts of water in admixture with acetonitrile has also been investigated and the beginning of passivation phenomena (i.e., formation of adsorbed oxygen) has been found to occur at water contents higher than 2 M (about 4 wt%). The oxidation of the platinum surface is also strongly dependent on the concentrations of Cl^- and Cl_2. This fact has been interpreted in terms of competitive reactions of hydration of Cl^- and formation of Cl_3^-.

2.1.2.2. The Cl_2/Cl^- Couple in Dimethylsulfoxide (DMSO). Solutions of HCl and LiCl in DMSO have recently been studied by the chronopotentiometric method [124] at platinum electrodes, and a single oxidation wave with a quarter-wave potential $E_{1/4} = 1.00$ V (SCE) in 0.1 M Et_4NClO_4 at 25°C has been observed. Hence, in contrast to the anodic behavior of Br^- ions, the trichloride ion Cl_3^- does not appear to be a reaction intermediate.

The anodic transfer coefficients are collected in Table 2.1.8. As in the case of the acetonitrile solutions, catalytic phenomena undoubtedly complicate the oxidation process for Cl^-; in fact, the $j\tau^{1/2}$ values collected in Table 2.1.9 substantially decrease as j increases. Taking into account that the $j\tau^{1/2}$ values are higher than the corresponding cathodic ones by more than one order of magnitude, the occurrence of a postchemical reaction seems to be very likely. This conclusion is further substantiated by controlled-

TABLE 2.1.8. Anodic Transfer Coefficients $(1 - \alpha)$ for the Oxidation of HCl in 0.1 M Et_4NClO_4 in DMSO at Platinum Electrodes at 25°C. Based on Chronopotentiometry [124]

$c_{HCl} \times 10^3$ (mole/liter)	$(1 - \alpha)^a$	$(1 - \alpha)^b$	$j \times 10^6$ (A/cm^2)
0.416	0.56	0.53	2.2
0.830	0.50	0.47	4.4
1.651	0.56	0.56	22
3.00	0.60	0.51	132
4.52	0.52	0.35	352
6.34	0.55	0.33	440

[a]Obtained from the $E_{t=0}$ vs log(j) plot at small j's (with transition times longer than 30 sec).

[b]Obtained from the $E_{t=0}$ vs log(c_{HCl}) plot at various j's ($E_{t=0}$ denotes the electrode potential value extrapolated to the beginning of the chronopotentiogram).

potential coulometry, where it was found that the number of electrons involved in the oxidation of 1 mole of HCl is a function of the electrolysis time and ranges from 8 to 20. Similar results were obtained with pyrolytic graphite [126].

2.1.2.3. The Cl_2/Cl^- Couple in Solvents other than Acetonitrile and DMSO. So far few solvents have been investigated other than ACN and DMSO; some data on the kinetics of Cl_2/Cl^- on platinum electrode are available in nitromethane [127], dimethylformamide [128], and sulfolane [129].

A vitreous carbon rotating microelectrode has been tested in dimethylsulfone at 127°C, the supporting electrolyte being $LiClO_4$ [130].

2.1.2.4. Higher-Valency Cl(ox)/Cl(red) Couples in Nonaqueous Solvents. Data can be found in Refs. 128 and 131-133 on this topic, to which little attention has been paid so far.

2.1.3. Fused Salts

2.1.3.1. The Cl_2/Cl^- Couple. The kinetic behavior of the Cl_2/Cl^- couple in fused chlorides has been studied because of its importance in the field of industrial electrochemistry, i.e., the production of sodium metal and characterization of chlorine electrodes for high-power-density and high-energy-density fused-salt batteries.

TABLE 2.1.9. Chronopotentiometric Data for the Oxidation of HCl in 0.1 M Et_4NClO_4 in DMSO at Platinum Electrodes at 25°C [124]

$C_{HCl} \times 10^3$ (mole/liter)	$j \times 10^6$ (A/cm^2)	$j\tau^{\frac{1}{2}} \times 10^6$ ($sec^{\frac{1}{2}} A/cm^2$)	$(\tau/\tau_D)^{\frac{1}{2}}$ [a]
0.416	220	647	9.7
0.416	308	616	9.3
0.416	440	572	8.6
1.651	1100	2477	9.4
1.651	1320	2402	9.1
1.651	1540	2348	8.8
3.97	3080	5812	9.2
3.97	3520	5658	9.0
3.97	4400	5359	8.5

[a] τ_D denotes the hypothetical diffusional transition time (sec).

A number of interesting papers have been published on this subject; Table 2.1.10 collects the relevant kinetic parameters obtained by a variety of experimental methods: galvanostatic pulses, steady-state polarization curves, linear-scan voltammetry, potentiostatic double pulses, and differential capacitance. A stoichiometric number $\nu = 1$ and a heat of activation of 7.1 kcal/mole in the reversible region were calculated [134].

A possible mechanism involves the reactions

$$Cl^-, \text{melt} = Cl_{ads} + e^- \quad (2.1.49a)$$

$$2Cl_{ads} = Cl_2, \text{gas} \quad (2.1.49b)$$

$$Cl^-, \text{melt} + Cl_{ads} = Cl_2, \text{gas} + e^- \quad (2.1.49c)$$

The usually low b-coefficients observed as well as other experimental data show that an anodic mechanism composed of a fast Cl^- discharge followed by a slow Cl_{ads} combination can explain the results, assuming either Langmuir-type or Temkin-type nonactivated adsorption [134, 134a, 135]. At high current densities an increase in b-coefficients was found for the case of graphite anodes [134a], and a mechanism implying a fast Cl^- discharge followed by a slow ion + atom reaction — Step (2.1.49c) — under nonactivated Temkin conditions appears to fit the results best.

With regard to the cathodic direction, it was concluded that the diffusion of the dissolved Cl_2 to the graphite-melt interface is undoubtedly the slow step at current densities greater

TABLE 2.1.10. Tafel b-coefficients (V/decade) and Exchange Current Densities j_0 (A/cm^2) for the Anodic Reaction $2Cl^-(melt) - 2e^- = Cl_2(g)$ at Various Temperatures

Electrode material	T (°C)	b	j_0	Refs.	Remarks
Spectroscopically pure graphite	656	0.093	0.13	[134]	LiCl melt
Porous graphite (30% porosity)	656	0.090	0.19	[134]	LiCl melt
Porous graphite (30% porosity)	714	0.093	0.24	[134]	LiCl melt
Porous graphite (30% porosity)	765	0.095	0.24	[134]	LiCl melt
Vitreous carbon	750	0.100 to 0.110	0.28	[134a]	NaCl + AgCl melt
Pyrolytic graphite (working surface coincident with the basal plane)	750	0.100 to 0.107		[134a]	NaCl + AgCl melt but low j's
Pyrolytic graphite (working surface coincident with the basal plane)	750	0.270 to 0.280		[134a]	NaCl + AgCl melt but high j's
Pyrolytic graphite (working surface perpendicular to the basal plane)	750	0.100 to 0.110		[134a]	NaCl + AgCl melt but low j's
Pyrolytic graphite (working surface perpendicular to the basal plane)	750	0.275 to 0.285		[134a]	NaCl + AgCl melt but high j's

than 1.2 A/cm^2 in molten LiCl at temperatures from 656 to 767°C. That the observed overpotential can be ascribed to diffusion is further substantiated by other published papers [136, 137].

However, the slope of $\eta_{cathode}$ vs $\log(jj_D/|j_D - j|)$ plots as well as the stoichiometric number would indicate [134] a partial kinetic control by the dissociation step Cl_2, gas = $2Cl_{ads}$ followed by fast ionization of Cl_{ads} to Cl^-, melt.

3. APPLIED ELECTROCHEMISTRY

3.1. CHLORINE PRODUCTION

Chlorine production represents the chief industrial electrochemical process. Its prominence and vitality as well as its role in today's industry is well illustrated by the 8% average annual growth rate [138] as the following approximate figures show for world production:

Year	1950	1955	1960	1965	1970
Millions of metric tons of chlorine	2.5	4.5	7.0	11.0	16.0

In earlier times the main product of brine electrolysis was sodium hydroxide, especially when the demand for rayon was booming (in the 1920s). During that period the by-product, chlorine, was used mainly as a bleaching agent (calcium hypochlorite); however, in the meantime the search for other chlorine-containing products continually increased, eventually resulting in a number of processes whose final products were chlorinated solvents, insecticides, and herbicides (e.g., Gammexane and chlordane). After World War II the petrochemical industries demanded ever-increasing amounts of chlorine for the production of ethylene oxide via the chlorohydrin process (although recently this process has been displaced by the direct-oxidation method).

Finally, the demand for allyl chloride, and mainly for polyvinyl chloride, has been responsible for the huge production of chlorine, so that chlorine is now the main product of the brine electrolysis whereas the profitable disposal of sodium hydroxide is the problem faced by the industry today.

Aqueous brines are currently electrolyzed in two basic types of cells: diaphragm cells and mercury-cathode cells.

A diaphragm cell is composed of graphite or dimensionally stable anodes (i.e., metallic anodes whose substrate is titanium activated by an electrocatalytic coating such as mixed oxides of noble metals) and iron cathodes separated by asbestos diaphragms. The relevant processes are

anode $2Cl^-, aq - 2e^- = Cl_2, g$

cathode $H_2O + 2e^- = H_2, g + 2OH^-, aq$

and the overall reaction results in the evolution of chlorine in the anode compartment and in the formation of hydroxide ions in the cathode one. The brine is fed continuously and flows from the anode to the cathode compartment; the flow rate is regulated near an optimum value to minimize back-diffusion and migration of OH^- ions. However, this flow always causes a

TABLE 3.1.1. Voltages and Currents in Diaphragm Cells

Year	1940	1950	1970
Total current per cell, A	25,000	25,000	70,000
Anodic current density, A/cm^2	0.09	0.12	0.13
Overall cell voltage	4.4	4.75	4.5
Power consumed per ton of Cl_2, kWh	3370	3640	3450

partial conversion of the fed NaCl. Consequently, the cathodic liquor requires a complicated and expensive treatment before it can be obtained and sold as pure caustic soda.

Some secondary reactions occur in the cell and they can be represented as

$$Cl_2, aq + 2OH^-, aq = ClO^-, aq + Cl^-, aq + H_2O$$

$$Cl_2, aq + H_2O = H^+, aq + Cl^-, aq + HClO, aq$$

$$3Cl_2, aq + 6OH^-, aq = ClO_3^-, aq + 5Cl^-, aq + 3H_2O$$

$$ClO^-, aq + 2HClO, aq = ClO_3^-, aq + 2H^+, aq + 2Cl^-, aq$$

During the last decade the electrolysis of aqueous NaCl in diaphragm cells has undergone impressive changes [138] with regard to current densities and overall cell currents (namely, cell productivities), as can be seen from the information in Table 3.1.1. The 1970 data in Table 3.1.1 refer to diaphragm cells equipped with dimensionally stable anodes (DSA).

A mercury-cathode cell consists of an inclined trough of mild steel on the baseplate of which mercury is flowed. The anodes (graphite or DSA) are constructed in rows parallel to the running mercury. The electrode processes are

<u>anode</u> $$Cl^-, aq - e^- = \frac{1}{2}Cl_2, g$$

<u>cathode</u> $$Na^+, aq + \{(1 - x_{Na})/x_{Na}\}Hg + e^- = [Na + \{(1 - x_{Na})/x_{Na}\}Hg]$$

where x_{Na} represents the mole fraction of sodium in the amalgam. Gaseous chlorine and sodium-mercury amalgam are the cell reaction products. The sodium amalgam flows, or is pumped, to a decomposer which is a horizontal trough just below the cell itself or to a vertical tower where the amalgam is mixed with water in the presence of graphite acting as a depolarizer. Actually this is a short-circuited galvanic element whose relevant reactions are

<u>anodic area</u> $$[Na + \{(1 - x_{Na})/x_{Na}\}Hg] - e^- = Na^+ + \{(1 - x_{Na})/x_{Na}\}Hg$$

<u>cathodic area</u> $$H_2O + e^- = \frac{1}{2}H_2, g + OH^-$$

TABLE 3.1.2. Characteristic Working Data for Mercury-Cathode Cells

Year	1950	1970
Total current per cell, A	30, 000	500, 000
Anodic current density, A/cm^2	0.37	1.3
Mercury hold-up, kg/1000-A cell	57	12
Power consumed per ton of Cl_2, kWh	3600	3400

Hence, the reaction products are mercury (containing little Na), hydrogen, and sodium hydroxide which is more concentrated and far purer than that obtained from the diaphragm cells. Mercury is recovered and recirculated.

The evolution of the mercury-cathode cells has been impressive [138]; the tremendous increase in anodic current density linked with the use of DSA, together with the introduction of silicon rectifiers that allow much higher whole-cell currents, determined a different structure of the plants with very low mercury holdup (cf. Table 3.1.2). A comprehensive review of the thermodynamic and kinetic factors relevant to the mercury cathode cells is given in Ref. 138.

A serious drawback for this type of cells is represented by the mercury losses (as high as 50 to 80 g/ton of Cl_2) which affect the economics of the process and also cause dangerous air-and-water pollution. Lately, however, an extensive research effort has led to a substantial improvement of the plant performances as regards mercury losses [139].

When comparing the two types of cells, the following items should be taken into account:

1. Electrical energy and other costs of cell operation.
2. Cost of purification and evaporation of the sodium hydroxide liquor.
3. Capital cost.

Inspection of the economics of the diaphragm- and mercury-cathode cells shows that the absence of mercury losses and the intrinsically lower electric power consumption of the first type is balanced by the high-purity sodium hydroxide together with the higher cell productivity, lower steam consumption, and lower operating costs which characterize the second type of cells. Consequently, conditions of high wages, relatively high-priced steam (usually applicable to the Western countries), and low-priced electric energy favor the mercury-cathode process. A point which must be borne in mind is represented by the commonly requested purity standards for the caustic soda, standards which make the sale of the hydroxide coming from the mercury-cell method much easier.

References 140-143 are recommended for further reading.

3.2. CHLORATES PRODUCTION

Chlorates are produced starting from chlorides via an electrochemical process whose basic stages are

anode $2Cl^- - 2e^- = 2Cl_2$ (3.2.1)

cathode $2H_2O + 2e^- = 2OH^- + H_2$ (3.2.2)

bulk solution $Cl_2 + H_2O = HClO + H^+ + Cl^-$ (3.2.3)

$HClO = H^+ + ClO^-$ (3.2.4)

$ClO^- + 2HClO = ClO_3^- + 2Cl^- + 2H^+$ (3.2.5)

Hence the resulting net overall process is

$Cl^- + 3H_2O = ClO_3^- + 3H_2$ (3.2.6)

Actually the situation is much more complicated because of a set of side reactions:

anode $6ClO^- + 3H_2O - 6e^- = 2ClO_3^- + 4Cl^- + 6H^+ + \frac{3}{2}O_2$ (2.1.27)

cathode $ClO^- + 2e^- + 2H^+ = Cl^- + H_2O$ (3.2.7)

$ClO_3^- + 6H^+ + 6e^- = Cl^- + 3H_2O$ (3.2.8)

Reaction (2.1.27), which was considered in some detail in Section 2.1.1.2, results in a decrease of the current efficiency for the chlorate formation and should be avoided in practice. The optimum operating conditions are characterized by slightly acidic pH (6.2 to 7) as well as high temperatures (60 to 110°C) which accelerate Step (3.2.5) more than Step (2.1.27). However, it has recently been shown that even in such a favorable situation the buildup of HClO and ClO^- concentrations can result in rather intense anodic evolutions of oxygen [144-146].

With regard to Reactions (3.2.7) and (3.2.8), the relevant current efficiency losses are unavoidable at the commonly used mild-steel cathodes unless sodium chromate is added to the solution. The mechanism by which the chromates can prevent the ClO^- and ClO_3^- reduction is unknown. It has tentatively been suggested that the cathode surface becomes covered by a mixed iron-chromium oxide acting as a semipermeable membrane. An alternative but expensive solution is represented by the use of noncatalytic cathodes such as nickel or nickel alloy.

The structure of a chlorate plant is the result of the fact that the process economics requires both high current efficiency and low cell voltage. These requirements are rather conflicting because high current efficiencies (that is, high conversions to ClO_3^- according to Reaction 3.2.5) are obtained only by using relatively large volumes, whereas low cell voltages can be attained only through minimum anode-cathode spacing. Hence, in practice the rows of electrodes are suspended in a large tank where liquid circulation is due to the cell gases (hydrogen); alternatively, cell and reactor are separated and the liquid (e.g., 190 g/l NaCl plus 330 g/l $NaClO_3$) is circulated by pumps [147, 148]. Graphite is used as anode material; recently, platinized titanium (or, more generally, activated titanium) has been introduced on an industrial scale [149]. Nowadays, commercial $NaClO_3$ cells operate at 80 to 87% current efficiencies with an average requirement of 6000 kWh (dc) of electrical energy per ton of product [149]. References 141, 150, and 151 are recommended for further details.

3.3. PERCHLORATE PRODUCTION

The anodic oxidation of chlorates leading to the production of perchlorates has been described in Section 2.1.1.4. Lead dioxide or platinum anodes are commonly used. It is noteworthy that PbO_2 electrodes do not withstand additions of sodium chromate to the electrolytic solution (e.g., 300 to 600 g/l $NaClO_3$ plus up to 400 g/l $NaClO_4$) because of the formation of an inactive film of lead chromate on the electrode surface. Hence, when using such anodes it is necessary to select nickel or nickel alloys for cathodes in order to prevent the electrochemical reduction of the chlorates.

The cell incorporating the battery of electrodes is often fitted with cooling devices to allow the temperature to remain within 40 to 60°C, since high current densities are used in perchlorate preparation with resultant intense thermal effects. The commercial cells operate with 95%, or even higher, current efficiencies and with energy consumptions ranging 3000 to 4400 kWh/ton of product.

Further details can be found in References 152-154.

REFERENCES

1. A. Cerquetti, P. Longhi, T. Mussini, and G. Natta, J. Electroanal. Chem., 20, 411 (1969).
2. D. D. Wagman, W. H. Evans, V. B. Parker, I. Halow, S. M. Bailey, and R. H. Schumm, Selected Values of Chemical Thermodynamic Properties, National Bureau of Standards Technical Note 270-3, Washington, D.C., 1968, pp. 24-30.
3. A. Cerquetti, P. Longhi, and T. Mussini, J. Chem. Eng. Data, 13, 458 (1968).
4. G. Faita, P. Longhi, and T. Mussini, J. Electrochem. Soc., 114, 340 (1967).
5. D. J. G. Ives and G. J. Janz, Reference Electrodes, Academic, New York, 1961, pp. 13 and 14.
6. R. G. Bates and V. E. Bower, J. Research Nat. Bur. Stand., 53, 282 (1954).
7. M. S. Sherrill and E. F. Izard, J. Amer. Chem. Soc., 53, 1667 (1931).
8. G. Bianchi, J. Electrochem. Soc., 112, 233 (1965).
9. G. Bianchi, A. Barosi, G. Faita, and T. Mussini, ibid., 112, 921 (1965).
10. G. Bianchi, G. Faita, and T. Mussini, J. Sci. Instrum., 42, 693 (1965).

11. R. G. Bates, E. A. Guggenheim, H. S. Harned, D. J. G. Ives, G. J. Janz, C. B. Monk, R. A. Robinson, R. H. Stokes, and W. F. K. Wynne-Jones, J. Chem. Phys., 25, 361 (1956).

12. G. Charlot, Oxidation-Reduction Potentials, Pergamon, Paris, 1958, p. 9.

13. A. J. de Bethune and N. A. Swendeman Loud, Standard Aqueous Electrode Potentials and Temperature Coefficients at 25°C, C. Hampel, Skokie, Illinois, 1964, pp. 12, 13, and 16.

14. W. M. Latimer, Oxidation Potentials, Prentice-Hall, Englewood Cliffs, New Jersey, 1952, pp. 55 and 56.

15. R. Parsons, Handbook of Electrochemical Constants, Butterworths, London, 1959, pp. 71-73.

16. B. E. Conway, Electrochemical Data, Elsevier, Amsterdam, 1952, p. 304.

17. I. E. Flis, Zhur. Fiz. Khim., 32, 573 (1958).

18. T. Naito, Kogyo Kagaku Zasshi, 65, 749, 1016 (1962).

19. V. M. Zolotukhin, I. E. Flis, and K. P. Mishchenko, Zhur. Prikl. Khim., 38, 369 (1965).

20. M. Pourbaix, Atlas d'Equilibres Electrochimiques à 25°C, Gauthier-Villars, Paris, 1963, pp. 590-603.

21. P. S. Buckley and H. Hartley, Phil. Mag., 8, 320 (1929).

22. Ref. 15, p. 73.

23. Ref. 16, p. 303.

24. G. Kortüm, Treatise on Electrochemistry, 2nd rev. Engl. ed., Elsevier, Amsterdam, 1965, p. 312.

25. A. McFarlane and H. Hartley, Phil. Mag., 13, 425 (1932).

26. J. C. Marchon and J. Badoz-Lambling, Bull. Soc. Chim. Fr., 1967, 4660.

27. J. C. Marchon, C. R. Acad. Sci., Paris, Ser. C, 267, 1123 (1968).

28. R. L. Benoit, M. Guay, and J. Desbarres, Can. J. Chem., 46, 1261 (1968).

29. G. Zimmermann and F. C. Strong, J. Amer. Chem. Soc., 79, 2063 (1957).

30. H. S. Harned and B. B. Owen, The Physical Chemistry of Electrolytic Solutions, 3rd ed., Reinhold, New York, 1958, pp. 466ff.

31. R. A. Robinson and R. H. Stokes, Electrolyte Solutions, 2nd rev. ed., Butterworths, London, 1965, pp. 481ff.

32. R. S. Greeley, W. T. Smith, M. H. Lietzke, and R. W. Stoughton, J. Phys. Chem., 64, 652, 1445 (1960).

33. H. S. Harned and E. C. Dreby, J. Amer. Chem. Soc., 61, 3116 (1939).

34. S. Lengyel, in Electrolytes (B. Pesce, ed.), Pergamon, Oxford, 1962, p. 208.

35. D. A. McInnes and J. A. Beattie, J. Amer. Chem. Soc., 42, 1117 (1920).

36. D. D. Perrin, Pure Appl. Chem., 20, 133 (1969).

37. R. L. Benoit and M. Guay, Inorg. Nucl. Chem. Lett., 4, 215 (1968).

38. I. V. Nelson and R. T. Iwamoto, J. Electroanal. Chem., 7, 218 (1964).

39. Ref. 33, p. 457 (for general bibliography), and pp. 717ff.

40. K. Schwabe and S. Ziegenbalg, Ber. Bunsenges. Phys. Chem. (Z. Elektrochem.), 62, 172 (1958).

41. K. Schwabe and W. Schwenke, ibid., 63, 441 (1959).

42. K. Schwabe and R. Hertzsch, ibid., 63, 445 (1959).

43. K. Schwabe and M. Kunz, ibid., 64, 1188 (1960).

44. K. Schwabe and N. Fu Nhuan, ibid., 65, 891 (1961).

45. K. Schwabe and K. Wankmüller, ibid., 69, 528 (1965).

46. K. Schwabe and R. Müller, ibid., 73, 178 (1969).

47. T. Mussini, P. Longhi, and P. Giammario, Chim. Ind. (Milan), 53, 1124 (1971).

48. D. A. Johnson and B. Sen, J. Chem. Eng. Data, 13, 376 (1968).

49. R. N. Roy and B. Sen, ibid., 13, 79 (1968).

50. H. Sadek, T. F. Tadros, and A. A. El-Harakany, Electrochim. Acta, 16, 339 (1971).

51. K. K. Kundu, P. K. Chattopadhyay, Debabrata Jana, and M. N. Das, J. Chem. Eng. Data, 15, 209 (1970).

52. D. Feakins and C. M. French, J. Chem. Soc., 1957, 2284.

53. B. Nayak and D. K. Sahu, Electrochim. Acta, 16, 1757 (1971).

54. J. S. Mayell and S. H. Langer, ibid., 9, 1411 (1964).

55. E. L. Littauer and L. L. Shreir, ibid., 11, 527 (1966).

56. A. N. Frumkin and G. A. Tedoradse, Z. Elektrochem., 62, 251 (1958).

57. G. A. Tedoradse, Zhur. Fiz. Khim., 33, 129 (1959).

58. S. F. Belevskii and S. V. Gorbachev, ibid., 36, 742 (1962).

59. B. M. Blavatnik and G. A. Tsiganov, Corrosion, 19, 421t (1963).

60. M. Enyo and T. Yokoyama, Electrochim. Acta, 15, 183 (1970).

61. S. Toshima and H. Okaniwa, Denki Kagaku, 34, 641 (1966).

62. S. Toshima and H. Okaniwa, ibid., 34, 958 (1966).

63. S. Toshima and H. Okaniwa, ibid., 35, 23 (1967).

64. T. Dickinson, R. Greef, and Lord Wynne-Jones, Electrochim. Acta, 14, 467 (1969).

65. M. Takahashi and T. Odashima, Denki Kagaku, 35, 805 (1967).

66. T. Mussini and G. Casarini, Chim. Ind. (Milan), 47, 600 (1965).

67. F. Chang and H. Wick, Z. Phys. Chem., 172, 448 (1935).

68. T. Yokoyama and M. Enyo, Electrochim. Acta, 15, 1921 (1970).

69. Chin Ying-Chech and M. A. Genshaw, J. Phys. Chem., 73, 3571 (1969).

70. V. S. Bagotzky, Yu. B. Vassilyev, J. Weber, and J. N. Pirtskhalava, J. Electroanal. Chem., 27, 31 (1970).

71. A. N. Chemodanov and Ya. M. Kolotyrkin, Proceedings of the 3rd European Symposium on Corrosion Inhibitors, Ferrara, Italy, 1970, p. 49.

72. L. I. Krishtalik, Electrochim. Acta, 13, 1045 (1968).

73. M. W. Breiter, ibid., 8, 927 (1963).

74. L. J. J. Janssen and J. G. Hoogland, ibid., 15, 941 (1970).

75. E. E. Littauer and L. L. Shreir, ibid., 12, 465 (1967).

76. B. V. Ershler, Zhur. Fiz. Khim., 18, 131 (1944).

77. K. J. Vetter, Electrochemical Kinetics, Academic, New York, 1967, p. 155.

78. K. Schwabe and Cl. D. Seiler, Chem. Ing.-Tech., 33, 366 (1961).

79. S. Toshima and H. Okinawa, Denki Kagaku, 35, 647 (1967).

80. V. I. Lyubushkin and K. G. Ilyin, Tr. Novocherk. Politekhn. Inst., 141, 79 (1964).

81. Chiou-Cheng Lu, S. Asakura, and T. Mukaibo, Denki Kagaku, 35, 796 (1967).

82. G. Raspi, F. Pergola, and D. Cozzi, J. Electroanal. Chem., 15, 35 (1967).

83. B. E. Conway, Theory and Principles of Electrode Processes, Ronald, New York, 1965, p. 202.

84. N. A. Balashova, Electrochim. Acta, 7, 559 (1962).

85. M. W. Breiter, ibid., 8, 925 (1963).

86. G. Faita, G. Fiori, and J. W. Augustynski, J. Electrochem. Soc., 116, 928 (1969).

87. G. Faita, G. Fiori, and A. Nidola, ibid., 117, 1333 (1970).

88. G. Faita and G. Fiori, J. Appl. Electrochem., 2, 31 (1972).

89. O. Schwarzer and R. Landsberg, J. Electroanal. Chem., 19, 391 (1968).

90. J. A. Harrison and Z. A. Khan, ibid., 30, 87 (1971).

91. L. Müller, Elektrokhimiya, 4, 199 (1968).

92. H. Wolf and R. Landsberg, J. Electroanal. Chem., 28, 295 (1970).

93. F. Foerster and E. Müller, Z. Elektrochem., 8, 515 (1902).

94. F. Foerster and E. Müller, ibid., 8, 665 (1902).

95. F. Foerster and E. Müller, ibid., 9, 171 (1903).

96. L. Hammar and G. Wranglén, Electrochim. Acta, 9, 1 (1964).

97. N. Konopik and E. Berger, Monatsh. Chem., 84, 666 (1953).

98. A. M. Hartley and A. C. Adams, J. Electroanal. Chem., 6, 460 (1963).

99. O. Schwarzer and R. Landsberg, ibid., 14, 339 (1967).

100. C. Raspi and F. Pergola, ibid., 20, 419 (1969).

101. L. M. Elina, T. I. Borisova, and T. S. Filippov, Zhur. Fiz. Khim., 30, 1282 (1956).

102. A. Stancu and L. Stancu, Bull. Inst. Politeh. Bucuresti, 20, 61 (1958).

103. M. M. Flisskii, Zhur. Fiz. Khim., 39, 186 (1965).

104. M. M. Flisskii, Elektrokhimiya, 2, 860 (1966).

105. M. M. Flisskii, ibid., 2, 942 (1966).

106. M. M. Flisskii, ibid., 3, 770 (1967).

107. Z. D. Rozhdestvenskaya and O. A. Songina, Zhur. Anal. Khim., 15, 138 (1960).

108. I. E. Veselovskaya, E. M. Kuchinskii, and L. V. Morochko, Zhur. Prikl. Khim., 37, 76 (1964).

109. W. C. Bray, Z. Phys. Chem., 54, 731 (1906).

110. I. M. Kolthoff and I. Hodara, J. Electroanal. Chem., 5, 2 (1963).

111. I. M. Kolthoff and I. Hodara, ibid., 5, 165 (1963).

112. G. Fiori, Chim. Ind. (Milan), 51, 1380 (1969).

113. T. Osuga, S. Fujii, K. Sugino, and T. Sekine, J. Electrochem. Soc., 116, 203 (1969).

114. K. Sugino and S. Aoyagi, ibid., 103, 166 (1956).

115. O. de Nora, P. Gallone, C. Traini, and G. Meneghini, ibid., 116, 146 (1969).

116. K. C. Narasimham, S. Sundarajan, and H. V. K. Udupa, ibid., 108, 798 (1961).

117. M. P. Grotheer and E. H. Cook, Electrochem. Technol., 6, 221 (1968).

118. E. Isabettini, Dissertation Thesis, University of Milan, 1970.

119. I. F. Franchuk, Izv. Akad. Nauk SSSR, Otd. Khim. Nauk, 1963(1), 63-66.

120. G. A. Rechnitz and H. A. Laitinen, Anal. Chem., 33, 1473 (1961).

121. E. V. Kasatkin, K. I. Rozental, and V. I. Veselovskii, Elektrokhimiya, 4, 1402 (1968).

122. E. V. Kasatkin, K. I. Rozental, A. A. Yakovleva, and V. I. Veselovskii, ibid., 5, 139 (1969).

123. I. M. Kolthoff and J. F. Coetzee, J. Amer. Chem. Soc., 79, 1852 (1957).

124. M. Michlmayr and D. T. Sawyer, J. Electroanal. Chem., 23, 387 (1969).

125. P. Longhi and G. Guerra, Chim. Ind. (Milan), 54, 205 (1972).

126. H. E. Zittel and F. J. Miller, Anal. Chim. Acta, 37, 141 (1967).

127. J. C. Marchon and J. Badoz-Lambling, Bull. Soc. Chim. Fr., 1967, 4660.

128. C. Sinicki, P. Desportes, M. Bréant, and R. Rosset, ibid., 1968, 829.

129. J. B. Headridge, D. Pletcher, and M. Callingham, J. Chem. Soc., A, 1967, 684.

130. B. Bry and B. Trémillon, J. Electroanal. Chem., 30, 457 (1971).

131. H. Schmidt and J. Noack, Z. Anorg. Allgem. Chem., 296, 262 (1958).

132. G. Cauquis and D. Serve, C. R. Acad. Sci., Paris, Ser. C, 262, 1516 (1966).

133. R. T. Foley, L. E. Helgan and L. S. Schubert, NASA Accession No. N66-14600, Rept. No. NASA-CR-62.23, (1965).

134. W. E. Triaca, C. Solomons, and J. O'M. Bockris, Electrochim. Acta, 13, 1949 (1968).

134a. R. Tunold, H. M. Bo, K. A. Paulsen, and J. O. Yttredd, ibid., 16, 2101 (1971).

135. P. Drossbach and H. Hoff, ibid., 14, 89 (1969).

136. D. A. J. Swinkels and R. N. Seefurth, J. Electrochem. Soc., 115, 994 (1968).

137. L. S. Leonova, Yu. M. Ryabukhin, and E. A. Ukshe, Elektrokhimiya, 4, 1245 (1968).

138. G. Bianchi, J. Appl. Electrochem., 1, 231 (1971).

139. R. Johnsen and E. Bohm, ibid., 1, 163 (1971).

140. R. Powell, Chlorine and Caustic Soda Manufacture, Noyes Development Corp., Park Ridge, New Jersey, 1969, pp. 115-127.

141. C. L. Mantell, Electrochemical Engineering, McGraw-Hill, New York, 1960.

142. Kirk-Othmer, Encyclopedia of Chemical Technology, Vol. 1, 2nd ed., Wiley (Interscience), New York, pp. 668-707.

143. A. T. Kuhn, Industrial Electrochemical Processes, Elsevier, Amsterdam, 1971, p. 106.

144. D. V. Kokoulina and L. I. Krishtalik, Elektrokhimiya, 7, 346 (1971).

145. D. Landolt and N. Ibl, J. Electrochem. Soc., 115, 713 (1968).

146. D. Landolt and N. Ibl, Electrochim. Acta, 15, 1165 (1970).

147. J. R. Newberry, W. C. Gardiner, A. J. Holmes, and R. F. Foyle, J. Electrochem. Soc., 116, 114 (1969).

148. T. R. Beck, ibid., 116, 1038 (1969).

149. A. Kuhn, Industrial Electrochemical Processes, Elsevier, Amsterdam, 1971, p. 92.

150. J. Fleck, Chem.-Ing. Tech., 43, 173 (1971).

151. Kirk-Othmer, Encyclopedia of Chemical Technology, Vol. 5, 2nd ed., Wiley (Interscience), New York, p. 50.

152. J. C. Schumacher, Perchlorates (American Chemical Society Monograph 146), Reinhold, New York, 1960.

153. A. Legendre, Chem.-Ing. Tech., 34, 379 (1962).

154. M. Ya. Fioshin and A. M. Mezheritskii, Khim. Prom., 43, 676 (1967).

155. L. I. Krishtalik and Z. A. Rothenberg, Zhur. Fiz. Khim., 39, 328 (1965).

156. L. I. Krishtalik and Z. A. Rothenberg, ibid., 39, 907 (1965).

157. R. G. Ehrenburg and L. I. Krishtalik, Elektrokhimiya, 4, 923 (1968).

158. V. V. Stender, Dokl. Akad. Nauk SSSR, 106, 487 (1956).

159. L. J. J. Janssen and J. G. Hoogland, Electrochim. Acta, 14, 1097 (1969).

160. O. Suzuki, A. Ikeda, and S. Abe, J. Electrochem. Soc. Japan, 27, E38 (1959).

161. L. J. J. Janssen and J. G. Hoogland, Electrochim. Acta, 15, 339 (1970).

Chapter I-2

BROMINE

T. MUSSINI and G. FAITA

Laboratory of Electrochemistry and Metallurgy
University of Milan
Milan, Italy

1. OXIDATION STATES AND STANDARD POTENTIALS

1.1. OXIDATION STATES

The more reliable oxidation numbers for which one can state the existence of bromine compounds are: **-1** for hydrobromic acid, HBr, and its salts, the bromides; +1 for hypobromous acid, HBrO, and its salts, the hypobromites; +5 for bromic acid, $HBrO_3$, and its salts, the bromates; and +7 for perbromic acid, $HBrO_4$, and its salts, the perbromates. There is evidence for the existence of the oxides Br_2O, BrO, and BrO_2 with oxidation numbers for bromine of +1, +2, and +4, respectively. Bromous acid (+3) certainly forms only as an intermediate step in several reactions.

Only recently was it possible to synthesize the perbromates [1, 2] and perbromic acid [2], corresponding to the oxidation number +7. This result is not only important because of the conspicuous absence that is removed (for one could reasonably accommodate the perbromates in their logical position between perchlorates and periodates, for homology reasons), but also because it clears away the strongly established belief of nonsynthesizability and nonexistence of perbromates, for which many structural argumentations have been alleged [3-9].

The synthesis of perbromates was realized by Appelman [1, 2] following three different methods:

1. The radiochemical method [1], based on a hot-atom process (the β-decay of radioactive ^{83}Se as selenate) according to the reaction

$$^{83}SeO_4^{2-} \longrightarrow {}^{83}BrO_4^- + \beta^- \quad (1.1.1)$$

By this method rubidium perbromate was obtained.

2. The electrolytic method [1], consisting of the oxidation of aqueous bromates at -15°C on rotating platinum microanodes at anodic current densities of about 10 A/cm^2, with separate electrode compartments (acidic catholyte and neutral anolyte).

3. The chemical method [1, 2] in two variations: (a) the oxidation of bromate to perbromate by aqueous xenon difluoride XeF_2 [1]; (b) the oxidation of bromate in alkaline solution by molecular fluorine [2], the only method which lends itself to large-scale preparations. Concentrated solutions of perbromic acid were obtained in this way.

The mass-spectral characterization of perbromic acid was carried out by Studier [10].

The thermodynamic properties of the perbromate ion were also determined [11, 12]. The values of the fundamental thermodynamic functions for the more significant species based on bromine at different oxidation states are collected in Table 1.1.1.

1.2. STANDARD POTENTIALS

1.2.1. Aqueous Solutions

1.2.1.1. The Bromine/Bromide Couple. The potential of the reversible bromine/bromide couple can be expressed in terms of either the electrode reaction

$$Br_2(l) + 2e^- = 2Br^-(aq) \quad (1.2.1)$$

or

$$Br_2(aq) + 2e^- = 2Br^-(aq) \quad (1.2.2)$$

A third form of electrode reaction must be considered, besides reactions (1.2.1) and (1.2.2), i.e.,

$$Br_3^-(aq) + 2e^- = 3Br^-(aq) \quad (1.2.3)$$

TABLE 1.1.1. Basic Thermodynamic Functions for Bromine at Different Oxidation States at 25° C[a]

Formula and description	State	ΔG_f° (kcal/mole)	ΔH_f° (kcal/mole)	S° (cal/deg mole)	Refs.
Br	g	19.701	26.741	41.805	[13]
Br_2	liq	0	0	36.384	[13]
Br_2, std state, m = 1	aq	0.94	-0.62	31.2	[13]
	aq	1.00	-0.22	32.32	[14]
	in CCl_4	0.36	0.71	37.6	[13]
	in CS_2		0.01		[13]
	in CH_3COOH		-0.25		[13]
	in $CHCl_3$		0.65		[13]
Br^-	g		-55.9		[13]
Br^-, std state, m = 1	aq	-24.85	-29.05	4.1[a]	[13]
HBr	g	-12.77	-8.70	47.463	[13]
HBr, std state, m = 1	aq	-24.85	-29.05	19.7	[13]
Br_3^-, std state, m = 1	aq	-25.59	-31.17	35.9[a]	[13]
	aq	-25.28	-31.39	34.1[a]	[14]
Br_5^-, std state, m = 1	aq	-24.8	-34.0	60.1[a]	[13]
BrO^-, std state, m = 1	aq	-8.0	-22.5	-5.6[a]	[13]
HBrO, undiss; std state, m = 1	aq	-19.7	-27.0	34.	[13]
BrO	g	25.87	30.06	56.75	[13]
BrO_2	c		11.6		[13]
BrO_3^-, std state, m = 1	aq	0.4	-20.0	23.4[a]	[13]
	aq	4.55	-15.95	23.0[a]	[11]
$HBrO_3$, std state, m = 1	aq	0.4	-20.0	39.0	[13]
BrO_4^-, std state, m = 1	aq	29.18	3.19	29.1[a]	[11]
$HBrO_4$, std state, m = 1	aq	29.18	3.19	44.7	[11]

[a]The entropy values of ions are quoted on the electrochemical scale, i.e., they are based on $S^\circ(H^+, aq) = \frac{1}{2}S^\circ(H_2, g) = 15.604$ cal/deg mole which follows from the electrochemical convention $E^\circ(H^+, aq + e^- = \frac{1}{2}H_2, g) = 0$ at all temperatures [15, 16]. Entropy values based on the convention $S^\circ(H^+, aq) = 0.000$, can be obtained by adding the quantity $15.604z_j$ cal/deg mole (where z_j is the ionic valency with sign) to the S° values listed in the table.

since bromine is known to react with bromide to form the tribromide ion Br_3^- according to the reaction

$$Br_2 + Br^- = Br_3^- \qquad (1.2.4)$$

At high concentrations, in addition to Br_3^-, the pentabromide ion Br_5^- might form according to

$$2Br_2 + Br^- = Br_5^- \qquad (1.2.5)$$

but this last case is of little or no importance for the determination of the standard potentials of the bromine-bromide couple being considered.

The more important contributions for the evaluation of these standard potentials are those of Lewis and Storch [17] and of Jones and Baeckström [18] in terms of both electrode Reactions (1.2.1) and (1.2.2) at 25° C, and of Mussini and Faita [14] in terms of the Reaction (1.2.2) over the temperature range 0 to 50° C. The results of the above investigations are in excellent agreement (see Table 1.2.1 and also Refs. 19-23), and one finds $E_m^\circ(Br_2, l/Br^-, aq) = 1.0652$ V and $E_m^\circ(Br_2, aq/Br^-, aq) = 1.0874$ V (abs) at 25° C, the "m" subscripts indicating standard potentials on the molal scale, and the signs of potentials being as prescribed by the Stockholm Convention of IUPAC [24, 25].

From the $E_m^\circ(Br_2, l/Br^-, aq)$ value the standard free energy of the bromide ion results as -24.562 kcal/mole, in good agreement with the thermal data (cf. Table 1.1.1).

The works of Lewis and Storch [17] and of Jones and Baeckström [18] were based on emf measurements at 25° C of the cell

$$Ag|AgBr|HBr, aq|Br_2|Pt\text{-}Ir \qquad (1.2.6)$$

and

$$Ag|AgBr|KBr, aq|Br_2|Pt\text{-}Ir \qquad (1.2.7)$$

respectively, the formation of the Br^- ion according to Reaction (1.2.4) being taken into account on the basis of previous knowledge of the equilibrium constant K for Reaction (1.2.4). Assuming a value of 17.9 kg/mole for this constant K_m (on the molal scale) at 25° C (see Refs. 14 and 26, and Section 1.3), the standard free energy of formation of the tribromide ion is -25.28 kcal/mole, and for the standard potential in terms of Reaction (1.2.3) one obtains $E_m^\circ(Br_3^-, aq/Br^-, aq) = 1.0503$ V at 25° C.

The simultaneous determination of E_m° and K_m for Reaction (1.2.4), starting from emf measurements of Cell (1.2.6), can be carried out by the following procedure. If there were no reaction forming tribromide ion, the emf of Cell (1.2.6) would be independent of the Br^- concentration and given simply by

$$E = E_m^\circ(Br_2, aq/Br^-, aq) - E_m^\circ(Ag/AgBr/Br^-, aq) + (k/2)\log(a_{Br_2}) \qquad (1.2.8)$$

where $k = 2.303RT/F$. But owing to the occurrence of Reaction (1.2.4), some Br^- ions are converted to Br_3^- ions in the compartment of the bromine/bromide electrode, and the molality of the Br^- ion in this compartment, $m^*_{Br^-}$, is different from that in the compartment of the silver/silver-bromide electrode, m_{Br^-}. This fact results in the appearance of a logarithmic term accounting for the $m^*_{Br^-}/m_{Br^-}$ ratio and of a liquid junction potential E_J between the two electrodes in the expression for the emf

TABLE 1.2.1. Standard Molal Potentials E_m° over a Range of Temperatures for Different Electrode Reactions Concerning the System $Br_2/Br^-/Br_3^-$ in Aqueous Solution, Together with the Isothermal Temperature Coefficient $(dE_m^\circ/dT)_{isoth}$ and the Second Isothermal Temperature Coefficient $(d^2E_m^\circ/dT^2)_{isoth}$. The Thermal Temperature Coefficient at 25°C Is Obtainable from $(dE_m^\circ/dT)_{therm} = (dE_m^\circ/dT)_{isoth} + 0.871$ mV/deg, the Constant 0.871 mV/deg Being the Thermal Temperature Coefficient for the SHE at 25°C

Reaction	T (°C)	E_m° (V)	$(dE_m^\circ/dT)_{isoth}$ (mV/deg)	$(d^2E_m^\circ/dT^2)_{isoth}$ (μV/deg²)	Comments
$Br_2(l) + 2e^- = 2Br^-(aq)$	25	1.0652	-0.629	-6.210	From Ref. 23. E_m° obtained from emf of cell Ag/AgBr/Br^-/Br_2/Pt-Ir; formation of Br_3^- taken into account: cf. Refs. 17, 18, and 20-23.
$Br_2(aq) + 2e^- = 2Br^-(aq)$	0	1.0978	-0.2870	-10.1	Data taken from, or calculated from, Ref. 14. Simultaneous determination of E_m° and K_m for the reaction $Br_2 + Br^- = Br_3^-$ over the same temperature range, from emf's of the cell Ag/AgBr/HBr/Br_2/Pt-Ir. Earlier value at 25°C: E_m° = 1.087 V (cf. Refs. 17-20 and 23). From entropy data in Table 1.1.1 one would calculate that $(dE_m^\circ/dT)_{isoth} = -0.51 \pm 0.01$ mV/deg at 25°C.
	10	1.0943	-0.3884		
	20	1.0900	-0.4899		
	25	1.0874	-0.5406		
	30	1.0846	-0.5913		
	40	1.0782	-0.6927		
	50	1.0707	-0.7941		
$Br_3^-(aq) + 2e^- = 3Br^-(aq)$	0	1.0601	-0.2787	-8.8	Data obtained from the above values of $E_m^\circ(Br_2, aq/Br^-, aq)$ and K_m for the reaction $Br_2 + Br^- = Br_3^-$ in Ref. 14. Earlier value at 25°C: E_m° = 1.05 V (Ref. 20).
	10	1.0566	-0.3664		
	20	1.0527	-0.4541		
	25	1.0503	-0.4980		
	30	1.0478	-0.5418		
	40	1.0418	-0.6296		
	50	1.0351	-0.7173		

$$E = E^\circ_m(Br_2, aq/Br^-, aq) - E^\circ_m(Ag/AgBr/Br^-, aq) + (k/2)\log(m_{Br_2}\gamma_{Br_2}) - k\log(m^*_{Br^-}\gamma^*_{Br^-}/\{m_{Br^-}\gamma_{Br^-}\}) + E_J \quad (1.2.9)$$

where m_{Br_2} is the molality of free bromine, γ_{Br_2} the corresponding molal activity coefficient, and the ratio $\gamma^*_{Br^-}/\gamma_{Br^-}$ can be safely equated to unity since the occurrence of the Reaction (1.2.4) causes no change in the ionic strength of the solution. The molality of free bromine, m_{Br_2}, can be expressed as a function of the equilibrium constant, K_m, and of the molality of total bromine in solution, m_{totBr_2}, by combining the relevant equations

$$K_m = \gamma_{Br_3^-}m_{Br_3^-}\ (\gamma_{Br_2}m_{Br_2}\gamma^*_{Br^-}m^*_{Br^-}) \quad (1.2.10)$$

and

$$m_{totBr_2} = m_{Br_2} + m_{Br_3^-} \quad (1.2.11)$$

Thereby, upon assuming $\gamma^*_{Br^-} = \gamma_{Br_3^-}$, one obtains

$$m_{Br_2} = m_{totBr_2}/(1 + K_m\gamma_{Br_2}m^*_{Br^-}) \quad (1.2.12)$$

and thus

$$E = E^\circ_m(Br_2, aq/Br^-, aq) - E^\circ_m(Ag/AgBr/Br^-, aq) + \frac{k}{2}\log\frac{\gamma_{Br_2}m_{totBr_2}}{1 + K_m\gamma_{Br_2}m^*_{Br^-}} - k\log\frac{m^*_{Br^-}}{m_{Br^-}} - E_J \quad (1.2.13)$$

If one considers the emf's E' and E'' corresponding to two different molalities $(m^*_{Br^-})'$ and $(m^*_{Br^-})''$ of bromide ion, at a same m_{totBr_2}, from Eq. (1.2.13) one obtains

$$E = E'' - E' = \frac{k}{2}\log\left(\frac{\gamma''_{Br_2} + \gamma'_{Br_2}\gamma''_{Br_2}m^{*\prime}_{Br^-}K_m}{\gamma'_{Br_2} + \gamma'_{Br_2}\gamma''_{Br_2}m^{*\prime\prime}_{Br^-}K_m}\right) - k\log\left(\frac{m^{*\prime\prime}_{Br^-}m'_{Br^-}}{m^{*\prime}_{Br^-}m''_{Br^-}}\right) + (E''_J - E'_J) \quad (1.2.14)$$

Now, as

$$m_{HBr} = m^*_{Br^-} + m_{Br_3^-} \quad (1.2.15)$$

from Eqs. (1.2.15) and (1.2.10) one obtains

$$m^*_{Br^-} = m_{HBr}/(1 + m_{Br_2}K_m\gamma_{Br_2}\gamma^*_{Br}/\gamma_{Br_3^-}) \quad (1.2.16)$$

Thus, if the molality of total bromine in solution is small (for instance, <0.001), it is evident that $m^*_{Br^-}$ differs from m_{HBr} (and hence from m_{Br^-}) by about 1% or less, as K_m is of the order of 17 kg/mole. Therefore the last logarithmic term in Eq. (1.2.14) can be dropped and the terms in E_J in Eqs. (1.2.13) and (1.2.14) are also negligible (cf. Ref. 18); moreover, m_{HBr} can be substituted for $m^*_{Br^-}$. If, in addition, m'_{HBr} and m''_{HBr} are not high enough to make γ_{Br_2} sensibly different from unity [27] — for instance $m'_{HBr} = 0.1$ and $m''_{HBr} = 0.01$ — Eq. (1.2.14) takes the final form

$$E'' - E' = \frac{k}{2}\log\left(\frac{1 + m'_{HBr}K_m}{1 + m''_{HBr}K_m}\right) \quad (1.2.17)$$

while Eq. (1.2.13) reduces to

$$E = E^\circ_m(Br_2, aq/Br^-, aq) - E^\circ(Ag/AgBr/Br^-, aq) - \frac{k}{2}\log\frac{m_{totBr2}}{1 + m_{HBr}K_m} \qquad (1.2.18)$$

Using Eq. (1.2.17), K_m is first obtained from the two emf values E' and E'', and subsequently it is introduced into Eq. (1.2.18) to obtain the required $E^\circ_m(Br_2, aq/Br^-, aq)$, the $E^\circ_m(Ag/AgBr/Br^-, aq)$ value being known from the recent redetermination by Hetzer, Robinson, and Bates [28]. The values of $E^\circ_m(Br_2, aq/Br^-, aq)$ obtained by Mussini and Faita [14] by the above method within the 0 to 50°C range are collected in Table 1.2.1 and can be reproduced accurately by the least-squares polynomial

$$E^\circ_m(Br_2, aq/Br^-, aq) = 0.797783 + 0.00248340T - (5.07121 \times 10^{-6})T^2 V \qquad (1.2.19)$$

(where T is the absolute temperature), in practice with all the significant figures: the maximum and the mean deviation of fit are 0.08 and 0.04 mV, respectively.

The values of $E^\circ_m(Br_3^-, aq/Br^-, aq)$, also given in Table 1.2.1, have been obtained by combining the above $E^\circ_m(Br_2, aq/Br^-, aq)$ values with the K_m values for Reaction (1.2.4) at the corresponding temperatures [14]. They can be reproduced by

$$E^\circ_m(Br_3^-, aq/Br^-, aq) = 0.808872 + 0.00211746T - (4.38612 \times 10^{-6})T^2 V \qquad (1.2.20)$$

the maximum and the mean deviation of fit being 0.18 and 0.06 mV, respectively. The first and the second isothermal temperature coefficients $(dE^\circ_m/dT)_{isoth}$ and $(d^2E^\circ_m/dT^2)_{isoth}$ can be readily computed from Eqs. (1.2.19) and (1.2.20) from which the standard entropy and heat capacity changes — ΔS° and ΔC_p, respectively — canbbe evaluated according to $(dE^\circ/dT)_{isoth} = \Delta S^\circ/nF$ and $(d^2E^\circ/dT^2)_{isoth} = \Delta C^\circ_p/nFT$.

1.2.1.2. The Hypobromite/Bromine and the Hypobromite/Bromide Couples. The standard potentials for these redox couples, like those discussed in Section 1.2.1.3 below, are not amenable to determinations by direct emf methods. Instead, the relevant E°'s can be calculated on the basis of thermodynamic data.

For the reactions

$$HBrO + H^+ + e^- = \frac{1}{2} Br_2(l) + H_2O \qquad (1.2.21)$$

and

$$BrO^- + H_2O + 2e^- = Br^- + 2\ OH^- \qquad (1.2.22)$$

the E° values in de Bethune's critical tabulation [23] agree well with those in Latimer's classical book [20] based on the values of the free energies of formation of HBrO and BrO^- taken from the equilibrium constant of bromine hydrolysis [29] according to

$$Br_2 + H_2O = HBrO + H^+ + Br^- \qquad (1.2.23)$$

and on the equilibrium constant of HBrO dissociation [30] according to

$$HBrO = H^+ + BrO^- \qquad (1.2.24)$$

Recalculation based on the free energies of formation in the more recent National Bureau of Standards compilation [13] lead to essentially the same results, as Table 1.2.2 shows.

TABLE 1.2.2. Standard Molal Potentials E_m° and Corresponding Isothermal Temperature Coefficients at 25°C for Redox Couples Other than Bromine/Bromide. Underlined Values Calculated from Thermodynamic Data in Table 1.1.1

Reaction	E_m° (V)	$(dE_m^\circ/dT)_{isoth}$ (mV/deg)	$(d^2E_m^\circ/dT^2)_{isoth}$ (μV/deg^2)	Refs.	Comments
$HBrO(aq) + H^+(aq) + e^-$	1.595	-	-	[23]	Cf. Refs. 20 and 21.
$= \frac{1}{2}Br_2(l) + H_2O$	$\underline{1.60}$	$\underline{-0.637}$	-		
$BrO^-(aq) + H_2O + 2e^-$	0.761	-	-	[23]	Cf. Refs. 20 and 21;
$= Br^-(aq) + 2OH^-(aq)$	$\underline{0.77}$	$\underline{-0.94}$	-		aqueous basic solutions.
$BrO_3^-(aq) + 6H^+(aq) + 5e^-$	1.52	-0.418	+1.871	[23]	Cf. Refs. 20 and 21.
$= \frac{1}{2}Br_2(l) + 3H_2O$	$\underline{1.51}$	$\underline{-0.422}$	-		
$BrO_3^-(aq) + 3H_2O + 6e^-$	0.61	-1.287	-6.748	[23]	Cf. Refs. 20 and 21;
$= Br^-(aq) + 6OH^-(aq)$	$\underline{0.614}$	$\underline{-1.29}$	-		aqueous basic solutions.
$BrO_4^-(aq) + 2H^+(aq) + 2e^-$	1.763	$\underline{-0.446}$	-	[11]	
$= BrO_3^-(aq) + H_2O$					

1.2.1.3. The Bromate/Bromine and the Bromate/Bromide Couples. The reactions involved are

$$BrO_3^- + 6H^+ + 5e^- = \frac{1}{2}Br_2(l) + 3H_2O \quad (1.2.25)$$

and

$$BrO_3^- + 3H_2O + 6e^- = Br^- + 6OH^- \quad (1.2.26)$$

and the values of the relevant E°'s from different sources [20, 23] show the same good agreement as those in Section 1.2.1.2. These values are compared in Table 1.2.2. Bromic acid is a stronger oxidant than chloric acid, and in high acidity it can oxidize water to oxygen.

1.2.1.4. The Perbromate/Bromate Couple. Only recently has the essential data become available [11] for this redox couple. The relevant reaction is

$$BrO_4^- + 2H^+ + 2e^- = BrO_3^- + H_2O \quad (1.2.27)$$

and the corresponding standard molal potential is $E_m^\circ(BrO_4^-/BrO_3^-) = 1.763$ V (see Table 1.2.2), which means that perbromate is a stronger oxidant than either perchlorate or periodate. This fairly high electrode potential plus the presence of a remarkable activation barrier [2] between Br(V) and Br(VII) can explain why the synthesis of perbromates has long been so difficult to attain [11].

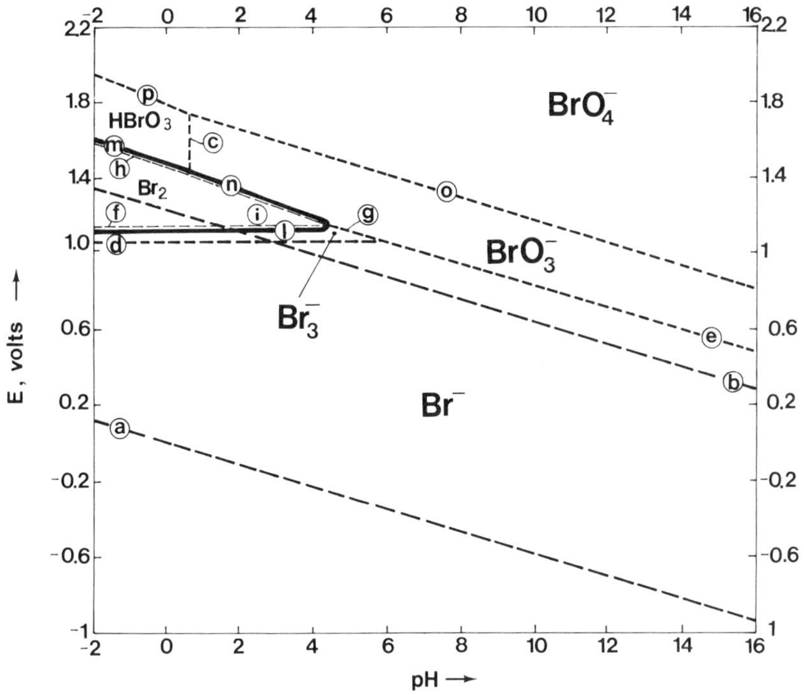

FIG. 1.2.1. Potential-pH diagram for the bromine-water system at 25°C. Equilibria considered: (a) H^+/H_2, g; (b) O_2, g/H_2O; (c) $HBrO_3/BrO_3^-$; (d) Br_3^-/Br^-; (e) BrO_3^-/Br^-; (f) Br_2, aq satd/Br_3^-; (g) BrO_3^-/Br_3^-; (h) $HBrO_3/Br_2$, aq satd; (i) BrO_3^-/Br_2, aq satd; (l) Br_2, l/Br_3^-; (m) $HBrO_3/Br_2$, l; (n) BrO_3^-/Br_2, l; (o) BrO_4^-/BrO_3^-; (p) $BrO_4^-/HBrO_3$.

1.2.1.5. Potential-pH Diagram for Bromine in Aqueous Solution. In his classical and fundamental treatise Pourbaix gave the potential-pH equilibrium diagram for bromine [31] under a variety of experimental conditions, with bromate as the highest oxidation state then known for bromine. The recent determination of the standard potential for the aqueous perbromate/bromate couple [11] makes it possible to trace an extended potential-pH diagram for bromine, as shown in Fig. 1.2.1. This diagram refers to the conditions: 25°C, 1 atm, and 1 gram-atom of total dissolved bromine, and is based on the set of important reactions and relevant functions listed below. These reactions are written consistently as reduction reactions; in any reaction the abbreviation aq for "aqueous" species in solution is omitted except for the species Br_2 to avoid misinterpretations; in the corresponding expressions of redox potentials or of thermodynamic functions the terms accounting for the water activities are omitted as usual, and the activities of the other species are indicated by round parentheses.

(a) $2H^+ + 2e^- = H_2(g)$ $\qquad E_a = (E^\circ)_a - 0.0592\ pH$

(b) $\frac{1}{2}O_2(g) + 2H^+ + 2e^- = 2H_2O$ $\qquad E_b = (E^\circ)_b - 0.0592\ pH$

(c) $HBrO_3 = H^+ + BrO_3^-$ $$\log \frac{(BrO_3^-)}{(HBrO_3)} = pK_{HBrO_3} + pH$$

(d) $Br_3^- + 2e^- = 3Br^-$ $$E_d = (E^\circ)_d + 0.0295 \log \frac{(Br_3^-)}{(Br^-)^3}$$

(e) $BrO_3^- + 6H^+ + 6e^- = Br^- + 3H_2O$ $$E_e = (E^\circ)_e + 0.0098 \log \frac{(BrO_3^-)}{(Br^-)} - 0.0592\ pH$$

(f) $3Br_2(aq\ satd) + 2e^- = 2Br_3^-$ $$E_f = (E^\circ)_f + 0.0295 \log \frac{(Br_2)^3}{(Br^-)^2}$$

(g) $3BrO_3^- + 18H^+ + 16e^- = Br_3^- + 9H_2O$ $$E_g = (E^\circ)_g + 0.0037 \log \frac{(BrO_3^-)^3}{(Br_3^-)} - 0.0665\ pH$$

(h) $2HBrO_3 + 10H^+ + 10e^- = Br_2(aq\ satd) + 6H_2O$

$$E_h = (E^\circ)_h - 0.0059 \log \frac{(HBrO_3)^2}{(Br_2)} - 0.0592\ pH$$

(i) $2BrO_3^- + 12H^+ + 10e^- = Br_2(aq\ satd) + 6H_2O$

$$E_i = (E^\circ)_i - 0.0059 \log \frac{(BrO_3^-)^2}{(Br_2)} - 0.0709\ pH$$

(l) $3Br_2(l) + 2e^- = 2Br_3^-$ $$E_l = (E^\circ)_l - 0.0592 \log (Br_3^-)$$

(m) $2HBrO_3 + 10H^+ + 10e^- = Br_2(l) + 6H_2O$ $$E_m = (E^\circ)_m - 0.0118 \log(HBrO_3) - 0.0592\ pH$$

(n) $2BrO_3^- + 12H^+ + 10e^- = Br_2(l) + 6H_2O$ $$E_n = (E^\circ)_n - 0.0118 \log (BrO_3^-) - 0.0709\ pH$$

(o) $BrO_4^- + 2H^+ + 2e^- = BrO_3^- + H_2O$ $$E_o = (E^\circ)_o - 0.0295 \log \frac{(BrO_4^-)}{(BrO_3^-)} - 0.0592\ pH$$

(p) $BrO_4^- + 3H^+ + 2e^- = HBrO_3 + H_2O$ $$E_p = (E^\circ)_p - 0.0295 \log \frac{(BrO_4^-)}{(HBrO_3)} - 0.0887\ pH$$

The values of K_{HBrO_3} and of (E°)'s for Reactions (a) to (n) are the same as in Ref. 31; the (E°) values for Reactions (o) and (p) are based on Ref. 11.

1.2.2. Nonaqueous Solutions

Contrary to the case of aqueous solutions, there is still a paucity of data of standard potentials for bromine-based redox couples in nonaqueous solutions, in spite of the growing

TABLE 1.2.3. Standard Molar Potentials E_c° for the Bromine/Bromide Couple in Various Nonaqueous or Mixed Solvents and at Different Temperatures

Reaction	Solvent	E_c° (V)	T (°C)	Refs.	Comments[a]
$Br_2 + 2e^- = 2Br^-$	Methanol	0.837	25	[32], [33], [34]	Referred to $E_c^\circ(H^+/H_2) = 0$
	Methanol	3.83	25	[35], [36]	Referred to $E_c^\circ(Rb^+/Rb) = 0$
	Ethanol	0.777	25	[32], [33], [37]	Referred to $E_c^\circ(H^+/H_2) = 0$
	Acetonitrile	0.47	25	[32]	Referred to $E_c^\circ(H^+/H_2) = 0$
	Acetonitrile	3.64	25	[35], [36]	Referred to $E_c^\circ(Rb^+/Rb) = 0$
	Liquid ammonia	3.76	-50	[35], [36], [38]	Referred to $E_c^\circ(Rb^+/Rb) = 0$
	HCOOH	3.97	25	[35], [36]	Referred to $E_c^\circ(Rb^+/Rb) = 0$
	20 wt% CH_3COOH-H_2O mixt	1.038	25	[39]	Apparently referred to $E_c^\circ(H^+/H_2) = 0$
	50 wt% CH_3COOH-H_2O mixt	0.986	25	[39], [40], [41]	Apparently referred to $E_c^\circ(H^+/H_2) = 0$
	80 wt% CH_3COOH-H_2O mixt	0.888	25	[39], [40], [41]	Apparently referred to $E_c^\circ(H^+/H_2) = 0$
	90 wt% CH_3COOH-H_2O mixt	0.836	25	[39], [40], [41]	Apparently referred to $E_c^\circ(H^+/H_2) = 0$
$Br_3^- + 2e^- = 3Br^-$	Sulfolane	-0.150	22	[42]	Referred to E_c°(ferricinium/ferrocene) = 0
	Methanol	0.39	22	[42]	Referred to E_c°(ferricinium/ferrocene) = 0
	Nitromethane	0.07	20	[43]	Referred to E_c°(ferricinium/ferrocene) = 0
	Acetonitrile	0.03	20	[43]	Referred to E_c°(ferricinium/ferrocene) = 0
$3Br_2 + 2e^- = 2Br_3^-$	Sulfolane	0.450	22	[42]	Referred to E_c°(ferricinium/ferrocene) = 0
	Methanol	0.60	22	[42]	Referred to E_c°(ferricinium/ferrocene) = 0
	Nitromethane	0.68	20	[43]	Referred to E_c°(ferricinium/ferrocene) = 0
	Acetonitrile	0.64	20	[43]	Referred to E_c°(ferricinium/ferrocene) = 0

[a]All the reference potentials are relative to the same solvent and the same temperature.

interest of electrochemists in the latter field of research. The available data essentially concern the bromine/bromide couple in terms of the electrode reactions

$Br_2 + 2e^- = 2Br^-$

$Br_3^- + 2e^- = 3Br^-$

$3Br_2 + 2e^- = 2Br_3^-$ (1.2.28)

A selection of values of standard potentials E_c° (molar scale) in various solvents is given in Table 1.2.3.

1.3. EQUILIBRIUM DATA

1.3.1. Aqueous Solutions

1.3.1.1. The Formation Constant for the Tribromide Ion. For the equilibrium constant for the reaction

$Br_2 + Br^- = Br_3^-$ (1.2.4)

in aqueous solutions some sparse data exist, mostly determined by the partition method. A more extensive set of data, covering the temperature range 0 to 50°C, was however obtained by the emf method [14] which was described in Section 1.2.1.1.

Table 1.3.1 provides a collection of data for the sake of comparison. Upon inspection, these data can be considered to agree to an acceptable extent, also taking into account the small contribution due to the use of two different scales (molal and molar).

1.3.1.2. The Formation Constant for the Pentabromide Ion. From Liebhafsky's work [29] the molar-scale equilibrium constant for the reaction

$2Br_2 + Br^- = Br_5^-$ (1.2.5)

of formation of the pentabromide ion in aqueous solution is found to be 40 and 20 liter^2 mole^{-2} at 0 and 25°C, respectively. From Jones and Baeckström's work [50], however, one finds 40 liter^2 mole^{-2} at 25°C, in considerable disagreement with the former values. Perhaps an independent approach — e.g., by the emf method — could help to settle this discrepancy. There would, however, be some difficulties: inter alia, the domain of high concentrations in which the presence of Br_5^- can be detected would make the evaluation of the activity coefficients relevant to the involved single species (Br_2, Br^-, Br_3^-, Br_5^-) highly problematic.

1.3.2. Nonaqueous Solutions

The available data concern the formation constant for the tribromide ion, essentially, and were obtained by a variety of methods (from voltammetry to UV spectroscopy).

TABLE 1.3.1. Values of the Equilibrium Constant on the Molal Scale (K_m) and the Molar Scale (K_c) for the Reaction of Formation (1.2.4) of the Tribromide Ion Br_3^- in Aqueous Solutions at Different Temperatures

	K_m (kg/mole)		K_c	
T (°C)	Ref. 14	Ref. 26	(l/mole)	Refs.
0	24.68	-	19.6	[44]
5	-	19.85	-	
10	21.91	-	-	
16.5	-	-	18.4	[27]
20	19.24	-	-	
21.5	-	-	17.8	[27]
25	17.90	16.85	17.2	[45]
25	-	-	16.1	[46]
25	-	-	16.4	[47]
26.5	-	-	15.9	[48]
30	16.77	-	-	
32.5	-	-	15.4	[49]
35	-	15.28	-	
40	14.89	-	-	
50	12.89	-	-	

Table 1.3.2 provides a selection of values of the above constant, on the molar scale, in the more popular nonaqueous or mixed solvents.

1.4. ACTIVITY COEFFICIENTS AND TRANSFERENCE NUMBERS

1.4.1. Activity Coefficients for Hydrobromic Acid

Aqueous hydrobromic acid is among those strong electrolytes whose activity coefficients were determined with the greatest accuracy [28, 60, 61]. This is the reason for selecting and proposing the value of the mean molal activity coefficient of 0.1 aqueous HBr as a standard value [28] for practical use: e.g., for pH-meter checks and for intercomparison of the potentials of bromide-reversible electrodes of different construction.

Extension of activity coefficient data for HBr to higher concentrations (up to 5.6 mole/kg) over the temperature range 0 to 70°C was recently obtained [62] from emf measurements of the cell

$$Pt|H_2(1\ atm)|HBr(m),\ in\ water|AgBr|Ag|Pt \qquad (1.4.1)$$

TABLE 1.3.2. Molar-Scale Equilibrium Constants (K_c) for the Reaction $Br_2 + Br^- = Br_3^-$ in Various Nonaqueous or Mixed Solvents at Different Temperatures

Solvent	K_c (l/mole)	T (°C)	Refs.
Methanol	200	22	[42]
	355	-15	[51]
	260	+5	[51]
	204	+18	[51]
	177	+25	[51]
93 wt% Methanol-water mixt	190	18	[51]
80 wt% Methanol-water mixt	165	18	[51]
75 wt% Methanol-water mixt	158	18	[51]
71.6 wt% Methanol-water mixt	148	18	[51]
Acetonitrile	$10^{7.0}$	20	[43]
Nitromethane	$10^{7.2}$	20	[52]
Sulfolane	$10^{6.8}$	22	[42], [53]
Acetone	2×10^9	25	[54]
Dimethylformamide	10^5	25	[55]
	2×10^6	25	[56]
Acetic acid	150	25	[57]
	55.5	30	[58]
	27.2	50	[58]
	13.8	70.5	[58]
90 wt% Acetic acid-water mixt	147	25	[39], [40], [41]
80 wt% Acetic acid-water mixt	100	25	[39], [40], [41]
60 wt% Acetic acid-water mixt	27.6	25	[39], [40], [41]
20 wt% Acetic acid-water mixt	20.2	25	[39]
Trifluoracetic acid	<0.9	25	[59]

taking into account the increasing solubility of AgBr in HBr with increasing HBr concentration. The hydrogen electrodes in Cell (1.4.1) were of the capillary-imbibition type described by Bianchi et al. [63-65] and the silver/silver bromide electrodes were of the electrolytic type [66].

Activity coefficients of HBr in nonaqueous or mixed solvents are available in only a few cases. They are based on emf measurements of either the cell

$$\mathrm{Pt}\,|\,H_2(1\text{ atm})\,|\,\mathrm{HBr}(m),\ \text{in S}\,|\,\mathrm{AgBr}\,|\,\mathrm{Ag}\,|\,\mathrm{Pt} \tag{1.4.2}$$

or

$$\mathrm{Pt}\,|\,H_2(1\text{ atm})\,|\,\mathrm{HBr}(m),\ \text{in S}\,|\,\mathrm{Hg_2Br_2}\,|\,\mathrm{Hg}\,|\,\mathrm{Pt} \tag{1.4.3}$$

where S indicates the nonaqueous or mixed aquo-organic solvent. The emf expression for both cells is of the type

$$^{S}E = {}^{S}E^{\circ}_{m} - 2k \log(m\,{}^{S}_{s}\gamma_{\pm})_{HBr} \tag{1.4.4}$$

where $^{S}E^{\circ}$ is the standard molal emf of cell and $^{S}_{s}\gamma_{\pm}$ the mean molal activity coefficient of HBr at the molality m. The superscript "s" indicates the solvent studied, the subscript "s" denoting reference to unity at infinite dilution in s. Once $^{S}E^{\circ}_{m}$ is obtained through the classical extrapolation procedure equivalent to

$$\lim_{m \to 0} \left| {}^{S}E + 2k \log(m)_{HBr} \right| = {}^{S}E^{\circ}_{m} \tag{1.4.5}$$

the $^{S}_{s}\gamma$'s can be determined from Eq. (1.4.4) at the molalities m's.

Cell (1.4.2) has been used by Kundu et al. [67, 68] and by Mussini et al. [69] for the cases of 10 to 100 wt% ethylene glycol-water and 5 to 82 wt% dioxane-water mixtures respectively. Cell (1.4.3) was used by Schwabe et al. [70, 71] in the cases of 20 to 99.5 wt% methanol-water and 20 to 95 wt% ethanol-water mixtures.

Table 1.4.1 shows how the increasing amount of the nonaqueous component in the solvent mixture affects the activity coefficient at a given HBr molality, 1,4-dioxane (lowest dielectric constant, zero dipole moment) and ethylene glycol (conspicuous dielectric constant, high dipole moment) being the nonaqueous component contrasted. At high dioxane percentages HBr behaves as a weak acid [69], as shown by the magnitudes of the medium effects [69]. Table 1.4.1 also includes the activity coefficients at 25°C for aqueous HBr for the sake of comparison.

1.4.2. Transference Numbers for Hydrobromic Acid

Recently the transference numbers of aqueous hydrobromic acid have been determined [72] over a wide range of concentrations (up to 5.5 mole/kg) at temperatures from 20 to 40°C, thus removing the previous lack of data. These transference numbers have been determined as

$$t_{Br^-} = dE_{tr}/dE \tag{1.4.6}$$

where E_{tr} is the emf of the cell with transference

$$Pt|H_2(1\ atm)|HBr(m_1)||HBr(m_2)|H_2(1\ atm)|Pt \tag{1.4.7}$$

and E is the emf of the corresponding cell without transference

$$Pt|H_2(1\ atm)|HBr(m_1)|AgBr|Ag|AgBr|HBr(m_2)|H_2(1\ atm)|Pt \tag{1.4.8}$$

with m_1 as fixed molality and m_2 as variable molality. The resulting transference numbers can be expressed as a function of HBr molality by the interpolation polynomial

TABLE 1.4.1. Mean Molal Activity Coefficients ($\gamma_\pm$) for HBr at Round Molalities in Various Dioxane-Water and Ethylene Glycol-Water Mixtures at 25°C. Data for HBr at Higher Molalities in 100% Water Are Also Included

m_{HBr} (mole/kg)	100% Water	1,4-Dioxane (wt%)							Ethylene glycol (wt%)					
		5	10	15	20	45	70	82	10	30	50	70	90	100
0.001	0.966	0.963	0.959	0.956	0.951	0.910	0.695	0.286	-	-	-	-	-	-
0.002	0.952	0.949	0.944	0.939	0.933	0.878	0.606	0.216	-	-	-	-	-	-
0.003	-	0.939	0.933	0.927	0.920	0.856	0.552	0.182	-	-	-	-	-	-
0.005	0.930	0.924	0.917	0.910	0.901	0.824	0.483	0.146	0.920	0.915	0.895	0.872	0.830	0.795
0.007	-	0.912	0.904	0.896	0.886	0.801	0.440	0.126	-	-	-	-	-	-
0.01	0.906	0.899	0.889	0.881	0.869	0.774	0.397	0.108	0.900	0.890	0.870	0.845	0.800	0.740
0.02	0.879	0.868	0.856	0.845	0.831	0.717	0.324	0.080	0.875	0.855	0.835	0.805	0.760	0.679
0.03	-	0.847	0.834	0.823	0.806	0.682	0.289	0.069	0.855	0.835	0.810	0.778	0.730	0.643
0.05	0.838	0.820	0.803	0.792	0.773	0.638	0.254	0.058	0.830	0.810	0.780	0.745	0.690	0.602
0.07	-	0.800	0.782	0.772	0.751	0.610	0.238	0.053	0.815	0.795	0.765	0.728	0.670	0.582
0.1	0.805	0.780	0.759	0.750	0.727	0.582	0.227	0.051	0.804	0.785	0.760	0.712	0.660	0.557
Refs.	[60]	[69]	[69]	[69]	[69]	[69]	[69]	[69]	[67]	[67]	[67]	[67]	[67]	[68]

Data at Higher HBr Molalities (in 100% Water)

m_{HBr} (mole/kg)	0.2	0.3	0.4	0.5	0.6	0.7	0.8	0.9	1.0	1.4	2.0	3.0	4.0	4.5	5.0	5.5
$\gamma_\pm$	0.782	0.777	0.781	0.789	0.801	0.815	0.832	0.850	0.871	0.969	1.168	1.674	2.300	2.75	3.4	5.3
Refs.	[60]	[60]	[60]	[60]	[60]	[60]	[60]	[60]	[60]	[60]	[60]	[60]	[62]	[62]	[62]	[62]

$$(t_{H^+})_{HBr} = (1 - t_{Br^-})_{HBr} = 0.79236 + (5.7807 \times 10^{-2})m_{HBr} - (4.6009 \times 10^{-4})m_{HBr}^2 \quad (1.4.9)$$

at 25°C.

Data for the transference numbers of hydrobromic acid in nonaqueous solutions are still lacking, contrary to the parallel case of hydrochloric acid for which accurate data are available over a wide range of concentrations and temperatures [73]. This lack of data is not peculiar to HBr, but there is a general and severe shortage of transference numbers for electrolytes in nonaqueous or mixed solvents, a shortage that seriously hinders the development of studies on structures and interactions within the very important field of nonaqueous solutions.

2. KINETIC STUDIES

2.1. KINETIC PARAMETERS AND MECHANISMS

2.1.1. Aqueous Solutions

2.1.1.1. The Br_2/Br^- Couple. Comparatively few experimental data can be found on the electrochemical kinetics, namely on the reaction mechanism and — within this — on the rate-determining step for the bromine/bromide couple.

The tables in this section collect the available data of transfer coefficients (or "symmetry coefficients"), reaction orders with respect to Br^- and Br_2 concentrations, stoichiometric numbers, and exchange currents (see also Fig. 2.1.1).

2.1.1.1.1. Iridium Electrode. The scanty data available (cf. Table 2.1.1) allow only rather general considerations [74]. The anodic and cathodic current/potential relationships clearly follow the Tafel equation and the sum of the anodic and the cathodic transfer coefficients is about 1: this experimental finding shows that the reaction rate is controlled by a charge-transfer stage. However, nothing can be said about the kinetic mechanism.

2.1.1.1.2. Platinum Electrode (Electrochemically Activated). The kinetic parameters collected in Tables 2.1.1, 2.1.2, and 2.1.3 have been used to define a reaction mechanism. On the basis of data provided by the faradic impedance method [75, 76], the following scheme has been suggested:

$$Br_2 + e^- \rightleftharpoons Br_2^- \quad \text{(rate-determining step)} \quad (2.1.1a)$$

$$Br_2^- = Br^- + Br_{ads} \quad (2.1.1b)$$

$$2Br_{ads} = Br_2 \quad (2.1.1c)$$

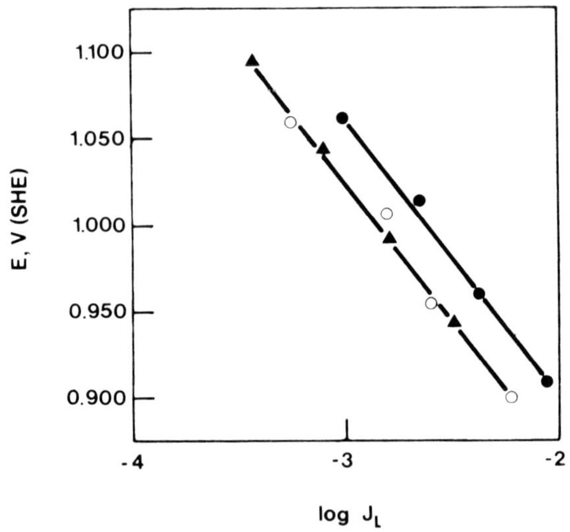

FIG. 2.1.1. Cathodic E/log(j_L) characteristics at 30°C, with aqueous 1 M $NaClO_4$ as a supporting electrolyte, pH = 2.0. (▲) 6×10^{-4} M Br_2 + 3×10^{-3} M NaBr; (o) 6×10^{-4} M Br_2 + 6×10^{-3} M NaBr; (●) 1.2×10^{-3} M Br_2 + 6×10^{-3} M NaBr. j_L is the limiting kinetic current, and SHE is the standard hydrogen electrode.

The equilibrium Reactions (2.1.1b) and (2.1.1c) follow or precede Reaction (2.1.1a) which is the rate-determining one. The behavior of the Br_2/Br^- couple has also been studied by using the rotating disk method [77] which allows a suitable elaboration of the experimental data in terms of 1/j vs $1/\omega^{1/2}$ functions [78]. In this way the influence of diffusion can be eliminated even at rather high reaction rates [79] and reliable kinetic parameters are consequently obtained.

The suggested mechanism consists of two successive charge-transfer stages which are characterized by slightly different exchange currents (cf. Table 2.1.3):

$$Br_2 + e^- \rightleftharpoons Br^- + Br_{ads} \qquad (2.1.2a)$$

$$Br_{ads} + e^- \rightleftharpoons Br^- \qquad (2.1.2b)$$

Reaction (2.1.2a) is substantially equivalent to Reaction (2.1.1a); in fact, it can be imagined as made up by two stages:

$$Br_2 + e^- \rightleftharpoons (Br_2^-)_{ads}$$

$$(Br_2^-)_{ads} = Br^- + Br_{ads} \quad \text{(fast chemical equilibrium)}$$

If the mechanism represented by the Stages (2.1.2a) and (2.1.2b) holds, the extrapolation of the log(j) vs potential straight lines leads to the exchange currents which characterize the two stages. More precisely, the extrapolation from the anodic data leads to the exchange currents for Stage (2.1.2b) whereas extrapolation from the cathodic data leads to the exchange currents for Stage (2.1.2a).

TABLE 2.1.1. Transfer Coefficients for the Reaction $Br_2 + 2e \rightleftharpoons 2Br^-$

	Quoted value	Method	Refs.	Comments
α (cathodic transfer coefficient)	0.47	Rot Ir dsk cathodic current/voltage curves	[80]	At 20°C; diffusion polarization taken into account.
	0.40	Rot Pt dsk cathodic current/voltage curves	[77]	At 30°C; electrochemical activation.[a]
	0.35	Rot Pt dsk cathodic current/voltage curves	[77]	At 30°C; unconditioned electrode.[b]
$1 - \alpha$ (anodic transfer coefficient)	0.25 to 0.40	Pt electrode farad. imp	[75]	25°C(?); electrochemically activated electrode.[a]
	0.58	Rot Ir dsk anodic current/voltage curves	[81]	At 20°C.
	0.40	Rot Pt dsk anodic current/voltage curves	[77]	At 30°C; electrochemically activated electrode.[a]

[a]Electrochemically activated electrode. Llopis and Vàzquez treatment [75]: same as Robertson's method [82] but followed by an anodic oxidation and a cathodic reduction (galvanostatic method). Faita, Fiori, and Mussini's treatment [77]: polishing of the Pt surface followed by a cyclic potentiostatic anodic/cathodic polarization.

[b]Unconditioned electrode. After the electrochemical activation, the electrodes were kept in contact with the test solution for about 3 hr.

In the present case the two exchange currents satisfy Vetter's equation [83]:

$$\left(\frac{\partial\eta}{\partial j}\right)_{j\to 0} = \frac{RT}{4F}\left\{\frac{1}{(j_0)_{2.1.2a}} + \frac{1}{(j_0)_{2.1.2b}}\right\} \qquad (2.1.3)$$

for two consecutive electrochemical stages. On the basis of Eq. (2.1.3) the charge-transfer resistance in both directions, anodic and cathodic, is the same, as it was experimentally found (cf. Table 2.1.3).

TABLE 2.1.2. Reaction Orders n and Stoichiometric Numbers ν for the Reaction Br^2 + 2e ⇄ $2Br^-$

	Quoted value	Refs.	Comments
$(n_{Br^-})_{cath}$	0	[75]	Farad. imp method; T = 25°C(?); Pt electrode, electrochemically activated.[a]
	0	[77]	Rot Pt dsk; cathodic current/voltage curves; activated and unconditioned electrodes[a, b]; T = 30°C.
$(n_{Br^-})_{an}$	1	[75]	Farad. imp method; T = 25°C(?); electrochemically activated Pt electrode.[a]
	1	[77]	Rot Pt dsk; anodic current/voltage curves; activated electrode[a]; T = 30°C.
$(n_{Br_2})_{cath}$	1	[75]	Farad. imp method; T = 25°C(?); electrochemically activated Pt electrode.[a]
	1	[77]	Rot Pt dsk; cathodic current/voltage curves; activated and unconditioned electrodes[a, b]; T = 30°C.
$(n_{Br_2})_{an}$	0.5	[75]	Farad. imp method; 25°C(?); electrochemically activated Pt electrode.[a]
	0	[77]	Rot Pt dsk; anodic current/voltage curves; activated electrodes[a]; T = 30°C.
$(\nu)_{cath}$	2.2	[77]	Rot Pt dsk; unconditioned electrodes[b]; T = 30°C.
	1.5 to 1.8	[77]	Rot Pt dsk; activated electrodes[a]; T = 30°C.
$(\nu)_{an}$	2.0	[77]	Rot Pt dsk; activated electrodes[a]; T = 30°C.

[a]Electrochemically activated electrode. Llopis and Vàzquez treatment [75]: same as Robertson's method [82] but followed by an anodic oxidation and a cathodic reduction (galvanostatic method). Faita, Fiori, and Mussini's treatment [77]: polishing of the Pt surface followed by a cyclic potentiostatic anodic/cathodic polarization.

[b]Unconditioned electrode. After the electrochemical activation, the electrodes were kept in contact with the test solution for about 3 hr.

TABLE 2.1.3. Exchange Current Densities j_0 and Charge-Transfer Resistance $R_t = (\partial\eta/\partial j)_{j=0}$ for the Reaction $Br_2 + 2e \rightleftharpoons 2Br^-$ at $m_{Br_2} = 0.00144$ and $m_{Br^-} = 0.00362$

R_t (Ω cm^2)	j_0 (A/cm^2)	Refs.	Supporting electrolyte	Comments
1.9	0.0136	[76]	2 m $HClO_4$	Farad. imp; activated Pt electrode[a]; 25°C(?).
2.4	0.0105	[76]	2 m $NaClO_4$	Farad. imp; activated Pt electrode[a]; 25°C(?).
8.4	0.003	[76]	2 m $NaClO_4$	Farad. imp; Pt electrode previously anodized[a]; 25°C(?).
-	0.0108	[76]	2 m $HClO_4$	Farad. imp; Pt electrode unconditioned[b]; 25°C(?).
27.7	0.0009	[76]	2 m $HClO_4$	Farad. imp; Pt electrode previously poisoned[c]; 25°C(?).
11.2	0.00215	[77]	1 m $NaClO_4$	pH = 2; T = 30°C; rot Pt dsk, electrochemically activated electrodes[a]; anodic.
12.2	0.0016	[77]	1 m $NaClO_4$	pH = 2; T = 30°C; rot Pt dsk, electrochemically activated electrodes[a]; cathodic.
69	0.00042	[77]	1 m $NaClO_4$	pH = 2; T = 30°C; rot Pt dsk, unconditioned electrodes[b]; cathodic.
-	0.000025	[80]	-	Rot Ir dsk; cathodic current/voltage curves; T = 20°C.
-	0.000036	[81]	-	Rot Ir dsk; anodic current/voltage curves; T = 20°C.

[a]Electrochemically activated electrode. Llopis and Vàzquez treatment [75]: same as Robertson's method [82] but followed by an anodic oxidation and a cathodic reduction (galvanostatic method). Faita, Fiori, and Mussini's treatment [77]: polishing of the Pt surface followed by a cyclic potentiostatic anodic/cathodic polarization.

[b]Unconditioned electrode. After the electrochemical activation, the electrodes were kept in contact with the test solution for about 3 hr.

[c]Poisoned electrodes. The electrodes were soaked for 10 min in a saturated Na_2S_x solution after the activation pretreatment, then rinsed with distilled water, and finally immersed into the test solution. Experimental data from Refs. 77, 80, and 81 were adapted to the unified values of m_{Br_2} and m_{Br^-} indicated in the table title for the sake of easier comparison.

The two suggested mechanisms may be incorporated into a more general scheme:

$$Br_2 + e^- \rightleftarrows (Br_2^-)_{ads} \quad (2.1.4a)$$

$$(Br_2^-)_{ads} = Br^- + Br_{ads} \quad \text{(fast chemical equilibrium)} \quad (2.1.4b)$$

$$Br_{ads} + e^- \rightleftarrows Br^- \quad (2.1.4c)$$

$$2Br_{ads} \rightleftarrows Br_2 \quad (2.1.4d)$$

This scheme is now discussed starting with the anodic reaction. The anodic kinetic parameters obtained by Llopis et al. (cf. Tables 2.1.2 and 2.1.3) show that, when the experimental conditions are very near equilibrium (as is the case in the faradic impedance method), the reaction rate of Stage (2.1.4d) is substantially higher than the one concerning Stage (2.1.4c). These stages are in parallel and the kinetics of the process is defined by the faster stage. Consequently the reaction orders are 1 with respect to the Br^- concentration and 1/2 with respect to the Br_2 concentration. Stage (2.1.4d), however, is just a chemical process and should be independent of the electrode potential. Stage (2.1.4c), on the contrary, is a typical electrochemical reaction and its rate increases as the electrode potential increases. Hence, when the experimental conditions are sufficiently far from equilibrium, as is the case for the extrapolation of the current vs potential curves, Stage (2.1.4c) is presumably faster than Stage (2.1.4d). If this situation holds, the reaction orders are 0 with respect to the Br_2 concentration and 1 with respect to the Br^- concentration. Therefore the conflict between Schemes (2.1.1) and (2.1.2) could be fictitious; as a matter of fact, it can be settled by assuming that a change of mechanism takes place when passing from a near-equilibrium situation to a substantially anodic overpotential.

As regards the cathodic reaction, the reaction orders obtained by the faradic impedance method and those obtained by the analysis of the current vs potential curves are the same (cf. Table 2.1.2). This experimental finding is explained on the basis of the general Scheme (2.1.4); in fact, at high cathodic overpotential only the kinetic parameters which characterize Step (2.1.4a) are obtained. Nothing can be said about the successive faster Steps (2.1.4c) and/or (2.1.4d), i.e., it is not possible to distinguish whether the reaction mechanism follows Path (2.1.1) or Path (2.1.2). However, one should keep in mind that the charge-transfer resistance data obtained by Faita et al. [77] are in fairly good agreement with the theoretical values obtained by using Eq. (2.1.3). This fact seems to support the validity of the two-consecutive-steps electrochemical mechanism at the low overpotentials as well, where the quantity $R_t = (\partial\eta/\partial j)_{j=0}$ is measured.

Activation energies for the reaction $2Br^- \longrightarrow Br_2$ have been evaluated at temperatures ranging from 20 to 80°C [84]: the 10.1 kcal/mole value obtained is in agreement with the hypothesis common to Schemes (2.1.1) and (2.1.2), i.e., the rate-determining step is electrochemical.

Thus far it has been assumed that the adsorption free energies for the reaction intermediates [e.g., $(Br_2^-)_{ads}$, Br_{ads}^-, and Br_{ads}] are independent of the degree of coverage; the adsorption processes in such a situation follow a Langmuir-type isotherm. Actually, the adsorption free energies should somehow depend on the degree of coverage, and a Temkin-type isotherm seems to be more appropriate for defining the adsorption processes.

Experimental data necessary for a reinterpretation of the kinetics of the Br_2/Br^- couple are scanty so far [85-88] and efforts in this direction would undoubtedly be welcome.

2.1.1.1.3. Platinum Electrode (Unconditioned or Spontaneously Oxidized). Llopis and Vàzquez [75] found that changes in the behavior of electrodes kept in contact with solutions containing Br^- and Br_2 can be explained simply by a decrease of the electroactive area without variations in the kinetic parameters. The same conclusions were drawn by Faita et al. [77] who obtained the same values of reaction orders and of Tafel coefficients for unconditioned and electrochemically activated platinum electrodes. The differences of behavior between the two types of electrodes is attributable to a partial oxidation of the electrode surfaces [75].

2.1.1.1.4. Platinum Electrode (Poisoned with Na_2S_x). The experimental data [75] show that the hypothesis of simple mechanical blockage of the platinum surface is rather rough. As a matter of fact, the adsorption and desorption rates of the intermediates seem to be considerably lowered.

2.1.1.1.5. Platinum Electrode (Electrochemically Oxidized). The data collected in Table 2.1.3 show that the Br_2/Br^- couple on a previously anodized platinum electrode is highly irreversible, the effect on the exchange current being particularly strong. Although a complete agreement on this point was not reached [89], the change of kinetic parameters, which is close to that observed for the Cl_2/Cl^- couple at platinum (see Chapter I-1, Chlorine), is understandable in terms of competitive adsorption of Br^- ions and oxygen on the electrode surface. This phenomenon has been studied and demonstrated by various authors.

Voltammetry and capacity measurements [90] in 1 M $HClO_4$ solutions containing various amounts of Br^- clearly show that Br^- adsorption begins in the hydrogen region during the anodic sweep; it increases in the double layer region and strongly delays the formation of adsorbed oxygen. Br^- ions are only substituted by oxygen atoms to a small extent; this substitution is less at higher Br^- concentrations. Concentrations of Br^- greater than 0.001 M completely hinder the formation of adsorbed oxygen under the voltammetric conditions used in Breiter's work [90] (scanning rate of 30 mV/sec).

Analogous conclusions were drawn from radiochemical measurements [91]. Here it was demonstrated that Br^- adsorption on a previously anodized electrode is extremely slow and becomes appreciable only after the oxygen layer has been partly reduced.

The data on the adsorption kinetics of the Br^- ion on a platinum surface lead to the following conclusions:

1. The Br^- ion is adsorbed quickly with high degrees of coverage on platinum electrodes which have been subjected to activation cycles, the final step being a cathodic treatment (e.g., electrochemically activated electrodes).

2. The degree of coverage slowly decreases as the electrode is kept in contact with a solution containing Br^- ions because of the formation of small quantities of adsorbed

oxygen (with consequent displacement of part of the adsorbed Br^-), e.g., on unconditioned or spontaneously oxidized electrodes.

3. If a platinum electrode has been subjected to a treatment ending with an anodic step (oxygen evolution), the steady state for the adsorption processes is rapidly reached, but the degree of coverage by Br^- is smaller by about one order of magnitude than the corresponding degree of coverage for an electrochemically activated electrode (e.g., anodized electrode). Only after a long contact time with the Br^- solution will the layer of adsorbed oxygen be destroyed and substituted by Br_{ads}^-.

2.1.1.1.6. Pyrolytic Graphite Electrode. Voltammetric data show that the anodic oxidation of Br^- contained in 0.5 M K_2SO_4 solution gives a well-defined wave whose half-wave potential is 1.08 ± 0.06 V (SCE) at 25°C, nearly pH-independent [87].

In addition, cyclic voltammetry shows that the reaction intermediate Br_{ads} is weakly adsorbed at the electrode surface. Moreover, the redox behavior of the Br_2/Br^- couple is less reversible than that of the I_2/I^- couple.

2.1.1.2. Higher-Valency Br^{ox}/Br^{red} Couples. The scarce available data on such couples as $HBrO/Br^-$ [92, 93], BrO_2^-/Br^- [93, 94], BrO_3^-/Br^- [93, 95-101], and BrO_4^-/BrO_3^- [1] are highly dispersed and do not allow one to draw any clear kinetic analysis.

2.1.2. Nonaqueous Solutions

2.1.2.1. The Br_2/Br^- Couple in Acetonitrile Solutions. The anodic and cathodic electrode potential vs current curves exhibit double waves [102-104], the related processes being

$$3Br^- \longrightarrow Br_3^- + 2e^- \quad \text{(1st anodic wave)} \tag{2.1.5a}$$

$E_{\frac{1}{2}}$ = 0.42 V vs (Ag/$AgNO_3$ 0.01 M in acetonitrile) electrode

$$2Br_3^- \longrightarrow 3Br_2 + 2e^- \quad \text{(2nd anodic wave)} \tag{2.1.5b}$$

$E_{\frac{1}{2}}$ = 0.71 V vs same electrode as above

and

$$3Br_2 + 2e^- \longrightarrow 2Br_3^- \quad \text{(1st cathodic wave)} \tag{2.1.6a}$$

$$Br_3^- + 2e^- \longrightarrow 3Br^- \quad \text{(2nd cathodic wave)} \tag{2.1.6b}$$

This behavior, which is peculiar to nonaqueous systems, is explained in terms of the equilibrium reaction

$$Br_2 + Br^- = Br_3^- \tag{2.1.7}$$

whose equilibrium constant is notoriously high in many organic solvents (cf. Table 1.3.2).

TABLE 2.1.4. Diffusion Coefficients for the Br^-, Br_2, and Br_3^- Species in 0.4 M $LiClO_4$ in Acetonitrile at 25°C

Species	Diffusion coefficients (cm^2/sec) $\times 10^5$	Method	Refs.
Br^-	1.17	Rot Pt dsk	[102]
Br_2	2.93	Rot Pt dsk	[102]
Br_3^-	1.97[a]	Rot Pt dsk	[102]
	1.89[b]	Rot Pt dsk	[102]

[a] From the cathodic polarization curves.

[b] From the anodic polarization curves.

TABLE 2.1.5. Observed Activation Energies $\Delta G^{\ddagger}$ for Various Steps in the Process $Br_2 + 2e^- \rightleftharpoons 2Br^-$. Experimental Conditions as in Table 2.1.4

Step	$\Delta G^{\ddagger}$ (kcal/mole)
$3Br^- \longrightarrow Br_3^- + 2e^-$	5.1 (±0.5)
$3Br_2 + 2e^- \longrightarrow 2Br_3^-$	3.4
$2Br_3^- \longrightarrow 3Br_2 + 2e^-$	3.6
$Br_3^- + 2e^- \longrightarrow 3Br^-$	4.5

Some kinetic parameters which characterize the mentioned reaction are collected in Tables 2.1.4, 2.1.5, and 2.1.6. The following reaction mechanisms have been suggested [102]:

$$Br^- \underset{K_{-1}}{\overset{K_1}{\rightleftharpoons}} Br_{ads} + e^- \tag{2.1.8a}$$

$$Br_{ads} + Br^- \underset{K_{-2}}{\overset{K_2}{\rightleftharpoons}} Br_2 + e^- \tag{2.1.8b}$$

$$Br_2 + Br^- \underset{K_{-3}}{\overset{K_3}{\rightleftharpoons}} Br_3^- \tag{2.1.8c}$$

and

$$Br^- \underset{K_{-1}}{\overset{K_1}{\rightleftharpoons}} Br_{ads} + e^- \tag{2.1.9a}$$

TABLE 2.1.6. Reaction Orders n_{Br^-} and n_{Br_2}, Tafel Coefficients b, Transfer Coefficients α (cathodic) and $1 - \alpha$ (anodic), and Exchange Currents j_0 for the $Br_2 + 2e^- \rightleftharpoons 2Br^-$ Reaction in 0.4 M $LiClO_4$ in Acetonitrile at 25°C. (All data from Ref. 102; rot Pt dsk method; data elaborated according to Frumkin and Tedoradse's treatment [78].)

Wave	c_{Br_2} (mole/liter)	c_{Br^-} (mole/liter)	n_{Br_2}	n_{Br^-}	b (mV/decade)	$1 - \alpha$	α	$1000j_0$ (A/cm^2)
1st anodic	0.0089	0.0203	-	1.00	130	0.45	-	0.796
	0.0058	0.0192	-	1.00	129	0.46	-	0.618
	0.0052	0.0158	-	1.05	112	0.52	-	0.945
1st cathodic	0.0074	-	1.13	-	129	-	0.46	0.708
	0.0102	-	-	-	122	-	0.48	1.200
	0.0034	-	-	-	134	-	0.44	0.815
2nd anodic	0.0066	0.0045	-	1.00	118	0.50	-	1.232
	0.0077	0.0067	-	1.15	125	0.47	-	1.451
	0.0156	0.0092	-	1.00	94	0.63	-	1.593
2nd cathodic	0.0058	0.0192	0.92	-	268	-	0.22	2.244
	0.0074	-	1.13	-	268	-	0.22	1.701
	0.0145	0.0247	0.98	-	275	-	0.21	0.984
	0.0102	-	1.16	-	256	-	0.23	1.635

$$Br_{ads} + Br_{ads} \underset{K_{-2}}{\overset{K_2}{\rightleftharpoons}} Br_2 \qquad (2.1.9b)$$

$$Br_2 + Br^- \underset{K_{-3}}{\overset{K_3}{\rightleftharpoons}} Br_3^- \qquad (2.1.9c)$$

These mechanisms are quite similar to those suggested for the Br_2/Br^- couple in aqueous solution, provided that due account is taken of Equilibrium (2.1.7). The analysis of the experimental data was carried out by adopting Christiansen's method extended by Bockris [105] to electrochemical reactions. The kinetic parameters are theoretically evaluated for each of the various combinations through which each reaction step may control the rate of the overall process (Table 2.1.7). Among the various postulated mechanisms, the one whose kinetic parameters best fit the experimental data is accepted as the true mechanism. The following conclusions can be drawn in the present case:

1. For the first cathodic wave the only satisfying mechanism is Mechanism (2.1.8) whose Step (2.1.8b) is the rate controlling one.
2. The same mechanism holds for the second cathodic wave and Step (2.1.8b) is the rate determining one. (The only difference between the two waves is connected with the $Br^- + Br_2 = Br_3^-$ equilibrium. One should, however, be aware that the experimental Tafel coefficient is $\simeq 4RT/F$, which cannot be explained by any of the proposed mechanisms.)
3. Mechanism (2.1.8) is also appropriate to explain the experimental anodic parameters.

2.1.2.2. Br_2/Br^- Couple in Dimethylsulfoxide. Solutions of HBr and KBr in dimethylsulfoxide (DMSO) have recently been studied by the chronopotentiometric method [106]. In addition to the main anodic wave (quarter-wave potential [$E_{1/4}$ = 0.88 V (SCE)], HBr and KBr exhibit small pre-waves [$E_{1/4}$ = 0.60 V (SCE)], the height of which is approximately 1/6 of the total transition time.

Anodic transfer coefficients for the main wave are collected in Table 2.1.8. In addition, a 0.5-value has been reported for the $n(1 - \alpha)$ product (n = number of electrons involved in the rate-determining step, $(1 - \alpha)$ – anodic transfer coefficient [107]).

The suggested reaction mechanism is the following one, which seems to be rather general for the oxidation of the halides in nonaqueous solvents:

$$HX = H^+ + X^- \qquad (2.1.10a)$$

$$3X^- \longrightarrow X_3^- + 2e^- \quad \text{(anodic pre-wave)} \qquad (2.1.10b)$$

$$X_3^- \longrightarrow 3/2X_2 + e^- \quad \text{(main anodic wave)} \qquad (2.1.10c)$$

The chronopotentiometric method, in addition, makes it clear that the oxidation process of the halides in DMSO is complicated by catalytic phenomena. Kinetic data which demonstrate this situation are collected in Table 2.1.9. The $j\tau^{1/2}$ values for the anodic wave of HBr are at least one order of magnitude higher than those of the cathodic wave: this implies a catalytic

TABLE 2.1.7. Theoretical Values for the Kinetic Parameters for the Reaction $Br_2 + 2e^- \rightleftarrows 2Br^-$: Tafel Coefficients b, Stoichiometric Numbers ν, and Br_{ads} coverage degree θ. Data from Ref. 102. $f = F/RT$[a]

Rate controlling step	Rate equation	b (mV/decade)	ν	θ
$Br^- \rightarrow Br_{ads} + e^-$	$j = F\underline{K}_1 c_{Br^-} \exp(\Delta Ef/2)$	120	1	$\rightarrow 0$
$2Br_{ads} \rightarrow Br_2$	$j = FK_1^2 \underline{K}_2 c_{Br^-}^{-2} \exp(2\Delta Ef)$	30	1	$\rightarrow 0$
$2Br_{ads} \rightarrow Br_2$	$j = F\underline{K}_{sat}(c_{Br,ads})_{sat}K_2$	∞	–	$\rightarrow 1$
$Br_{ads} + Br^- \rightarrow Br_2 + e^-$	$j = 2FK_1\underline{K}_2 c_{Br^-}^{-2} \exp(3\Delta Ef/2)$	40	1	$\rightarrow 0$
$Br_{ads} + Br^- \rightarrow Br_2 + e^-$	$j = 2F\underline{K}_2 c_{Br^-} \exp(\Delta Ef/2)$	120	3	$\rightarrow 1$
$Br_2 + e^- \rightarrow Br_{ads} + Br^-$	$j = F\underline{K}_{-2} c_{Br_2} \exp(-\Delta Ef/2)$	120	1	$\rightarrow 0$
$Br_2 + e^- \rightarrow Br_{ads} + Br^-$	$j = F\underline{K}_{-2}(1/K_3) c_{Br_3^-}/c_{Br^-} \exp(-\Delta Ef/2)$	120	1	$\rightarrow 0$
$Br_{ads} + e^- \rightarrow Br^-$	$j = 2FK_2 K_3 \underline{K}_{-1} c_{Br_3^-}/c_{Br^-}^{-2} \exp(-3\Delta Ef/2)$	40	1	$\rightarrow 0$
$Br_{ads} + e^- \rightarrow Br^-$	$j = 2F\underline{K}_{sat}\underline{K}_{-1} \exp(-\Delta Ef/2)$	120	3	$\rightarrow 1$

[a] The anodic and cathodic transfer coefficients have been taken as 0.5. K_1, K_2, and K_3 are the "equilibrium constants" for Reactions (2.1.8a), (2.1.8b), (2.1.8c). The K's are rate constants for single steps; K_{sat} accounts for the saturation coverage for Br_{ads}.

process with regeneration of HBr [106, 107]. Such a conclusion is supported also by controlled-potential coulometry [106]. The number of electrons involved in the oxidation of HBr depends on the electrolysis time and varies between 8 and 20 per mole of HBr. This finding is not surprising since it is well known that solutions of the halogens in organic solvents, such as acetonitrile, nitromethane, and dimethylformamide, are rather unstable. In the present case the catalytic postchemical reaction can be represented by

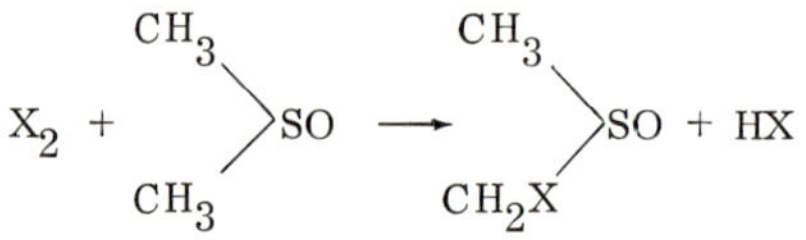

$$X_2 + (CH_3)(CH_3)SO \rightarrow (CH_3)(CH_2X)SO + HX \tag{2.1.11}$$

This cyclic mechanism is consistent with the large $(\tau/\tau_D)^{1/2}$ ratios observed (cf. Table 2.1.9) and with the coulometric results.

TABLE 2.1.8. Anodic Transfer Coefficients $(1 - \alpha)$ for the Oxidation of HBr in 0.1 M Et_4NClO_4 in DMSO at Platinum Electrodes at 25°C. All Data from Ref. 106. Based on Chronopotentiometry

$c_{HBr} \times 10^3$ (mole/liter)	$(1 - \alpha)^a$	$(1 - \alpha)^b$	$j \times 10^6$ (A/cm^2)
0.405	0.60	0.65	2.2
0.808	0.62	0.59	4.4
1.603	0.52	0.59	22
2.77	0.58	0.67	88
3.16	0.64	0.53	132
3.54	0.61	0.48	220

[a]Obtained from the $E_{t=0}$ vs log(j) relationship at small j's.

[b]Obtained from the $E_{t=0}$ vs $\log(c_{HBr})$ relationship at various j's. ($E_{t=0}$ denotes the electrode potential value extrapolated to the beginning of the chronopotentiogram.)

TABLE 2.1.9. Chronopotentiometric Data for the Oxidation of HBr in 0.1 M Et_4NClO_4 in DMSO at Platinum Electrodes at 25°C. All Data from Ref. 106

$c_{HBr} \times 10^3$ (mole/liter)	$j \times 10^6$ (A/cm^2)	$j\tau^{1/2} \times 10^6$ ($sec^{1/2}$ A/cm^2)	$(\tau/\tau_D)^{1/2}$ [a]
0.405	88	370	5.0
0.405	132	361	4.9
0.405	176	352	4.8
1.603	352	1400	4.8
1.603	440	1390	4.8
1.603	528	1375	4.7
3.54	1320	3050	4.8
3.54	1760	2970	4.6
3.54	2200	2900	4.5

[a]τ_D denotes the theoretical diffusional transition time.

2.1.2.3. The Br_2/Br^- Couple in Solvents Other Than Acetonitrile and Dimethylsulfoxide. There is little data available in the literature for sulfolane (tetramethylenesulfone). The cathodic and the anodic waves for the Br_2/Br^- couple are very similar to those for the I_2/I^- couple [108, 109]. A reaction scheme such as that in Eq. (2.1.11) has been suggested; the height of the waves is proportional to the reactant concentration, and the shape of the waves seems to indicate a certain degree of irreversibility for the cathodic reactions [108].

The half-wave potentials $E_{1/2}$ vs E(Ag/AgCl/Cl$^-$) in sulfolane saturated with AgCl and Et_4NCl are 1.34 and 1.88 V in 0.1 M $HClO_4$ at 40°C [109], respectively. The analysis is complicated by the instability of Br_2 solutions in sulfolane. Chronopotentiometric data on the couple Br_2/Br^- in dimethylformamide are available in Ref. 106; again, the presence of catalytic reactions complicates the analysis.

2.1.3. Fused Salts

2.1.3.1. The Br_2/Br^- Couple in the KNO_3-$NaNO_3$ Eutectic Melt at 250°C. Br^- is electrochemically oxidized to Br_2 at a platinum surface [110]: the half-wave potential $E_{1/2}$ is 0.644 ± 0.010 V vs E(Ag/Ag$^+$), and the limiting current is linear with the concentration of the Br^- ions. Inspection of the wave shape suggests that the reaction $2Br^- \rightleftharpoons Br_2 + 2e^-$ is reversible; the theoretically predicted shift in the half-wave potential is also observed. Difficulties arise in trying to interpret the results for the reduction of the electrochemically generated bromine because of its low solubility in the melt.

2.1.3.2. The $Br_2(g)/Br^-$ System as a Reference Electrode at High Temperatures. The observed reversibility of the process $Br_2(g) + 2e^- = 2Br^-$ makes the $Br_2(g)/Br^-$ couple useful as a bromine reference electrode for electrochemical measurements at high temperatures [111-114], both in fused salts and in ionic solid conductors.

Graphite has usually been employed as the inert conductor for this high-temperature bromine electrode [113]. The electrode assemblage has recently been redesigned by Leonardi and Brenet [115]. The temperature range is 420 to 810°C and the bias potential is 0.1 mV.

3. APPLIED ELECTROCHEMISTRY

3.1. BROMATES PRODUCTION

The process by which the bromates are produced starting from alkali bromide solutions is rather similar to the chlorate manufacture. Such a process consists of a set of electrochemical and chemical reactions:

anode	$6Br^- + 6e^- = 3Br_2$
cathode	$6H_2O + 6e^- = 3H_2 + 6OH^-$
bulk solution	$2Br_2 + 2OH^- = 2HBrO + 2Br^-$
	$Br_2 + 2OH^- = BrO^- + H_2O + Br^-$
	$2HBrO + BrO^- = BrO_3^- + 2H^+ + 2Br^-$

so that the overall net reaction is

$$Br^- + 3H_2O = 3H_2 + BrO_3^-$$

Graphite electrodes are commonly used. PbO_2 and activated titanium anodes have recently been proposed for commercial cells in order to obtain a higher-quality bromate. Graphite electrodes lead to the formation of yellow mud which tends to discolor the product.

Ref. 116 is recommended for further reading.

REFERENCES

1. E. H. Appelman, J. Amer. Chem. Soc., 90, 1900 (1968).
2. E. H. Appelman, Inorg. Chem., 8, 223 (1969).
3. L. Pauling, Chem. Eng. News, 25, 2970 (1947).
4. R. Ferreira, Bull. Soc. Chim. Fr., 1950, 131.
5. Z. Z. Hugus, Jr., J. Amer. Chem. Soc., 74, 1076 (1952).
6. B. Lakatos, Acta Chim. Acad. Sci. Hung., 8, 219 (1955).
7. R. S. Nyholm, Proc. Chem. Soc., 1961, 273.
8. D. S. Urch, J. Inorg. Nucl. Chem., 25, 77 (1963).
9. W. E. Dasent, Nonexistent Compounds, Dekker, New York, 1965, pp. 117 and 120-124.
10. M. H. Studier, J. Amer. Chem. Soc., 90, 1901 (1968).
11. G. K. Johnson, P. N. Smith, E. H. Appelman, and W. N. Hubbard, Inorg. Chem., 9, 119 (1970).
12. J. R. Brand and S. A. Bunck, J. Amer. Chem. Soc., 91, 6500 (1969).
13. D. D. Wagman, W. H. Evans, V. B. Parker, I. Halow, S. M. Bailey, and R. H. Schumm, Selected Values of Chemical Thermodynamic Properties, National Bureau of Standards (U.S.) Technical Note 270-3, Washington, D.C., 1968, pp. 32-34.
14. T. Mussini and G. Faita, Ric. Sci., 36, 175 (1966).
15. G. Faita, P. Longhi, and T. Mussini, J. Electrochem. Soc., 114, 340 (1967).
16. D. J. G. Ives and G. J. Janz, Reference Electrodes, Academic, New York, 1961, pp. 13 and 14.
17. G. N. Lewis and H. Storch, J. Amer. Chem. Soc., 39, 2544 (1917).
18. G. Jones and S. Baeckström, ibid., 56, 1524 (1934).
19. G. Charlot, Oxidation-Reduction Potentials, Pergamon, Paris, 1958, p. 8.

20. W. M. Latimer, Oxidation Potentials, Prentice-Hall, Englewood Cliffs, New Jersey, 1952, p. 60.

21. R. Parsons, Handbook of Electrochemical Constants, Butterworths, London, 1959, p. 71.

22. B. E. Conway, Electrochemical Data, Elsevier, Amsterdam, 1952, p. 293.

23. A. J. de Bethune and N. A. Swendeman Loud, Standard Aqueous Electrode Potentials and Temperature Coefficients at 25°C, C. Hampel, Skokie, Illinois, 1964, pp. 12, 13, and 16.

24. J. A. Christiansen and M. Pourbaix, Compt. Rend. Conf. Union Intern. Chim. Pure et Appliq., 17th Conf. Stockholm, p. 83 (1953).

25. Ref. 16, p. 26.

26. D. B. Scaife and H. J. V. Tyrrell, J. Chem. Soc., 1958, 386.

27. R. O. Griffith, A. McKeown, and A. G. Winn, Trans. Faraday Soc., 28, 101 (1932).

28. H. B. Hetzer, R. A. Robinson, and R. G. Bates, J. Phys. Chem., 66, 1423 (1962).

29. H. A. Liebhafsky, J. Amer. Chem. Soc., 56, 1500 (1934).

30. E. A. Shilov, ibid., 60, 490 (1938).

31. M. Pourbaix, Atlas d'Equilibres Electrochimiques à 25°C, Gauthier-Villars, Paris, 1963, p. 604.

32. Ref. 21, p. 73.

33. Ref. 22, p. 303.

34. P. S. Buckley and H. Hartley, Phil. Mag., 8, 320 (1929).

35. H. Strehlow, Z. Elektrochem., 56, 827 (1952).

36. G. Kortüm, Treatise on Electrochemistry, 2nd rev. Engl. ed., Elsevier, Amsterdam, 1965, p. 312.

37. A. McFarlane and H. Hartley, Phil. Mag., 13, 425 (1932).

38. V. A. Pleskov, Acta Physicochim. URSS, 20, 578 (1945).

39. L. G. Lavrenova, T. V. Zegzhda, and V. M. Shulman, Elektrokhimiya, 7, 83 (1971).

40. R. M. Keefer and L. J. Andrews, J. Amer. Chem. Soc., 78, 3637 (1956).

41. T. W. Nakagawa, L. J. Andrews, and R. M. Keefer, J. Phys. Chem., 61, 1007 (1957).

42. R. L. Benoit, M. Guay, and J. Desbarres, Can. J. Chem., 46, 1261 (1968).

43. J. C. Marchon, C. R. Acad. Sci. Paris, Ser. C, 267, 1123 (1968).

44. G. Jones and M. L. Hartmann, Trans. Amer. Electrochem. Soc., 30, 295 (1916).

45. M. S. Sherrill and E. F. Izard, J. Amer. Chem. Soc., 50, 1671 (1928).

46. G. N. Lewis and M. Randall, ibid., 38, 2348 (1916).

47. G. N. Lewis and H. Storch, ibid., 39, 2551 (1917).

48. G. A. Linhart, ibid., 40, 158 (1918).

49. F. P. Worley, J. Chem. Soc., 87, 1107 (1905).

50. G. Jones and S. Baeckström, J. Amer. Chem. Soc., 56, 1517 (1934).

51. J. E. Dubois and H. Herzog, Bull. Soc. Chim. Fr., 1963, 57.

52. J. C. Marchon and J. Badoz-Lambling, ibid., 1967, 4660.

53. R. L. Benoit and M. Guay, Inorg. Nucl. Chem. Lett., 4, 215 (1968).

54. I. V. Nelson and R. T. Iwamoto, J. Electroanal. Chem., 7, 218 (1964).

55. C. Sinicki and M. Bréant, Bull. Soc. Chim. Fr., 1967, 3080.

56. A. J. Parker, J. Chem. Soc., A, 1966, 220.

57. P. B. D. de la Mare, O. M. H. el-Dusouqui, J. G. Tillet, and M. Zeltner, ibid., 1964, 5306.

58. K. Nozaki and R. A. Ogg, Jr., J. Amer. Chem. Soc., 64, 697 (1942).

59. P. Alcais, F. Rothenberg, and J. E. Dubois, J. Chim. Phys., 64, 1818 (1967).

60. R. A. Robinson and R. H. Stokes, Electrolyte Solutions, 2nd rev. ed., Butterworths, London, 1965, pp. 481 and 491.

61. H. S. Harned and B. B. Owen, The Physical Chemistry of Electrolytic Solutions, 3rd ed., Reinhold, New York, 1958, p. 727.

62. G. Faita, T. Mussini, and R. Oggioni, J. Chem. Eng. Data, 11, 162 (1966).

63. G. Bianchi, J. Electrochem. Soc., 112, 233 (1965).

64. G. Bianchi, A. Barosi, G. Faita, and T. Mussini, ibid., 112, 921 (1965).

65. G. Bianchi, G. Faita, and T. Mussini, J. Sci. Instrum., 42, 693 (1965).

66. Ref. 16, pp. 203-207.

67. S. K. Banerjee, K. K. Kundu, and M. N. Das, J. Chem. Soc., A, 1967, 161.

68. K. K. Kundu, P. K. Chattopadhyay, Debabrata Jana, and M. N. Das, J. Chem. Eng. Data, 15, 209 (1970).

69. T. Mussini, C. Massarani-Formaro, and P. Andrigo, J. Electroanal. Chem., 33, 177 (1971).

70. K. Schwabe, R. Urlass, and A. Ferse, Ber. Bunsenges. Phys. Chem., 68, 46 (1964).

71. K. Schwabe and A. Ferse, ibid., 70, 849 (1966).

72. J. W. Augustynski, G. Faita, and T. Mussini, J. Chem. Eng. Data, 12, 369 (1967).

73. H. S. Harned and E. C. Dreby, J. Amer. Chem. Soc., 61, 3113 (1939).

74. K. J. Vetter, Electrochemical Kinetics, Academic, New York, 1967, p. 470.

75. J. Llopis and M. Vàzquez, Electrochim. Acta, 6, 167 (1962).

76. J. Llopis and M. Vàzquez, ibid., 6, 177 (1962).

77. G. Faita, G. Fiori, and T. Mussini, ibid., 13, 1765 (1968).

78. A. N. Frumkin and G. A. Tedoradse, Z. Elektrochem., 62, 251 (1958).

79. G. Faita, G. Fiori, and J. W. Augustynski, Electrochim. Metallorum, 2, 437 (1967).

80. M. Loshkarev and O. Essin, Acta Physicochim. USSR, 8, 189 (1958).

81. F. Chang and H. Wick, Z. Phys. Chem., A172, 448 (1935).

82. W. D. Robertson, J. Electrochem. Soc., 100, 194 (1953).

83. Ref. 74, p. 155.

84. S. F. Belevskii and S. V. Gorbachev, Zh. Fiz. Khim., 36, 742 (1962).

85. Ya. Veber, Dzh. Pirtskhalava, Yu. B. Vasilev, and V. S. Bagotskii, Electrokhimiya, 5, 1037 (1969).

86. R. A. Osteryoung, Anal. Chem., 35, 1100 (1963).

87. G. Dryhurst and P. J. Elving, ibid., 39, 606 (1967).

88. Chin-Ying-Chech and M. A. Genshaw, J. Phys. Chem., 73, 3571 (1969).

89. W. D. Cooper and R. Parsons, Trans. Faraday Soc., 66, 1698 (1970).

90. M. W. Breiter, Electrochim. Acta, 8, 925 (1963).

91. N. A. Balashova, ibid., 7, 559 (1962).
92. D. C. Johnson and S. Bruckenstein, J. Electrochem. Soc., 117, 460 (1970).
93. A. A. Sakharov and N. N. Martynova, Uch. Zap. Petrozavodsk. Gos. Univ., 14, 89 (1966).
94. A. F. Krivis and G. R. Supp, Anal. Chem., 40, 2063 (1968).
95. P. Delahay and M. Kleinerman, J. Amer. Chem. Soc., 82, 4509 (1960).
96. Yousif Abulahad Yonan, Dissertation Abstr., B27, 3010 (1967).
97. S. Roffia and M. Lavacchielli, J. Electroanal. Chem., 22, 117 (1969).
98. P. G. Desideri and L. Lepri, ibid., 22, 265 (1969).
99. T. Osuga and K. Sugino, J. Electrochem. Soc., 104, 448 (1957).
100. S. Sundararajan et al., Chem. Process Eng., 43, 438, 447 (1962).
101. E. A. Dzhafarov and Sh. M. Efendieva, Azerb. Khim. Zh., 5, 166 (1967).
102. T. Iwasita and M. C. Giordano, Electrochim. Acta, 14, 1045 (1969).
103. I. M. Kolthoff and J. F. Coetzee, J. Amer. Chem. Soc., 79, 1852 (1957).
104. A. I. Popov and D. H. Geske, ibid., 80, 1340, 5346 (1958).
105. J. O'M. Bockris, J. Chem. Phys., 24, 817 (1956).
106. M. Michlmayr and D. T. Sawyer, J. Electroanal. Chem., 23, 387 (1969).
107. H. E. Zittel and F. J. Miller, Anal. Chim. Acta, 37, 141 (1967).
108. J. B. Headridge, D. Pletcher, and M. Callingham, J. Chem. Soc., A, 1967, 684.
109. R. L. Benoit, M. Guay, and J. C. Desbarres, Can. J. Chem., 46, 1261 (1968).
110. H. S. Swofford, Jr., and J. H. Propp, Anal. Chem., 37, 974 (1965).
111. E. J. Salstrom and J. H. Hildebrand, J. Amer. Chem. Soc., 52, 4650 (1930).
112. J. G. Murgulescu and O. J. Marchidan, Rev. Chim. (Bucharest), 3, 1 (1958).
113. H. A. Laitinen and J. W. Pankey, J. Amer. Chem. Soc., 81, 1053 (1959).
114. Ref. 16, pp. 577, 587, and 593.
115. J. Leonardi and J. Brenet, C. R. Acad. Sci., Paris, Ser. C, 264, 2090 (1967).
116. A. Kuhn, Industrial Electrochemical Processes, Elsevier, Amsterdam, 1971, p. 121.

Chapter I-3

IODINE and ASTATINE

PIER GIORGIO DESIDERI, LUCIANO LEPRI, and DANIELA HEIMLER

Institute of Analytical Chemistry
University of Florence
Florence, Italy

1. STANDARD AND FORMAL POTENTIALS

1.1. AQUEOUS SOLUTION

The standard and formal potentials in aqueous solution have been determined mainly by means of thermodynamic and potentiometric methods. For some redox couples the potentials

TABLE 1.1.1. Standard and Formal Potentials in Aqueous Solution

Half-reaction	Standard or formal potential (V)	Conditions	Refs.
$I_2(s) + 2e^- \rightleftharpoons 2I^-$	+0.536	-	[4]-[10]
$I_2(aq) + 2e^- \rightleftharpoons 2I^-$	+0.621	-	[7], [8], [11]
	+0.628	0.5 M H_2SO_4	[12]
$I_3^- + 2e^- \rightleftharpoons 3I^-$	+0.536	-	[13]
	+0.545	0.5 M H_2SO_4	[12]
$HIO + H^+ + 2e^- \rightleftharpoons I^- + H_2O$	+0.987	-	[3]
	+0.99	-	[1]
$2HIO + 2H^+ + 2e^- \rightleftharpoons I_2 + 2H_2O$	+1.354	-	[3]
$IO^- + 2H^+ + 2e^- \rightleftharpoons I^- + H_2O$	+1.313	-	[3]
$IO^- + H_2O + 2e^- \rightleftharpoons I^- + 2OH^-$	+0.49	-	[1]
$ICN + 2e^- \rightleftharpoons I^- + CN^-$	+0.30	KOH	[14]
$2ICN + 2H^+ + 2e^- \rightleftharpoons I_2 + 2HCN$	+0.63	-	[15], [16]
$2ICl + 2e^- \rightleftharpoons I_2 + 2Cl^-$	+1.19	-	[1]
$2ICl_2^- + 2e^- \rightleftharpoons I_2 + 4Cl^-$	+1.06	-	[1]
$2IBr + 2e^- \rightleftharpoons I_2 + 2Br^-$	+1.02	-	[1]
$2IBr_2^- + 2e^- \rightleftharpoons I_2 + 4Br^-$	+0.87	-	[1]
$2ICl_3(s) + 4e^- \rightleftharpoons 2ICl(s) + 4Cl^-$	+1.31	-	[1]
$2ICl_3(s) + 6e^- \rightleftharpoons I_2 + 6Cl^-$	+1.28	-	[1]
$HIO_3 + 5H^+ + 6e^- \rightleftharpoons I^- + 3H_2O$	+1.077	-	[3]
$2HIO_3 + 10H^+ + 10e^- \rightleftharpoons I_2 + 6H_2O$	+1.169	-	[3]
$IO_3^- + 6H^+ + 6e^- \rightleftharpoons I^- + 3H_2O$	+1.085	-	[3]
$IO_3^- + 3H_2O + 6e^- \rightleftharpoons I^- + 6OH^-$	+0.26	-	[1]
$2IO_3^- + 12H^+ + 10e^- \rightleftharpoons I_2 + 6H_2O$	+1.195	-	[1]
	+1.178	-	[3]
$HIO_4 + H^+ + 2e^- \rightleftharpoons IO_3^- + H_2O$	+1.603	-	[3]
$HIO_4 + 2H^+ + 2e^- \rightleftharpoons HIO_3 + H_2O$	+1.626	-	[3]
$IO_4^- + 2H^+ + 2e^- \rightleftharpoons IO_3^- + H_2O$	+1.653	-	[3]
$H_5IO_6 + H^+ + 2e^- \rightleftharpoons IO_3^- + 3H_2O$	~+1.6	-	[1]

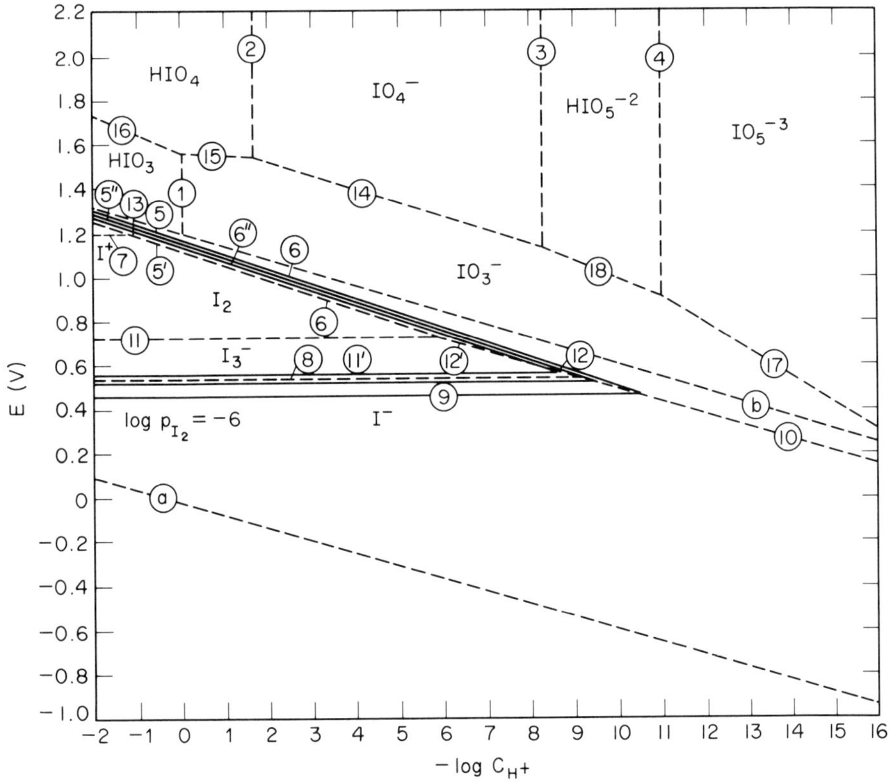

FIG. 1.1.1. Potential-pH diagram of iodine [3]. The numbers refer to the data of Tables 1.1.2 and 1.1.3. The lines (a) and (b) have been obtained on the basis of the following equations:

Line (a) $E = 0.000 - 0.0591$ pH

Line (b) $E = 1.228 - 0.0591$ pH

Such equations correspond to the reactions:

$H_2 \rightarrow 2H^+ + 2e^-$

$2H_2O \rightarrow O_2 + 4H^+ + 4e^-$

when $P_{H_2} = P_{O_2} = 1$ atm.

have been drawn from the standard potential of a different system and the equilibrium constant of this system and the unknown one or on the basis of the chemical properties. These data are shown in Table 1.1.1.

These data have been taken from Latimer [1], Charlot [2], and Pourbaix [3] by selecting the redox couples which, in our opinion, have a special interest from an electrochemical point of view.

The potential-pH diagram for the species in the different oxidation states is given in Fig. 1.1.1. The diagram, which refers to an iodine concentration of 0.5 M, has been constructed considering the pH values limiting the region in which the different species exist (Table 1.1.2) and the potential variations as the pH changes (Table 1.1.3) for several couples.

TABLE 1.1.2. Limits of the Region in Which the Different Species Exist

Couple		pH
1	HIO_3/IO_3^-	0.80
2	HIO_4/IO_4^-	1.70
3	IO_4^-/HIO_5^{2-}	8.30
4	HIO_5^{2-}/IO_5^{3-}	11.00

TABLE 1.1.3. Potential Variations of the Different Redox Couples as the pH Changes

Couple		Potential (V)
5	I_2/HIO_3	1.166 - 0.0591 pH + 0.0118 log C - 0.0059 log P
5'	I_2/HIO_3 and $I_2(s)$	1.154 - 0.0591 pH
5"	$I_2(s)/HIO_3$	1.186 - 0.0591 pH + 0.0118 log C
6	I_2/IO_3^-	1.175 - 0.0709 pH + 0.0118 log C - 0.0059 log P
6'	I_2/IO_3^- and $I_2(s)$	1.165 - 0.00709 pH
6"	$I_2(s)/IO_3^-$	1.195 - 0.0709 pH + 0.0118 log C
7	I_2/I^+ and $I_2(s)$	1.205
8	I^-/I_3^-	0.532 - 0.0591 log C
9	I^-/I_2	0.636 - 0.0591 log C + 0.0295 log P
10	I^-/IO_3^-	1.085 - 0.0591 pH
11	I_3^-/I_2 and $I_2(s)$	0.714
11'	$I_3^-/I_2(s)$	0.562 - 0.0591 log C
12	I_3^-/IO_3	1.154 - 0.0665 pH + 0.0074 log C
12'	I_3^-/IO_3^- and $I_2(s)$	1.090 - 0.0591 pH
13	I^+/HIO_3 and $I_2(s)$	1.143 - 0.0739 pH
14	IO_3^-/IO_4^-	1.653 - 0.0591 pH
15	IO_3^-/HIO_4	1.603 - 0.0295 pH
16	HIO_3/HIO_4	1.626 - 0.0591 pH
17	IO_3^-/IO_5^{3-}	2.223 - 0.1182 pH
18	IO_3^-/HIO_5^{2-}	1.898 - 0.0886 pH

TABLE 1.2.1. Electrode Potentials in Nonaqueous Solution

Solvent	Half-reaction	Potential (V)	Reference electrode	Conditions	Refs.
Acetonitrile	$3I_2 + 2e^- \rightleftharpoons 2I_3^-$	+0.396	Ag/Ag^+ 10^{-2} M [17]	0.1 M $NaClO_4$	[18]
		+0.31	Ferrocene/ferricene	0.1 M $LiClO_4$	[19]
		+0.65	Aq SCE	0.1 M $NaClO_4$	[20]
	$I_2 + 2e^- \rightleftharpoons 2I^-$	+0.26	Aq SCE	0.1 M Et_4NClO_4	[20]
	$I_3^- + 2e^- \rightleftharpoons 3I^-$	+0.06	Aq SCE	0.1 M $NaClO_4$	[20]
		-0.248	Ag/Ag^+ 10^{-2} M	0.1 M $LiClO_4$	[18]
		-0.33	Ferrocene/ferricene	0.1 M $LiClO_4$	[19]
Nitromethane	$3I_2 + 2e^- \rightleftharpoons 2I_3^-$	+0.36	Ferrocene/ferricene	0.1 M Et_4NClO_4	[19]
		+0.71	Aq SCE	0.1 M Et_4NClO_4	[20]
	$I_2 + 2e^- \rightleftharpoons 2I^-$	+0.31	Aq SCE	0.1 M Et_4NClO_4	[20]
	$I_3^- + 2e^- \rightleftharpoons 3I^-$	+0.11	Aq SCE	0.1 M Et_4NClO_4	[20]
		-0.28	Ferrocene/ferricene	0.1 M Et_4NClO_4	[19]
Acetone	$3I_2 + 2e^- \rightleftharpoons 2I_3^-$	+0.86	Aq SCE	0.1 M Et_4NClO_4	[20]
	$I_2 + 2e^- \rightleftharpoons 2I^-$	+0.36	Aq SCE	0.1 M Et_4NClO_4	[20]
	$I_3^- + 2e^- \rightleftharpoons 3I^-$	+0.11	Aq SCE	0.1 M Et_4NClO_4	[20]
DMF	$3I_2 + 2e^- \rightleftharpoons 2I_3^-$	+0.270	Hg/Hg^{2+} [21]	0.1 M $HClO_4$	[22]
	$I_3^- + 2e^- \rightleftharpoons 3I^-$	-0.370	Hg/Hg^{2+}	0.1 M $HClO_4$	[22]
Ac_2O	$3I_2 + 2e^- \rightleftharpoons 2I_3^-$	-0.065	$Ag/AgClO_4$ 10^{-2} M	0.1 M $LiClO_4$	[23]
	$I_3^- + 2e^- \rightleftharpoons 3I^-$	-0.78	$Ag/AgClO_4$ 10^{-2} M	0.1 M $LiClO_4$	[23]
AcOH	$3I_2 + 2e^- \rightleftharpoons 2I_3^-$	+0.344	NHE	1 M AcONa + 0.5 M $NaClO_4$	[24]
		+0.339	NHE	0.5 M $NaClO_4$	[24]
	$I_2 + 2e^- \rightleftharpoons 2I^-$	+0.273	Ag/AgCl 0.3 M LiCl	-	[25]
	$I_3^- + 2e^- \rightleftharpoons 3I^-$	-0.112	NHE	1 M AcONa + 0.5 M $NaClO_4$	[24]
		-0.118	NHE	0.5 M $NaClO_4$	[24]

1.2. NONAQUEOUS SOLUTION

The redox potentials of iodine in nonaqueous solution have been widely studied. The redox couples considered are I_2/I_3^-, I_3^-/I^-, and, to a lesser extent, I_2/I^-.

From the data of Table 1.2.1 we can deduce that, among the different solvents, acetonitrile was the most widely studied.

1.3. MOLTEN SALTS

The redox potentials of iodine in molten salts have been recently determined and are mostly obtained with emf measurements in molten $LiNO_3$-KNO_3 in the temperature range 162-217°C [26]. Beyond this temperature the oxidation of iodine by nitrate increases remarkably and is inconsistent with the emf data. The data are collected in Table 1.3.1. Other thermodynamic data are reported in Table 1.3.2. ΔG° refers to 177°C.

It is interesting to compare the value of E° found with emf measurements for the reaction

$$I_2(\text{soln}) + 2e^- \rightleftharpoons 2I^-$$

with that obtained by Swofford and Propp [27] at 250°C with voltammetric measurements. Converting the concentration of the reference electrode from 0.07 M to 1 mole/kg and considering the density difference between molten $LiNO_3$-KNO_3 and $NaNO_3$-KNO_3, the value -0.195 V is obtained with respect to the value -0.221 V drawn by extrapolating the data of Tables 1.3.1 and 1.3.2. The agreement between the two values is quite good in view of the nonequilibrium nature of the measurements of Swofford and Propp and the assumption that the diffusion coefficients of iodide and iodine were the same.

TABLE 1.3.1. Electrode Potentials in Molten Salts

Half-reaction	Potential (V)	Conditions	Reference electrode	Refs.
$I_2(g) + 2e^- \rightleftharpoons 2I^-$	-0.372	$LiNO_3$-KNO_3, 177°C	Ag/Ag^+ 1 mole/kg	[26]
$I_2(\text{soln}) + 2e^- \rightleftharpoons 2I^-$	-0.245	$LiNO_3$-KNO_3, 177°C	Ag/Ag^+ 1 mole/kg	[26]
	+0.126	$NaNO_3$-KNO_3, 250°C	Ag/Ag^+ 0.07 M	[27]
$I_3^- + 2e^- \rightleftharpoons 3I^-$	-0.332	$LiNO_3$-KNO_3, 177°C	Ag/Ag^+ 1 mole/kg	[26]
$3I_2(g) + 2e^- \rightleftharpoons 2I_3^-$	-0.451	$LiNO_3$-KNO_3, 177°C	Ag/Ag^+ 1 mole/kg	[26]
$3I_2(\text{soln}) + 2e^- \rightleftharpoons 2I_3^-$	-0.072	$LiNO_3$-KNO_3, 177°C	Ag/Ag^+ 1 mole/kg	[26]

TABLE 1.3.2. Thermodynamic Data for the Electrode Reactions in $LiNO_3$-KNO_3 at 177°C [26]

Half-reaction	dE°/dT (mV/deg^{-1})	ΔG° (kcal mole^{-1})	ΔH° (kcal mole^{-1})	ΔS° (cal mole^{-1} deg^{-1})
$I_2(g) + 2e^- \rightleftharpoons 2I^-$	-0.1531 ∓ 0.0075	17.1643 ∓ 0.0065	13.99 ∓ 0.15	-7.06 ∓ 0.35
$I_2(soln) + 2e^- \rightleftharpoons 2I^-$	0.338 ∓ 0.056	11.324 ∓ 0.080	18.4 ∓ 1.2	15.6 ∓ 2.6
$I_3^- + 2e^- \rightleftharpoons 3I^-$	0.435 ∓ 0.015	15.328 ∓ 0.017	24.37 ∓ 0.29	20.08 ∓ 0.67
$3I_2(g) + 2e^- \rightleftharpoons 2I_3^-$	-1.329 ∓ 0.026	20.836 ∓ 0.033	-6.77 ∓ 0.52	-61.3 ∓ 1.2
$3I_2(soln) + 2e^- \rightleftharpoons 2I_3^-$	0.15 ∓ 0.13	3.32 ∓ 0.18	6.4 ∓ 2.7	6.8 ∓ 6.0

2. VOLTAMMETRIC CHARACTERISTICS

2.1. POLAROGRAPHIC CHARACTERISTICS

2.1.1. Aqueous Solution

2.1.1.1. Reduction of Periodate. The periodate ion is directly reduced on a dropping mercury electrode in spite of the high redox potential of the periodate/iodate couple. Two waves corresponding successively to the reduction of periodate to iodate and to iodide are observed.

As regards the first reduction step, according to Coe and Rogers [28], the reduction wave is not well-defined and its half-wave potential is not measurable because it coincides with that of the anodic dissolution of mercury. Souchay [29], on the contrary, claimed that the half-wave potential of the first wave could be measured by changing the reference electrode to a mercury-mercurous sulfate electrode from a saturated calomel electrode. The shape of the first wave is greatly influenced by the pH value. For more negative potentials than -0.2 V/SCE the limiting current decreases as the pH increases. The drop of the limiting current observed in the case of periodate is common to many anions which give a diffusion current at the zero-charge point. Kryukova [30] demonstrated that this minimum is greatly influenced by the concentration of the supporting electrolyte. Since the polarographic behavior of periodate is similar to that of other anions, it may probably be attributed to the same effects observed for the reduction of persulfate [31], i.e., the influence of the double layer structure on slow electrode processes. The pH influence on the half-wave potential is not constant. In H_2SO_4 (pH = 0) the half-wave potential is +0.385 V/SCE and changes to 22 mV per pH unit up to pH = 5. For pH > 5 the variation is 45 mV. Proportionality between the height of the first reduction wave and the periodate concentration was found in the concentration range 0.8×10^{-4} to 10^{-2} M. In this concentration range Souchay [29] set up a method for the quantitative determination of periodate in the presence of iodate.

2.1.1.2. Reduction of Iodate. According to Rylich [32], the electrode process is

$$IO_3^- + 6e^- \rightarrow I^- + 3O^{2-}$$

This scheme is in accord with the independence of the decomposition potential of iodate with pH value in alkaline solution.

In acidic solution, on the contrary, the decomposition potential shifts to more positive values with decreasing pH of the solution and for this reason the overall electrode process seems to be in accord with the scheme proposed by Orlemann and Kolthoff [33]:

$$IO_3^- + 6H^+ + 6e^- \rightarrow I^- + 3H_2O$$

The half-wave potentials for iodate reduction using different supporting electrolytes and at different pH values are gathered in Table 2.1.1.

TABLE 2.1.1. Polarographic Characteristics of Iodate [33]

Conditions	pH	$E_{1/2}$ (V/SCE)	$C_{IO_3^-} \times 10^3$ M
2 M $HClO_4$	0.2	+0.040	0.500
1M HNO_3	0.55	0.000	0.500
0.12 M $HClO_4$	0.90	-0.050	0.525
0.1 M H_2SO_4 + 0.2 M KNO_3	1.10	-0.080	0.500
0.1 M biphthalate + 0.12 M KCl	3.20	-0.305	1.000
0.1 M acetate + 0.12 M KCl	4.20	-0.425	1.000
0.1 M acetate + 0.12 M KCl	4.90	-0.500	1.000
0.1 M acetate	5.95	-0.650	0.500
1.5 M acetate	6.10	-0.640	0.500
0.2 M phosphate	6.40	-0.790	0.500
0.2 M phosphate	7.10	-1.050	0.500
0.2 M KCl	-	-1.240	1.000
0.2 M KNO_3 + 0.5 M borax	9.2	-1.200	1.000
0.12 M KCl + 0.1 M NaOH	12.7	-1.210	1.000

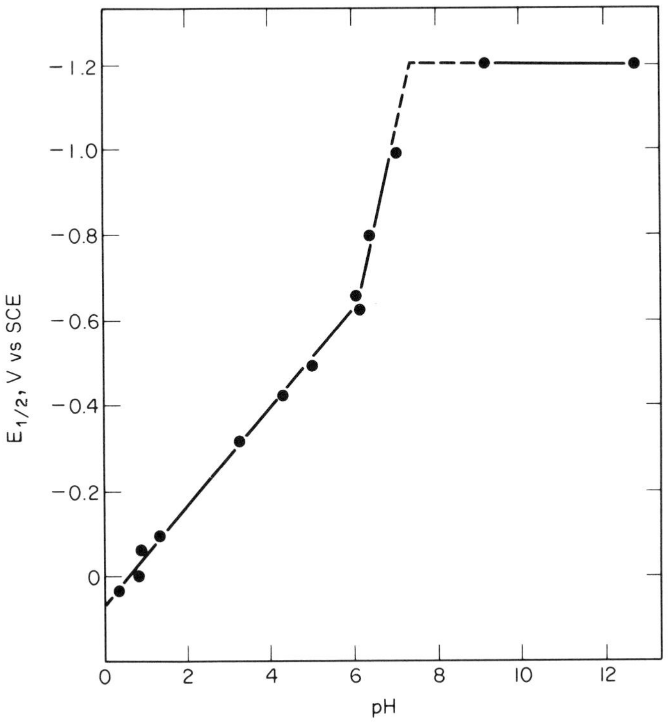

FIG. 2.1.1. Variation of the half-wave potential ($E_{1/2}$) of iodate with pH in buffered solutions at 25°C [33].

Since the iodate reduction on the DME is irreversible, the half-wave potential variation with pH (see Fig. 2.1.1) is not in accord with what one would predict from the net stoichiometric reaction. The half-wave potential, indeed, is shifted about 120 mV per unit change in pH to more negative potentials till pH = 6.0. Such a shift is practically twice that expected for the case of a reversible process. For pH > 6.0 the half-wave potential tends to a limiting value (about -1.2 V/SCE) which is reached at pH ≥ 8. In the pH range 6-8 the increase is about 400 mV per pH unit.

In unbuffered weakly acidic solution two waves are obtained when the ratio of hydrogen ion to iodate is less than 2. The first wave refers to the reduction of iodic acid and the second to that of iodate.

Cernak [34] also obtained two waves in buffered solution in the pH range 5.6-6.8, and he determined the rate of association of iodic acid from the currents relative to the reduction of iodic acid and of iodate (see Table 2.1.2).

The concentration and the kind of supporting electrolyte have a large influence on the reduction potential of iodate. Rylich [32] was the first to notice the influence of the charge of the supporting electrolyte cation on the reduction potential (see Table 2.1.3).

The shift of the reduction potential to less negative values (as the cation charge increases) was attributed by Heyrovsky to the formation of "ion pairs" between the cation and the iodate; such an effect has been studied and closely examined by Zykov and Zhdanov [35, 36] and

TABLE 2.1.2. Rate Constant for the Association of Iodic Acid[a]

pH	$\bar{i}_{HIO_3}/\bar{i}_{IO_3^-}$	k x 10^{-11} (mole^{-1} liter sec^{-1})
5.75	0.79	3.78
6.0	0.625	2.43
6.23	0.5	2.57
6.75	0.225	2.26
7.0	0.145	2.50

[a]K_A = 1.67 x 10^{-1}; drop time = 2.5 sec.

TABLE 2.1.3. Reduction Potentials[a] of Iodate in Different Supporting Electrolytes

Supporting electrolyte	Potential, V vs NCE
0.1 N KCl	-1.05
0.1 N NaCl	-1.05
0.1 N LiCl	-1.05
0.1 N K_2SO_4	-1.05
0.1 N $CaCl_2$	-0.85
0.1 N $BaCl_2$	-0.85
0.1 N $SrCl_2$	-0.85
0.01 N $LaCl_3$	-0.40

[a]The reduction potentials were determined by the 45° tangent method at a concentration of 0.001 M at room temperature.

Bernard and co-workers [37]. The former found that the log(i/i_d - i) vs E plot was linear for every kind of electrolyte; the slope of the lines ($2.3RT/\alpha n_a F$), on the contrary, decreases, in the case of ions of the same valence, with increasing ion size (i.e., the slope decreases in changing from Li^+ to Cs^+). Bernard and co-workers noticed that the Ilkovic equation, verified by Lingane and Kolthoff [38] in the case of iodate in 0.1% HCl, was not obeyed in the presence of salts of polyvalent cations as supporting electrolytes. The deviation of the polarographic behavior of iodate from the Ilkovic equation was explained by the authors by assuming that the following equilibria exist in solution:

$$M^{z+} + IO_3^- \rightleftharpoons [M(IO_3)]^{(z-1)+}$$

$$[M(IO_3)]^{(z-1)+} + IO_3^- \rightleftharpoons [M(IO_3)_2]^{(z-2)+}$$

2.1.1.3. Reduction of Iodine and Hypoiodite. Iodine in acid solution gives a reduction wave which starts from zero applied emf owing to the more positive redox potential of the couple I_2/I^- than of the couple Hg_2^{2+}/Hg. In alkaline solution iodine disproportionates to hypoiodite and iodide.

Souchay [29] studied the reduction of hypoiodite in alkaline solution. The reduction wave starts at the dissolution potential of mercury and corresponds to the reduction to iodide. From the decrease of the diffusion current it is possible to measure the rate of disproportionation of hypoiodite to iodate and iodide.

2.1.1.4. Oxidation of Iodide. The oxidation potential of mercury in the presence of iodide is shifted to more negative values owing to the formation of Hg_2I_2 on the electrode. Kolthoff and Miller [40] obtained a well-defined wave with iodide concentrations smaller than 5×10^{-4} M. For concentrations $>5 \times 10^{-4}$ M the limiting current is very poorly defined and undergoes irregular fluctuations owing to the formation of an Hg_2I_2 film on the electrode surface. Since the formation of this film has a notable influence on the growth of the mercury drops, the irregularities become smaller when a more rapid drop time is used or when a small amount of gelatin is added.

Successively, it has been pointed out [41] that iodide gives a pre-wave corresponding to the formation of a monomolecular layer of mercurous iodide. Takemori and Tachi [42] reported only a single ac polarographic wave for the oxidation of iodide ion, while Breyer and Hacobian [43] and, more recently, Biegler [44] found two ac waves. The more negative wave was rather broad and was not exclusively due to a redox process, but it depended, to a great extent, on the adsorption of iodide ions on the electrode. This first wave corresponded to the potential region in which the dc polarographic step occurs, as Fig. 2.1.2 shows. The second wave was associated with the presence of irregular fluctuations in the dc polarogram owing to the formation of an Hg_2I_2 film.

Perchard and co-workers [45] used the formation of the film of Hg_2I_2 for analytical purposes; they determined the iodide concentration by reducing the Hg_2I_2 once formed by means of cathodic redissolution chronoamperometry at a dropping mercury electrode. Practical analytical applications of the anodic iodide wave for the determination of iodide in blood and biological fluids have been described by Zimmerman and Layton [46].

2.1.2. Nonaqueous Solution

2.1.2.1. Oxidation of Iodide. Iodide facilitates mercury oxidation more in nonaqueous solution than in aqueous solution. On this basis it is to be expected that the dissociation constants and the solubility products of the salts of Hg(I) with iodide are much smaller in organic solvents than in water. Iodide gives two anodic waves in acetonitrile [47], dimethylformamide (DMF) [48, 49], isobutyronitrile (IBN) [50], and acetone [51]. In IBN the first wave has been attributed by Coetzee and Hedrick [50] to the electrode reaction

$$Hg + 3I^- \rightleftharpoons HgI_3^- + 2e^-$$

and the second to the electrode reaction

$$Hg + 2HgI_3^- \rightleftharpoons 3HgI_2 + 2e^-$$

The equation for the first wave (neglecting the activity coefficients) is

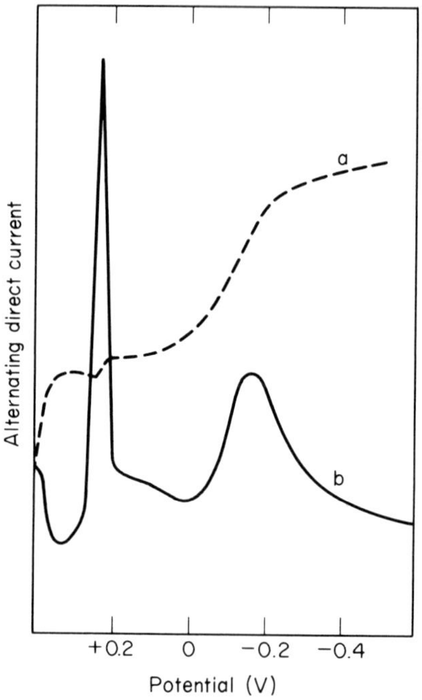

FIG. 2.1.2. Polarograms, (a) dc and (b) ac, of 1.48 x 10^{-4} M iodide in 0.93 M $HClO_4$ [44].

$$E_{DME} = E_{1/2} + \frac{0.059}{2} \log \frac{i}{(i_d - i)^3}$$

and for the second wave:

$$E_{DME} = E_{1/2} + \frac{0.059}{2} \log \frac{i^3}{(i_d - i)^2}$$

Plots of the potential of the dropping mercury electrode as a function of the quantities $\log [i/(i_d - i)^3]$ and $\log [i^3/(i_d - i)^2]$ are linear with a slope of 0.030 and 0.035 for $C_{I^-} = 5 \times 10^{-4}$ M. For $C_{I^-} = 1 \times 10^{-3}$ M the slopes of the plots are 0.037 and 0.030, respectively. These values indicate that the two waves are quasi-reversible.

The polarographic behavior of iodide in acetone [51] is similar to its behavior in IBN.

In DMF a plot of E_{DME} vs $\log [i/(i_d - i)^4]$ for the first wave gives a straight line whose slope is 0.029; this was attributed by Given and Peover [48] to the electrode reaction

$$Hg + 4I^- \rightleftharpoons HgI_4^{2-} + 2e^-$$

The overall equilibrium constant of the reaction

$$Hg^{2+} + 4I^- \rightleftharpoons HgI_4^{2-}$$

is approximately 6.2 x 10^{38} at 20°C in 0.1 M $LiClO_4$.

TABLE 2.1.4. Polarographic Characteristics of Iodide in Nonaqueous Solvents

Solvent	Conditions	$E_{1/2}$ (V/SCE)		I (A)		Remarks	Refs.
DME	0.1 M $LiClO_4$	+0.32	+0.10	1.24	0.50	20°C $C_{I^-} = 1.96 \times 10^{-3}$ M	[48]
Acetone	0.1 M Et_4NClO_4	-0.37	+0.20	-	-	-	[51]
IBN	0.05 M $NaClO_4$	-0.42	+0.19	-	-	$C_{I^-} = 1 \times 10^{-3}$ M	[50]

The second wave corresponds, according to the authors, to the electrode reaction

$$Hg + 3I^- \rightleftharpoons HgI_3^- + 2e^-$$

More recently Matsui and co-workers [49] came to the conclusion that, in DMF also, the electrode reactions corresponding to the two anodic waves are the same as those proposed for IBN and acetone.

The polarographic characteristics of iodide in different solvents are listed in Table 2.1.4.

2.1.3. Molten Salts

In the fused $NaNO_3$-KNO_3 eutectic mixture (250°C) Swofford and Holifield [107] noticed that the mercury oxidation is facilitated, as has been observed in water and in nonaqueous solution, by the presence of iodide. The resulting wave is not well defined and exhibits a maximum followed by a minimum; behind the minimum the wave merges with the normal oxidation of mercury. A study of the wave shows that the oxidation process is irreversible. The height of the anodic wave at both maximum and minimum current values is proportional to the KI concentration in the range 4×10^{-4} to 3×10^{-4} M. The half-wave potential (+0.465 V vs Ag/Ag^+ 0.07 M reference electrode) is shifted 80 mV as the KI concentration increases from 1.09×10^{-3} to 4.88×10^{-3} M. The irreversibility of the anodic process seems attributable to a blockage of the electrode surface due to the formation of a film of insoluble mercury iodide and to the adsorption of iodide and nitrate on the electrode.

Francini and co-workers [108, 109], by means of oscillographic polarography at linear voltage sweep in the same medium at 232°C, found that the oxidation of iodide was reversible at the dropping mercury electrode when the scanning potential rate was 5-10 V/sec. In the case of iodide, indeed, the three equations for an anodic reversible process are obeyed:

$$(i_p)_r = 0.447nFACD^{1/2}\left(\frac{nFV}{RT}\right)^{1/2} \qquad (2.1.1)$$

$$(E_p)_r = E_{1/2}{}^r + \frac{RT}{nF} \qquad (2.1.2)$$

$$(E_{p/2})_r - (E_p)_r = 2.20\frac{RT}{nF} \qquad (2.1.3)$$

where C = iodide concentration; A = electrode area; V = scanning potential rate; and $E_{1/2}{}^r$ = half-wave potential of conventional polarography. The authors further noticed that the diffusion coefficients of the three halides changed in the same way with the temperature and that the electrochemical oxidation of iodide (and also of bromide and chloride) was associated with the reaction

$$nHg + I^- \rightleftharpoons (HgI) \cdot nHg + e^-$$

Mercurous iodide, which in molten salts is generally stable in the form Hg_2I_2, is more soluble in the metal than in the molten salt.

2.2. VOLTAMMETRIC CHARACTERISTICS

2.2.1. Aqueous Solution

2.2.1.1. Oxidation of Iodide without Complexing Agents of I_2 and/or I^+. The voltammetric behavior of iodide on solid electrodes (platinum, graphite, and carbon) has been widely investigated in the last 20 years. Iodide gives two anodic waves on a Pt electrode, the first of which corresponds to the oxidation of iodide to iodine [52-54, 57, 59] and the second to the oxidation of iodine to iodate [54, 59-61] rather than to I^+ [52] or HIO_2 [53].

The first wave is always well defined in acid solution, while the second shows a maximum followed by a minimum. On the pyrolytic graphite electrode three waves are found which have been attributed to the successive oxidation of iodide to iodine, I^+, and iodate [55, 56, 58].

In Table 2.2.1 the voltammetric characteristics of iodide on solid electrodes in different supporting electrolytes are presented.

Kolthoff and Jordan [52], using a rotating platinum electrode, found that two waves are obtained when iodide is oxidized in acid solution. The first wave, which corresponds to the process

$$2I^- \rightleftharpoons I_2 + 2e^-$$

is well defined and reversible for $pH \leq 1$; for $pH > 1$ the wave becomes irreversible. No mechanism has been proposed by the authors to explain this occurrence. In alkaline solution nowave has been found. The anodic diffusion current of iodide in 0.1 M $HClO_4$ measured at +0.7 V/SCE is proportional to the concentration in the range 10^{-5} to 10^{-3} M.

Newson and Riddiford [62], in a further study of iodide oxidation in acid solution, found that the formation of the I_3^- ion has a remarkable influence on the reversibility of the oxidation reaction of iodide to iodine. According to these authors, the oxidation reaction is reversible only for $C_{I^-} \leq 2.50 \times 10^{-4}$ M. The variation of the reversibility with pH (observed by Kolthoff and Jordan [52]) was attributed to an alteration of the surface characteristics of the electrode.

It should be noted that Anson and Lingane [63], in a chronopotentiometric study, have shown the influence of platinum oxide films on the oxidation of iodide to iodine.

The wave corresponding to the oxidation of iodide to iodine is not uniform [64] since, until a current value corresponding to 50% of the voltammetric curve is reached, it is primarily controlled by the reaction

$$2I^- \rightleftharpoons I_2 + 2e^-$$

while, beyond this point, it is controlled by the reaction

$$3I^- \rightleftharpoons I_3^- + 2e^-$$

According to whether the first or the second reaction is prevailing, the iodide/iodine system approaches reversibility to a greater or lesser extent. The oxidation process is more reversible when the iodide concentration is decreased because the I_3^- formation is associated with the bulk concentration of iodide. Toren and Driscoll [64] observed the appearance of an anodic maximum on the voltammetric wave for $C_{I^-} \geq 2.5 \times 10^{-3}$ M due to the formation of solid iodine on the electrode surface.

The presence of adsorbed iodine on the electrode surface for $C_{I^-} = 5 \times 10^{-4}$ M has been shown by Raspi, Pergola, and Cozzi [54]; there is no evidence, however, of the influence of adsorbed iodine on the reversibility of the iodide/iodine couple. On a platinum electrode, with periodic renewal of the diffusion layer, the oxidation of iodide to iodine is also reversible for $C_{I^-} = 5 \times 10^{-4}$ M and is independent of the electrode surface.

The influence of the I_3^- formation on the reversibility of the iodide/iodine couple has been studied by Guidelli and Piccardi [65]. They set up a quantitative treatment which takes the instability constant of I_3^- into account. The value of this constant is 1.27×10^{-3} moles/l at 25°C [99] in water, and for this reason the condition required by the treatment ($K \gg 3/2C$, where K = instability constant) is not satisfied. The theoretical curve corresponding to $C_{I^-} = 1 \times 10^{-4}$ M was found to be not very different from that corresponding to $K = \infty$. For this concentration the E vs log $[(i_d - i)/i^2]$ plot is linear; the slope is 30 mV and it agrees with the slope predicted for a reversible oxidation. For $C_{I^-} = 1 \quad 10^{-3}$ M a straight line is obtained whose slope, however, is 35 mV. Such a value shows that at this iodide concentration the oxidation process is slightly irreversible on a platinum electrode.

Three waves were found by Miller and Zittel [55] on a stationary pyrolytic graphite electrode in the oxidation of iodide. The first two waves tend to merge at pH > 10. The resulting wave has $E_p = \sim +0.85$ V/SCE. The first wave ($E_p = \sim +0.5$ V/SCE) corresponds to the oxidation of iodide to iodine and its peak potential does not change with pH.

Dryhurst and Elving [56], using a rotating pyrolytic graphite electrode, noticed that the half-wave potential of the first wave is independent not only of the pH value but also of the bulk iodide concentration. On this electrode the oxidation to iodine is reversible when single-scan voltammetry is employed, while it is slightly reversible for $C_{I^-} = 5 \times 10^{-4}$ M, by cyclic voltammetry.

TABLE 2.2.1. Voltammetric Characteristics of Iodide in the Absence of Complexing Agents of I_2 and/or I^+

Product	Conditions	Electrode	Technique	Potential (V/SCE)[a]	Rev	Remarks	Refs.
I_2	0.1 $\underline{M}$ $HClO_4$	Rot Pt wr	E swp	+0.5	Rev	D_{I_2} = 1.11 × 10^{-5} cm^2/sec	[52]
						D_{I^-} = 2.05 × 10^{-5} cm^2/sec	[52]
I_2	0.1 $\underline{M}$ $HClO_4$	Stat Pt	E swp	+0.55	Rev	C_{I^-} = 1 × 10^{-3} $\underline{M}$	[53]
I_2	1 $\underline{M}$ $HClO_4$	Pt[b]	E swp	+0.488	Rev	C_{I^-} = 5 × 10^{-4} $\underline{M}$	[54]
I_2	1 $\underline{M}$ $NaClO_4$	Stat PGE[c]	E swp	~+0.5	-	C_{I^-} = 2 × 10^{-3} $\underline{M}$	[55]
I_2	0.5 $\underline{M}$ K_2SO_4	Rot PGE[c]	E swp	+0.511	Rev	3 × 10^{-5} < C_{I^-} < 2 × 10^{-3} $\underline{M}$	[56]
I_2	0.5 $\underline{M}$ K_2SO_4 or						
	0.5 $\underline{M}$ K_2SO_4 + 0.5 $\underline{M}$ H_2SO_4	Stat PGE[c]	E swp	+0.510	-	C_{I^-} = 5 × 10^{-4} $\underline{M}$	[56]
I_2	0.5 $\underline{M}$ K_2SO_4 or						
	0.5 $\underline{M}$ K_2SO_4 + 0.5 $\underline{M}$ H_2SO_4	PGE[c]	CV	+0.47	Sl rev	Scan rate = 0.012 V/Sec	[56]
I_2	0.1 $\underline{M}$ $HClO_4$ + 0.4 $\underline{M}$ KNO_3	Rot G-A dsk[d]	E swp	+0.5	-	D_{I_2} = 0.96 × 10^{-5} cm^2/sec	[57]
						D_{I^-} = 1.82 × 10^{-5} cm^2/sec	[57]
I_2	0.1 $\underline{M}$ $NaClO_4$ (pH = 5.5)	PGE[c]	i stp	+0.524	Sl rev	C_{I^-} = 1.1 × 10^{-3} $\underline{M}$	[58]
I_2	0.1 $\underline{M}$ $NaClO_4$ (pH = 5.5)	GCE[e]	i stp	+0.513	Sl rev	C_{I^-} = 1.1 × 10^{-3} $\underline{M}$	[58]
I_2	0.1 $\underline{M}$ $HClO_4$	Rot Pt dsk	E swp	+0.49	Rev	D_{I_2} = 1.04 × 10^{-5} cm^2/sec	[59]
						D_{I^-} = 1.73 × 10^{-5} cm^2/sec	[59]
I^+	0.1 $\underline{M}$ $HClO_4$	Rot Pt wr	E swp	~+1.1	Irr	-	[52]

HIO_2	0.1 M $HClO_4$	Stat Pt	E swp	~+1.2	Irr	$C_{I^-} = 1 \times 10^{-3}$ M	[53]
I^+	1 M $NaClO_4$	Stat PGE[c]	E swp	~+0.9	-	$C_{I^-} = 2 \times 10^{-3}$ M	[55]
I^+	0.5 M K_2SO_4	Rot PGE[c]	E swp	+0.73	Irr	$C_{I^-} = 4 \times 10^{-4}$ M	[56]
I^+	0.5 M K_2SO_4	Stat PGE[c]	E swp	+0.85	Irr	$C_{I^-} = 5.1 \times 10^{-4}$ M	[56]
I^+	0.5 M K_2SO_4 + 0.5 M H_2SO_4	Stat PGE[c]	E swp	+0.97	Irr	$C_{I^-} = 5.1 \times 10^{-4}$ M	[56]
I^+	0.5 M K_2SO_4	PGE[c]	CV	+0.78	Irr	$C_{I^-} = 5 \times 10^{-4}$ M	[56]
I^+	0.5 M K_2SO_4 + 0.5 M H_2SO_4	PGE[c]	CV	+0.97	Irr	$C_{I^-} = 5 \times 10^{-4}$ M	[56]
IO_3^-	1 M $NaClO_4$	Stat PGE[c]	E swp	~+1.2	-	$C_{I^-} = 2 \times 10^{-3}$ M	[55]
IO_3^-	0.5 M K_2SO_4	Rot PGE[c]	E swp	~+1.28	Irr	-	[56]
IO_3^-	0.5 M K_2SO_4	PGE[c]	CV	+1.33	Irr	$C_{I^-} = 5 \times 10^{-4}$ M	[56]
IO_3^-	0.5 M K_2SO_4 + 0.5 M H_2SO_4	PGE[c]	CV	+1.50	Irr	$C_{I^-} = 5 \times 10^{-4}$ M	[56]

[a] The potentials relative to voltammetry at stationary pyrolytic graphite electrode (PGE) and glassy-carbon electrode (GCE) are peak potentials (E_p).

[b] Platinum microelectrode with periodical renewal of the diffusion layer.

[c] PGE = pyrolytic graphite electrode.

[d] G-A = graphite-azobenzene electrode.

[e] GCE = glassy-carbon electrode.

TABLE 2.2.2. Voltammetric Determination of Iodide in the Presence of Bromide and Chloride in 0.5 M K_2SO_4 + 0.5 M H_2SO_4 Supporting Electrolyte

KI (mM)	KBr (mM)	KCl (mM)	Iodide found, average (mM)
0.03006	3.027	2.97	0.0372
0.1002	3.027	2.97	0.1067
1.002	3.027	2.97	0.984

A pyrolytic graphite electrode has been employed for the quantitative determination of iodide alone or in the presence of bromide and chloride. From the first voltammetric wave iodide may be determined up to a concentration of 1×10^{-5} M with an accuracy of 4-5%. In the presence of chloride and bromide the accuracy and sensitivity decrease as the data of Table 2.2.2 show.

According to Kolthoff and Jordan [52], the second wave observed in acid solution on the rotating platinum electrode in the oxidation of iodide is due to the irreversible reaction

$$I_2 + 2xH_2O \rightarrow 2[I(H_2O)_x]^+ + 2e^-$$

The current, measured at the maximum of the second wave, is twice as high as the first wave. At pH = 8 the two waves coincide and the resulting wave corresponds to the oxidation of iodide to I^+. The same wave is observed starting from solutions of iodine.

According to Wakkad and co-workers [53], on the other hand, the second wave corresponds to the formation of iodous acid according to

$$I^- + 2H_2O \rightarrow HIO_2 + 3H^+ + 4e^-$$

The second wave observed on a platinum electrode with periodic renewal of the diffusion layer has been attributed by Raspi and co-workers [54] to the oxidation of iodine to iodate, in agreement with Anson and Lingane [60] and Zakharov and Songina [61].

From the value of the ratio $(n_1 + n_2)/n_1$ (n_1 = the number of electrons involved in the first step; n_2 = the number of electrons involved in the second step) Anson and Lingane [60] concluded that the second step refers to the oxidation of iodine to iodate. The value of this ratio, which may be found from the measurements of the transition times according to the relation

$$\frac{i_2(\tau_1 + \tau_2)^{1/2}}{i_1\tau_1^{1/2}} = \frac{n_1 + n_2}{n_1}$$

is 6 only for $C_{I^-} \geq 1 \times 10^{-3}$ M. For $C_{I^-} < 1 \times 10^{-3}$ M higher values have been obtained. The increase of the transition time of the second step, and therefore of the number of electrons involved as the iodide concentration decreases, has been attributed by these authors

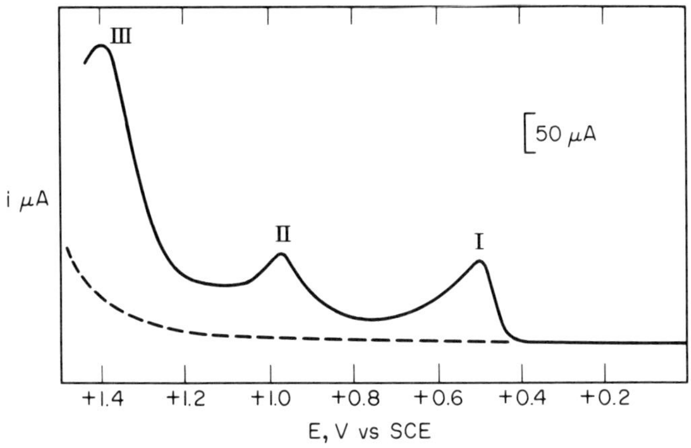

FIG. 2.2.1. Typical voltammogram of 2×10^{-3} M iodide in 1 M $NaClO_4$. Scan rate = 0.5 V/min; area of PGE = 0.4 cm^2 [55].

to the surface oxidation of the platinum electrode associated with the formation of a platinum oxide film. The oxidation of water may also give a positive contribution, but the possibility of the chemical reaction iodate-iodide must be noted; the iodine, once formed, would diffuse to the bulk of the solution.

Beran and co-workers [59], using a ring-disk electrode, confirmed the iodate formation in the second step of the oxidation of iodide in acid solutions. The authors, oxidizing I^- to I_2 and successively to I^{n+} on a platinum disk electrode, suppose that the I^{n+} species, diffusing to a ring electrode (which surrounds the disk electrode at a distance of ~2 mm), may react with the excess iodide in solution:

$$I^{n+} + nI^- \rightleftharpoons 1/2(n+1)I_2$$

Such a reaction is monitored by reducing the iodine, once formed, at the ring electrode. By keeping the ring electrode at the reduction potential of iodine and varying the potential of the disk electrode from the value corresponding to the oxidation of I^- to I_2 to that of I_2 to IO_3^-, the n value may be determined from the ratio of the two limiting currents on the ring electrode. The n value is 5 because the value of the ratio of the two limiting currents is 6. Such a value is obtained with a rotation rate of the electrode smaller than 400 rpm. When the rotation rate is greater than 400 rpm the value of the ratio of the two currents decreases noticeably. This occurrence was explained by the authors by assuming that the rate of the iodate-iodide reaction is the step controlling the overall process.

On graphite electrodes, unlike platinum electrodes, it is possible to study the oxidation process of iodine to I^+ corresponding to the second wave in the oxidation of iodide.

The voltammetric waves on a stationary pyrolytic graphite electrode are shown in Fig. 2.2.1 [55].

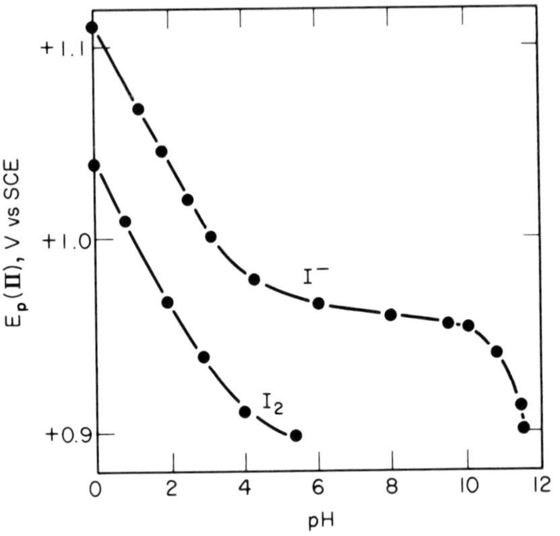

FIG. 2.2.2. Peak potential (E_p) vs pH for 5×10^{-4} M iodine and 1×10^{-3} M iodide. Scan rate = 1.0 V/min; area of PGE = 0.4 cm^2 [55].

The second wave may be described by the following possible reactions:

$$I_2 + 2H_2O \rightarrow HIO + 2H^+ + 2e^- \qquad E \text{ (vs NHE) } +1.45 - 0.059 \text{ pH} \qquad (2.2.1)$$

$$I^- + H_2O \rightarrow HIO + H^+ + 2e^- \qquad +0.99 - 0.029 \text{ pH} \qquad (2.2.2)$$

$$I_3^- + 3H_2O \rightarrow 3HIO + 3H^+ + 4e^- \qquad +1.22 - 0.044 \text{ pH} \qquad (2.2.3)$$

According to the authors [55], the third wave is associated with the reaction

$$HIO + 2H_2O \rightarrow IO_3^- + 5H^+ + 4e^- \qquad (2.2.4)$$

At pH > 10 the two waves coalesce to form a single broad peak. The variation of the peak potential of the second wave along with the pH starting from iodide and iodine solutions is reported in Fig. 2.2.2.

The value of $\Delta E_p/\Delta$pH (see Fig. 2.2.2.) is 0.034 per unit change in pH; this value is intermediate between the values of Eqs. (2.2.2) and (2.2.3). However, since the slopes $\Delta E_p/\Delta$pH are the same with both iodide and iodine solutions, the authors assume that the second wave is associated with reaction (2.2.1).

These same authors, in a later chronopotentiometric study [58], found that the experimental data obtained with this technique were in disagreement with the preceding ones. The values of the slopes, which refer to the steps corresponding to the formation of I^+ and IO_3^-, in particular, disagree with what could be predicted from Reactions (2.2.1) and (2.2.4). This was attributed to the irreversibility of the two reactions and/or to the lack of reaching electromechanical equilibrium. Also, the value of τ_2/τ_1, which according to Berzins and Delahay [67] should be 3 if $n_1 = n_2$, is smaller than 2. Since Evans [68] on the basis of the relation

$$\frac{\tau_2}{\tau_1} = 4\frac{n_2^2 + n_1n_2}{n_1^2 + 2n_1n_2}$$

had ascertained that the value of the ratio τ_2/τ_1 must be 1.78, if alternate path reactions occur, the authors assume that iodide or iodine, alternately, are oxidized. Neither the adsorption of iodide and/or iodine on the electrode nor the possible disproportionation of I^+ to give I^- and IO_3^- are excluded.

Dryhurst and Elving [56], using a rotating pyrolytic graphite electrode, found that $E_{1/2}$ for the wave attributed to the formation of I^+ is connected with the pH value by

$$E_{1/2} = 0.99 - 0.049 \text{ pH}$$

This relation agrees with the expected value for a one electron-one proton or two electrons-two protons process. Because by cyclic voltammetry the oxidation product was found to be unstable, the authors suggest a reaction similar to the one proposed by Miller and Zittel [55].

On a rotating graphite-azobenzene electrode Landsberg and co-workers [69] found that the iodide oxidation occurs similarly to what was pointed out successively by Miller and Zittel [55]. On this electrode Landsberg and co-workers [57, 70] closely examined the oxidation of iodide to hypoiodite and iodate; they found that at pH > 12 only one wave, corresponding to the direct oxidation of iodide to iodate, is obtained. With regard to the wave attributed to the formation of I^+, its limiting current in the pH range 2-7 is controlled by the rate of

$$I_2 + H_2O \rightleftarrows IOH + H^+ + I^- \qquad (2.2.5)$$

Beyond pH 9 the wave is diffusion-controlled since the slow Reaction (2.2.5) is replaced by the fast reaction

$$I_2 + OH^- \rightarrow IOH + I^- \qquad (2.2.6)$$

The wave related to iodate formation is diffusion-controlled in the pH range 3-12. Its height, furthermore, increases with pH up to pH 11, showing that the hydroxyl ions take part directly in the electrochemical process according to

$$OH_{aq}^- \rightleftarrows OH_{ads}^-$$

$$I^- + OH_{ads}^- \rightarrow IOH_{ads}^- + e^-$$

The formation of IO_3^- from IOH_{ads}^-, however, is not explained.

2.2.1.2. Oxidation of Iodide in the Presence of Complexing Agents of I_2 and/or I^+. The voltammetric characteristics of iodide in the presence of complexing agents of I_2 and/or I^+ are reported in Table 2.2.3.

Kolthoff and Jordan [52] noticed that the presence of chloride does not affect the shape of the wave relative to the oxidation of iodide to iodine. The second wave, on the contrary,

TABLE 2.2.3. Voltammetric Characteristics of Iodide in the Presence of Complexing Agents of I_2 and/or I^+

Product	Conditions	Electrode	Technique	Potential (V/SCE)	Rev	$C_{I^-} \times 10^3$ M	Refs.
I_2	0.1 $\underline{M}$ HCl	Rot Pt wr	E swp	+0.48	Rev	0.2	[52]
I_2	0.1 $\underline{M}$ HBr	Rot Pt wr	E swp	+0.62	Rev	0.2	[52]
I_2	0.5 $\underline{M}$ K_2SO_4 + 3 × 10^{-3} $\underline{M}$ KBr	PGE[a]	CV	+0.52	Rev	1	[56]
I_2	0.5 $\underline{M}$ HCl + 0.5 $\underline{M}$ $HClO_4$	Pt[b]	E swp	+0.488	Rev	0.5	[54]
I_2	0.1 $\underline{M}$ HCl + 0.4 $\underline{M}$ NaCl	Rot Pt ring-dsk	E swp	+0.48	Rev	1	[59]
ICl	0.1 $\underline{M}$ HCl + 0.4 $\underline{M}$ $NaClO_4$	Rot Pt ring-dsk	E swp	+0.79	Sl rev	1	[59]
ICl	0.1 $\underline{M}$ HCl + 0.4 $\underline{M}$ NaCl	Rot Pt ring-dsk	E swp	+0.72	Sl rev	1	[59]
IBr	0.5 $\underline{M}$ K_2SO_4 + 3 × 10^{-3} $\underline{M}$ KBr	PGE[a]	CV	+0.69	Irr	1	[56]
IBr_2^-	1$\underline{M}$ $HClO_4$ + 5 × 10^{-3} $\underline{M}$ HBr	Pt[b]	E swp	+0.737	Rev	0.5	[54]
ICN	0.05 $\underline{M}$ KCN + 0.1 $\underline{M}$ $HClO_4$	Rot Pt wr	E swp	+0.5	Irr	0.2	[52]
$[Ipy_2]^+$	Acetate buffer (pH = 6)	Rot Pt wr	E swp	+0.56	Irr	0.12	[52]

[a]PGE = pyrolytic graphite electrode.

[b]Platinum microelectrode with periodical renewal of the diffusion layer.

appears well-defined in the presence of chloride. The oxidation scheme of iodine in 0.1 M HCl is

I wave $2I^- \rightleftharpoons I_2 + 2e^-$

II wave $I_2 + 2Cl^- \rightarrow 2ICl + 2e^-$

In 0.1 M HBr iodide gives a single anodic wave corresponding to the formation of iodine. Since bromide is more easily oxidized than iodine, the second wave is not observed. In the presence of hydrocyanic acid iodide gives a single oxidation wave; such a wave, however, corresponds to a two-electron oxidation and is associated with

$$I^- + HCN \rightarrow ICN + H^+ + 2e^-$$

In the presence of pyridine in weakly acid solution, a single wave is obtained attributed to

$$I^- + 2py \rightarrow [I(py)_2]^+ + 2e^-$$

The same behavior is observed starting with iodine solutions instead of iodide solutions. In 4 M HCl iodine, which exists in solution as I_2Cl^-, gives an anodic wave ($E_{1/2} = +0.56$) corresponding to

$$I_2Cl^- + 3Cl^- \rightarrow 2ICl_2^- + 2e^-$$

This is somewhat analogous to the behavior of iodide in 1 M KCl + 0.1 M HCl.

The oxidation of iodine to I^+ starting from iodide or iodine solutions is irreversible at rotating platinum electrodes. At a stationary pyrolytic graphite electrode, Dryhurst and Elving [56], employing cyclic voltammetry, observed two waves in the oxidation of iodide ($C_{I^-} = 1 \times 10^{-4}$ M) in the presence of a large amount of bromide ($C_{Br^-} = 3 \times 10^{-3}$ M). The second irreversible wave was attributed to

$$\frac{1}{2}I_2 + Br^- \rightarrow IBr + e^-$$

At a platinum electrode with periodic renewal of the diffusion layer [54], on the contrary, the oxidation of iodine to iodine chloride and iodine bromide are both reversible. The second oxidation wave of iodide in the presence of bromide has been obtained with $C_{I^-} = 5 \times 10^{-4}$ M and $C_{Br^-} = 5 \times 10^{-3}$ M in 1 M $HClO_4$. This second wave is described by

$$I_2 + 4Br^- \rightarrow 2IBr_2^- + 2e^-$$

The prevailing species in the diffusion layer, when iodide is oxidized in the presence of chloride, are I_2Cl^-, ICl, and ICl_2^-.

The value of the equilibrium constant

$$\frac{[ICl][I^-]}{[I_2Cl^-]} = K$$

is 3.3×10^{-9} moles/l [71].

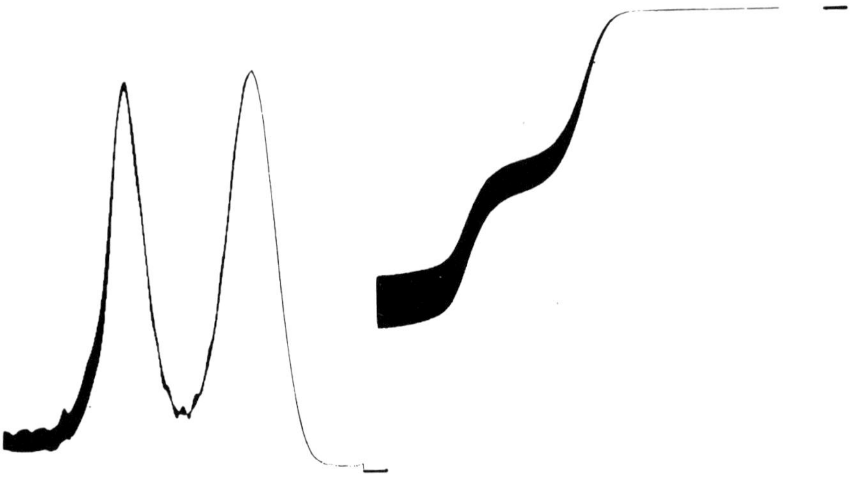

FIG. 2.2.3. Ac and dc voltammetric curves of 5×10^{-4} M potassium iodide in 1 M HCl on smooth platinum electrode at 25°C. Starting potential = +0.25 V/SCE; electrode area = 6.3×10^{-2} cm^2; scan rate = 0.055 V/min [76].

The value of the analogous constant calculated by oxidizing iodide in the presence of bromide is 3.9×10^{-7} moles/l [72].

On the periodic renewal of the electrode with diffusion layer, the heights of the two waves corresponding to the oxidation of iodide to iodine and to iodine chloride, respectively, are not the same, as many other authors [52, 73-75] have found on rotating and stationary platinum electrodes.

In order to explain the different heights of the two waves, whose ratio changes with the rotation rate of the electrode, Kolthoff and Jordan [52] assumed that part of the iodine formed in the first electrochemical step can escape from the electrode before being oxidized to ICl.

According to Morgan [73], the decrease of the second wave with respect to the first one at stationary platinum electrodes can be explained by the reaction of iodine chloride with incoming iodide to form slower-diffusing iodine [52, 57, 59]. This assumption has been accepted by Beilby and Crittenden [74].

Beran and co-workers [59], on the contrary, assumed that the lowering of the second anodic wave was attributable to the formation of a platinum oxide film on the electrode surface. As regards the reversibility of the oxidation reaction of iodine to ICl_2^-, they noticed that the wave is reversible only until 30% of the limiting current is reached. Beyond this value the voltammetric curve is irreversible owing to the formation of the platinum oxide film.

On a platinum electrode with periodic renewal of the diffusion layer, Desideri and coworkers [76] obtained two well-formed ac peaks (see Fig. 2.2.3). The similarity of the two heights shows that the oxidation process of iodide to iodine chloride is reversible in both steps. The slopes of the dc waves (30 mV in both cases) also agree with a reversible process. The different heights of the two dc waves may be attributed, according to Beran and co-workers [59], to the formation of a platinum oxide film in the potential region in which the second wave occurs. Such a film, whose formation seems inhibited in ac voltammetry, has no influence, in opposition to Beran [59], on the reversibility of the process, but only decreases the active surface of the electrode.

The decrease of the second dc wave was attributed by Guidelli [83] to the different value of the diffusion coefficients of iodide, iodine, I_2Cl^-, ICl, and ICl_2^-; the influence of a possible reaction between iodide and iodine chloride to form iodine is excluded.

2.2.1.3. Reduction of Iodate. A reduction wave of iodate in 5 M H_2SO_4 on a rotating platinum electrode was first observed by Shain and Crittenden [77]. Anson [78], in a chronopotentiometric study, examined the effect of oxidation of the electrode surface on the voltammetric behavior of iodate; iodate is reduced with a smaller overpotential on platinum electrodes which were preoxidized than on prereduced ones. Such behavior was attributed to the reduction of the surface platinum oxide with the formation of an active surface which catalyzes the reduction of iodate.

Badoz-Lambling and Guillaume [79], using rotating platinum electrodes, observed that iodate reduction was favored by an increase of the iodate concentration and of the acidity of the solution. They noticed that if the $HClO_4$ concentration changed from 1 to 5 M (keeping the ionic strength equal to 5 with $NaClO_4$) well-defined waves were obtained and the limiting currents did not vary with acidity. Iodate reduction is favored by the addition of iodine to the solution, since iodine supplies iodide ions which may react chemically with iodate to give iodine. The assumption of Anson [78] of the importance of the chemical reduction of iodate in the electrochemical reduction of iodate was confirmed.

Desideri [80], in a study of the influence of sulfuric acid concentraion on the reduction of iodate, noticed that for $C_{H_2SO_4} > 3$ M the wave splits in two steps, the second of which is smaller and badly-defined. The half-wave potential of the first step is shifted from +0.385 V/SCE in 0.75 M H_2SO_4 to +0.446 V/SCE in 3 M H_2SO_4. For $C_{H_2SO_4} > 3$ M, $E_{1/2}$ decreases until a value of +0.294 V/SCE in 10 M H_2SO_4 is reached. The shifting of $E_{1/2}$ to less positive values, after the maximum value, is related to the reduction of ionic species more complex than HIO_3. The split of the wave in two steps was confirmed by coulometric measurements to be associated with the reduction of polymeric species of HIO_3 rather than with a two-step reduction. With regard to the reduction mechanism of iodate, Desideri demonstrated that iodate is directly reduced at the electrode until the supply of iodide ions is sufficient to make the chemical reduction of iodate possible. The shift of the reduction potential of iodate in the presence of complexing agents of iodide [i.e., Hg(II)] confirms this mechanism.

The reduction scheme for iodate in acid solution is therefore

$$HIO_3 + 5H^+ + 6e^- \xrightarrow{\text{slow}} I^- + 3H_2O$$

$$\frac{1}{5}HIO_3 + I^- + H^+ \xrightarrow{\text{fast}} \frac{3}{5}I_2 + \frac{3}{5}H_2O$$

$$I_2 \xrightarrow{\text{fast}} 2I^- + 2e^-$$

With regard to the effect of electrode pretreatment on iodate reduction, Desideri [81] noticed that the alterations of the electrode surface only influence the first step of the oxidation process. The shape of the reduction waves is the same on preoxidized, prereduced, and polished electrodes. The slope, on the other hand, decreases according to the sequence: preoxidized electrode > prereduced electrode > polished electrode.

At an electrode prereduced for less than 2 min, the shape of the chronopotentiometric curves is different from that obtained at an electrode preoxidized for the same length of time. The shape of the chronopotentiometric wave is analogous both at prereduced and preoxidized electrodes if the treatment time is greater than 2 min. The differences observed for treatment times of less than 2 min may be eliminated by the addition of iodine to the solution.

Iodate reduction in hydrochloric acid has recently been studied by Beran and co-workers [59] and by Desideri [82].

According to Beran and co-workers [59] the half-wave potential noticeably varies as the iodate concentration changes in solution at constant ionic strength (I = 0.5). There is a proportionality between the limiting current and the concentration of iodate only for $2 \times 10^{-4}\ \underline{M} < C_{IO_3^-} < 1 \times 10^{-3}\ \underline{M}$. The reaction scheme proposed by these authors is analogous to that proposed for the iodate reduction in other acid media.

Desideri [82], on the contrary, observed that the reduction of iodate occurs at more positive potentials in HCl than in $HClO_4$ or in H_2SO_4 (while keeping the H^+ concentration constant).

Iodate is reduced in HCl through an autocatalytic cycle similar to what is found in other acid media; such a cycle, however, concerns iodine chloride instead of iodine as the electroactive species. The proposed mechanism is

$$IO_3^- + 6H^+ + 6e^- \xrightarrow{\text{slow}} I^- + 3H_2O$$

$$2I^- + IO_3^- + 3Cl^- + 6H^+ \xrightarrow{\text{fast}} 3ICl + 3H_2O$$

$$ICl + 2e^- \xrightarrow{\text{fast}} I^- + Cl^-$$

The voltammetric characteristics of iodate in different media are collected in Table 2.2.4.

2.2.1.4. Reduction of Iodine Chloride. The voltammetric characteristics of iodine monochloride are reported in Table 2.2.5.

The reduction of ICl was studied by Kolthoff and Jordan [52] in 4 M HCl at a rotating platinum electrode. Two cathodic waves are obtained which are associated with

$$2ICl_2^- + 2e^- \rightarrow I_2Cl^- + 3Cl^- \qquad (2.2.7)$$

$$I_2Cl^- + 2e^- \rightarrow 2I^- + Cl^- \qquad (2.2.8)$$

TABLE 2.2.4. Voltammetric Characteristics of Iodate in Different Acid Media

Product	Conditions	Electrode	Technique	Potential (V/SCE)	Rev	Remarks $C_{IO_3^-} \times 10^3$ M	Refs.
I^-	1 M H_2SO_4	Bub Pt	E swp	+0.410	Irr	5	[80]
I^-	1 M H_2SO_4	Pt[a]	i stp	+0.465	Irr	5	[81]
I^-	1 M H_2SO_4	Pt[b]	i stp	+0.575	Irr	5	[81]
I^-	0.2 M HCl + 0.3 M NaCl	Rot Pt dsk	E swp	+0.491	Irr	1	[59]
I^-	5 M $HClO_4$	Rot Pt	E swp	+0.43	Irr	1	[79]
I^-	1 M HCl	Pt[c]	E swp	+0.450	Irr	0.833	[82]

[a]Preoxidized electrode for $2 \leq t \leq 60$ min.

[b]Prereduced electrode for $15 \leq t \leq 60$ min. For a reduction time of 2 min. $E_{\tau/4} = +0.460$ V/SCE.

[c]Platinum microelectrode with periodical renewal of the diffusion layer.

TABLE 2.2.5. Voltammetric Characteristics of Iodine Monochloride

Product	Conditions	Electrode	Technique	Potential (V/SCE)	Rev	$C_{ICl} \times 10^3$ M	Refs.
I_2Cl^-	0.1 M HCl	Pt[a]	E swp	+0.768	Rev	0.5	[76]
I_2	0.1 M HCl + 0.4 M $NaClO_4$	Rot Pt dsk	E swp	+0.78	Rev	1	[59]
I_2	0.5 M HCl + 0.5 M $HClO_4$	Pt[a]	E swp	+0.682	Rev	0.5	[54]
I_2Cl^-	1 M HCl	Pt[a]	E swp	+0.652	Rev	0.5	[76]
I_2Cl^-	4 M HCl	Rot Pt wr	E swp	+0.56	-	0.2	[52]
I_2Cl^-	10 M HCl	Pt[a]	E swp	+0.452	Rev	0.5	[76]
I^-	0.1 M HCl	Pt[a]	E swp	+0.555	Rev	0.5	[76]
I^-	0.1 M HCl + 0.4 M $NaClO_4$	Rot Pt dsk	E swp	+0.49	Rev	1	[59]
I^-	0.5 M HCl + 0.5 M $HClO_4$	Pt[a]	E swp	+0.477	-	0.5	[54]
I^-	1 M HCl	Pt[a]	E swp	+0.517	Rev	0.5	[76]
I^-	4 M HCl	Rot Pt wr	E swp	+0.36	-	0.2	[52]
I^-	10 M HCl	Pt[a]	E swp	+0.336	-	0.5	[76]

[a]Platinum microelectrode with periodical renewal of the diffusion layer.

TABLE 2.2.6. Slopes and Half-Wave Potentials of the Two Cathodic Waves of Iodine Monochloride at Different H^+ and Cl^- Concentrations

C_{H^+} (M)	C_{Cl^-} (M)	Slope 1[a] (V)	Slope 2[a] (V)	$E_{1/2}^1$ (V/SCE)	$E_{1/2}^2$ (V/SCE)
0.1	0.1	0.032	0.032	+0.78	+0.49
0.1	0.5	0.032	0.032	+0.70	+0.48
0.3	0.1	0.032	0.040	+0.76	+0.48

[a] Of $\log[i_l - i/i]$ vs E plot.

On a platinum electrode with periodical renewal of the diffusion layer, Raspi and co-workers [54] observed that iodine monochloride gives rise to two cathodic waves in 0.5 M HCl + 0.5 M $HClO_4$. The half-wave potential of the first wave is analogous to that of the oxidation wave of iodine to iodine chloride in the same medium.

Beran and Bruckenstein [59] studied the reversibility of the two reduction waves in solutions with different H^+ and Cl^- concentrations. The slopes for the two waves are reported in Table 2.2.6.

On the basis of these data the reduction of ICl to iodine is reversible, while the reduction to iodide is reversible only for $C_{H^+} < 0.3$ M. From the shifting of $E_{1/2}$ as the concentrations of H^+ and Cl^- change, the authors found that ICl_2^-, $HICl_2$, and, to a lesser extent, ICl are the prevailing species in solution.

Desideri and co-workers [76] found that the reversibility of the reduction process of iodine chloride to iodine is noticeably influenced both by the hydrochloric acid concentration and by the electrode surface.

On smooth platinum, as the ac voltammetric curves of Fig. 2.2.4 show, reduction process (2.2.7), which is quasi-reversible for $C_{HCl} \leq 1$ M, becomes entirely irreversible for $C_{HCl} = 4$ M. The formation of a poorly defined peak is already observed for $C_{HCl} \geq 2$ M.

Reaction (2.2.7) is reversible at a platinized platinum electrode only for a platinizing time of 5 sec at a current density of 240 mA/cm^2 (see Table 2.2.7).

For a platinizing time of 5 sec, the heights of the two ac peaks are similar because of the increased reversibility of the ICl_2^-/I_2Cl^- couple. Since on a platinized (t = 5 sec) platinum electrode the reduction of iodine chloride to iodine is reversible in the whole HCl concentration range explored, the stability constants of the chloride complexes of I^+ are obtainable at this electrode. By applying the De Ford-Hume method [85], I^+ was found to exist in hydrochloric solution in the following forms: ICl, ICl_2^-, and ICL_3^{2-}. The percentage distribution of these forms, at different chloride concentrations, is shown in Fig. 2.2.5.

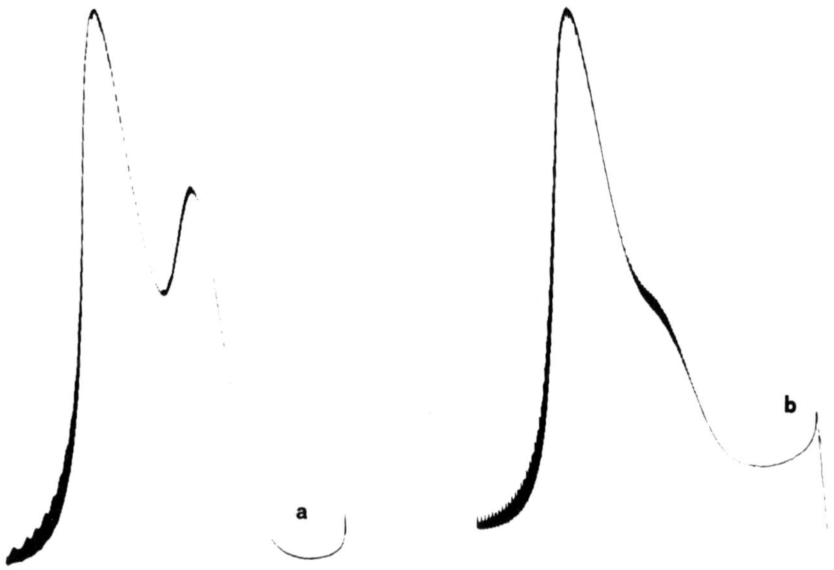

FIG. 2.2.4. Ac voltammetric curves of 5×10^{-4} M iodine monochloride in (a) 1 M HCl and (b) 4 M HCl on smooth platinum electrode at 25°C. Starting potential = +0.75 V/SCE; electrode area = 6.3×10^{-2} cm^2; scan rate = 0.055 V/min [76].

TABLE 2.2.7. Slopes of Log(i_l - i/i) vs E for the Wave Relative to the Reaction (2.2.7) as the Platinizing Time of the Electrode Changes

Platinizing time (sec)	Slope (V)
1	0.039
3	0.035
5	0.030

The overall stability constants obtained at different ionic strengths and the extrapolated values at ionic strength = 0 are collected in Table 2.2.8.

2.2.1.5. Reduction of Iodine and Triiodide. A cathodic wave with a well-defined limiting current is obtained when iodine is reduced at rotating platinum electrodes in the pH range 1-8 [52]. The wave, which is reversible in 0.1 M $HClO_4$, deviates from reversibility as the pH increases and it is completely irreversible at pH > 4. The limiting current is proportional to the iodine concentration between 5×10^{-6} and 5×10^{-4} M. The diffusion

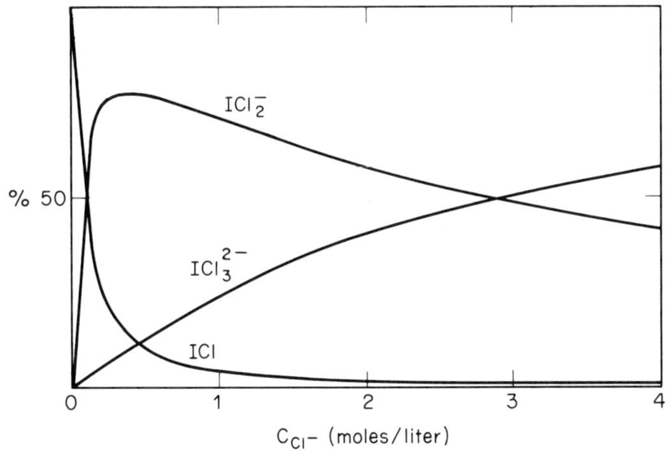

FIG. 2.2.5. Percentage distribution of I^+ chlorocomplexes vs the chloride concentration at ionic strength = 5 [76].

TABLE 2.2.8. Overall Stability Constants as the Ionic Strength Changes

I	log β_1	log β_2	log β_3
0[a]	2.88	4.2	3.74
1	2.93	4.24	3.85
3	3.11	4.27	3.78
5	3.17	4.37	3.91

[a]Extrapolated values at ionic strength = 0.

coefficient of iodine at infinite dilution is similar to that of triiodide (i.e., 1.11 × 10^{-5} cm^2/sec at 25°C.

When $HClO_4$ is replaced by HCl no changes are observed in the half-wave potential until $C_{HCl} \leq 0.5$ M [59]; for $C_{HCl} > 0.5$ M a potential shift is observed toward less positive values. In 4 M HCl $E_{1/2}$ is shifted from +0.5 to +0.36 V/SCE [52].

Newson and Riddiford [86], in a study of triiodide reduction at a rotating platinum disk cathode, determined the limiting currents as a function of the triiodide concentration, the angular rotation rate of the electrode, and the temperature (25.0–44.7°C). In Fig. 2.2.6 the limiting currents are plotted as a function of the triiodide concentration at three different angular rotation rates of the disk at 25°C. In each case proportionality between the triiodide concentration and the limiting current is observed. The reduction reaction at the rotating platinum electrode was investigated in depth by Barbasheva and co-workers [87] with

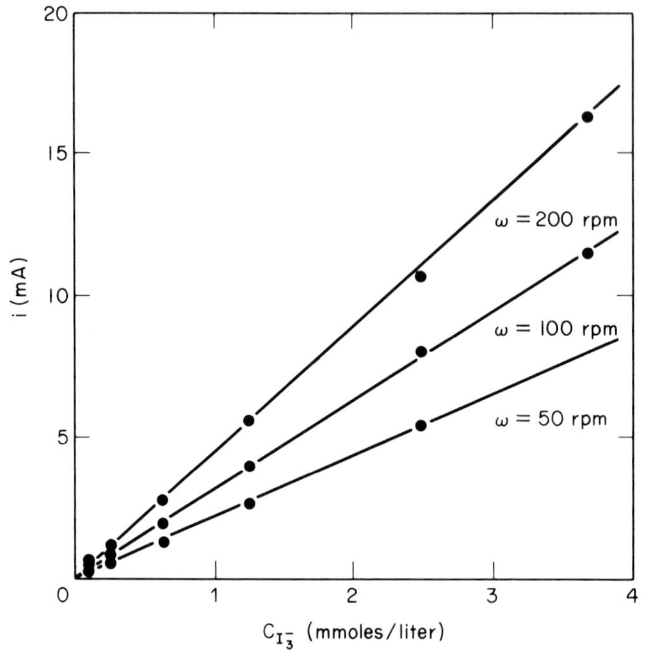

FIG. 2.2.6. Limiting currents vs triiodide concentration in 0.1 M potassium iodide at three different angular velocities of the platinum disk electrode. Apparent area of the electrode = 7.64 cm^2 [86].

triiodide solutions more concentrated than those employed by Newson and Riddiford [86]. The limiting currents fall off gradually with the passage of time. This effect is more pronounced as the concentration of iodine (with respect to that of iodide) and of the supporting electrolyte increase. Such an occurrence was attributed to a decrease in the number of active centers on the electrode surface.

2.2.2. Nonaqueous Solution

2.2.2.1. Oxidation of Iodide in the Absence of Complexing Agents of I_2 and/or I^+.

The voltammetric characteristics of iodide in different solvents are collected in Table 2.2.9.

Acetonitrile (ACN) was the first solvent employed in the study of the voltammetric behavior of iodide. In this solvent two anodic waves, the second of which was twice as high as the first, were observed at a rotating platinum electrode by Kolthoff and Coetzee [47]. The two waves are associated with

$$6I^- \rightleftharpoons 2I_3^- + 4e^- \quad (2.2.9)$$

$$2I_3^- \rightleftharpoons 3I_2 + 2e^- \quad (2.2.10)$$

In this solvent the oxidation of iodide to iodine occurs in two steps because of the formation of I_3^-, whose stability is higher in acetonitrile than in water.

This oxidation scheme was confirmed by Popov and Geske [89]; two other waves, however, were observed by these authors at more positive potentials than those corresponding to the formation of molecular iodine. They noted that a third wave was observed when oxidizing iodide or triiodide, but not iodine, solutions. For this reason the third wave was attributed to the formation of I^+ through a mechanism involving nascent iodine. With regard to the fourth wave, no acceptable assumption has been proposed. The electrode processes corresponding to the first two waves are reversible, though the slopes differ from the theoretical ones because of the influence of the electrode surface.

Guidelli and Piccardi [65], in a quantitative study, also concluded that the two waves are reversible, though the reversibility depends strictly on the pretreatment of the platinum electrode.

At pyrolytic graphite electrodes [66] three waves are found whose i_p/C ratio is 2:1:1, and this suggests that a similar ratio of electrons are lost in the iodide oxidation sequence. The two first waves, indeed, are analogous to those observed at platinum electrodes (the second wave is reversible) while the third wave is associated with the formation of I^+:

$$I_2 \rightarrow 2I^+ + 2e^-$$

Two oxidation waves corresponding to Reactions (2.2.9) and (2.2.10) have been found at platinum electrodes in acetone [88, 20], 1,2-dimethoxyethane [88], acetic anhydride [88], dimethylsulfoxide [88, 92], dimethylformamide [22, 93], tetrahydrofuran (buffer solution at pH = 5.4) [94], nitromethane [20, 90], and in acetone/water (1:1) [88] and ethylene glycol/1,2-dimethoxyethane [88] mixtures. In pyridine [88], due to its stabilizing properties of I^+, two waves are observed which are associated with

$$2I^- \rightarrow I_2 + 2e^-$$

$$I_2 + 2py \rightarrow 2[Ipy]^+ + 2e^-$$

In buffered tetrahydrofuran solutions (pH 1.5-3) [94] only one wave, corresponding to the oxidation of triiodide to iodine, is observed because iodine is chemically oxidized:

$$3I^- + 2H^+ \rightarrow H_2 + I_3^-$$

A single wave corresponding to the oxidation of iodide to iodine (similar to what is observed in water) has been found in methanol, propanol, butanol, and acetic acid by the chronopotentiometric method [88] and in liquid ammonia with the voltammetric method [95]. In this last solvent only iodide is stable, while I_2 and I^+ are reduced by the solvent as the pH increases. In acetic acid, however, two waves attributed to Reactions (2.2.9) and (2.2.10) were found by the voltammetric method by Guidelli and Piccardi [96] and, more recently, by Durand and Trémillon [24].

The instability constants of I_3^- in the different solvents have been determined by many authors. The data are collected in Table 2.2.10.

Two waves were found by Zittel and Miller [91] in dimethylsulfoxide at pyrolytic graphite electrodes. The two waves were of similar height and were attributed, unlike what is found at platinum electrodes [92, 101, 102], to the reactions

TABLE 2.2.9. Voltammetric Characteristics of Iodide in Nonaqueous Solution in the Absence of Complexing Agents of I_2 and/or I^+

Product	Solvent	Conditions	Electrode	Technique	Potential (V/SCE)	Rev	Remarks, C_{I^-}	Refs.
I_3^-	ACN	0.1 M $NaClO_4$	Rot Pt	E swp	+0.3	-	5×10^{-4} M	[88]
I_3^-	ACN	0.1 M $LiClO_4$	Rot Pt	E swp	+0.29	Rev	1×10^{-3} M	[89]
I_3^-	ACN	0.1 M $NaClO_4$	Stat Pt	i stp	+0.25	Rev	-	[88]
I_3^-	ACN	0.1 M Et_4NClO_4	Rot PGE[a]	E swp	+0.357	-	5.78×10^{-4} M	[56]
I_3^-	ACN	0.1 M Et_4NClO_4	Stat PGE	CV	+0.377	-	5×10^{-4} M	[56]
I_3^-	Acetone	0.1 M Et_4NClO_4	Rot Pt	E swp	+0.28	-	1.5×10^{-3} M	[20]
I_3^-	Acetone	0.1 M $NaClO_4$	Stat Pt	i stp	+0.22	-	-	[88]
I_3^-	1,2-DME[b]	0.1 M $NaClO_4$	Stat Pt	i stp	+0.39	-	-	[88]
I_3^-	Ac_2O	0.1 M $NaClO_4$	Stat Pt	i stp	+0.20	-	-	[88]
I_3^-	Ac_2O	0.1 M $LiClO_4$	Rot Pt	E swp	-0.5[e]	Irr	4×10^{-3} M	[23]
I_3^-	DMSO	0.1 M $NaClO_4$	Stat Pt	i stp	+0.48	-	-	[88]
I_3^-	NM[c]	0.1 M Et_4NClO_4	Rot Pt	E swp	+0.33	Irr	2.9×10^{-4} M	[20]
I_3^-	NM[c]	0.1 M $LiClO_4$	Stat Pt	i stp	+0.220	-	5.62×10^{-3} M	[90]
I_3^-	Acetone/water (1:1)	0.1 M $NaClO_4$	Stat Pt	i stp	+0.42	-	-	[88]
I_3^-	EG[d]/1,2-DME[b]	0.1 M $NaClO_4$	Stat Pt	i stp	+0.37	-	-	[88]
I_3^-	EG[d]	0.1 M $NaClO_4$	Stat Pt	i stp	+0.42	-	-	[88]
I_3^-	DMF	0.1 M $HClO_4$	Rot Pt	E swp	-0.185[f]	-	5×10^{-3} M	[22]
I_2	ACN	0.1 M $NaClO_4$	Rot Pt	E swp	+0.6	-	5×10^{-4} M	[47]
I_2	ACN	0.1 M $LiClO_4$	Rot Pt	E swp	+0.61	Rev	1×10^{-5} M	[89]
I_2	ACN	0.1 M $NaClO_4$	Stat Pt	i stp	+0.55	Rev	-	[88]
I_2	ACN	0.1 M Et_4NClO_4	Rot PGE[a]	E swp	+0.707	Rev	5.78×10^{-4} M	[56]
I_2	ACN	0.1 M Et_4NClO_4	Stat PGE[a]	CV	+0.707	-	5×10^{-4} M	[56]
I_2	Acetone	0.1 M Et_4NClO_4	Rot Pt	E swp	+0.77	-	1.5×10^{-3} M	[20]

I_2	Acetone	0.1 M $NaClO_4$	Stat Pt	i stp	+0.62	-	-	[88]
I_2	1,2-DME[b]	0.1 M $NaClO_4$	Stat Pt	i stp	+0.65	-	-	[88]
I_2	Ac_2O	0.1 M $NaClO_4$	Stat Pt	i stp	+0.60	-	-	[88]
I_2	Ac_2O	0.1 M $LiClO_4$	Rot Pt	E swp	-0.10[e]	Sl rev	4×10^{-3} M	[23]
I_2	DMSO	0.1 M $NaClO_4$	Stat Pt	i stp	+0.70	-	-	[88]
I_2	DMSO	0.1 M But_4NClO_4	Stat Pt	CV	+0.84	Irr	1.5×10^{-3} M	[91]
I_2	NM[c]	0.1 M Et_4NClO_4	Rot Pt	E swp	+0.60	Q-rev	2.9×10^{-4} M	[20]
I_2	NM[c]	0.1 M $LiClO_4$	Stat Pt	i stp	+0.530	-	5.62×10^{-3} M	[90]
I_2	Acetone/water (1:1)	0.1 M $NaClO_4$	Stat Pt	i stp	+0.54	-	-	[88]
I_2	EG[d]/1,2-DME[b]	0.1 M $NaClO_4$	Stat Pt	i stp	+0.56	-	-	[88]
I_2	EG[d]	0.1 M $NaClO_4$	Stat Pt	i stp	+0.52	-	-	[88]
I_2	CH_3OH	0.1 M $NaClO_4$	Stat Pt	i stp	+0.40	-	-	[88]
I_2	1-Butanol	0.2 M $LiClO_4$	Stat Pt	i stp	+0.42	-	-	[88]
I_2	2-Propanol	0.2 M $LiClO_4$	Stat Pt	i stp	+0.42	-	-	[88]
I_2	AcOH	0.1 M $NaClO_4$	Stat Pt	i stp	+0.44	-	-	[88]
I_2	DMF	0.1 M $HClO_4$	Rot Pt	E swp	+0.20[f]	-	5×10^{-3} M	[22]
I^+	ACN	0.1 M $LiClO_4$	Rot Pt	E swp	+1.35	-	1×10^{-3} M	[89]
I^+	ACN	0.1 M Et_4NClO_4	Rot PGE[a]	E swp	+1.337	-	5.78×10^{-4} M	[56]
I^+	ACN	0.1 M Et_4NClO_4	Stat PGE[a]	CV	+1.487	-	5×10^{-4} M	[56]
I^+	DMSO	0.1 M But_4NClO_4	Stat Pt	CV	+1.10	Irr	1.5×10^{-3} M	[91]

[a]PGE = pyrolytic graphite electrode.
[b]1,2-DME = 1,2-dimethoxyethane.
[c]NM = nitromethane.
[d]EG = ethylene glycol.
[e]Potential value vs $Ag/AgClO_4$ 10^{-2} M reference electrode.
[f]Potential value vs Hg/Hg(II) reference electrode [21].

TABLE 2.2.10. Instability Constants of Triiodide Ion in Different Solvents at 25 °C

Solvent	pK	Refs.
Acetonitrile	7.4	[18]
	6.6	[20]
Dimethylformamide	7.35	[22]
	6.83	[97]
	4.89	[98]
Acetic anydride	8.3	[100]
	7.1	[96]
Acetic acid	5.24	[24]
	4.40	[96]
Acetone	8.3	[20]
Nitromethane	6.7	[20]
Dimethylsulfoxide	7-8	[101]
Water	2.9	[99]

$$2I^- \rightarrow I_2 + 2e^-$$

$$I_2 \rightarrow 2I^+ + 2e^-$$

2.2.2.2. Oxidation of Iodide in the Presence of Complexing Agents of I_2 and/or I^+. Addition of chloride in acetonitrile [89] gives rise to a remarkable increase of the current beyond that due to the oxidation of iodide to iodine. This current increase is probably due to

$$\frac{1}{2} I_2 + Cl^- \rightarrow ICl + e^-$$

Only a single, not well-defined, wave is obtained as the chloride concentration increases. The current at the inflection point is about twice as high as what could be expected for the oxidation of iodide to iodine in a noncomplexing medium.

In nitromethane [90], in the presence of 0.05 M tetramethylammonium chloride, iodide oxidation occurs according to

$$I^- + Cl^- \rightarrow ICl + 2e^-$$

In acetonitrile, in the presence of heterocyclic amines (pyridine, 2-picoline, 2,6-lutidine, 2,2'-bipyridine, 4,4'-bipyridine, 2,2',2''-tripyridine, and 1,10-phenantroline), three waves are observed in the oxidation of iodide. The first two waves correspond to those observed in the same solvent in the absence of complexing agents, while the third wave is attributed to the oxidation of iodine to the complex $[I\text{-Amine}]^+$ [89, 139]. The height of the third wave increases with the amine concentration. The height of the wave is proportional to the iodide concentration for a given amine concentration.

TABLE 2.2.11. Voltammetric Characteristics of Iodine in Different Solvents at the Rotating Platinum Electrode

Product	Solvent	Conditions	Technique	Potential (V/SCE)	Rev	Remarks, C_{I_2}	Refs.
IAcO	Ac_2O	0.1 M $LiClO_4$	E swp	+1.25[a]	Irr	4×10^{-3} M	[23]
$IAcO^{n+}$	Ac_2O	0.1 M $LiClO_4$	E swp	+1.7[a]	Irr	4×10^{-3} M	[23]
?	ACN	0.1 M $LiClO_4$	E swp	+2.09[b]	Irr	-	[89]
I_3^-	ACN	0.1 M $LiClO_4$	E swp	+0.62	Rev	4×10^{-4} M	[89]
I_3^-	Ac_2O	0.1 M $LiClO_4$	E swp	-0.15[a]	Sl rev	4×10^{-3} M	[23]
I_3^-	DMF	0.1 M $HClO_4$	E swp	+0.155[c]	Rev	2.5×10^{-3} M	[22]
I^-	ACN	0.1 M $LiClO_4$	E swp	+0.14	Q-rev	4×10^{-4} M	[89]
I^-	Ac_2O	0.1 M $LiClO_4$	E swp	-0.9[a]	Irr	4×10^{-3} M	[23]
I^-	DMF	0.1 M $HClO_4$	E swp	-0.510[c]	-	2.5×10^{-3} M	[22]

[a] Potential value vs $Ag/AgClO_4$ 10^{-2} M reference electrode.

[b] Reverse trace.

[c] Potential value vs Hg/Hg(II) reference electrode [21].

2.2.2.3. Voltammetric Behavior of Iodine. The voltammetric characteristics of iodine are reported in Table 2.2.11. Poorly defined waves are obtained both in acetonitrile [89] and in tetrahydrofuran [94] when iodine is oxidized at a platinum electrode; for this reason it was not possible to determine the oxidation state of the products formed.

Iodide is oxidized directly to iodine in fluorosulfonic acid; therefore, only iodine oxidation has been studied in it [103]. Iodine has been studied in a neutral unbuffered medium and in basic medium in this solvent. In both cases two waves, corresponding to the following reactions, are observed:

1st wave $I_2 \rightarrow I_2^+ + e^-$

2nd wave $I_2^+ \rightarrow 2I^{3+} + 5e^-$

In a basic medium I_2^+, which is unstable, disproportionates to I_2 and I^{3+}.

Several authors [23, 104, 24, 105] have studied the oxidation of iodine in acetic anydride and in acetic acid.

Plichon and co-workers [23] observed two anodic waves in acetic anhydride, the first of which is associated with

$$I_2 \rightarrow 2I^+ + 2e^-$$

$$2I^+ + 2CH_3COOH \rightarrow 2CH_2ICOOH + 2H^+$$

The second wave has been attributed to a further oxidation of monoiodoacetic acid to give an unstable product ($IAcO^{n+}$). The final oxidation state of the iodine has not been reported.

Guidelli and Piccardi [104], in a recent pulse-polarographic study, have found that, in the product formed along the second wave, the oxidation state of the iodine is +5. The process is irreversible.

Durand and Trémillon [24] observed two anodic waves and a peak before the waves in anhydrous acetic acid. The trend is similar both in an acidic medium (0.5 M $NaClO_4$ + 1 M $HClO_4$ as supporting electrolyte) and a basic one (0.5 M $NaClO_4$ + 1 M AcONa as supporting electrolyte). The peak was attributed to iodine adsorbed on the platinum surface; this assumption was confirmed by the influence on the peak of the nature of the electrode, of its pretreatment, and of the presence of iodide ions which are strongly adsorbed on platinum.

The first oxidation wave is attributed to the formation of monoiodoacetic acid (similarly to what is found in acetic anhydride). The second wave, which in acid medium tends to coalesce with that corresponding to the formation of monoiodoacetic acid, was attributed to the oxidation of CH_2ICOOH according to

$$CH_2ICOOH \rightarrow CH_2I^{\cdot} + CO_2 + H^+ + e^-$$

$$CH_2I^{\cdot} + H^+ \rightarrow CH_3^{\cdot} + I^+$$

The $CH_3^{\cdot}$ radical reacts further to give hydrocarbons.

At a smooth platinum electrode with periodic renewal of the diffusion layer a single anodic wave is observed in the same solvent with 0.3 M $LiClO_4$ as supporting electrolyte [105]. This wave was attributed by Guidelli and Piccardi [105] to the formation of I(V).

One or two cathodic waves are obtained when iodine is reduced, depending on whether the oxidation of iodide in that solvent occurs in one or two steps. The reactions associated with the cathodic waves represent the reverse of the corresponding anodic reactions. In a buffered solution of tetrahydrofuran [94] at pH = 5.4 iodine gives rise to two cathodic waves, while in a buffered solution at pH = 2 only a single wave, corresponding to the reduction to I_3^-, is obtained.

2.2.2.4. Voltammetric Behavior of Triiodide and Interhalogen Species. Triiodide gives rise to a cathodic wave corresponding to its reduction to I^- and to one or more anodic waves corresponding to its oxidation to iodine and to species with an oxidation degree higher than 0 [89, 92].

The voltammetric characteristics of triiodide and interhalogen species in acetonitrile [106] are shown in Table 2.2.12.

The cathodic voltammetric behavior of the diiodobromide ion is analogous to that of triiodide. Their stability constants should be similar because the half-wave potentials, relative to the reduction to I^-, are very close (see Table 2.2.12).

TABLE 2.2.12. Voltammetric Characteristics of Interhalogen Species in Acetonitrile at a Rotating Platinum Electrode [106]. (0.1 M $LiClO_4$ supporting electrolyte)

Substance	Product	Potential (V/SCE)	I[b] (μA/mM)
I_3^-	I^-	+0.12[a]	9.52
I_3^-	I_2	+0.62	-5.08
I_3^-	I^+	+1.38	-1.11
I_3^-	?	+2.10	-24.6
I_2Br^-	I^-	+0.18[a]	7.80
I_2Br^-	I_2	+0.64	-2.38
I_2Br^-	$IBr + Br_2$	+1.05	-2.77
I_2Br^-	?	+2.18[c]	-9.5
IBr	?	+0.19[a]	2.43
IBr	?	+0.56[a]	1.53
IBr	?	+0.83[a]	2.10
ICl	?	+0.15[a]	1.88
ICl	?	+0.56[a]	1.81
ICl	?	+0.89[a]	2.53
IBr_2^-	I_2Br^-	~+0.4	~1.5
IBr_2^-	I^-	+0.22[a]	1.47
IBr_2^-	$IBr + Br_2$	+1.11	-4.86
IBr_2^-	?	+1.72	-0.6
ICl_2^-	I_2Cl^-	+0.09[a]	1.07
ICl_2	I^-	+0.42[a]	1.09
ICl_2^-	$ICl + Cl_2$	+1.44[a]	-3.82
$IBrCl^-$	I^-	~+0.27[a]	~3.4
$IBrCl^-$	$ICl + Br_2$	+1.27	-4.53
$IBrCl^-$	?	+1.79	-0.35

[a]Data obtained on forward scan.

[b]$I = i_e/C$; anodic process given a negative sign.

[c]Data for reverse scan.

On the other hand, the anodic behavior seems slightly different and fairly complex. In order to explain this, Popov and Geske [106] assumed that I_2Br^- disproportionates to I_3^- and IBr_2^- and attributed the first two anodic waves to

$$I_3^- \rightarrow \frac{3}{2}I_2 + e^-$$

$$IBr_2^- \rightarrow IBr + \frac{1}{2}Br_2 + e^-$$

The process associated with the third anodic wave, which is analogous to the corresponding oxidation wave of triiodide and of iodine, has not been clarified.

With regard to the interhalogen species in which iodine has an oxidation state of +1, Popov and Geske [106] noticed that the current-potential curves for the reduction of ICl and IBr are very similar, as both involve a three-step wave followed by a current decrease at more negative potentials. The overall electrode reaction is

$$IX + 2e^- \rightarrow I^- + X^-$$

The voltammetric behavior of ICl_2^- is similar to that of IBr_2^- and $IBrCl^-$. The anodic wave is associated with

$$IX_2^- \rightarrow IX + \frac{1}{2}X_2 + e^-$$

while the two cathodic waves are associated with

$$IX_2^- + e^- \rightarrow \frac{1}{2}I_2X^- + \frac{3}{2}X^-$$

$$\frac{1}{2}I_2X^- + e^- \rightarrow I^- + \frac{1}{2}X^-$$

2.2.3. Molten Salts

The voltammetric characteristics of iodide and iodine in molten salts are reported in Table 2.2.13.

The first study concerning the oxidation of iodide in molten AgI was by Mugulescu and co-workers [112]. They concluded that the electrochemical discharge of iodide occurs at carbon electrodes without appreciable activation overvoltage (2–3 mV) unlike the behavior in water. The absence of an overvoltage was attributed both to the high temperatures (450–600°C) at which the process occurs and to the presence of iodide in a nonhydrated state. A remarkable depolarization is observed at temperatures higher than 600°C and is related to the anodic process and particularly to the desorption and diffusion of iodine from the carbon electrode.

Two anodic waves, the first of which is due to the oxidation of iodide to iodine, were found by Novik and Lyalikov [113, 114] in molten KNO_3-$NaNO_3$ at a rotating platinum electrode. The second wave, which occurs in the potential region where the oxidation of nitrite occurs, is higher than that observed in the same medium in the absence of iodide. The increase has been attributed by the authors to an increase of the nitrite concentration:

$$KNO_3 + 2KI \rightarrow I_2 + KNO_2 + K_2O$$

TABLE 2.2.13. Voltammetric Characteristics of Iodine (250°C) and Iodide (150°C) in Molten Salts at the Rotating Platinum Electrode

Substance	Product	Medium	Reference Electrode	Technique	Potential	Rev	Refs.
I^-	I_2	$(Na, K)NO_3$	Ag/Ag^+ 0.07 M	E swp	+0.14[a]	Rev	[27]
I^-	I_2	$(Na, K)NO_3$	Ag/Ag^+ 0.07 M	E swp	+0.2	Rev	[110]
I^-	I^+	$(Na, K)NO_3$	Ag/Ag^+ 0.07 M	E swp	+0.72	Irr	[27]
I^-	I^+	$(Na, K)NO_3$	Ag/Ag^+ 0.07 M	E swp	+0.75	Irr	[110]
I_2	I^-	$(Li, Na, K)NO_3$	Platinum	E swp	+1.17	Irr	[111]

[a]Potential value at $C_{I^-} = 1 \times 10^{-3}$ M.

The increase of the current of the second wave was attributed by Swofford and Propp [27] to the irreversible oxidation of iodine formed at the electrode. Swofford [115], using potentiometric measurements, has found that iodide is stable in this eutectic at 250°C.

As regards the first wave, the anodic limiting current is proportional to the iodide concentration, and its height does not change with time; this confirms the iodide stability. A study of this wave suggests that

$$2I^- \rightleftharpoons I_2 + 2e^-$$

is reversible. The shift of $E_{1/2}$ as a function of the iodide concentration is predicted by theory (0.052 V with every tenfold change in iodide concentration). The stability constant of I_3^- is ~250 moles/l.

The results obtained for the first wave have been confirmed by Fulton and Swofford [116] in a chronopotentiometric study.

The characteristics of the second wave have been closely examined by Fulton and Swofford [110]. With controlled potential electrolysis [+0.85 V vs Ag/Ag(I) 0.07 M reference electrode] the second step was ascertained to be a one-electron oxidation of iodine to I^+. However, I^+ is unstable in the melt and is reduced to iodine by nitrate:

$$I^+ + NO_3^- \rightarrow \frac{1}{2}I_2 + NO_2 + \frac{1}{2}O_2$$

The limiting current of the second wave is higher than that of the first one due to a contribution of a catalytic current, the current that is necessary to oxidize the iodine produced by the chemical reaction in the following cycle:

$$\frac{1}{2}I_2 \rightleftharpoons I^+ + e^-$$

$$I^+ + NO_3^- \rightleftharpoons \frac{1}{2}I_2 + NO_2 + \frac{1}{2}O_2$$

I^+ may also be reduced by iodide:

$$I^+ + I^- \rightleftharpoons I_2$$

This reaction becomes more important at high iodide concentrations but does not produce a catalytic current. In fact, at high iodide concentrations the catalytic cycle is diminished and the limiting current ratio (i_2/i_1) approaches unity.

3. KINETIC PARAMETERS AND DOUBLE-LAYER PROPERTIES

3.1. KINETIC PARAMETERS

3.1.1. Aqueous Solution

3.1.1.1. Reduction of Iodate. A wide study of the kinetic parameters of iodate reduction has been carried out by Delahay and Strassner [84]. The authors noticed that at a dropping mercury electrode the rate constant for this reaction changes with the overvoltage and the gelatin concentration. The free energy of activation and the quantity αn_a decrease with increasing the gelatin concentration until $C_{gelatin}$ = 0.01%; at higher concentrations there is practically no change in the values of ΔG_{act} and αn_a. These parameters do not vary with the iodate concentration; they are, on the contrary, greatly influenced by the pH of the supporting electrolyte as the data of Table 3.1.1 show.

TABLE 3.1.1. ΔG_{act} and αn_a for the Reduction of 1×10^{-3} M Potassium Iodate in the Presence of 0.01% Gelatin

pH of the solution	t (sec)	ΔG_{act} (kcal)	αn_a
0.94	3.61	34.6	1.13
1.91	3.59	25.3	0.59
4.06	3.57	25.4	0.53
6.03	3.24	21.4	0.28
7.82	2.99	23.5	0.31
9.69	2.75	33.1	0.61
11.6	2.78	32.7	0.63
13.7	2.62	31.0	0.61

TABLE 3.1.2. Kinetic Parameters for the Iodate Reduction in 0.05 M NaOH + 0.1 M Na_2SO_4 as Supporting Electrolytes

Electrode	Technique	Rate constant (cm/sec)	αn_a	Remarks, % gelatin	Refs.
DME	E swp	$k_c^\circ = 1.4 \times 10^{-16}$	0.77	0.002	[117]
Pt/Hg[a]	E swp	$k_c^\circ = 5.6 \times 10^{-12}$	0.44	0.002	[118]

[a]Platinum electrode, coated with a mercury film, with periodic renewal of the diffusion layer.

The values of αn_a are in good agreement with those found by De Mars and Shain [39] by means of voltammetry with linearly varying potential at a hanging mercury electrode. On the basis of these data Delahay and Strassner, supposing that the transfer coefficient did not appreciably vary as the concentration of the supporting electrolyte changes and that its value was approximately 0.3, deduced that the step determining the rate of the iodate reduction involves one electron in the pH range 6-8 and two electrons beyond pH 8 and between pH 2 and 6. This α value is lower than that found by Delahay and Mamantov [134] in the chronopotentiometric reduction of 4×10^{-3} M KIO_3 in 1 M KOH (0.5).

The kinetic parameters for iodate reduction in alkaline solution are reported in Table 3.1.2.

The rather high value of αn_a obtained at a dropping mercury electrode agrees with the results of Breiter and co-workers [141] in 0.02 N KOH (0.81) and confirms the assumption that in strongly alkaline solution the discharge step occurs through the exchange of two electrons [142].

At a platinum electrode coated with a mercury film under the same experimental conditions, a k_c° value higher than that found by Delahay and Mattax [117] was found by Roffia and co-workers [118]. From the potential dependence of log k_c, iodate reduction at a solid electrode was found to be more irreversible than at a dropping electrode. According to the authors, this difference may be explained by supposing that the adsorption equilibrium of gelatin is slow; for this reason the measurements were carried out under equilibrium adsorption conditions at the solid electrode where surface renewal is missing, in opposition to the nonequilibrium nature of the measurements at a dropping mercury electrode.

The influence of surface platinum oxide on the rate of the reduction of iodate and of iodic acid was studied by Davis [135] and Müller [119]. Davis found that the presence of platinum oxide greatly influences the quantity αn_a in an acetic buffer solution (pH = 4.7). In the absence of oxide (i.e., electrode reduced below ~ +0.25 V) there is an abrupt shift in αn_a from 0.78 to 0.35. The value of αn_a on the reduced platinum electrode is similar to the values found by other authors [39, 84] at a mercury electrode in the same pH range.

Müller [119] concluded that the IO_2—OH bond is catalytically broken or weakened in acid solution owing to the presence of platinum oxide. He pointed out [120] that platinum oxide accelerates the initial step of the electrochemical reduction of iodate and that such a reaction consists of the dissociative chemisorption of iodate:

$$IO_2{-}OH \rightleftharpoons IO_2^{\cdot}{}_{chem} + OH^{\cdot}_{chem} \quad (3.1.1)$$

In the potential region 1.1-0.75 V vs SCE the order of the reaction is 0.5. According to Müller the radical $IO_2^{\cdot}$ may react fast with water:

$$IO_2^{\cdot} + H_2O \rightleftharpoons IO_2^- + H^+ + OH^{\cdot} \quad (3.1.2)$$

Two $OH^{\cdot}$ radicals which are formed from an iodic acid molecule are successively reduced at the electrode

$$OH^{\cdot} + e^- \rightarrow OH^- \quad (3.1.3)$$

The IO_2^- formed in Reaction (3.1.2) is unstable and disproportionates at a high rate:

$$3IO_2^- \rightleftharpoons 2IO_3^- + I^- \quad (3.1.4)$$

The rate of the reaction of iodate with iodide at a rotating platinum electrode in $HClO_4$ was studied by Beran and Bruckenstein [121]; in order to start the catalytic reduction of iodate, a small amount of iodine was added to the iodate solution. ICl was added in HCl to start the catalytic process. A relation similar to that proposed by other authors [122, 123] was found in $HClO_4$:

$$dC_{I_2}/dt = k_1[H^+]^2[IO_3^-][I^-] + k_2[H^+]^2[IO_3^-][I^-]^2$$

where $k_1 = 4 \times 10^5\ l^4\ moles^{-4} sec^{-1}$ and $k_2 = 5.1 \times 10^9\ l^5 moles^{-5} sec^{-1}$.

The value of k_2 is in good agreement with the results of Myers and Kennedy [122], while the value of k_1 is higher. In HCl the relation is:

$$dC_{I_2}/dt = k[H^+]^3[IO_3^-][I^-][Cl^-]$$

where $k = 1.7 \times 10^9\ l^5 moles^{-5} sec^{-1}$.

3.1.1.2. The I^-/I_3^- System. The kinetic parameters of the I^-/I_3^- system can not be determined with polarization curves at stationary platinum electrodes since both the anodic and the cathodic processes are essentially diffusion controlled [12]. For this reason most studies are relative to ac methods (i.e., faradaic impedence method) [124, 125, 128, 130]; many authors [62, 126, 129], however, employ polarization curves at rotating platinum electrodes. In the latter case it is necessary to work at conditions such that the inconvenience found by Vetter [12] at a stationary platinum electrode may be removed as much as possible.

The kinetic parameters of the I^-/I_3^- system are gathered in Table 3.1.3. A comparison of the data obtained with the different techniques shows that the kinetic parameters do not agree with one another. Such differences are attributable to the remarkable variation of

TABLE 3.1.3. Kinetic Parameters for the I^-/I_3^- System

Conditions	Electrode	Technique	Rate constant	α	$1 - \alpha$	Remarks	Refs.
1 N H_2SO_4 + 1 × 10^{-3} M I_3^- + 0.1 M I^-	Pt	Farad imp	$j_0 = 0.4 \times 10^{-2}$ A/cm^2	-	0.78	-	[124]
0.5 M K_2SO_4 + 1 × 10^{-3} M I_3^- + 0.1 M I^-	Pt	Farad imp	$j_0 = 2.1 \times 10^{-2}$ A/cm^2	-	~0.6	-	[125]
1.24 × 10^{-3} M I_3^- + 0.1 M KI	Rot Pt dsk	E swp	$k_c = 2.63 \times 10^{-7}$ cm/sec	0.65	-	133.3 rpm	[62]
			$j_0 = \sim 4 \times 10^{-4}$ A/cm^2				[62]
0.1 M $HClO_4$ + 1 × 10^{-3} M I^- [e]	Stat Pt[a]	E swp	$k_a^\circ = 1.5 \times 10^{-2}$ cm^4/mole sec	-	0.83	-	[126]
1 M KI + 2.5 × 10^{-3} M I_2	Pt	i stp[b]	$j_0 = 2 \times 10^{-4}$ A/cm^2	-	-	0.01% citric acid	[127]
0.005 M H_2SO_4	Pt wr	Farad imp	$j_0^\circ = 0.4$ A/cm^2 [c]	-	~0.5	E_a = 15 kcal/mole[d]	[128]
0.5 M K_2SO_4 + 0.2 M I^- + 10^{-2} M I_2	Rot Pt wr	E swp	$j_0 = 1.9 \times 10^{-2}$ A/cm^2	1.2	0.4	-	[129]
2 M KCl + 0.5 M KI 2 × $10^{-3} \le C_{I_2} \le 6.5 \times 10^{-2}$ M	Pt dsk	Farad imp	$k_c = (1\text{–}5) \times 10^{-2}$ cm/sec	0.47	-	-	[130]

[a] A rotated electrolysis cell is used.

[b] Pulsed galvanostatic method.

[c] Standard exchange current density, i.e., at $C_{I^-} = C_{I_2} = 1$ mole/liter.

[d] This energy of activation pertains to the reversible potential $\epsilon_r = 0.574$ V.

[e] The kinetic parameters refer to the I^-/I_2 system.

TABLE 3.1.4. Kinetic Parameters of the Anodic and Cathodic Reactions in the KI/I_2 System on Pt at 25°C [129]

C_{I^-} (moles/l)	C_{I_2} (moles/l)	$C_{Na_2SO_4}$ (moles/l)	j° [a] (mA/cm^2)	$j^{\circ}_{a,c}$ [b] (mA/cm^2)	j°_a [c] (mA/cm^2)	j°_c [d] (mA/cm^2)	α_a [c]	α_c [d]
1.0	1×10^{-2}	-	30	29	65.6	21.4	0.4	1.3
0.5	1×10^{-2}	-	18.4	20.9	41.6	17.8	0.36	0.8
0.2	1×10^{-2}	-	9.1	12.4	21.4	6.17	0.3	1.3
5×10^{-2}	1×10^{-2}	-	2.9	4.3	5.5	2.9	0.4	0.7
0.2	5×10^{-2}	-	9.8	18.2	25.1	14.1	0.3	0.5
0.2	2×10^{-3}	-	7.1	6.8	14.0	5.0	0.4	1.1
0.2	1×10^{-2}	0.2	14.1	15	24.8	11.2	0.4	1.3
0.2	1×10^{-2}	0.5	19	22	39	17.4	0.4	1.2
0.2	1×10^{-2}	1.0	29	35	47	22	0.4	1.8
0.2	1×10^{-2}	2.0	36	42	59	25	0.6	1.5

[a] j° is determined from the slope of the straight line i_∞ vs η at low overvoltages (2-10 mV).

[b] $j^{\circ}_{a,c}$ is the exchange current determined from the point of intersection of anodic and cathodic partial log i vs η curves at $\eta = 0$.

[c] j°_a and α_a are determined from the linear portion of the anodic log i vs η curve.

[d] j°_c and α_c are determined from the linear portion of the cathodic log i vs η curve.

the kinetic parameters of the anodic and cathodic processes with the overvoltage and the bulk concentration of iodide and triiodide [62, 127, 129]. If a rotating electrode is used, the kinetic parameters also depend on the rotation rate of the electrode [62, 136].

The anomalies observed in the data obtained with the same technique (i.e., faradaic impedence method) under the same experimental conditions are due to the strong dependence of the kinetic parameters on the number of iodine atoms adsorbed at the electrode surface [127]. In order to compare such data, therefore, the degree of coverage of the electrodes must be exactly the same. Since the electrode coverage depends on the concentrations of iodide, triiodide, and possible foreign electrolytes, the dependence of the kinetic parameters on the bulk concentration of such ions can be explained.

In Table 3.1.4 the kinetic parameters determined at rotating platinum electrodes under different experimental conditions (i.e., for a coverage degree of the electrode included between 0 and 0.5) are reported [129]. The values of $\alpha_c + \alpha_a$ are included between 0.8 and 2.1. In general $\alpha_c + \alpha_a = \sim 2$ when $C_{I^-} \gg C_{I_2}$, while it approaches 1 as the iodide concentration approaches that of iodine.

TABLE 3.1.5. Kinetic Parameters of Electrode Reactions in the I^-/I_2 System at an Iridium Electrode [131]

C_{I^-} (N)	C_{I_2} (N)	α_a [a]	α_c [a]	j_a^o [b] (mA/cm^2)	j_c^o [b] (mA/cm^2)	j^o [c] (mA/cm^2)
1.0	1×10^{-2}	0.55	0.73	44	19	32
0.5	1×10^{-2}	0.52	0.70	38.8	17.2	28
0.2	1×10^{-2}	0.42	0.75	25.5	13.1	16.3
0.05	1×10^{-2}	0.52	0.75	6	4	4.7
0.2	5×10^{-2}	0.48	0.75	31	20	29
0.2	2×10^{-3}	0.44	0.70	18	13	8.9

[a] Values calculated from the anodic and cathodic log i vs η curves.

[b] Data obtained extrapolating the Tafel curves to $\eta = 0$.

[c] Exchange current determined over the range $\eta = \mp 10$ mV from the i, η curves.

An effect similar to that obtained by increasing the iodide concentration is observed by adding sodium sulfate.

The lack of coincidence of the anodic and cathodic exchange current and the nonattainment of the value of 1 for the sum $\alpha_c + \alpha_a$ indicate that the overall electrode reaction proceeds through an intermediate step or that the electrode processes proceed simultaneously according to several reactions [133]. The nature of the electrode also has a noticeable influence on the kinetic parameters since iodine adsorption, and therefore the degree of coverage, changes with the metal employed [131].

The kinetic parameters determined at an iridium electrode are reported in Table 3.1.5. The α_c values in this case do not exceed unity; for this reason the degree of coverage with electroactive iodine approaches zero at this electrode.

3.1.2. Nonaqueous Solution

3.1.2.1. Kinetic Parameters of the I^-/I_2 System. The kinetic parameters of the I^-/I_2 system have been determined with polarization curves at rotating platinum electrodes in dimethylsulfoxide [92, 101, 138] and acetonitrile [140]. The values are gathered in Tables 3.1.6 and 3.1.7, respectively.

In both solvents the transfer coefficients change very little with the bulk concentration of iodine and iodide. The anodic and cathodic exchange current densities, as expected, increase as the depolarizer concentration increases.

The diffusion coefficients of iodide, iodine, and triiodide in different solvents are reported in Table 3.1.8.

TABLE 3.1.6. Kinetic Parameters of the I^-/I_2 System in Dimethylsulfoxide

Reaction	Conditions	Electrode	Technique	Rate constant	α	$1-\alpha$	Remarks	Refs.
$I_3^- + 2e^- \rightarrow 3I^-$	0.8 $\underline{M}$ $NaClO_4$ + 2.9 × 10^{-3} $\underline{M}$ I_2	Rot Pt dsk	E swp	$j_c^\circ = 0.28$ mA/cm^2	0.27	-	$C_{I^-} \gg C_{I_2}$; t = 24.9°C	[138]
$I_3^- + 2e^- \rightarrow 3I^-$	0.4 $\underline{M}$ $NaClO_4$ + 0.905 × 10^{-2} $\underline{M}$ I_3^- + 0.0132 $\underline{M}$ NaI	Pt	i stp	$k_c = 1.13 \times 10^{-4}$ cm/sec	0.43	-	-	[92]
$I_2 + 2e^- \rightarrow 2I^-$	0.8 $\underline{M}$ $NaClO_4$ + 2.85 × 10^{-3} $\underline{M}$ I_2	Rot Pt dsk	E swp	$j_c^\circ = 0.22$ mA/cm^2	0.44	-	-	[101]
$3I^- \rightarrow I_3^- + 2e^-$	0.8 $\underline{M}$ $NaClO_4$ + 7.4 × 10^{-3} $\underline{M}$ I^-	Rot Pt dsk	E swp	$j_a^\circ = 0.38$ mA/cm^2	-	0.30	t = 27.2°C; E_a = 4 kcal/mole	[138]
$2I_3^- \rightarrow 3I_2 + 2e^-$	0.8 $\underline{M}$ $NaClO_4$ + 2.5 × 10^{-3} $\underline{M}$ I_3^-	Rot Pt dsk	E swp	$j_a^\circ = 0.16$ mA/cm^2	-	0.5	t = 27.2°C; E_a = 3 kcal/mole	[138]

TABLE 3.1.7. Kinetic Parameters of the I^-/I_2 System in Acetonitrile [140]. Iodine Concentration = 1.51×10^{-2} M

Reaction	Conditions	Electrode	Technique	$j^\circ \times 10^{-3}$ (A/cm^2)	α	$1-\alpha$	Activation energy (kcal/mole)
$3I_2 + 2e^- \rightarrow 2I_3^-$	0.4 M $NaClO_4$ + 8.1×10^{-3} M I^-	Rot Pt dsk	E swp	$j_c^\circ = 1.05$	0.51	-	4.41
$I_3^- + 2e^- \rightarrow 3I^-$	0.4 M $NaClO_4$ + 2.5×10^{-2} M I^-	Rot Pt dsk	E swp	$j_c^\circ = 0.83$	0.29	-	4.6
$3I^- \rightarrow I_3^- + 2e^-$	0.4 M $NaClO_4$ + 2.5×10^{-2} M I^-	Rot Pt dsk	E swp	$j_a^\circ = 0.84$	-	0.42	5.2
$2I_3^- \rightarrow 3I_2 + 2e^-$	0.4 M $NaClO_4$ + 8.1×10^{-3} M I^-	Rot Pt dsk	E swp	$j_a^\circ = 1.32$	-	0.52	4.86

TABLE 3.1.8. Diffusion Coefficients of Iodide, Iodine, and Triiodide in Different Solvents

Solvent	Conditions	$D \times 10^5$ (cm^2/sec) I^-	I_3^-	I_2	Refs.
Nitromethane	0.1 M $LiClO_4$	-	1.6	-	[90]
Acetonitrile	0.1 M $NaClO_4$	-	3.05	-	[90]
	0.4 M $NaClO_4$	1.68	1.91	1.55	[140]
Dimethylformamide	0.5 M $NaNO_3$	0.8	0.6	0.6	[137]
Dimethylsulfoxide	0.8 M $NaClO_4$	0.37	0.29	0.37	[101]
	0.8 M $KClO_4$	0.69	0.37	-	[138], [92]

3.2. IODIDE AND IODINE ADSORPTION AND ITS INFLUENCE ON THE DOUBLE-LAYER PROPERTIES

Iodide adsorption at a dropping mercury electrode and its influence on the double-layer properties is included in the more general field of anion adsorption at such electrodes, and

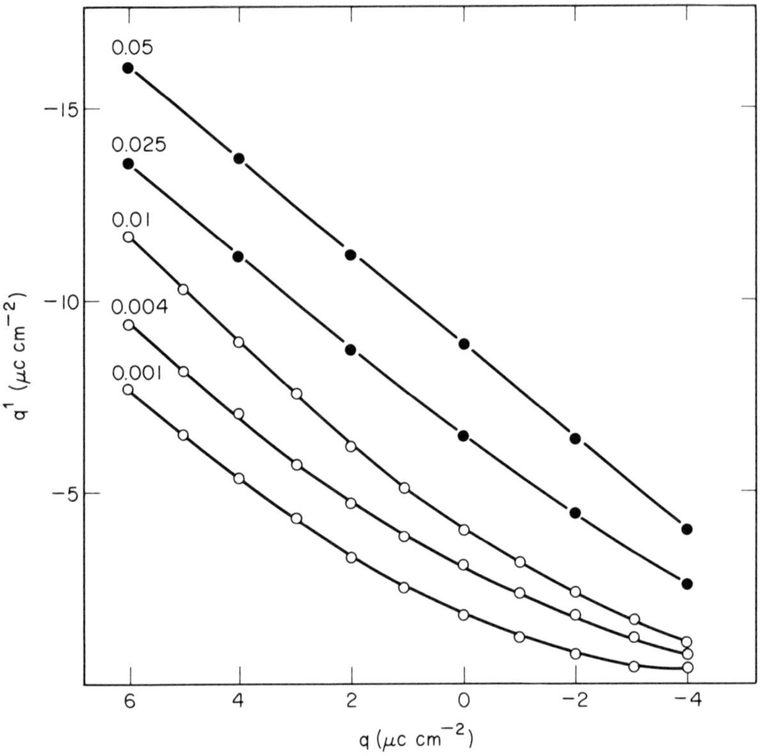

FIG. 3.2.1. Amount of specifically adsorbed iodide vs charge density on the electrode for different KI concentrations: o [145]; • [143, 144].

for this reason it is beyond our treatment. For a discussion of this problem, see the extensive study of Devanathan and Canagaratna [166].

It is important to recall, however, that from measurements of the zero-charge potential and of the double-layer capacity the quantity of adsorbed iodide may be estimated. Such studies have been carried out by many authors [143-145] and the data obtained are reported in Fig. 3.2.1.

In the case of solid electrodes, the adsorption of iodide and iodine has a greater importance owing to its influence on the mechanism and the kinetics of the electrochemical reaction concerning the iodide/iodine couple. On these electrodes, on which the accurate determination of the zero-charge point is very difficult, adsorption measurements are generally carried out with techniques different from those employed in the case of the dropping mercury electrode.

The influence of the iodide adsorption on the zero-charge potential at a platinized platinum electrode has been studied by Balashova and co-workers [146]. The zero-charge point in an acid solution of sodium iodide falls at ~-0.5 V/NHE; that is, about 0.7 V more negative than in an acid solution of sodium sulfate. The influence of pH on the zero-charge point is very small since in an alkaline solution of sodium iodide [147] no appreciable difference is observed with respect to the value found in acid solution.

Iodide adsorption at platinum electrodes has been investigated by several authors [148, 149]. Osteryoung and co-workers [151], by means of integration of single-sweep oscillopolarograms, concluded that iodine was the prevalent adsorbed species with respect to iodide and triiodide at a platinum electrode. These results, however, were later corrected by Kazarinov and Balashova [150] and by the same authors [152, 153]; with radiotracers measurements iodide, rather than iodine, was found to be adsorbed to a greater extent. Between the two adsorbed species, however, only part of the iodine is still electroactive. The amount of electroinactive adsorbed iodide or iodine is about 2×10^{-9} moles/cm^2; that of electroactive iodine is 1×10^{-9} moles/cm^2. Such values agree with those found by Schuldiner and Presbrey [127]; these authors claimed that every iodine atom is bonded with two platinum atoms.

The presence of iodine adsorbed on the surface of platinum electrodes, dipped in iodide solutions, has been pointed out by Toth [154]. The adsorption is small at platinum treated with aqua regia, or heated in an oxidizing flame, or in the presence of concentrated HNO_3. On the other hand, at platinum cathodically prepolarized or saturated with hydrogen, the adsorption is high and saturation is reached in 20-30 min and corresponds to 0.7 μg/cm^2 of adsorbed iodine. The pH also has a noticeable influence on the adsorption, i.e., it is highest at pH < 3 and progressively decreases as the pH increases. At pH ≈ 8, in fact, the amount of adsorbed iodine is only 20% of the amount at pH < 3. The same author [155] later proposed the following scheme for the adsorption of iodine at a preoxidized platinum electrode:

$$Pt_2{-}Pt_1{-}O + 2H^+ + 2I^- \rightarrow [Pt_1IX_n]^{4-(n+1)} + H_2O + I + Pt_2$$

$$Pt_2 + I \rightarrow Pt_2\text{-}I$$

where Pt_2 is an inner layer of the electrode, Pt_1 an outer one, and $[PtIX_n]^{4-(n+1)}$ represents platinum-iodide complexes.

Comparing the value of the surface of the preoxidized electrode with the amount of adsorbed iodine, the author deduced that saturation corresponds to the formation of a monomolecular iodine layer and that each iodine atom is bonded to one platinum atom. At a prereduced electrode the oxidation of iodide to iodine and the adsorption of iodine occur according to:

$$Pt(H) + H^+ + I^- \rightarrow Pt(I) + H_2$$

The amount of adsorbed iodine is 15-20% lower than at preoxidized electrodes.

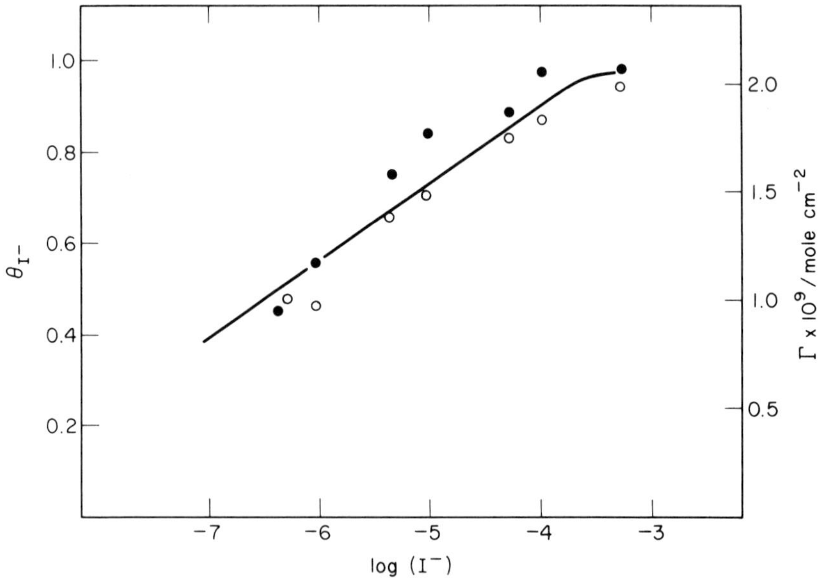

FIG. 3.2.2. Adsorption isotherm of iodide on smooth platinum at φ_t = 0.1 V measured from (o) a decrease in oxygen adsorption and (•) a decrease in hydrogen adsorption [156].

An extensive study on iodide adsorption on smooth platinum electrodes has been carried out by Bagotzky and co-workers [156]. They found that 90% of the surface is covered and that the amount of adsorbed iodide is the same as that reported by Osteryoung and Anson [152]; such a value, however, may also be reached with iodide concentrations smaller than 1×10^{-7} M (instead of 1×10^{-4} M) if the electrode potential is shifted from 0.1 to 0.3 V. The adsorption isotherm of iodide in 0.5 M H_2SO_4 at a potential of 0.1 V is reported in Fig. 3.2.2. The results show that iodide adsorption follows the Temkin isotherm:

$$\theta_{I^-} = A(\varphi) + \frac{1}{f'} \ln C_{I^-}$$

where $A(\varphi)$ is a function of the potential and the adsorptivity f' is 13.8. The adsorption of iodide is time dependent and is controlled by the adsorption rate rather than by diffusion.

The kinetic isotherm of iodide (θ_{I^-} vs ln τ) is linear to a first approximation and may be described by the following relation:

$$\theta_{I^-} = B + \frac{1}{\alpha f'} \ln \tau$$

The quantity $\alpha f'$ is 6.35.

Iodide adsorption depends mostly on the electrode surface inhomogeneity and only slightly on the electrostatic repulsion forces. In fact, the value f' = 13.8 for iodide is equal to that found for the adsorption of hydrogen or of neutral particles.

In the presence of anodic iodine films which have a high electric resistance, the oxidation current of iodide shows kinetic characteristics owing to the dissolution rate of iodine into iodide to give triiodide. The rate constant of the dissolution rate of the iodine film determined in 1×10^{-2} M KI solutions is 6.1×10^{-3} cm/sec [157].

Iodide adsorption as a function of pH has been studied on silver electrodes by Medvedeva and Kolotyrkin [158] with polarization and radiochemical measurements. The extent of adsorption increases when the electrode potential is shifted to more positive values and decreases sharply in alkaline solutions due to the adsorption of OH^- ions.

Finally, we include the study of the adsorption of iodide on iron electrodes [159], that on graphite macroelectrodes in the temperature range 85-400°C [160], and that of iodine adsorption on active carbon electrodes in fused KI + AgI [161].

3.3. EFFECT OF THE DOUBLE LAYER ON REACTION KINETICS

Frumkin and his school [162, 163] have demonstrated that the overall rates of electrode reactions are strongly dependent on double-layer properties. With regard to iodine compounds, the influence of the double-layer properties has been pointed out and discussed in the case of iodate reduction at a dropping mercury electrode. Several authors [32, 33, 38] had already observed that the overvoltage of the reduction of iodate depends not only on the current density, but also on the pH value and on the cationic charge of the supporting electrolyte (see Section 2.1). Other authors [35, 117, 164] later pointed out that, in strongly alkaline solution (where the discharge rate is pH independent), the rate of the electrode reaction depends on the concentration and on the kind of supporting electrolyte. In double-layer terms, the observed salt effect can be accounted for by the resulting changes in the potential of the outer Helmholtz plane.

Aramata and Delahay [165] noticed that the overvoltage for iodate reduction decreases, changing the supporting electrolyte cation from Li^+ to Cs^+, and that the Li^+ ion was able to cause anomalous overvoltages; this occurrence was explained by postulating that supporting electrolyte ions are able to affect the hydration shell of iodate and render the reduction easier.

4. ELECTROCHEMICAL STUDIES

4.1. MECHANISM OF THE ELECTRODE REACTIONS FOR THE IODIDE/IODINE SYSTEM

4.1.1. Aqueous Solution

The mechanism of the electrochemical reaction of the iodide/triiodide couple at a platinum electrode has been studied by many authors [62, 124, 126, 129, 136, 152, 153, 167, 169].

TABLE 4.1.1. Order of the Anodic and Cathodic Reactions for the Iodide/Iodine System

Anodic			Cathodic			Refs.
I^-	I_2	I_3^-	I^-	I_2	I_3^-	
1	0	–	–1	1	–	[129]
1	–	–	–	–	1	[136]
–	–	–	–	–	0.5	[62]
2	–	–	–	1	–	[126]
1	–	0	–0.5	–	0.5	[124]

In order to have a complete picture of such studies the reaction orders relative to the oxidation of iodide and to the reduction of iodine or triiodide are reported in Table 4.1.1.

Vetter [124], from the determination of the dependence of the exchange current density on the iodide and triiodide concentration, concluded that the anodic reaction order was 1 with respect to iodide, while the cathodic order was 0.5 with respect to triiodide. On the basis of such results the following mechanism has been proposed:

$2(I^- \rightleftharpoons I + e^-)$ slow charge-transfer step   (4.1.1)

$2I \rightleftharpoons I_2$ chemical reactions (here in equilibrium)   (4.1.2)

$I_2 + I^- \rightleftharpoons I_3^-$ (4.1.3)

It is not clear, however, if Reaction (4.1.2), and consequently Reaction (4.1.1), take place among iodine atoms adsorbed at the electrode surface or occur homogeneously.

Newson and Riddiford [62] studied the electroreduction of triiodide at platinum electrodes and confirmed the mechanism proposed by Vetter [124]. In the proposed mechanism, however, the iodine atoms involved in Reactions (4.1.1) and (4.1.2) are considered to be adsorbed at the electrode surface:

$I^- + S \underset{}{\overset{slow}{\rightleftharpoons}} I\text{-}S + e^-$

$I\text{-}S + I\text{-}S \underset{}{\overset{fast}{\rightleftharpoons}} I_2 - 2S$

$I_2 + I^- \underset{}{\overset{fast}{\rightleftharpoons}} I_3^-$

where S is an active center on the electrode surface.

Iodide oxidation in the absence of iodine was studied by Jordan and Javick [126, 167] with hydrodynamic voltammetry. The process was found to be controlled by a second-order reaction with respect to iodide. This result was explained by

$$2I^- + S \rightarrow \left[S \begin{smallmatrix} \cdot^{e^-} \cdots I \\ | \\ \cdot_{e^-} \cdots I \end{smallmatrix} \right]^{2-} \rightarrow S(2e^-) + I_2$$

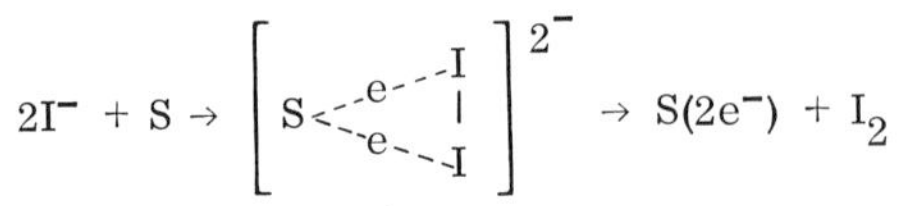

The disagreement with the preceding data was explained by considering that the previous researches had been carried out with iodine solution in the presence of large amounts of iodide where, therefore, the triiodide ion was the prevailing species.

Osteryoung [170] observed that the exchange currents at a platinum electrode in iodide/iodine solutions were greater than the apparent rates of the electron exchange between adsorbed iodine (or iodide) and dissolved iodine (or iodide). Osteryoung and Anson [152] later demonstrated that iodine was adsorbed on a surface covered with electroinactive iodide ions which are irreversibly adsorbed; for this reason both the second reaction proposed by Newson and Riddiford [62] and the other mechanisms based on the assumption that the charge-transfer occurs among adsorbed species [126, 167, 168] can be criticized.

Thomas and Brodd [125] explained their results by the mechanism proposed by Vetter with slow electron transfer to an iodine atom as the rate-determining step. With regard to the possibility that the reaction may occur at the platinum surface, they decided that even though the adsorbed iodide and iodine in the first layer do not exchange with the solution species [152, 171], it is still possible to have electron transfer through the very thin adsorbed layer.

As a matter of fact the interpretation of transients given by Osteryoung and co-workers [152, 153] is far from being reliable, since it does not take into account the charge separation and recombination that may take place at an interphase without the flow of external current [169].

More recently Dané and co-workers [128] have considered the possible reactions into which the electrode process may be divided:

$$I^- \rightleftharpoons I_{ads} + e^- \quad \text{(Volmer reaction)} \tag{4.1.4}$$

$$I_{ads} + I^- \rightleftharpoons I_2 + e^- \quad \text{(Heyrovsky reaction)} \tag{4.1.5}$$

$$2I_{ads} \rightleftharpoons I_2 \quad \text{(Tafel reaction)} \tag{4.1.6}$$

$$I_2 + I^- \rightleftharpoons I_3^- \quad \text{(complex reaction)} \tag{4.1.7}$$

They demonstrated that the process consists of Reactions (4.1.4), (4.1.5), and (4.1.7), and that the exchange step is determined by Reaction (4.1.5). This scheme has been confirmed by Povarov and co-workers [129]. The authors, however, assume that Reactions (4.1.4) and (4.1.5) proceed at commensurate rates.

On the basis of the results reported, the different authors prefer an electrode process which occurs at the platinum surface. There is considerable disagreement, however, with regard to the rate-determining step of such processes.

4.1.2. Nonaqueous Solution

The mechanism of the iodide/iodine couple in nonaqueous solution has been studied by Arvia and co-workers [101, 138, 140]. The order of the anodic (oxidation of I^- and I_3^-)

and cathodic (reduction of I_2 and I_3^-) reactions in acetonitrile [140] and dimethylsulfoxide [101, 138] is 1 with respect to I^- and I_3^- and to I_2 and I_3^-, respectively. The mechanism proposed for the electrochemical processes for the two anodic and cathodic waves in both solvents are the following:

1st anodic wave

$I^- + S \rightleftharpoons I\text{-}S + e^-$ slow

$I^- + I\text{-}S \rightleftharpoons I_2 + S + e^-$

$(I_2 + I^- \rightleftharpoons I_3^-)$

2nd anodic wave

$2(I_3^- \rightleftharpoons I_2 + I^-)$

$I^- + S \rightleftharpoons I\text{-}S + e^-$

$I\text{-}S + I^- \rightleftharpoons I_2 + S + e^-$ slow

1st cathodic wave

$I_2 + S + e^- \rightleftharpoons I\text{-}S + I^-$ slow

$I\text{-}S + e^- \rightleftharpoons I^- + S$

$2(I_2 + I^- \rightleftharpoons I_3^-)$

2nd cathodic wave

$(I_3^- \rightleftharpoons I_2 + I^-)$

$2I_2 + S + e^- \rightleftharpoons I\text{-}S + I^-$ slow

$I\text{-}S + e^- \rightleftharpoons I^- + S$

5. APPLIED ELECTROCHEMISTRY

5.1. ELECTROSYNTHESIS OF PERIODATE

Periodate is very important in some industrial processes in which it is employed as an oxidizing agent. Since in oxidation processes it is reduced to iodate, the oxidation of iodate to periodate has a practical interest. The reaction is carried out on lead dioxide electrodes [172]; only on such electrodes, in fact, is the electrochemical oxidation of iodate fast. This occurrence has been explained by several authors [173, 174].

Allen [173] assumed that the reaction was catalyzed by the formation of "active" oxygen on the electrode, while Hickling and Richards [174] postulated a mechanism of several steps wherein hydrogen peroxide and lead oxides enter into the oxidation process.

Lancaster and Conway [175] derived a mathematical equation which takes the experimental conditions into account. The purpose of their investigation was not to elucidate the actual mechanism but to find a mathematical expression which could be used over a reasonable range of experimental conditions.

6. ASTATINE

6.1. ELECTROCHEMICAL PROPERTIES OF ASTATINE

Astatine is a synthetic unstable element and can be worked with only at concentrations of between 10^{-11} and 10^{-15} M by tracer scale experiments. Though the more stable isotope is ^{210}At (8.3 h half-life) [176], ^{211}At (7.5 [176] or 7.2 [177, 132] h half-life) is generally employed in the study of chemical properties. Astatine-211 is prepared by the bombardment of bismuth with α-particles with energy between 21 and 29 MeV. Johnson and co-workers [176] observed that when astatine is dissolved in sulfuric acid and electrolyzed, no deposition at the cathode is observed until dichromate ion is added to the cell.

The critical deposition potentials of astatine from various solutions are reported in Table 6.1.1.

The following oxidation states of astatine have been characterized: -1, 0, +1, and +5. No evidence has been found for a +7 state.

On the basis of the potential of the redox couples employed by Johnson and co-workers [176] in order to change from one oxidation state of astatine to another, Latimer [1] constructed the following potential diagrams (V vs NHE):

$$\text{Acidic solution:}\quad At^- \xrightarrow{+0.2} At_2 \xrightarrow{+0.7} HAtO \xrightarrow{+1.4} HAtO_3$$

$$\text{Basic solution:}\quad At^- \xrightarrow{+0.2} At_2 \xrightarrow{0.0} AtO^- \xrightarrow{+0.5} AtO_3^-$$

Appleman [177] constructed the following potential scheme in 0.1 M acid with suitably chosen redox couples

$$At^- \xrightarrow{+0.3} At_2 \xrightarrow{+1.0} HAtO \xrightarrow{+1.5} HAtO_3$$

At^-, in fact, is oxidized to At_2 by the iodide/iodine couple (+0.60 V) and, at pH < 4, by the As(III)/As(V) couple (emf > +0.30). At_2 is oxidized to At^+ by the bromide/bromine couple (+1.04 V) even at very low bromine concentrations, while At^+ is oxidized to AtO_3^- by the iodate/periodate couple (~+1.6 V). The astatate, furthermore, is reduced to At^+ by the chloride/chlorine couple (+1.40 V), while At^+ is reduced to At_2 in the dark by the VO^{2+}/VO_2^+ couple (+0.95 V). In light, however, the vanadium couple oxidizes At_2 to At^+. Finally, At_2 is reduced to At^- at pH > 4 by the As(III)/As(V) couple (emf < +0.3 V).

In Table 6.1.2 the astatine potentials are compared with those of other halogens. The astatine values are in agreement with the trends exhibited by the lighter members of the series, although astatate is a stronger oxidizing agent than the corresponding halates.

TABLE 6.1.1. Critical Deposition Potentials of Astatine Solutions [176]

Solution	Astatine concentration ($\underline{M} \times 10^{-3}$)	Potential, V vs NHE
0.066 $\underline{M}$ HNO_3	2.8	1.225
1.0 $\underline{M}$ HNO_3	0.5	1.24
0.075 $\underline{M}$ H_2SO_4 + 0.1 $\underline{M}$ $Na_2Cr_2O_7$	6	1.20
0.066 $\underline{M}$ HNO_3 + 3 mg Au	1	1.22
0.066 $\underline{M}$ HNO_3	0.4	1.22

TABLE 6.1.2. Comparison of Halogen Potentials in 0.1 $\underline{M}$ Acid (vs NHE)

Halogen	$X^-/X_2(aq)$	$X_2(aq)/HXO$	HXO/HXO_3
Cl	+1.40	+1.53	+1.35
Br	+1.09	+1.51	+1.42
I	+0.62	+1.31	+1.07
At	+0.3	+1.0	+1.5

REFERENCES

1. W. M. Latimer, Oxidation Potentials, 2nd ed., Prentice-Hall, New York, 1952, p. 64.
2. G. Charlot, Oxidation-Reduction Potentials, Pergamon, London, 1958.
3. M. Pourbaix, Atlas d'équilibries electrochemiques à 25°C, Gautier Villars, Paris, 1963.
4. R. G. Bates and W. C. Vorsburgher, J. Amer. Chem. Soc., 59, 1188 (1937).
5. J. N. Pearce and A. R. Fortsch, ibid., 45, 2852 (1923).
6. G. Jones and B. B. Kaplan, ibid., 50, 2066 (1928).
7. H. D. Murray, J. Chem. Soc., 127, 882 (1925).

8. G. Jones and W. C. Schumb, Proc. Amer. Acad. Arts Sci., 56, 199 (1921).
9. A. MacKeown, Trans. Faraday Soc., 17, 517 (1922).
10. W. Maitland, Z. Elektrochem., 12, 263 (1906).
11. A. L. Rotinyan and I. I. Appenin, J. Chem. Gen. USSR, 10, 1524 (1940).
12. K. J. Vetter, Z. Phys. Chem., 199, 22 (1952).
13. G. N. Lewis and M. Randall, Thermodynamics, McGraw-Hill, New York, 1923, p. 526.
14. L. Kovach, Z. Phys. Chem., 80, 107 (1910).
15. D. F. Bowersox, E. A. Butler, and E. H. Swift, Anal. Chem., 28, 221 (1956).
16. R. Gauguin, Bull. Soc. Chim. Fr., 15, 1052 (1948).
17. V. A. Pleskov, J. Phys. Chem. USSR, 22, 351 (1948).
18. J. Desbarres, Bull. Soc. Chim. Fr., 1961, 502.
19. M. G. C. Marchon, ibid., 1968, 1123.
20. I. V. Nelson and R. T. Iwamoto, J. Electroanal. Chem., 7, 218 (1964).
21. J. K. Gorman, Univ. Michigan 58-7254; Dissertation Abstr., 19, 1930 (1959).
22. H. Breant and C. Sinicki, Bull. Soc. Chim. Fr., 1965, 5016.
23. V. Plichon, J. Badoz-Lambling, and G. Charlot, ibid., 1964, 287.
24. G. Durand and B. Trémillon, Anal. Chim. Acta, 49, 135 (1970).
25. R. Guidelli and G. Pezzatini, Collect. Czech. Chem. Commun., 36, 855 (1971).
26. G. A. Sacchetto, G. G. Bombi, and M. Fiorani, J. Electroanal. Chem., 20, 89 (1969).
27. H. S. Swofford, Jr. and J. H. Propp, Anal. Chem., 37, 974 (1965).
28. R. H. Coe and L. B. Rogers, J. Amer. Chem. Soc., 70, 3276 (1948).
29. P. Souchay, Anal. Chim. Acta, 2, 17 (1948).
30. T. A. Kryukova, Dokl. Akad. Nauk. SSSR, 65, 517 (1949).
31. G. M. Florianovich and N. Frumkin, Zh. Fiz. Khim., 29, 1827 (1955).
32. A. Rylich, Collect. Czech. Chem. Commun., 7, 288 (1935).
33. E. F. Orlemann and I. M. Kolthoff, J. Amer. Chem. Soc., 64, 1044 (1942).
34. V. Cernak, Collect. Czech. Chem. Commun., 21, 1344 (1956).
35. V. I. Zykov and S. I. Zhdanov, Zh. Fiz. Khim., 32, 644 (1958).
36. V. I. Zykov and S. I. Zhdanov, ibid., 32, 791 (1958).
37. M. L. Bernard, A. Barbet, J. Brègeon, and M. Lezay, Bull. Soc. Chim. Fr., 1965, 6101.
38. J. L. Lingane and I. M. Kolthoff, J. Amer. Chem. Soc., 61, 825 (1939).
39. R. D. De Mars and I. I. Shain, ibid., 81, 2654 (1959).
40. I. M. Kolthoff and C. S. Miller, ibid., 63, 2732 (1941).
41. I. M. Kolthoff and Y. Okinaka, ibid., 83, 47 (1961).
42. Y. Takemori and I. Tachi, Bull. Chem. Soc. Japan, 28, 151 (1955).
43. B. Breyer and S. Hacobian, Aust. J. Sci. Res., A4, 610 (1951).
44. T. Biegler, J. Electroanal. Chem., 6, 373 (1963).
45. J. P. Perchard, M. Buvet, and R. Molina, ibid., 14, 57 (1967).
46. W. J. Zimmerman and W. M. Layton, J. Biol. Chem., 181, 141 (1949).

47. I. M. Kolthoff and J. F. Coetzee, J. Amer. Chem. Soc., 79, 1952 (1957).
48. P. H. Given and N. E. Peover, J. Chem. Soc., 1959, 1602.
49. J. Matsui, J. Kurosaki, and J. Date, Bull. Chem. Soc. Japan, 43, 1707 (1970).
50. J. F. Coetzee and J. L. Hedrick, J. Amer. Chem. Soc., 67, 221 (1963).
51. J. F. Coetzee and W. S. Siao, Inorg. Chem., 2, 14 (1963).
52. I. M. Kolthoff and J. Jordan, J. Amer. Chem. Soc., 75, 1571 (1953).
53. S. E. S. El Wakkad, S. E. Khalafalla, and A. M. Shams El Din, Rec. Trav. Chim. Pays-Bas, 76, 30 (1957).
54. G. Raspi, F. Pergola, and D. Cozzi, J. Electroanal. Chem., 15, 35 (1967).
55. F. J. Miller and H. E. Zittel, ibid., 11, 85 (1966).
56. G. Dryhurst and P. J. Elving, Anal. Chem., 39, 607 (1967).
57. W. Geissler, R. Nitzche, and R. Landsberg, Electrochim. Acta, 11, 389 (1966).
58. H. E. Zittel and F. J. Miller, J. Electroanal, Chem., 13, 193 (1967).
59. P. Beran and S. Bruckenstein, Anal. Chem., 40, 1045 (1968).
60. F. C. Anson and J. J. Lingane, J. Amer. Chem. Soc., 79, 1015 (1957).
61. V. A. Zakharov and O. A. Songina, Zh. Fiz. Khim., 36, 1226 (1962).
62. J. D. Newson and A. C. Riddiford, J. Electrochem. Soc., 108, 699 (1961).
63. F. C. Anson and J. J. Lingane, J. Amer. Chem. Soc., 79, 4901 (1957).
64. E. C. Toren and C. P. Driscoll, Anal. Chem., 38, 872 (1966).
65. R. Guidelli and G. Piccardi, Electrochim. Acta, 12, 1085 (1967).
66. G. Dryhurst and P. J. Elving, J. Electroanal. Chem., 12, 416 (1966).
67. T. Berzins and P. Delahay, J. Amer. Chem. Soc., 75, 4205 (1953).
68. D. H. Evans, J. Electroanal. Chem., 6, 419 (1963).
69. R. Landsberg and R. Nitzche, Z. Chem., 6, 100 (1964).
70. R. Landsberg, R. Nitzche, and W. Geissler, Electrochim. Acta, 11, 495 (1966).
71. G. Piccardi and R. Guidelli, J. Phys. Chem., 72, 2782 (1968).
72. R. Guidelli and F. Pergola, J. Inorg, Nucl. Chem., 31, 1373 (1969).
73. E. Morgan, Thesis, University of Washington, 1956.
74. A. L. Beilby and A. L. Crittenden, J. Phys. Chem., 64, 177 (1960).
75. S. V. Gorbachev and Yu. A. Korostelin, Zh. Fiz. Khim., 39, 1469 (1965).
76. P. G. Desideri, L. Lepri, and D. Heimler, Unpublished Data.
77. I. Shain and A. L. Crittenden, Anal. Chem., 26, 281 (1954).
78. F. C. Anson, J. Amer. Chem. Soc., 81, 1554 (1959).
79. J. Badoz-Lambling and C. Guillaume, Proceedings of the Second International Polarographic Congress, Cambridge, 1959.
80. P. G. Desideri, J. Electroanal. Chem., 9, 218 (1965).
81. P. G. Desideri, ibid., 9, 229 (1965).
82. P. G. Desideri, ibid., 17, 229 (1968).
83. R. Guidelli, Anal. Chem., 43, 1715 (1971).
84. P. Delahay and J. E. Strassner, J. Amer. Chem. Soc., 73, 5219 (1951).
85. D. D. De Ford and D. N. Hume, ibid., 73, 5321 (1951).

86. J. D. Newson and A. C. Riddiford, J. Electrochem. Soc., 108, 695 (1961).

87. I. E. Barbasheva, Yu. M. Povarov, and P. D. Lukovtsev, Elektrokhimiya, 6, 175 (1970).

88. R. T. Iwamoto, Anal. Chem., 31, 955 (1959).

89. A. I. Popov and D. H. Geske, J. Amer. Chem. Soc., 80, 1340 (1958).

90. J. D. Voorhies and E. J. Schurdak, Anal. Chem., 34, 939 (1962).

91. H. E. Zittel and F. J. Miller, Anal. Chim. Acta, 37, 141 (1967).

92. M. C. Giordano, J. C. Bazán, and A. J. Arvia, Electrochim. Acta, 12, 723 (1967).

93. I. E. Barbasheva, Yu. M. Povarov, and P. D. Lukovtsev, Elektrokhimiya, 3, 1149 (1967).

94. J. Perichon and R. Buvet, Bull. Soc. Chim. Fr., 1967, 3707.

95. J. Badoz-Lambling and M. Herlem, in Polarography 1964 (G. J. Hills, ed.), Macmillan, London, 1966, 1113.

96. R. Guidelli and G. Piccardi, Anal. Lett., 1, 779 (1968).

97. A. J. Parker, J. Chem. Soc., 1966, 220.

98. Yu. M. Povarov and I. E. Barbasheva, Elektrokhimiya, 3, 745 (1967).

99. G. Jones and B. B. Kaplan, J. Amer. Chem. Soc., 50, 1600 (1928).

100. V. Plichon, Bull. Soc. Chim. Fr., 1964, 287.

101. A. J. Arvia, M. C. Giordano, and J. J. Podestà, Electrochim. Acta, 14, 389 (1969).

102. M. C. Giordano, J. C. Bazán, and A. J. Arvia, ibid., 11, 741 (1966).

103. G. Adhami and M. Herlem, J. Electroanal. Chem., 26, 363 (1970).

104. R. Guidelli and G. Piccardi, Anal. Chem., 43, 1639 (1971).

105. R. Guidelli and G. Piccardi, ibid., 43, 1646 (1971).

106. A. I. Popov and D. H. Geske, J. Amer. Chem. Soc., 80, 5346 (1958).

107. H. S. Swofford, Jr. and C. L. Holifield, Anal. Chem., 37, 1513 (1965).

108. M. Francini and S. Martini, Electrochim. Metal., 3(2), 136 (1968).

109. M. Francini, S. Martini, and C. Monfrini, Electrochim. Metal., 2(3), 325 (1967).

110. R. B. Fulton and H. S. Swofford, Jr., Anal. Chem., 41, 2027 (1969).

111. Yu. K. Delimarsky and G. V. Shilina, Electrochim. Acta, 10, 973 (1965).

112. I. G. Mugulescu, S. Sternberg, L. Medintev, and C. Mustetea, ibid., 8, 65 (1963).

113. R. M. Novik and Yu. S. Lyalikov, Polyarografiya Rasplaven Solci, Akad. Nauk. Ukr. SSSR Obshch. i Neorgam. Khim., 1962, 41; Chem. Abstr., 57, 11857h (1962).

114. R. M. Novik and Yu. S. Lyalikov, Zh. Anal. Khim., 13, 691 (1958).

115. H. S. Swofford, Jr., Anal. Chem., 37, 610 (1965).

116. R. B. Fulton and H. S. Swofford, Jr., ibid., 40, 1373 (1968).

117. P. Delahay and C. C. Mattax, J. Amer. Chem. Soc., 76, 5314 (1954).

118. S. Roffia, F. Busi, and M. Lavacchielli, Gazz. Chim. Ital., 98, 1343 (1968).

119. L. Müller, J. Electroanal. Chem., 16, 531 (1968).

120. L. Müller, ibid., 16, 67 (1968).

121. P. Beran and S. Bruckenstein, J. Phys. Chem., 72, 3630 (1968).

122. O. E. Myers and J. W. Kennedy, J. Amer. Chem. Soc., 72, 897 (1950).

123. K. J. Morgan, M. G. Peard, and C. F. Cullis, J. Chem. Soc., 1951, 1965.

124. K. J. Vetter, Z. Phys. Chem., 199, 285 (1952).

125. A. B. Thomas and R. J. Brodd, J. Phys. Chem., 68, 3363 (1964).

126. J. Jordan and R. A. Javick, Electrochim. Acta, 6, 23 (1962).

127. S. Schuldiner and C. M. Presbrey, Jr., J. Electrochem. Soc., 111, 457 (1964).

128. L. M. Dané, L. J. J. Janssen, and J. G. Hoogland, Electrochim. Acta, 13, 507 (1968).

129. Yu. M. Povarov, I. E. Barbasheva, and P. D. Lukovtsev, Electrokhimiya, 3, 306 (1970).

130. V. A. Tyagai and G. Ya. Kolbasov, ibid., 6, 473 (1970).

131. A. M. Trukhan, Yu. M. Povarov, and P. D. Lukovtsev, ibid., 6, 744 (1970).

132. C. Duval, Chim. Anal. (Paris), 1963, 557.

133. K. J. Vetter, Electrochemical Kinetics, Academic, New York, 1967.

134. P. Delahay and G. Mamantov, Anal. Chem., 27, 478 (1955).

135. D. G. Davis, Jr., Talanta, 3, 335 (1960).

136. I. E. Barbasheva, Yu. M. Povarov, and P. D. Lukovtsev, Electrokhimiya, 6, 92 (1970).

137. Yu. M. Povarov, I. E. Barbasheva, and P. D. Lukovtsev, ibid., 3, 1202 (1967).

138. M. C. Giordano, J. C. Bazán, and A. J. Arvia, Electrochim. Acta, 11, 1553 (1966).

139. G. Pezzatini and R. Guidelli, ibid., 16, 1415 (1971).

140. V. A. Macagno, M. C. Giordano, and A. J. Arvia, ibid., 14, 335 (1969).

141. M. Breiter, M. Kleinerman, and P. Delahay, J. Amer. Chem. Soc., 80, 5111 (1958).

142. L. Gierst, in Transaction of the Symposium on Electrode Processes (E. Yeager, ed.), Wiley, New York, 1961, p. 109.

143. D. C. Grahame, J. Amer. Chem. Soc., 80, 4201 (1958).

144. D. C. Grahame, Technical Report of the Office of Naval Research, No. 5, 2nd Series, Contract N-onr-2309(01), August 1, 1957.

145. P. Delahay and D. Y. Kelsh, J. Electroanal. Chem., 18, 194 (1968).

146. N. A. Balashova and V. E. Kazarinov, Elektrokhimiya, 1, 512 (1965).

147. V. E. Kazarinov, Z. Phys. Chem., 226, 167 (1964).

148. N. A. Balashova, Zh. Fiz. Khim., 32, 2266 (1958); Z. Phys. Chem., 207, 340 (1957).

149. A. N. Frumkin, in Transaction of the Symposium on Electrode Processes (E. Yeager, ed.), Wiley, New York, 1959, p. 1.

150. V. E. Kazarinov and N. A. Balashova, Dokl. Akad. Nauk. SSSR, 134, 864 (1960).

151. R. A. Osteryoung, G. Lauer, and F. C. Anson, J. Electrochem. Soc., 110, 926 (1963).

152. R. A. Osteryoung and F. C. Anson, Anal. Chem., 36, 975 (1964).

153. A. T. Hubbard, R. A. Osteryoung, and F. C. Anson, ibid., 38, 692 (1966).

154. G. Toth, Radiokhimiya, 5, 411 (1963).

155. G. Toth and L. Zsinka, Acta Chim. Akad. Sci. Hung., 61, 289 (1969).

156. V. S. Bagotzky, Yu. B. Vassilyev, J. Weber, and J. N. Pirtskhalava, J. Electroanal. Chem., 27, 31 (1970).

157. A. Ya. Gokhstein, Elektrokhimiya, 1, 906 (1965); ibid., 3, 32 (1967); ibid., 1, 1052 (1965).

158. L. A. Medvedeva and Ya. M. Kolotyrkin, Dokl. Akad. Nauk. SSSR, 143, 1384 (1962).

159. I. Ammor, S. Darwish, and M. Etman, Electrochim. Acta, 12, 485 (1967).

160. P. Connor, J. B. Lewis, and W. J. Thomas, Proceeding of the Fifth Conference on Carbon, Pergamon, London, 1962.

161. S. Sternberg and I. Galasiu, Rev. Roumaine Chim., 11, 447 (1966).

162. A. N. Frumkin, Zh. Fiz. Khim., 24, 244 (1950); Z. Elektrochem., 59, 807 (1955); J. Chem. Phys., 26, 1552 (1957).

163. T. Kalish and A. N. Frumkin, Zh. Fiz. Khim., 28, 473 (1954).

164. L. Gierst, "Cinétique d'approache et Réactions d'électrodes irréversibles," Thesis, University of Brussels, 1958.

165. A. Aramata and P. Delahay, J. Amer. Chem. Soc., 66, 2710 (1962).

166. M. A. V. Devanathan and S. G. Canagaratna, Electrochim. Acta, 8, 77 (1963).

167. J. Jordan and R. A. Javick, J. Amer. Chem. Soc., 80, 1264 (1958).

168. J. Llopis, J. Biarge, and M. Fernandez, Electrochim. Acta, 1, 130 (1959).

169. P. Delahay, J. Phys. Chem., 70, 2067 (1966); ibid., 70, 2373 (1966).

170. R. A. Osteryoung, Anal. Chem., 35, 1100 (1963).

171. R. A. Osteryoung, G. Lauer, and F. C. Anson, ibid., 34, 183 (1962).

172. H. F. Conway and E. B. Lancaster, Electrochem. Technol., 2, 46 (1964).

173. M. J. Allen, Organic Electrode Processes, Reinhold, New York, 1958.

174. A. Hickling and S. H. Richards, J. Chem. Soc., 1940, 256.

175. E. B. Lancaster and H. F. Conway, Electrochem. Technol., 1, 253 (1963).

176. G. L. Johnson, R. F. Leiniger, and E. Segrè, J. Chem. Phys., 17, 1 (1949).

177. E. M. Appleman, J. Amer. Chem. Soc., 83, 805 (1961).

Chapter I-4

CADMIUM

ROGER J. LATHAM
Department of Chemistry
Loughborough University of Technology
Leicestershire, England

and

NOEL A. HAMPSON
School of Chemistry
City of Leicester Polytechnic
Leicestershire, England

1. STANDARD AND FORMAL POTENTIALS

1.1. AQUEOUS SOLUTION

The determination of accurate values for standard state potentials by extrapolation to infinite dilution of various functions is not feasible if association occurs to appreciable but unknown extents between ions which are involved in the net cell reaction. The cell Cd-Hg(2-phase amalgam) $|CdX_2(m)|AgCl|Ag$, because of the existence and moderate stability of the CdX^+ ion in dilute solutions, has been used for determinations of equilibrium potentials. Thus cells of this type were investigated by Harned and Fitzgerald [1] who evaluated the stability constant K of the $CdCl^+$ ion, and the standard state potential E° by special methods which they described.

A publication by Kielland [2] outlines the simultaneous evaluation, from electromotive force data, of K and E° at 25°C by the following alternative method. If conventional choices of standard state for Ag, AgCl, and aqueous CdX_2 are made, and the two-phase equilibrium amalgam as the standard state environment for the cadmium metal is chosen, the reversible potential of the cell E may be expressed as

$$E = E^\circ - (RT/2F)\,\ell n(m - m_{CdX^+})\,(2m - m_{CdX^+})^2\,\gamma_{Cd^{2+}}\,\gamma_{X^-}^2 \qquad (1.1.1)$$

where m is the analytical molality of CdX_2, m_{CdX^+} is the actual molality of the CdX^+ ion, and the other symbols have their usual meaning. The stability constant K of the CdX^+ ion is then represented by the equation

$$K = \frac{m_{CdX^+}\,\gamma_{CdX^+}}{(m - m_{CdX^+})\,(2m - m_{CdX^+})\,\gamma_{Cd^{2+}}\,\gamma_{X^-}} \qquad (1.1.2)$$

This treatment assumes that the species CdX_2, CdX_3^-, or CdX_4^{2-} are of negligible importance. However, this may not always be the case and it has been reported by Trueman and Ferris [3] that an upward shift in the reported values of the stability constant of CdX^+ may well confirm this contention. Of the reported values for E° it is noteworthy that a significant difference exists between E° for the amalgam electrode and that of the solid metal electrode. This was shown by Parks and La Mer [4] who measured the potential difference between amalgam and solid electrodes.

In Table **1.1.1** the main features of the equilibrium data are shown in summary form [5].

A scrutiny of the reported data indicates that the data of Harned and Fitzgerald [1] is certainly the most comprehensive and appears to be quite reliable. A comprehensive collection of data is shown in Table **1.1.2**.

Harned and Fitzgerald [1] and Bates [6] have shown that the variation of E° with temperature may be represented by an equation of the type:

$$E_t^\circ = E_{25}^\circ + \alpha_0(t - 25) + b(t - 25)^2 \qquad (1.1.3)$$

TABLE 1.1.1. Summary of Equilibrium Potential Data[a]

Electrode	E°	$(dE^\circ/dT)_{th}$ mV/K	$(dE^\circ/dT)_{isoth}$ mV/K	$(d^2E^\circ/dT^2)_{isoth}$ μV/K^2
$Cd^{2+} + 2e = Cd$	-0.4029	+0.778	-0.093	-
		+0.788[b]	-0.053[b]	
$Cd^{2+} + 2e + Hg = Cd(Hg)$	-0.3516	+0.62	-0.250[b]	+1.6[b]
	-0.3515	-	-0.252[b]	-0.56[b]
	-0.3514	-	-0.292[b]	+0.20[b]
$CdS + 2e = Cd + S^{2-}$	-1.175	0.0	-0.87	-
$Cd(CN)_4^{2-} + 2e = Cd + 4CN^-$	-1.028	-	-	-
$Cd(OH)_2 + 2e = Cd + 2OH^-$	-0.809	-0.143	-1.014	-
$CdCO_3 + 2e = Cd + CO_3^{2-}$	-0.74	-0.361	-1.232	-
$Cd(NH_3)_4^{2+} + 2e = Cd + 4NH_3(aq)$	-0.613	-	-	-

[a]The thermal temperature coefficient $(dE^\circ/dT)_{th}$ has a positive value if the hot electrode is the (+) terminal in a thermal cell of two such electrodes at two different temperatures. The thermal liquid junction potential is taken as zero in KCl solutions. The first and second isothermal temperature coefficients have positive values if $d(E' - E)/dT$ and $d^2(E' - E)/dT^2$ are positive, respectively, in the isothermal cell: (E)Cd|SHE||electrolyte|electrode|Cd'(E'). Values given are calculated. The thermodynamic relations used are

$$E^\circ = \Delta G^\circ(\text{oxidn})/nF$$

where ΔG° is that for the oxidation reaction

$$R \rightarrow O + ne^-$$

and

$$(dE^\circ/dT)_{isoth} = \Delta S^\circ(\text{redn})/nF$$

$$(dE^\circ/dT)_{th} = (dE^\circ/dT)_{isoth} + 0.871 \text{ mV}/^\circ\text{C}$$

$$(d^2E^\circ/dT^2)_{isoth} = \Delta C_p^\circ(\text{redn})/nFT$$

where ΔS° and ΔC_p° are those for the reduction reaction.

[b]Experimental.

TABLE 1.1.2. Standard and Formal Potentials

Half-reactions	Standard or formal potential (V)	Conditions	Refs.
$Cd^{2+} + 2e = Cd(Hg)$	-0.58151	0°C, 1 N $CdCl_2$	[1]
	-0.58039	5°C, 11% Cd(Hg)	
	-0.57900	10°C, Ag/AgCl reference	
	-0.57755	15°C	
	-0.57581	20°C	
	-0.57300	25°C	
	-0.57175	30°C	
	-0.56955	35°C	
	-0.56730		
$Cd^{2+}/Cd(Hg)$ 11%	-0.3452	0°C, 11% Cd(Hg)	[1]
	-0.3465	5°C, 11% Cd(Hg)	
$Cd^{2+} + 2e = Cd(Hg)$	-0.3477	10°C, 11% Cd(Hg)	
	-0.3491	15°C, 11% Cd(Hg)	
	-0.3503	20°C, 11% Cd(Hg)	
	-0.3515	25°C, 11% Cd(Hg)	
	-0.3526	30°C, 11% Cd(Hg)	
	-0.3539	35°C, 11% Cd(Hg)	
	-0.3553	40°C, 11% Cd(Hg)	
$Cd^{2+} + 2e = Cd(s)$	-0.4006	0°C	[1]
	-0.4010	5°C	
	-0.4013	10°C	
	-0.4016	15°C	
	-0.4018	20°C	
	-0.4020	25°C	
	-0.4021	30°C	
	-0.4020	35°C	
	-0.4026	40°C	
$Cd^{2+} + 2e = Cd(Hg)$	-0.4026		[3]
$Cd^{2+} + 2e = Cd(Hg)$	-0.4250	5°C, 10% Cd(Hg)	[6]
	-0.4248	10°C	
	-0.4243	15°C	
	-0.4236	20°C	
	-0.4227	25°C	
	-0.4215	30°C	
	-0.4201	35°C	
	-0.4185	40°C	
Cd^{2+} (a = 1)/Cd(Hg)	-0.3451	5°C	[6]
	-0.3468	10°C	
$Cd^{2+} + 2e = Cd(Hg)$	-0.3483	15°C	
	-0.3499	20°C	
	-0.3514	25°C	
	-0.3528	30°C	
	-0.3541	35°C	
	-0.3554	40°C	

(continued)

TABLE 1.1.2 (continued)

Half-reactions	Standard or formal potential (V)	Conditions	Refs.
Cd(Hg) = Cd(s) + Hg	-0.05538	0°C, 0.5 M $CdSO_4$	[4]
	-0.05359	10°C, satd amalgam	
	-0.05148	20°C	
	-0.04950	30°C	
Cd^{2+} + 2e = Cd(Hg)	-0.587	0.5 M Na_2SO_4, SCE	[7]
Cd^{2+} + 2e = Cd(s)	-0.4035	Dil $HClO_4$	[8]
Cd^{2+} + 2e = Cd(s)	-0.395	Single crystal	[9]

1.2. NONAQUEOUS SOLUTION

The standard potential of the cadmium electrode in nonaqueous systems has been the subject of but few investigations.

Cadmium chloride is sufficiently insoluble in formamide for a Cd|$CdCl_2$(s)|Cl^- electrode of the second kind to function satisfactorily. The standard potential of this electrode was evaluated by Andrews et al. [10] from measurements on cells of the type Cd|$CdCl_2$(s)|Cl^-|AgCl|Ag.

A complicating factor was that cadmium chloride is sufficiently soluble in formamide (about 0.02 mole/kg at 298°K) for a cadmium electrode to be set up in solutions unsaturated with respect to cadmium chloride.

However, measurements were made on the cell, Cd:Hg (2-phase)|$CdCl_2$(m)|AgCl|Ag, with $0 < m < 0.02$ mole/kg, with the object of determining the standard potential of the Cd^{2+}/Cd electrode in formamide. Values of emf were corrected for the two-phase nature of cadmium amalgam and the corrected values of emf are given in Table 1.2.1.

These values lead to a formal potential of the Cd^{2+}/Cd electrode of -0.412 V at 298.15°K. The authors showed, however, that there is extensive complex formation in formamide, the dominant species being the $CdCl_4^{2-}$ ion. The final potential of the Cd^{2+}/Cd electrode in formamide has been measured by Paulopoulos and Strehlow [11] using the cell Cd|$CdCl_2$(s)|KCl (saturated)|$Cd(NO_3)_2$|Cd, and obtained the value -0.414 ± 0.009 V in a molality scale. Their uncertainty is large owing to the presence of a liquid-junction potential.

TABLE 1.2.1. Emf of the Cell $Cd|CdCl_2(m)|AgCl|Ag$ at 298.15°K

m (mole kg^{-1})	Emf (V)
0.000 114	0.944 2
0.000 260	0.915 3
0.000 301	0.903 8
0.000 545	0.889 7
0.001 131	0.876 4
0.002 023	0.861 2
0.003 050	0.855 9
0.004 027	0.851 5
0.004 160	0.849 1
0.005 008	0.841 8
0.006 781	0.835 8
0.007 140	0.832 8
0.008 702	0.826 0
0.008 950	0.825 1
0.009 954	0.824 5
0.012 989	0.817 0
0.014 750	0.813 7
0.017 060	0.811 5

1.3. MOLTEN SALTS

Measurements of cell emf's in which the electrodes are reversible to the main ionic constituents of the melt, or in which the electrode potential of an electrode responsive to a minor constituent is measured with respect to a more or less standard reference electrode have been reported. Many solvents for molten salt systems have still to be explored, and thus emf studies are a necessary precursor of kinetic and mechanistic studies of electrode processes. The melt systems must be characterized by the determination of the ranges of coexistence of metals and melts, and cell emf's may be compared with thermal data to see whether the ranges of electrochemical stability are as wide as expected. An important difficulty is that suitable reference electrodes are often not known or completely established. Consequently some effort has gone to developing reference electrodes which are isolated but not insulated from the rest of the system. Formidable problems have been overcome in many of the molten salt systems (for example, fluorides) now attracting attention.

The number of reliable, systematic determinations of cell emf's in molten salts is relatively small. Workers using the same solvent do not always use the same reference electrode. However, in some cases comparison may be made where the potential of the particular reference electrode used is measured with respect to other available reference electrodes. Comparison of results is also made difficult since there is no standard temperature established for emf measurements in molten salts; this is complicated also by the wide range of melting points of salts commonly used. The choice of standard states for the establishment of standard emf's in molten salts is largely influenced by the methods used to obtain the measurements.

The thermodynamic standard states of the pure substance in its stable form under 1 atm pressure at the temperature of measurement can be employed for the metal. The choice of a standard state for any species which is included in the solvent or is linked to the solvent

TABLE 1.3.1. Standard and Formal Electrode Potentials in Molten Salts

Half-reaction	Standard or formal potential (V)	Conditions	Refs.
$Cd^{2+} + 2e = Cd$	-1.3415		[12]
$Cd^{2+} + 2e = Cd$	-1.338		[13]
$Cd^{2+} + 2e = Cd$	-1.312	475°C, $HgCl_2$-NaCl-KCl	[14]
$Cd^{2+} + 2e = Cd$	-0.620	700°C, NaCl-KCl	[15]
$Cd^{2+} + 2e = Cd$	-0.580	800°C, NaCl-KCl	[15]
$Cd^{2+} + 2e = Cd(s)$	-0.143	600°C, $CdCl_2$-LiCl-KCl eutectic melt vs Cd-Au	[16]
$Cd^{2+} + 2e = Cd\text{-}Au$	-1.1498	633°C, 1.00 N $CdCl_2$ vs Cl_2, graphite	[16]
$Cd^{2+} + 2e = Cd$	-1.2909	633°C, 1.00 N $CdCl_2$ vs Cl_2, graphite	[16]
$Cd^{2+} + 2e = Cd$	-1.73	218°C, $AlCl_3$-NaCl-KCl	[17]
$Cd^{2+} + 2e = Cd(Hg)$	-0.375	495°C vs Zn/Zn(II) 1 M (Li, Na, K) acetates	[18]

by a chemical equilibrium cannot be independent of the choice of the solvent itself; the standard state for such a species is chosen as the actual state of the species in the specific solvent molten salt under 1 atm pressure at the temperature of measurement. The definition of standard state with regard to solute ionic species, Cd(II), is more complex. The partial free energy of such a species, μ_+, is given by

$$\mu_+ = RT \ln C_+ + RT \ln \gamma_+ + \mu_+^\circ$$

where C_+ is the concentration of the Cd(II) ionic species, γ_+ is its activity coefficient, and μ_+° is its standard partial free energy. Since $\mu = -nFE$, this equation becomes

$$E_+^\circ = E_+ - RT/2F \ln C_+ - RT/2F \ln \gamma_+$$

If the other species taking part in the half reaction are included, this equation is the well-known Nernst equation. The cell generally is most conveniently constructed so that these other species are present in their standard states, i.e., $E_+ = E$ and $E_+^\circ = E^\circ$. If γ_+ is defined so that γ_+ approaches zero as C_+ approaches unity, then $E^\circ = E - RT/nF \ln C_+$. A linear relationship is observed between the measured values of E and ln C_+ (within experimental error) whenever C_+ is less than approximately 0.1 m; this indicates that C_+ remains unity over this concentration range. Thus E° may be calculated from this relationship. However, this requires that $E = E^\circ$ when C_+ is unity on the concentration scale chosen. The concept of the standard state of an ionic species is one which is physically unreal, for it is a solution of unit concentration which has certain of the thermodynamic properties of an infinitely dilute solution. The value of E° is usually calculated from several sets of $E - RT/nF \ln C_+$ data, and the error assigned is usually the standard deviation in E°. These E° values are significant in solutions of concentrations below about 0.1 m, because C_+ does not remain unity at higher concentrations.

A list of data is presented in Table 1.3.1.

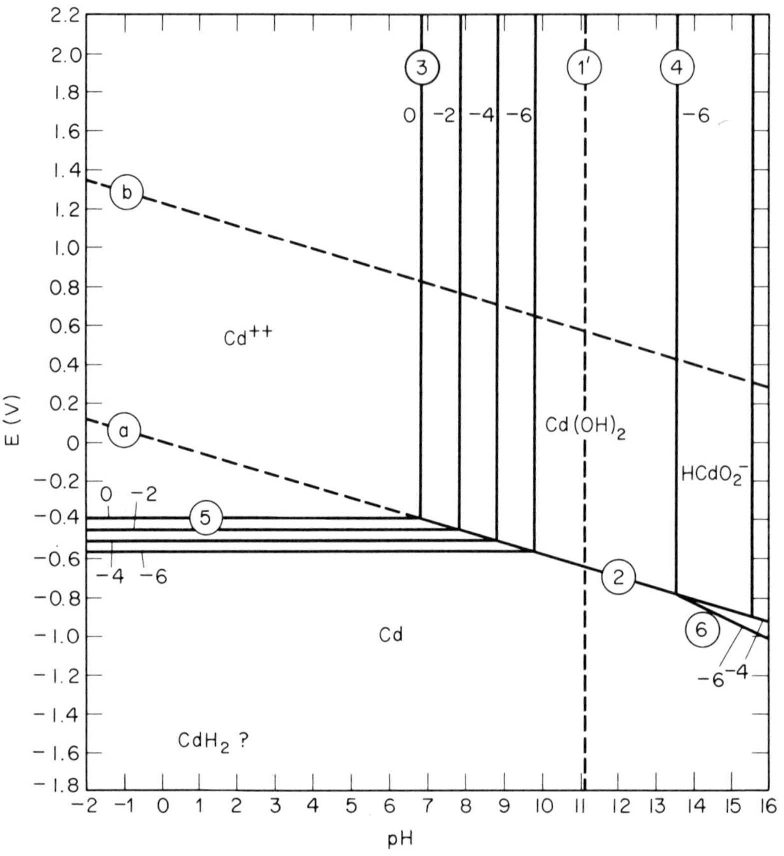

FIG. 1.4.1. Potential-pH diagram for cadmium in aqueous solution. (Data from Ref. 19, p. 417, by permission of Pergamon Press.)

1.4. EQUILIBRIUM DATA

Cadmium forms Cd^{2+} ions and complexes which are based on these ions. On the basis of species considered likely to be present in the cadmium-aqueous solution system (namely Cd, CdO-3 hydrated forms, dissolved substances based on Cd^{2+} — cadmous, cadmite, and bicadmite ions, and gaseous CdH_2) Deltombe, Pourbaix, Zoubov [19] have constructed a potential-pH diagram (Fig. 1.4.1) for the cadmium-water system at 25°C. This diagram was constructed with the following equations:

$$Cd^{2+} + 2H_2O = HCdO_2^- + 3H^+ \qquad \log \frac{[HCdO_2^-]}{[Cd^{++}]} = -33.34 + 3\ pH \qquad (1.4.1)$$

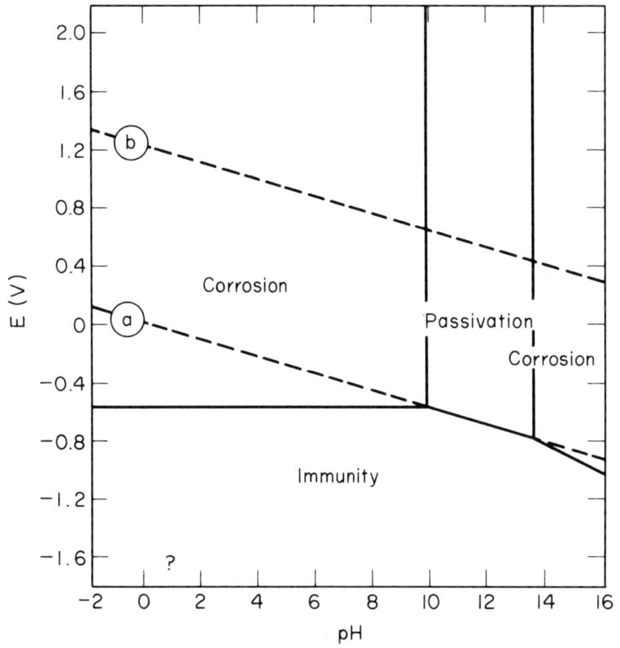

FIG. 1.4.2. Theoretical conditions of corrosion, immunity, and passivation of cadmium at 25°C. (Data from Ref. 19, p. 418, by permission of Pergamon Press.)

The limit of the region in which Cd^{2+} exists as a simple dissolved ion:

$Cd^{2+}/HCdO_2^-$	pH = 11.1	
$CdO + 2H^+ + 2e = Cd + H_2O$	a. $E_0 = 0.005 - 0.0591$ pH	(1.4.2)
	b. $E_0 = 0.023 - 0.0591$ pH	
	c. $E_0 = 0.063 - 0.0591$ pH	
$Cd^{2+} + H_2O = CdO + 2H^+$	a. $\log [Cd^{2+}] = 13.81 - 2$ pH	(1.4.3)
	b. $\log [Cd^{2+}] = 14.39 - 2$ pH	
	c. $\log [Cd^{2+}] = 15.76 - 2$ pH	
$CdO + H_2O = HCdO_2^- + H^+$	a. $\log [HCdO_2^-] = -19.54 +$ pH	(1.4.4)
	b. $\log [HCdO_2^-] = -18.96 +$ pH	
	c. $\log [HCdO_2^-] = -17.59 +$ pH	
$Cd^{2+} + 2e^- = Cd$	$E_0 = -0.403 + 0.0295 \log [Cd^{2+}]$	(1.4.5)
$HCdO_2^- + 3H^+ + 2e^- = Cd + 2H_2O$	$E_0 = 0.583 - 0.0886$ pH $+ 0.0295 \log [HCdO_2^-]$	(1.4.6)
$Cd + H^+ + e^- = CdH$	$E_0 = -2.417 - 0.0591$ pH $+ 0.0591 \log P_{CdH}$	(1.4.7)

In a similar manner these authors have constructed a simplified active-passive diagram (Fig. 1.4.2).

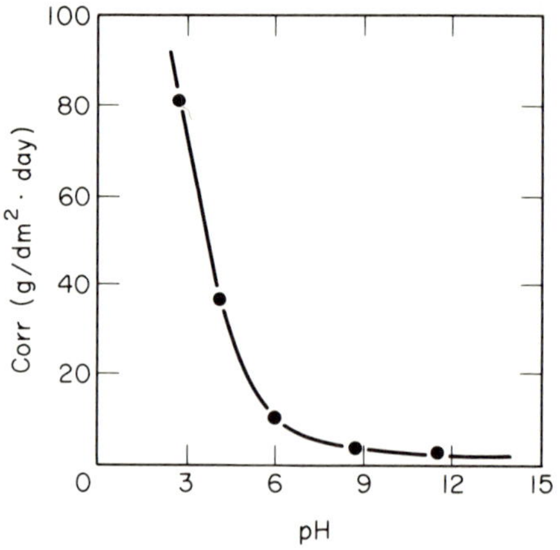

FIG. 1.4.3. Influence of pH on the corrosion rate of cadmium. (Data from Ref. 21.)

These diagrams are valid only in the absence of substances with which cadmium can form soluble complexes or insoluble salts. According to Charlot [20], most anions form complexes with Cd^{2+}, e.g., Cl^-, Br^-, I^-, NO_3^-, SO_4^{2-}, and $S_2O_3^{2-}$. Complexes are also formed with NH_3 and HCN. In general cadmium complexes are not very stable. A large number of salts and basic salts of cadmium are sparingly soluble or very sparingly soluble, e.g., the carbonate, the cyanide, the phosphate, and the sulfide.

From Figs. 1.4.1 and 1.4.2 Deltombe et al. [19] point out that the upper limit of stability of cadmium practically coincides with the lower limits of stability of water at atmospheric pressure. It is clear that cadmium can therefore be considered as being at the dividing line between noble and base metals; it is less noble than lead, appreciably more noble than zinc, and slightly more noble than iron.

Figures 1.4.1 and 1.4.2 show that cadmium can corrode in acid and very alkaline solutions, but corrosion is appreciable only in the presence of oxidizing or complexing substances; in the absence of such substances pure cadmium corrodes only very slightly on account of the large hydrogen overpotential. Cadmium can become covered with a film of hydroxide in alkaline solutions. According to Figs. 1.4.1 and 1.4.2, this film should form readily at pH's of about 10-13. The hydroxide film constitutes a protective layer when it is obtained by the anodic passivation of the metal. The region of stability of $Cd(OH)_2$ is therefore a region of real passivation of the metal; however, the region of passivation of cadmium has the disadvantage of being relatively narrow. Figure 1.4.3 represents some experimental results of Chatalov [21] concerning the influence of pH on the corrosion rate of cadmium.

The existence of a white solid cadmium hydride having a formula $\sim CdH_2$ and thermodynamically unstable in respect of its components H_2 and Cd, mentioned by Hurd [22], has been included in the diagram although the regions are considered by Deltombe et al. [19]

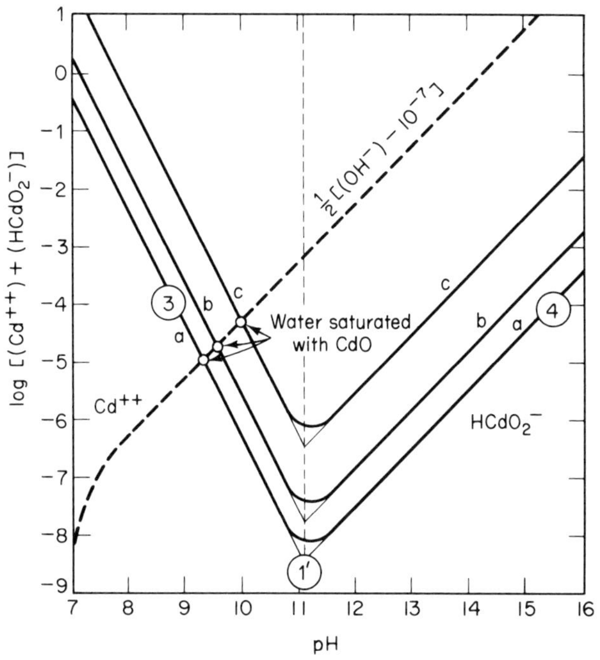

FIG. 1.4.4. Influence of pH on the solubility of cadmium at 25°C: (a) "inactive" $Cd(OH)_2$; (b) "active" $Cd(OH)_2$; (c) "anhydrous" CdO. (Data from Ref. 19, p. 419, by permission of Pergamon Press.)

to be only a rough guide. In the absence of complexing solutions cadmium can easily be cathodically protected by lowering its electrode potential to below -0.60 V in acid or neutral solutions. In agreement with Figs. 1.4.1 and 1.4.2, the electrodeposition of cadmium is generally very easy.

Figure 1.4.4 shows that the characteristics of pure water saturated with various varieties of the oxide and hydroxides of cadmium are:

	pH	$\log[Cd^{2+}]$	mg Cd/liter
$Cd(OH)_2$, "inactive"	9.40	-4.95	1.26
$Cd(OH)_2$, "active"	9.55	-4.75	2.00
CdO, "anhydrous"	10.00	-4.30	5.63

The colorless bivalent ion Cd^{2+} is converted into the white hydroxide $Cd(OH)_2$ on increasing the pH of the solution or on the oxidation of cadmium in acid or neutral solution. Cadmium hydroxide dissolves in solutions of high pH's to give the colorless bicadmite ion $HCdO_2^-$ which is the predominant dissolved form of cadmium at pH's above 11.1; it is approximately at this pH that $Cd(OH)_2$ has its minimum solubility. The hydroxide is soluble in both acid and alkaline media; the amphoteric character of this compound has been described by Piater [23].

Of the two forms of cadmium hydroxide reported in the literature, the "active" $Cd(OH)_2$ turns into "inactive" $Cd(OH)_2$ by the process of aging [24, 25]. Very alkaline oxidizing solutions should dissolve cadmium to give bicadmite ions $HCdO_2^-$.

2. VOLTAMMETRIC CHARACTERISTICS

2.1. POLAROGRAPHIC CHARACTERISTICS

2.1.1. Aqueous Solution

Cadmium is especially suitable for polarographic determination. Well-defined reduction waves are observed in neutral, alkaline, and acid media (nitric, hydrochloric, or perchloric acids). Table 2.1.1 gives a number of supporting electrolytes for the polarographic determination of cadmium. Cadmium can be easily determined in the presence of zinc, since the cadmium wave precedes that of zinc to an appreciable extent. Hence polarography can be used for the determination of even trace quantities of cadmium-zinc containing materials.

Several complex ions of cadmium are suitable for polarographic determination. For example, well-defined, diffusion-controlled waves are obtained in acid or neutral solutions of tartrates up to pH 9 in ammoniacal media and also in cyanide media. The waves obtained in the reduction of the cadmium iodide complex and in the reduction of cadmium in strongly alkaline tartrate solutions are not suitable. Cadmium can be determined polarographically in a range of materials. Often it is not necessary to separate the cadmium prior to its determination. Copper, bismuth, and lead give reduction waves that precede cadmium, and in these cases it is necessary to remove the elements by a controlled cathode potential electrodeposition. Copper can be complexed with cyanide and thus will not interfere with the reduction wave of cadmium.

Details of the usual analytical methods can be found in standard texts, for example, Kolthoff and Elving [26].

2.1.2. Nonaqueous

Auerbach and McGuire [37] have measured stability constants of cadmium nitrate complexes in molten dimethylsulfone at 125° C. Half-wave potentials were measured as a function of nitrate concentration. The ionic strength was maintained constant by addition of lithium perchlorate. A normalized curve-fitting method was used to evaluate the stability constants (log K values are 2.12, 3.72, 5.2, and 6.8).

TABLE 2.1.1. Polarographic Data for Cd in Various Aqueous Electrolytes[a]

Substance	Product	Conditions	$E_{1/2}$ (V vs SCE)	I	$E_{3/4}-E_{1/4}$ (mV)	Remarks	Refs.
Cd(II)	Cd(Hg)	0.5 M NaF + 0.5 M	-0.5607	-	-31 ± 1	drpt = 0.32	[27]
	Cd(Hg)	0.5 M NaF + 0.5 M	(Ag/AgCl)	-			[27]
	Cd(Hg)	0.5 M NaF + 0.5 M	-0.5610	-	-31 ± 1	drpt = 0.24	[27]
	Cd(Hg)	0.5 M NaF + 0.5 M	-0.5609	-	-31 ± 1	drpt = 0.16	[27]
Cd(II)	Cd(Hg)	1 M KCl, 0.25 X 10^{-3} M Cd^{2+}	-	-	-	$D_s = 0.72 \times 10^{-5}$	[28]
Cd(II)	Cd(Hg)	KNO_3	-	-	-	$D = 6.3 \times 10^{-6}$	[29]
		KCl	-	-	-	$D = 0.87 \times 10^{-5}$	
Cd(II)	Cd(Hg)	1 M $HClO_4$	-0.62			0.01% gelatin	[30]
Cd(II)	Cd(Hg)	1 M HNO_3	-0.59			0.01% gelatin	[30]
Cd(II)	Cd(Hg)	0.1 M KNO_3	-0.578				[30]
Cd(II)	Cd(Hg)	0.5 M H_2SO_4	-0.59				[30]
Cd(II)	Cd(Hg)	0.4 M Ac, pH 4.7	-0.61				[30]
$Cd(Cl)_x^{(2-x)+}$	Cd(Hg)	0.1 M KCl	-0.60			0.01% gelatin	[30]
$Cd(Cl)_x^{(2-x)+}$	Cd(Hg)	4 M NaCl	-0.69				[30]
$CdBr_2^{(2-x)+}$	Cd(Hg)	0.5 M KBr	-0.65				[30]
$CdBr_2^{(2-x)+}$	Cd(Hg)	3 M KBr	-0.70				[30]
CdI_4^{2-}	Cd(Hg)	0.1 M KI	-0.65				[30]

(continued)

TABLE 2.1.1 (continued)

Substance	Product	Conditions	$E_{1/2}$ (V vs SCE)	I	$E_{3/4}-E_{1/4}$ (mV)	Remarks	Refs.
CdI_4^{2-}	Cd(Hg)	3 M KI	-0.80				[30]
$Cd(CNS)_x^{(2-x)+}$	Cd(Hg)	KNO_3, 0.1 M KNCS, ionic strength 2	-0.58				[30]
$Cd(CNS)_x^{(2-x)+}$	Cd(Hg)	2 M KNCS	-0.664				[30]
$Cd(S_2O_3)_3 4-$	Cd(Hg)	0.1 M KNO_3, 1 M $Na_2S_2O_3$	-0.78			0.01% gelatin	[30]
$Cd(NH_3)_4^{2+}$	Cd(Hg)	0.1 M NH_4NO_3, 0.1 M NH_3	-0.674				[30]
$Cd(NH_3)_4^{2+}$	Cd(Hg)	1 M NH_4Cl, 1 M NH_3	-0.81				[30]
Cd(II) complexed	Cd(Hg)	0.4 M Ac^-, 0.1 M nitrilotriacetic acid	-0.87				[30]
$Cd(CN)_4^{2-}$	Cd(Hg)	0.1 M KNO_3, 1 M KCN	-1.16			0.01% gelatin	[30]
Cd^{2+}		0.01 M KCl				$D = 7.15 \times 10^{-6}$	[30]
Cd^{2+}		0.1 M KCl				$D = 7.15 \times 10^{-6}$	[30]
Cd^{2+}		0.1 M KNO_3				$D = 6.9 \times 10^{-6}$	[30]
Cd^{2+}		1.0 M KNO_3				$D = 6.8 \times 10^{-6}$	[30]
Cd^{2+}		1.0 M KCl				$D = 7.9 \times 10^{-6}$	[30]
Cd^{2+}		3.0 M KCl				$D = 7.9 \times 10^{-6}$	[30]
Cd(II)	Cd(Hg)	2 M HOAc, 2 M NH_4OAc	-0.653	2.3	-28	0.01% gelatin	[31]
Cd(II)	Cd(Hg)	0.5 M HOAc, 0.5 M NaOAc	-0.615w		-40	Satd with phenol	[31]
Cd(II)	Cd(Hg)	1 M NH_3, 1 M NH_4Cl	-0.81w	3.68	-28		[31]
Cd(II)	Cd(Hg)	2 M K_2CO_3, pH 11.0	-0.74				[31]

Cd(II)	Cd(Hg)	0.1 M KCl, LiCl, Me_4NCl, NH_4Cl	-0.60	3.51	-28		[31]
Cd(II)	Cd(Hg)	1 M HCl	-0.642w	3.58	-28		[31]
Cd(II)	Cd(Hg)	5 M $CaCl_2$	-0.80				[31]
Cd(II)	Cd(Hg)	8 M HCl	-0.80			Irr	[31]
Cd(II)	Cd(Hg)	Satd citric acid	-0.514w		-28		[31]
Cd(II)	Cd(Hg)	0.5 M K_3citrate, pH 3	-0.611w		-27		[31]
Cd(II)	Cd(Hg)	0.1 M $(NH_4)_3$citrate, pH 6.1	-0.700w		-28		[31]
Cd(II)	Cd(Hg)	0.1 M NH_3, 0.1 M $(NH_4)_3$citrate pH 8.5	-0.706w		-42		[31]
Cd(II)	Cd(Hg)	1 M KCN	-1.18w			Irr	[31]
Cd(II)	Cd(Hg)	10 M NaCN, 1 M NaOH	-1.216w		-33		[31]
Cd(II)	Cd(Hg)	0.5 M en, 0.5 M K_3PO_4	-0.95w				[31]
Cd(II)	Cd(Hg)	0.05 M EDTA	-0.65fi			$Cd^{2+}/Cd(EDTA)^-$	[31]
		0.8 M HOAc, pH 2	-0.85w				
Cd(II)	Cd(Hg)	0.1 M EDTA	-1.27i				
		0.8 M NaOAc, pH 2.7					
Cd(II)	Cd(Hg)	0.05 M glutamic acid, 0.2 M $NaClO_4$,	-0.84				[31]
		0.2 M borate buffer, pH 9.5					
Cd(II)	Cd(Hg)	Satd hydrazine dichloride	-0.729fw		-27		[31]
Cd(II)	Cd(Hg)	1 M N_2H_4, 1 M NH_3, 1 M NH_4Cl	-0.823fi		-26		[31]
Cd(II)	Cd(Hg)	1 M NaOH	-0.783w		-38	Low solubility	[31]

(continued)

TABLE 2.1.1 (continued)

Substance	Product	Conditions	$E_{1/2}$ (V vs SCE)	I	$E_{3/4}-E_{1/4}$ (mV)	Remarks	Refs.
Cd(II)	Cd(Hg)	10 M NaOH	-0.913fi		-25	Large maximum	[31]
Cd(II)	Cd(Hg)	Satd hydroxylamine hydrochloride	-0.714w		-30		[31]
Cd(II)	Cd(Hg)	Satd malonic acid	-0.519w		-27		[31]
Cd(II)	Cd(Hg)	0.1 M NH_3, 0.1 M $(NH_4)_2$malonate	-0.676w		-28		[31]
Cd(II)	Cd(Hg)	0.25 M $Na_2C_2O_4$, pH 3-4	-0.63				[31]
Cd(II)	Cd(Hg)	0.1 M NH_3, 0.1 M $(NH_4)_2C_2O_4$	-0.705w		-57		[31]
Cd(II)	Cd(Hg)	0.1 M $HClO_4$ or $NaClO_4$	-0.59	3.06	-28	1 M HNO_3, 0.01% gelatin	[31]
Cd(II)	Cd(Hg)	7.3 M H_3PO_4	-0.77w	-	-28		[31]
Cd(II)	Cd(Hg)	0.1 M pyridine, 0.1 M pyridinium chloride	-0.617w	-	-29		[31]
Cd(II)	Cd(Hg)	0.1 M NH_3, 0.1 M $(NH_4)_2$tartrate	-0.727w	3.40	-36		[31]
Cd(II)	Cd(Hg)	0.25 M Na_2tartrate, 2.0 M NaOH	-0.856w	-	-26		[31]
Cd(II)	Cd(Hg)	1 M KSCN	0.651fw/1.27	-	-26/-		[31]
Cd(II)	Cd(Hg)	0.3 M triethanolamine, 0.1 M NaOH	-0.82	-	-		[31]
Cd(II)	Cd(Hg)	0.5 M total ionic strength, pH 2.9-4.4, containing					[32]
		0.1 M $CaCl_2$, pH 3.4	-0.653				
		2×10^{-3} M Cd^{2+}, pH 3.7	-0.654				
		4×10^{-3} M Cd^{2+}					

		pH 3.00	-0.642				
		pH 3.70	-0.650				
		pH 4.20	-0.646				
Cd(II)	Cd(Hg)	0.1 M KCl	-0.630 (vs AgCl)			$\alpha = 0.5$, reversible	[33]
Cd(II)	Cd(Hg)	1 M KNO_3, 1.0 mM Cd^{2+}	-0.64			25 ± 0.1°C, small amount polyoxyethylene laurylether	[34]
Cd(II)	Cd(Hg)	2 M $NaClO_4$/NaF, pH 5.0 ± 0.2					
		0.0 MF	-0.999				
		0.05 MF	-1.003				
		0.10 MF	-1.006				
		0.15 MF	-1.008				
		0.20 MF	-1.011				
		0.25 MF	-1.012				
		0.30 MF	-1.014				
		0.50 MF	-1.018				
		0.60 MF	-1.020				
Cd(II)	Cd(Hg)	1 M KNO_3, 2 mM Cd^{2+}	-	-	-	$D = 0.68 \times 10^{-5}$	[36]
		0.1 M KNO_3, 2 mM Cd^{2+}				$D = 0.75 \times 10^{-5}$	
		0.1 M KCl, 2 mM Cd^{2+}				$D = 0.70 \times 10^{-5}$	
						$D_{calc} = 0.72 \times 10^{-5}$	

[a] Data mainly from Kolthoff and Elving [26] and Meites [31].

The reference electrode was contained in a fine porosity frit immersed in the solution. At the start of each experiment the electrode was constructed in situ from a 1-mm silver wire spiral in contact with a 0.01-M silver perchlorate—2-M lithium perchlorate solution in molten DMS. The electrode was reversible according to the Nernst criterion and stable over the period required for any of the experiments; reproducibility of $E_{1/2}$ measurements was ±0.005 V for any solution composition.

Values of stability constants are given in Tables 2.1.2 and 2.1.3

TABLE 2.1.2. Stability Constants in Dimethylsulfone at 125°C and Ionic Strength 2

Log	$\frac{[Cd(NO_3)_i^{2-i}]}{[Cd^{2+}][NO_3^-]^i}$
β_1	2.12 ± 0.11
β_2	3.72 ± 0.14
β_3	5.2 ± 0.1
β_4	6.8 ± 0.1

TABLE 2.1.3. Observed Values of $\Delta E_{1/2}$ and Values Calculated from Eq. (2.1.1). Ion Concentration = 2×10^{-4} M

	$\Delta E_{1/2}$ (mV)	
$\lvert NO_3^- \rvert$ (M $\times 10^3$)	Observed	Calculated
3.99	9	8
7.78	18	15
11.4	25	21
15.6	32	29
26.3	50	45
36.3	66	59
52.4	87	77
77.2	108	99
115	131	131
149	146	140
291	181	184
443	221	213
650	251	239
861	275	258

The values were calculated using a relationship developed by DeFord and Hume [38].

$$F_0 = \text{antilog}\left[\frac{0.434nF}{RT}\,\Delta E_{1/2} + \log\frac{I_{x=0}}{I_x}\right] \tag{2.1.1}$$

$$= 1 + \beta_1 X + \beta_2 X^2 + \beta_3 X^3 + \beta_4 X^4 + \cdots \tag{2.1.2}$$

where x = nitrate concentration; $\Delta E_{1/2} = (E_{1/2})_{x=0} - (E_{1/2})_x$; $I_{x=0}$ = diffusion current constant of the metal ion in 2 M lithium perchlorate; I_x = same, observed at nitrate concentration x; and β_i = overall stability constant of the species $Cd(NO_3)_i^{(2-i)}$, and a normalized curve-fitting approach of Sillen [39].

Jones and Fritsche [40] have made potential sweep chronoamperometry in dimethyl sulfoxide at the hanging mercury drop electrode in experiments using a commercial fiber junction SCE as the reference electrode. A Randles–Sevcik plot of peak current against the square root of the voltage scan was linear. In millimolar cadmium solutions half-wave potentials were -0.66 and -0.63 V at the DME in 1.0 and 0.10 M $KClO_4$, respectively. The polarographic wave plots obtained were linear and had slopes of 0.029 V in 1.0 M $KClO_4$ and 0.034 V in 0.10 M $KClO_4$.

Polarograms were well formed in all cases. The influence of the depolarizer concentration and that of the carrier electrolyte concentration is shown in Table 2.1.4. These data record peak potentials of the carrier electrolyte concentration.

TABLE 2.1.4. Influence of Carrier Electrolyte Concentration of the Peak Potential of Cadmium in DMSO.

Molarity	E_p(V) 1.00×10^{-3} M Cd^{2+}	5.00×10^{-4} M Cd^{2+}
10^{-3}	-0.71	-0.72
10^{-2}	-0.65	-0.68
10^{-1}	-0.65	-0.67
10^{0}	-0.66	-0.65

The reduction of millimolar cadmium in 0.10 M $KClO_4$ in DMSO is probably diffusion-controlled at the DME since values of the ratio of the diffusion current to the square root of the height of the mercury column were found to be constant to within ±4% of the mean. This finding agrees with that of Burrus [41] who showed that a plot of wave height vs concentration is linear up to 1.0×10^{-3} M Cd^{2+} in 1 M $KClO_4$, 0.1 M $KClO_4$, and 0.1 M $(n-C_4H_9)_4NClO_4$. Slopes of the plots in these electrolytes were reported as 3.18, 5.36, and 5.56 μA/mmole.

Although with lithium chloride or lithium nitrate supporting electrolyte it has been reported that pyridine solutions are unsatisfactory for nonaqueous polarography, Table 2.1.5 shows that in a wide variety of solvents Cd^{2+} and related complexes undergo simple reductions at mercury electrodes [42].

TABLE 2.1.5. Polarographic Data for Cd in Various Nonaqueous Solutions [42]

Substance	Conditions		$E_{1/2}$	
$Cd(ClO_4)_2$	MeCN	0.1 M TBAP[a]	-0.55	Ag/0.01 M $AgNO_3$
$Cd(ClO_4)_2 \cdot 6H_2O$	MeCN	0.1 M $NaClO_4$	-0.27	
$Cd(ClO_4)_2$	PhCN	0.1 M TEAP[b]	-0.17	
$Cd(ClO_4)_2$	PhCN	0.1 M $LiClO_4$	-0.24	
$Cd(ClO_4)_2$	EtCN	0.1 M TEAP	-0.25	
$Cd(ClO_4)_2$	$CH_2 = CH - CH_2CN$	0.1 M TEAP	-0.19	
$Cd(ClO_4)_2$	$PhCH_2CN$	0.1 M TEAP	-0.18	
Amalgam	DMF	0.1 M $NaClO_4$	-0.17	Ag/AgCl(s)
$Cd(ClO_4)_2$	DMF	0.1 M TEAP	-0.17	Ag/AgCl(s)
$Cd(NO_3)_2$	DMF	0.1 M $NaNO_3$	+1.453	Na-Hg(s)/$NaClO_4$(s)
$3Cd(SO_4)_2 \cdot 8H_2O$	$HCONH_2$	0.2 M $NaClO_4$	-0.69	
$CdCl_2$	$HCONH_2$	0.2 M LiCl	-0.672	Ag/AgCl(s)
$Cd(NO_3)_2$	DMSO	0.1 M $NaNO_3$	+0.384	Zn-Hg(s)/$Zn(ClO_4)_2$(s)
$Cd(NH_3)_6(NO_3)_2$	NH_3	0.1 M KNO_3	-0.45	Pb/0.1 N $Pb(NO_3)_2$
$Cd(ClO_4)_2 \cdot H_2O$	PC[e]	0.1 M TEAP	-0.11	
Cd(II)	EDA[f]	0.29 M LiCl	-0.696	Hg pool
$CdCl_2$	EDA	0.1 M $NaNO_3$	-0.64	NCE
$Cd(NO_3)_2$	EDA	0.1 M $NaNO_3$	-0.50	NCE
$Cd(ClO_4)_2$	EDA	0.1 M $NaNO_3$	-0.70	NCE
$CdCl_2$	EDA	0.1 M LiCl	-0.74	NCE
$Cd(NO_3)_2$	EDA	0.1 M LiCl	-0.60	NCE
$Cd(ClO_4)_2$	EDA	0.1 M LiCl	-0.80	NCE
$CdCl_2$	EDA	0.1 M TEAN[c]	-0.46	NCE
$Cd(NO_3)_2$	EDA	0.1 M TEAN	-0.60	NCE
$Cd(ClO_4)_2$	EDA	0.1 M TEAN	-0.50	NCE
$CdCl_2$	EDA	0.1 M $NaNO_3$	-0.64	NCE
$CdCl_2$	MOR[g]	0.1 M TBAI[d]	-0.95	NCE
$CdCl_2$	MeOH	0.2 M LiCl	-0.595	Ag/AgCl(S)

(continued)

TABLE 2.1.5 (continued)

Substance	Conditions		$E_{1/2}$	
$CdCl_2$	[$HOCH_2$ / CH_2OH]	0.2 M LiCl	-0.600	Ag/AgCl(S)
$CdCl_2$	EtOH	0.2 M LiCl	-0.590	Ag/AgCl(S)
$CdCl_2$	n-PrOH	0.2 M LiCl	-0.585	Ag/AgCl(S)
$CdCl_2$	n-BuOH	0.2 M LiCl	-0.581	Ag/AgCl(S)
$CdCl_2$	HOAc	0.25 M NH_4OAc	-0.83	
$Cd(ClO_4)_2 \cdot 2H_2O$	HOAc	1 M $LiClO_4$	-0.295	
$Cd(ClO_4)_2 \cdot 2H_2O$	HOAc	0.25 M NH_4OAc	-0.685	
$Cd(ClO_4)_2 \cdot 2H_2O$	HOAc	1 M LiCl	-0.290	
$CdCl_2$	HCOOH	0.5 M NaOOCH	-0.337	SCE in HCOOH
$CdCl_2$	HCOOH	0.4 M NaBr + 0.5 M NaOOCH	0.621	SCE in HCOOH
$CdCl_2$	HCOOH	0.5 M K_2SO_4	-0.356	SCE in HCOOH
$Cd(ClO_4)_2 \cdot 2H_2O$	EtCOOH	1 M $LiClO_4$	-0.270	
$Cd(ClO_4)_2 \cdot 2H_2O$	i-PrCOOH	1 M $LiClO_4$	-0.290	
$Cd(ClO_4)_2 \cdot 2H_2O$	$CH_2 = CHCO_2H$	1 M $LiClO_4$	-0.325	
$CdCl_2$	HCOOH	2 M NaOOCH	-0.356	SCE in HCOOH
$CdCl_2$	HCOOH	0.2 M NaF + 0.5 M NaOOCH	-0.344	SCE in HCOOH
$CdCl_2$	HCOOH	0.4 M NaF + 0.5 M NaOOCH	-0.349	SCE in HCOOH
$CdCl_2$	HCOOH	0.2 M NaBr + 0.5 M NaOOCH	0.585	SCE in HCOOH
$Cd(ClO_4)_2$	SUL[h]	0.1 M $NaClO_4$	+0.33	Ag/AgCl-TEAC(s)

[a]Tetra-n-butylammonium perchlorate.

[b]Tetraethylammonium perchlorate.

[c]Tetraethylammonium nitrate.

[d]Tetra-n-butylammonium iodide.

[e]Propylene carbonate.

[f]Ethylene diamine.

[g]Morpholine.

[h]Sulfolane.

2.1.3. Fused Salts

Francini, Martini, and Geiss [43] have investigated the electroreduction of cadmium ion in alkaline thiocyanate based melts. The investigation utilized conventional and single linear potential sweep (oscillographic) polarography employing a dropping mercury electrode as cathode and the Ag/Ag^+ electrode as reference anode in the temperature range 137-210° C.

The results are shown in Tables 2.1.6 to 2.1.9.

The linear relationships between peak current and concentration in the range 0.1-2.5 M are shown in Fig. 2.1.1.

TABLE 2.1.6. Linear Potential Sweep Data for the Cd^{2+} Reduction in Thiocyanate Melts

T (°K)	C (m$\underline{M}$/1)	$E_{\frac{1}{2}}^{obsd}$ (V)	E_p^{th} (V)	E_p^{obsd} (V)
410	0.420	-0.290	-0.309	-0.305
410	0.760	-0.287	-0.306	-0.304
410	0.810	-0.290	-0.309	-0.306

TABLE 2.1.7. Linear Potential Sweep Data for the Cd^{2+} Reduction in Thiocyanate Melts

$\left[\frac{2.3RT}{nF}\right]_{th}$ (mV)	$\left[\frac{2.3RT}{nF}\right]_{obsd}$ (mV)	$\left[E_{p/2}-E_p\right]_{th}$ (mV)	$\left[E_{p/2}-E_p\right]_{obsd}$ (mV)
42.8	45	41.0	44

TABLE 2.1.8. Diffusion Coefficients for Cd^{2+} in Thiocyanate Melts

$D^{160°}_{\text{Randles-Sevcik}}$ ($cm^2 sec^{-1}$)	$D^{160°}_{\text{Ilkovic}}$ ($cm^2 sec^{-1}$)
0.52×10^{-6}	0.61×10^{-6}

TABLE 2.1.9. Energies of Diffusion of Cd^{2+} in Thiocyanate Melts

$E^{\text{Randles-Sevcik}}$ (kcal/mole)	E^{Ilkovic} (kcal/mole)
-10.3	-10.3

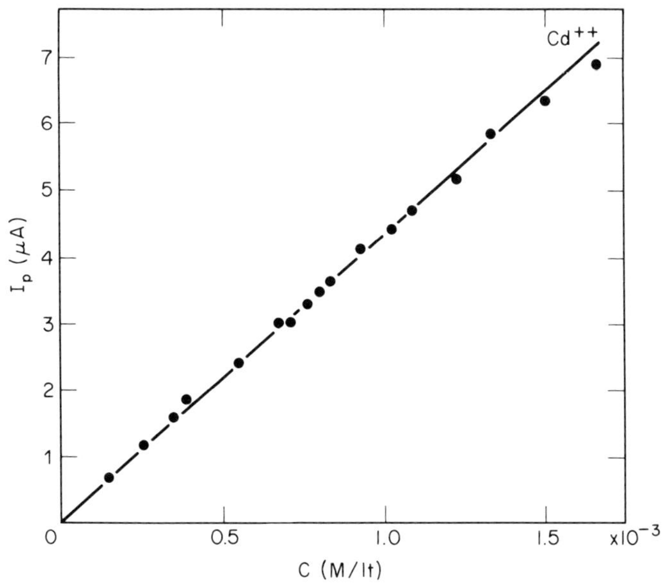

FIG. 2.1.1. Linear plot of peak current against $[Cd^{2+}]$: T = 137°C; m = 0.643 mg/sec; Υ = 6.5 sec; sweep rate = 0.364 V/sec. (Data from Ref. 43 by permission of Electrochim. Metallorum.)

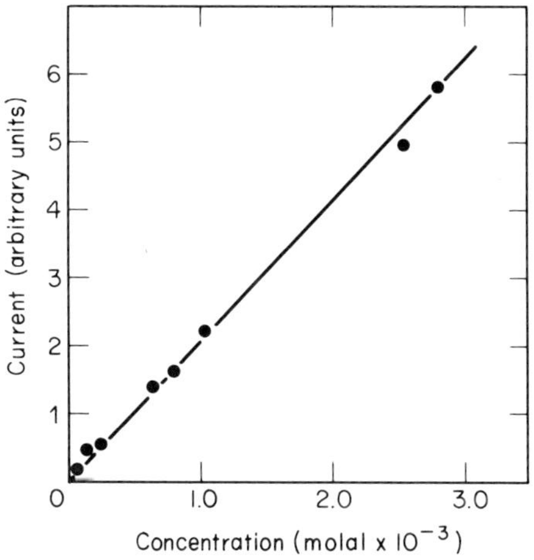

FIG. 2.1.2. Plot of i_d against $[Cd^{2+}]$ in $LiNO_3$ + $NaNO_3$ + KNO_3 melt. (Data from Ref. 44, p. 3019, by permission of the Faraday Society.)

Inman, Lovering, and Narayan [44] have made polarographic studies for the reduction of Cd(II) at a DME in molten $LiNO_3$ + $NaNO_3$ + KNO_3 at 145°C. Figure 2.1.2 shows plots of diffusion current vs concentration.

Table 2.1.10 gives the relevant details.

TABLE 2.1.10. Polarographic Data for Cd^{2+} in $LiNO_3 + NaNO_3 + KNO_3$ Melts[a]

	Average $E_{\frac{1}{2}}$ (V)	Probable error (mV)	Typical slope (mV)	Temp range (°C)	No. observations	Min concn ($\times 10^{-3}$ m)	Max concn ($\times 10^{-3}$ m)	D (cm^2/sec) $\times 10^{-6}$	No. observations	Temp (°C)
1[b]	-0.483	±2	42	148-150	6	0.075	1.24	1.02 ± 0.20	6	149 ± 1
2[c]	-0.499	±7	41	140-142	8	0.64	4.4	-	-	-

[a]These data are compared in Table 2.1.11 with some earlier measurements.

[b]These melts contained very small concentrations of Cl^- and H_2O, as indicated by their residual currents.

[c]These melts contained higher concentrations of Cl^- and other impurities, as indicated by their residual currents (nonbatch-selected $LiNO_3 \cdot 3H_2O$ was employed in these experiments).

TABLE 2.1.11. Polarographic Data for Cd^{2+} in Various Melts

Solvent	T (°C)	Reference electrode	Micro-electrode	Method	Half-wave potential (V)	Reversibility	Maxima	D (cm^2/sec) $\times 10^{-6}$	Refs.
$LiNO_3 + NaNO_3 + KNO_3$	160	Hg/Hg_2Cl_2	dme	Dc pol	-0.549	Good	None	1.5	[45]
	156	Hg/Hg_2CL_2		Sq wave	-0.540			0.72	[46]
	162	Ag/AgCl (satd)		E swp	-0.392			1.2	[47]
$NaNO_3 + KNO_3$	250	Ag/AgI		Dc pol	-0.665			-	[48]
$LiCl + KCl + AlCl_3$	160	Hg/Hg_2Cl_2			-0.370			-	[49]
$NaCl + KCl + AlCl_3$	140	Al			+0.480			-	[50]
LiAc + NaAc + KAc	197	Ag/AgCl			-1.10			-	[51]
$NH_4 \cdot O \cdot CO \cdot H$	125	Hg-pool			-0.170			-	[52]
0.1 M KNO_3(aq)	25	Hg/Hg_2Cl_2			-0.58			6.9	-

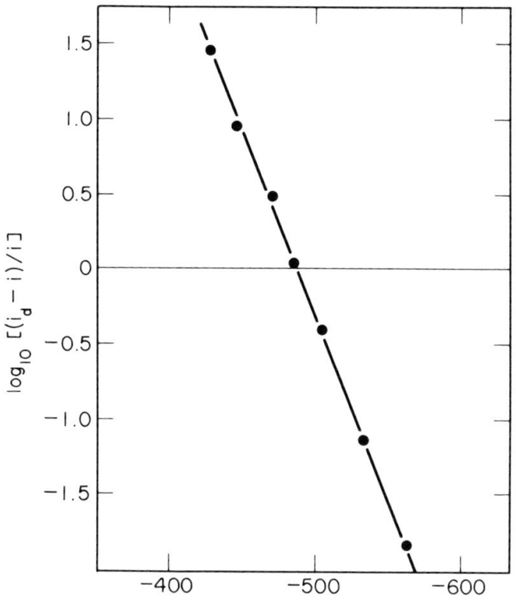

FIG. 2.1.3. Plot of $\log_{10}(i_d - i)/i$ against E for 0.15 Cd^{2+}. (Data from Ref. 44, p. 3030, by permission of the Faraday Society.)

The log plots of the polarograms indicate that these polarographic reductions are best described by the Heyrovsky-Ilkovic equation:

$$E_{DME} = E_{1/2} + 2.303\frac{RT}{nF}\log\left(\frac{i_d - i}{i}\right) \qquad (2.1.3)$$

where $E_{1/2}$ is the half-wave potential, i_d the limiting current, and i the current corresponding to the potential, E_{DME}, on the reduction wave.

The slopes of these plots indicate that the reduction of Cd^{2+} ions in nitrate solvent is reversible, forms mercury-soluble products, and involves the transfer of two electrons in the unit step (Fig. 2.1.3).

The authors [53] also determined stability constants of halide complexes under similar conditions. Results are shown in Table 2.1.12.

Stability constants are compared with the values reported by other workers in Table 2.1.13. The results are similar to those previously reported and there does not appear to be any change in structure.

TABLE 2.1.12. Overall Stability Constants

$Cd^{2+} + Cl^-$	Data 1	Data 2
	$C_M = 0.001043$ m	$C_M = 0.000146$ m
Computer analysis	$\beta_0 = 0.93 \pm 0.20$	$\beta_0 = 1.12 \pm 0.26$
	$\beta_1 = 197 \pm 46$	$\beta_1 = 108 \pm 47$
	$\beta_2 = 48 \times 10^2 \pm 23 \times 10^2$	$\beta_2 = 51 \times 10^2 \pm 22 \times 10^2$
	$\beta_3 = 20 \times 10^4 \pm 3.7 \times 10^4$	$\beta_3 = 25 \times 10^4 \pm 2.3 \times 10^4$
	$\beta_4 = 72.5 \times 10^4 \pm 15.2 \times 10^4$	4-degree fit not obtainable

β_0 assumed as unity

de Ford and Hume graphical analyses	$\beta_1 = 180$ (+ 20)	$\beta_1 = 100$ ($\pm$ 20)	The probable errors on β_1 were estimated graphically.
	$\beta_2 = 59 \times 10^2$	$\beta_2 = 80 \times 10^2$	
	$\beta_3 = 11 \times 10^4$	β_3 and β_4 not obtainable	
	$\beta_4 = 2 \times 10^6$		

$Cd^{2+} + Br^-$

	$C_M = 0.000089$ m		
Computer analysis	$\beta_0 = 1.20 \pm 0.03$	Graphical analysis	β_0 assumed as unity
	$\beta_1 = 642 \pm 29$		$\beta_1 = 660$ ($\pm$ 30)
	$\beta_2 = 91 \times 10^3 \pm 7 \times 10^3$		$\beta_2 = 78 \times 10^3$
	$\beta_3 = 73 \times 10^5 \pm 5 \times 10^5$		$\beta_3 = 80 \times 10^5$
	$\beta_4 = 93 \times 10^6 \pm 9 \times 10^6$		$\beta_4 = 87 \times 10^6$

TABLE 2.1.13. Stepwise Stability Constants

Solvent melt	Temp (°C)	Method	Stability constants	Refs.
		Chloro-cadmium$^{(2+)}$ Complexes		
$LiNO_3 + KNO_3$	180	dc pol	$K_1 = 200$[a]	[54]
			$K_2 = 15$	
			$K_3 = 40$	
			$K_4 = 5$	
$LiNO_3 + KNO_3$	152	emf	$K_1 = 156, 176$	[55], [53]
			$K_2 = 64$	
$LiNO_3 + KNO_3$	160	Isotopic tracer	$K_1 = 832$	[56]
			$K_2 = 3$	
			$K_3 = 52$	
			$K_4 = 5$	
$LiNO_3 + KNO_3$	160	emf	$K_1 = 157$	[57]
			$K_2 = 78$	
$LiNO_3 + NaNO_3 + KNO_3$	160	Cathode-ray polarography (E swp)	$K_1 = 277$	[58]
	180		$K_1 = 202$	
	204		$K_1 = 157$	
$LiNO_3 + NaNO_3 + KNO_3$	145	dc pol	$K_1 = 155$	[53]
			$K_2 = 30$	
		Bromo-cadmium$^{(2+)}$ Complexes		
$LiNO_3 + KNO_3$	171	emf	$K_1 = 681$	[57]
			$K_2 = 281$	
	240		$K_1 = 272$	
			$K_2 = 111$	
$LiNO_3 + NaNO_3 + KNO_3$	164	Cathode-ray polarography (E swp)	$K_1 = 350$	[58]
	189		$K_1 = 249$	
	199		$K_1 = 236$	
	204		$K_1 = 220$	
$LiNO_3 + NaNO_3 + KNO_3$	145	dc pol	$K_1 = 640$	[53]
			$K_2 = 140$	
			$K_3 = 80$	
			$K_4 = 15$	

[a] This value has recently been recalculated by Braunstein et al. [57] as 128.

The order of stability of the complexes of the metal ion, (F <<) Cl^- < Br^- (<<CN^-), is consistent with the formation of discrete complex ions in which substantial electron delocalization along the metal-halogen bonds occurs. The series is in agreement with the predications of Irving [59].

During this work maxima were observed on dc polarograms in these melts. They did not in general resemble the classical kind observed in aqueous solutions. They are related to adsorption processes involving complex ions present in the bulk solution. The electrode processes may behave in an ideally reversible manner in the region of the maxima.

These results may be explained in terms of the theory formed by Delahay [60]. However, the assumptions and the conjectural arguments involved in interpreting classical polarographic data in this way should be recognized.

It is possible that the addition of bromide ions to the solutions containing the metal ions increases the rate of electron transfer and the metal-metal ion electrode reaction becomes ideally reversible. In this case, according to Delahay [60], charging of the double layer will occur entirely by charge combination, $M^{2+} + 2e \rightarrow M$, without a flow of current in the external circuit. Thus current only flows in the external circuit because the anion excess left at the electrode surface when the rapid charge combination occurs has to be balanced.

According to Berglund [61], ammonium sulfamate melts at 125°C without decomposition. Above 160°C it decomposes, yielding ammonia gas and ammonium imidodisulfate. However, it is reported [62] that ammonium sulfamate melts at 132.8°C and that at 160°C decomposes slightly (only a few units per cent). It should, therefore, be a suitable solvent for polarographic studies. Bartocci, Marrassi, and Pucciarelli [63] have studied the polarographic behavior in molten ammonium sulfate at 162 ± 2°C. In spite of thermal decomposition of the solvent the measurements were made at this temperature because more stable half-wave potential values have been obtained than at lower temperatures. The decomposition did not significantly alter the nature of the solvent during experiments. The data are shown in Table 2.1.14.

Plots of log $[i/i_d - i]$ vs E showed the applicability of the Heyrovsky-Ilkovic equation. The reciprocal of the slopes of the curve for Cd^{2+} was in good agreement with the theoretical value (86/n mV), with a deviation of less than 5%. All polarographic waves were well defined. Potential measurements were made against a mercury pool electrode.

Francini, Martini, and Monfrini [64] have examined the oscillopolarographic behavior of Cd^{2+} ions in $AlCl_3 \cdot NaCl$ melt. The reference electrode was a very pure Al wire immersed in a small tube sealed at one end by a fritted disk; it was filled with the melt of the cell after bubbling with inert gas. Table 2.1.15 shows that the peak potentials do not depend on concentration or on the voltage sweep rate.

The difference between peak and half-peak potentials depends on the absolute temperature and number of electrons of the electrode reaction only. The obtained values are shown in Table 2.1.16 and are in agreement with theory in the investigated temperature range 150-230°C.

TABLE 2.1.14. Polarographic Data for Cd^{2+} at 435°K

C (m$\underline{M}$/l)	$E_{1/2}$ (mV)	I (μA/m$\underline{M}$) (1 $mg^{-2/3}$ $sec^{1/2}$)	D X 10^6 (cm^2/sec)	2.303 RT/nF (mV)
0.61	(-170)	1.08	0.77	43
0.67	-190	1.09	0.78	42
0.73	-77	1.03	0.70	45
1.32	-184	1.06	0.74	42
1.42	-180	1.04	0.71	44
1.44	-180	1.06	0.74	43
1.84	-186	1.06	0.74	42
2.06	-183	1.05	0.72	42
2.14	-185	1.03	0.70	46
2.48	-186	1.09	0.78	43
2.68	-185	1.05	0.72	43
2.80	-189	1.10	0.79	44
2.84	-187	1.03	0.70	44
3.51	-186	1.02	0.68	44
Av	-184	1.06	0.73	43
Av dev	±3	1.75%	4.01%	2.14%

TABLE 2.1.15. Peak Potentials at Various [Cd^{2+}] in $AlCl_3$.NaCl

T (°C)	C (m$\underline{M}$/l)	E_p^{obsd} (V)
176	0.199	+0.468
176	0.397	+0.468
176	0.595	+0.468
176	0.794	+0.468
176	0.993	+0.468

TABLE 2.1.16. Relationship between Observed and Theoretical Peak Potential

T (°C)	$(E_{p/2} - E_p)_{th}$ (mV)	$(E_{p/2} - E_p)_{obsd}$ (mV)
150	40.1	40
160	41.0	41
170	41.9	42
180	42.9	43
190	43.9	44
200	44.8	45
210	45.7	46
220	46.7	47
230	47.6	48

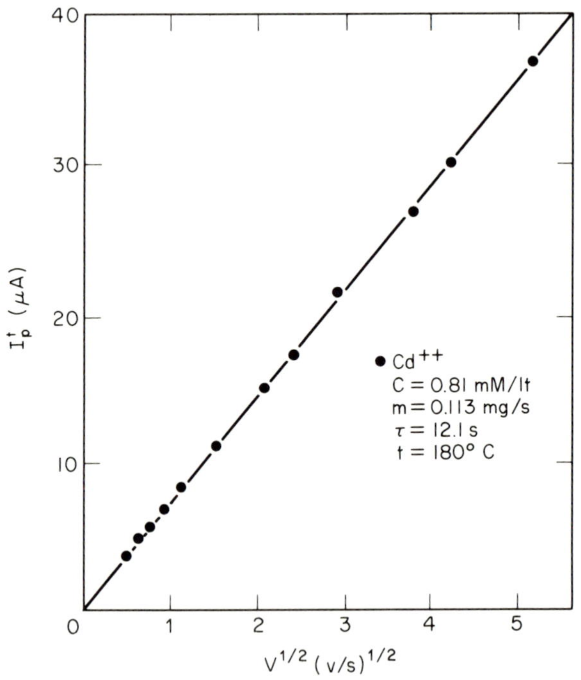

FIG. 2.1.4. Plot of current peak height against (sweep rate)$^{1/2}$ for Cd^{2+} in $AlCl_3 \cdot NaCl$ melt. (Data from Ref. 64, p. 3, by permission of Electrochim. Metallorum.)

The relationship between current heights and square root of the voltage sweep rate is linear up to approximately 10 V/sec (Fig. 2.1.4).

The deviations from linearity at higher sweep rates are related to a partial kinetic control of the current; this is confirmed by the corresponding progressive increase of the ($E_{p/2}$ - E_p) values given in Table 2.1.17. This value increases as the deviation from linearity becomes greater.

Table 2.1.18 gives the diffusion coefficient value at 160° C as well as the corresponding activation energy (E*).

Francini and Martini [65] have made similar measurements in alkali nitrate melts. For sweep rates up to 10 V/sec the peak currents are proportional to sweep rate. The values of the differences between half-peak and peak potentials at various sweep rates are given in Table 2.1.19. At the same time a distortion of the wave occurs which becomes smaller and rounded as predicted by theory for nonreversible reactions.

TABLE 2.1.17. $E_{p/2} - E_p$ at a Series of Sweep Rates ($AlCl_3$ Melts) ($E_{p/2}^{th} - E_p^{th} = 44.8$ mV)

V (V/sec	0.250	0.416	0.630	0.910	1.39	2.38	4.25	5.78	8.33	11.7	17.5	26.3
$E_{p/2}^{obsd} - E_p^{obsd}$ (mV)	44	44	44	44	44	44	44	44	44	44	46	49

TABLE 2.1.18. Diffusion Coefficient and Activation Energy ($AlCl_3$ Melts)

D, 160° C (cm^2/sec)	E* (kcal/mole)
$1.8^4 \times 10^{-6}$	-5.940

TABLE 2.1.19. $E_{p/2} - E_p$ at a Series of Sweep Rates (Alkali Nitrate Melts) ($E_{p/2}^{th} - E_p^{th} = 42.2$ mV)

V (V/sec)	0.157	0.244	0.400	0.609	0.925	1.38	2.22	3.55	5.40	10.4	15.1	23.8	38.4
$E_{p/2}^{obs} - E_p^{obs}$ (mV)	42	42	42	42	42	42	42	42	42	46	49	51	68

Oscillographic polarography is much more sensitive than conventional polarography as many electrodes reactions that appear fully reversible by conventional polarography do not completely obey the theoretical equations. This can be seen in Table 2.1.20 where the experimental values of the slopes 2.3RT/nF are compared with the theoretical ones, showing that for conventional polarography the reaction for Cd^{2+} in the ternary melt is reversible.

TABLE 2.1.20. Theoretical and Observed Tafel Slopes in Melts

$LiNO_3$-$NaNO_3$-KNO_3 eutectic			$NaNO_3$-KNO_3 eutectic		
	2.3 RT/nF			2.3 RT/nF	
T (°C)	Theoretical (mV)	Observed (mV)	T (°C)	Theoretical (mV)	Observed (mV)
162	43.1	43	232	50	50

It has been demonstrated that the experimental values of the peak potentials agree with those predicted by theory related to the half-wave potentials, and are relatively unaffected by small deviations from reversibility. This is also true in the presence of relatively high quantities of a ligand as shown in Table 2.1.21 where the values experimentally measured at different chloride ion concentrations are reported.

TABLE 2.1.21. Effect of Melt Composition on Polarographic Waves [66]

$n_{Cl^-} = \frac{n_{Cl^-}}{n_{NO_3^-}} \times 10^5$	$(E_{1/2} - E_p)_{obsd}$
None	18
7.30	19
14.6	18
43.7	19
73.0	20
291	20
582	18
1160	18

The authors emphasize, however, that in connection with ternary melts the base melt is important when comparing the results of different investigators. The results obtained with solvents containing Li^+ are compared with pure solvent and the influence of the Li^+ ion on the equilibrium parameters as well as on the kinetics of the electrode process is shown (ion-solvent interaction).

2.2. VOLTAMMETRIC DATA IN MOLTEN SALTS

Laitinen and Gaur [67] investigated the reduction of cadmium ions in a fused eutectic mixture of potassium chloride and lithium chloride at 450° C. Platinum microelectrodes of different areas and geometry were used. Providing that the dimensions of the electrode were considerably greater than the thickness of the diffusion layer, linear diffusion theory was obeyed. The transition time constant for cadmium ions was found to be $0.83 \pm 0.02 \times 10^3$ A cm $sec^{1/2}$/mole. The diffusion coefficient was calculated to be 2.08×10^{-5} cm^2/sec. Naryshkin, Yurkinskii, and Yavich [68] have made oscillographic linear potential sweep experiments in fused salts using an equimolar mixture of sodium and potassium chlorides.

Kal'voda and Tamanova [69] have studied the Cd(II)/Cd reaction using linear potential sweep voltammetry in potassium chloride and lithium chloride in the ratio 1:1 (temperature 460° C). A platinum disk electrode was employed in conjunction with a stationary platinum reference electrode. The data is shown in Table 2.2.1.

TABLE 2.2.1. Potential Sweep Data in KCl/LiCl Melt

Potential peak on the (dE/dt)-E curve (V)		Potential peak on the I-E curve (V)
Cathodic	Anodic	
-0.75	-0.80	-0.85
-1.00	-1.00	-1.15

Similar measurements were made in potassium and sodium nitrates in the molar ratio 1:1 (temperature 250° C). The experiments were complicated by the fact that the background curve shows various peaks which depend on the potential range in which the electrode is polarized and also on the magnitude of the initial polarization potential.

The results of a number of other investigations are given in Table 2.2.2.

TABLE 2.2.2. Potential Sweep Data in Melts

Reaction	Conditions	Method	Values	Refs.
$Cd^{2+} + 2e = Cd$	KCl-LiCl, 450°C 0.015-0.073 MCd^{2+}	chronopot	$D = 2.08 \times 10^{-5}$	[67]
$Cd^{2+} + 2e - Cd$	KCl-LiCl, 450°C 0.002-0.79 MCd^{2+}	chronopot	$D = 1.68 \times 10^{-5}$	[70]
$Cd^{2+} + 2e = Cd$	KCl-LiCl, 395-809°C	chronopot	$D = 1.2 \pm 0.2 \times 10^{-5}$ $\Delta H = 6.5 \pm 0.3$ kcal/mole	[71]
$Cd^{2+} + 2e = Cd$	$LiNO_3$-KNO_3-$NaNO_3$	pol	$D = 1.5 \pm 0.2 \times 10^{-6}$ $\Delta H = 13$kcal/mole	[44]

3. KINETIC PARAMETERS AND DOUBLE-LAYER PROPERTIES

3.1. KINETIC PARAMETERS

3.1.1. Aqueous Solution

3.1.1.1. Amalgam Electrodes. Table 3.1.1 shows the results of all the reported investigations to date. No attempt has been made to categorize the data. It is not clearly understood why the cadmium exchange at an amalgam electrode should be so fast, and this reaction has attracted much attention with regard to the marked effect of anions on the kinetics and the unusual value of the transfer coefficient. The exchange has not yet been fully described mechanistically. It is concluded from all the available evidence that the formation of a specifically adsorbed, partially charged ion occurs during the reaction, and it may be that the formation of this entity is the rate-determining step, although this does not account for all the observed facts. The electrochemistry of the Cd(II)/Cd(Hg) system has been complicated by a number of observations which are difficult to explain. It has been indicated that the electrode mechanism may be more complex than an electron-transfer combined with solely diffusive mass transfer. Results obtained in different investigations have not generally been in agreement, and in similar systems anomalies have sometimes been reported. However, reported values for the kinetic parameters (rate constant k, transfer coefficient α) have generally differed even in the absence of anomalies. It has been proposed by Bauer [88] that all reported observations may be explained by taking into account that mass transport is influenced by electrical migration of cadmium ions within the diffuse double layer. The ratio of measured to apparent (corrected for electrostatic effects) rate constants depends on the potential ϕ° at the outer Helmholtz plane in the manner shown in Fig. 3.1.1. For a particular value of ϕ°, the measured rate constant is infinite; for smaller values of ϕ° it is positive (phase angle $< 45^{\circ}$ in faradaic-admittance studies) and for larger values it is negative (phase angle $> 45^{\circ}$).

ϕ° has a critical value which depends on the true exchange current density (i.e., on the concentrations of the oxidant and reductant and on the transfer coefficient) and on the nature and concentration of the supporting electrolyte. Therefore, in a system where ϕ° can be close to this critical value, the reduction of a positively-charged species may be expected to show certain of the following characteristics:

1. Increasingly negative values of ϕ° lead to large, infinite, and, finally, negative measured rate constants; thus the use of electrolytes containing anions of increasing surface activity will cause such a change.

2. At a fixed value of ϕ°, changes in the true exchange current density can produce similar effects.

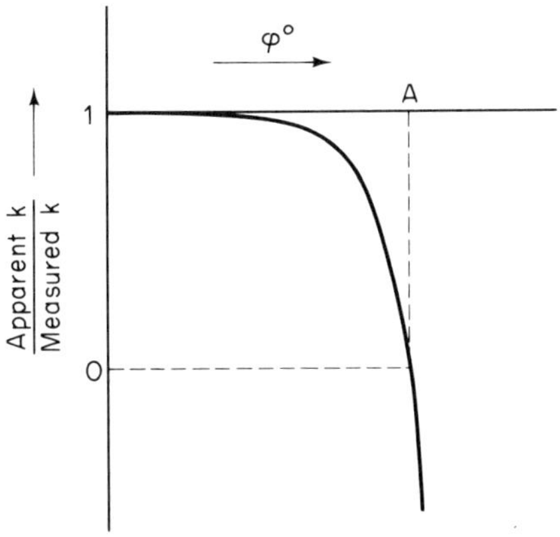

FIG. 3.1.1. Variation of measured rate constant as a function of potential across the diffuse double layer for attraction of the depolarizer. (Data from Ref. 88, p. 65, by permission of Elsevier Ltd.)

3. Since variations of ϕ° of the order of millivolts or even less, or corresponding variations in the exchange current density, can have such profound effects (steep slope of the curve in Fig. 3.1.1 at the critical value of ϕ°) it will be difficult to obtain reproducible results.

The behavior of cadmium fits the expectations for mass transport influenced by electrical migration. The low value of the cathodic transfer coefficient observed in the majority of the reported investigations has prompted the suggestion that the charge transfer step in the cadmium exchange consists of two steps:

$$Cd^{2+} + e = Cd^{+} \qquad (3.1.1)$$

$$Cd^{+} + e + Hg = Cd(Hg) \qquad (3.1.2)$$

Galvanostatic investigations [89] have been interpreted assuming this mechanism with a following chemical disproportionation of the intermediate. Although the general behavior of the experimental data appeared to be satisfactory, the magnitude of the exchange current density was two orders less than the generally accepted values. Chemical contamination may well be the cause of these and other effects which have also been observed in potential step experiments. The deviations from simple Randles behavior of the Faradaic impedance analog observed by Despic et al. [90] was interpreted in terms of a Cd^{+} intermediate. However, these were only observed at very high reactant concentrations under conditions when the electrode surface may very well be covered with cadmium.

TABLE 3.1.1. Kinetic Parameters of the Exchange Reaction at an Amalgam Electrode

Reaction	Conditions	Electrode
$Cd^{2+} + Hg + 2e \rightarrow Cd(Hg)$	1 M KNO_3, 10^{-3} M Cd^{2+}, 10^{-3} M Cd(Hg)	DME
$Cd^{2+} + Hg + 2e \rightarrow Cd(Hg)$	0.5 M Na_2SO_4	hng drp
$Cd(CN_3) + Hg + 3e \rightarrow Cd(Hg) + 3CN^-$	Excess cyanide	hng drp
$Cd(CN)_2 + Hg + 2e \rightarrow Cd(Hg) + 2CN^-$	Little cyanide	hng drp
$Cd^{2+} + Hg + 2e \rightarrow Cd(Hg)$	1 M Na_2SO_4 ($1.0\text{–}10.0 \times 10^{-7}$ M Cd^{2+}), 4×10^{-4} M Cd(Hg)	hng drp
$Cd^{2+} + Hg + 2e \rightarrow Cd(Hg)$	0.5 M Na_2SO_4 (0.5–3 mM Cd^{2+})	hng drp
$Cd^{2+} + Hg + 2e \rightarrow Cd(Hg)$	0.5 M HCl	DME
$Cd^{2+} + Hg + 2e \rightarrow Cd(Hg)$	0.5 M K_2SO_4	DME
	1 M $NaClO_4$, pH 2	DME
	1 M KNO_3, pH2	DME
	1 M $HClO_4$	DME
	1 M KCl, pH 2	DME
$Cd^{2+} + Hg + 2e \rightarrow Cd(Hg)$	0.5 M Na_2SO_4 (0.5 mM Cd^{2+})	DME
	1 M Na_2SO_4 (0.2 mM)	DME
	1 M Na_2SO_4 (1.0 mM)	DME
$Cd^{2+} + Hg + 2e \rightarrow Cd(Hg)$	1 M KNO_3 (1.22 mM Cd) (0 mM Cd^{2+})	DME
	1 M KNO_3 (1.5 mM Cd) (1.0 mM Cd^{2+})	DME
	1 M KNO_3 (1.08 mM Cd^{2+})	DME
	0.5 M KCl (0.5 mM Cd^{2+})	DME
	0.5 M HCl (1.08 mM Cd^{2+})	DME
	1.0 M HCl (1.0 mM Cd^{2+})	DME
$Cd^{2+} + Hg + 2e \rightarrow Cd(Hg)$	1 M $NaClO_4$	DME
$Cd^{2+} + Hg + 2e \rightarrow Cd(Hg)$	0.5 M Na_2SO_4	DME
	1 M KCl	DME
$Cd^{2+} + Hg + 2e \rightarrow Cd(Hg)$	1 M KNO_3, 5×10^{-4} M Cd^{2+}	DME
	2×10^{-4} M Cd^{2+}	DME

in Aqueous Solutions

Technique	Rate constant	α	Remarks	Refs.
farad. imp	~0.6	-	$\Delta H = 21$ kJ/mole $A = 10^3$ cm/sec	[72]
farad. imp	4.2×10^{-2}	0.17	20°C	[73]
farad. imp	-	0.3	20°C	[73]
farad. imp	-	0.25	20°C	
i stp	4.47×10^{-2}	0.23		[74]
V stp	2.6×10^{-2}	0.25	20°C	[75]
ac pol	$0.5 \times 10^{-1} \pm 15\%$		23.7-25.4°C	[76]
sq wave pol	$k_0 \sim 0.08$	~0.08	22°C	[77]
sq wave pol	k_0 2.3	0.07	22°C	
sq wave pol	k_0 6.3	0.15	22°C	
sq wave pol	$k_0 \sim 1.8$	~0.08	22°C	
sq wave pol	k_0 2.9	0.78	22°C	
ac pol	0.25	-		[78]
ac pol	0.30	-		
ac pol	0.10	-		
ac pol	1.0	-		[78]
ac pol	1.0	-		
ac pol	0.4	-		
ac pol	1.5	-		
ac pol	~0.5	-		
ac pol	0.6	-		
farad. imp	1.8	0.08		[79]
farad. imp	0.085	0.12		[79]
farad. imp	5.0	0.7		
ac pol	6×10^{-1}	-	-	[80]
ac pol	6×10^{-1}			

(continued)

Table 3.1.1 (continued)

Reaction	Conditions	Electrode
$Cd^{2+} + Hg + 2e \rightarrow Cd(Hg)$	1 $\underline{M}$ KCl, 0.02-0.08 m$\underline{M}$ Cd^{2+}	DME
		DME
$Cd^{2+} + Hg + 2e \rightarrow Cd(Hg)$	0.5 $\underline{M}$ Na_2SO_4 (2 m$\underline{M}$ Cd^{2+})	DME
$Cd^{2+} + Hg + 2e \rightarrow Cd(Hg)$	1 $\underline{M}$ $HClO_4$(0.27-0.54 m$\underline{M}$ Cd^{2+})	DME
	0.1 $\underline{M}$ $HClO_4$, 0.9 $\underline{M}$ $NaClO_4$	DME
$Cd^{2+} + Hg + 2e \rightarrow Cd(Hg)$	1 $\underline{M}$ $HClO_4$	DME
	0.1 $\underline{M}$ $HClO_4$, 0.9 $\underline{M}$ $NaClO_4$	DME
	1 $\underline{M}$ $HClO_4$ + n-BuOH (θ_{ads} 0.35)	DME
	0.1 $\underline{M}$ $HClO_4$ + 0.9 $\underline{M}$ $NaClO_4$ +	DME
	Leucoriboplasm (θ_{ads} 0.45)	
$Cd^{2+} + Hg + 2e \rightarrow Cd(Hg)$	0.5 $\underline{M}$ Na_2SO_4, 0.5 m$\underline{M}$ Cd^{2+}	hng drp
	0.5 m$\underline{M}$ Cd^{2+}	hng drp
	2.5 m$\underline{M}$ Cd^{2+}	hng drp
	200 m$\underline{M}$ Cd^{2+}	
	0.5 $\underline{M}$ $HClO_4$ + 0.5 $\underline{M}$ $NaClO_4$, 0.01 m$\underline{M}$ Cd^{2+}	hng drp
	0.1 m$\underline{M}$ Cd^{2+}	
	0.25 m$\underline{M}$ Cd^{2+}	
	0.5 m$\underline{M}$ Cd^{2+}	
$Cd^{2+} + Hg + 2e \rightarrow Cd(Hg)$	1 $\underline{M}$ Na_2SO_4, pH 4, 2.5-50 m$\underline{M}$ Cd(Hg)	hng drp
	0.5-10 m$\underline{M}$ Cd^{2+}	
$Cd^{2+} + Hg + 2e \rightarrow Cd(Hg)$	1 $\underline{M}$ $NaClO_4$, pH 4, 1.19-2.5 m$\underline{M}$ Cd(Hg)	hng drp
	0.12-2.0 m$\underline{M}$ Cd^{2+}	
$Cd^{2+} + Hg + 2e \rightarrow Cd(Hg)$	1 $\underline{M}$ KCl, pH 4, 0.5-9.7 m$\underline{M}$ Cd	hng drp
	0.44-2.15 m$\underline{M}$ Cd^{2+}	
$Cd^{2+} + Hg + 2e \rightarrow Cd(Hg)$	1 $\underline{M}$ $NaClO_4$	stat hng drp

Technique	Rate Constant	α	Remarks	Refs.
farad. imp	$k_{sh} > 12$		$D_{Cd^{2+}} = 8.9 \times 10^{-6}$, 25°C	[81]
farad. imp	$k_{sh} = 4-8$		$D_{Cd^{2+}} = 4.02 \times 10^{-6}$, 0°C	
V stp	3.6×10^{-2}	0.29		[82]
farad. imp	0.35 ± 0.03	0.14		[83]
farad. imp	0.46 ± 0.03	0.14		
farad. imp	0.35 ± 0.02	0.14 ± 0.01		[84]
farad. imp	0.45 ± 0.02	0.09 ± 0.01		
farad. imp	0.075 ± 0.002	0.13 ± 0.02		
farad. imp	0.113 ± 0.005	0.14 ± 0.03		
coul	4.8×10^{-3} A cm^{-2}	0.5	$C_L = 23.7$ μF cm^{-2}	[85]
coul	16.7×10^{-3}			
coul	36.7×10^{-3}			
	184×10^{-3}			
coul	7.7×10^{-3} A cm^{-2}	0.5	$C_L = 29.9$ μF cm^{-2}	
	11.5×10^{-3}			
	18.1×10^{-3}			
	34.7×10^{-3}			
i stp	6.3×10^{-2} cm/sec	0.13 ± 0.02	$C_L = 21$ μF cm^{-2}	[86]
			$D_O = (5.3 \pm 0.5) \times 10^{-6}$	
			$D_R = (11 \pm 2.0) \times 10^{-6}$	
i stp	0.55 ± 0.1	0.24 ± 0.03	$D_O = (8.2 \pm 0.5) \times 10^{-6}$	[86]
			$D_R = (9.5 \pm 1.0) \times 10^{-6}$	
			$C_L = 29.5$ μF cm^{-2}	
i stp	5/cm	-	$C_L = 25$ μF cm^{-2}	[86]
			$D_O = (8.7 \pm 0.03) \times 10^{-6}$	
			$D_R = (8.5 \pm 0.5) \times 10^{-6}$	
farad. imp	0.44	0.68 ± 0.05	$D = 9.8 \pm 1.0 \times 10^{-6}$, 23°C	[87]

Radiolysis experiments [91] have established, beyond doubt, the existence of Cd^+, and the standard potential of the Cd^{2+}/Cd^+ couple has been estimated as -1.9 to -2.5 V. It is clear from this that Cd^+ will only be formed from Cd^{2+} by reduction with the solvated electron and certainly not by hydrogen atoms. The standard free energy of activation of cadmium with Cd^+ in solution as an intermediate has been estimated by Biegler et al. [92] to be ~ 200 kJ/mole. Biegler et al. [92] conclude that it is obviously inconsistent for Cd^+ to occur as an intermediate without very considerable stabilization due to adsorption. It is difficult to estimate whether or not such stabilization could be achieved since these latter authors calculate that an adsorption coefficient of the order of 10^3 cm would be required, which is unknown in the systems involving the strong adsorption of other ions on mercury.

The formation of the dimeric intermediate $Cd_2{}^{2+}$ has been established in the case of molten alumino-chlorides [93, 94]. It is not possible to say from the published data whether $Cd_2{}^{2+}$ in aqueous solution would be stable but it has been suggested [92] that such a species may be more stable than Cd^+. Anbar [91] suggests on the basis of radiolysis experiments that Cd^+ may be removed, dimerized, and subsequently disproportionated. Biegler et al. [92] consider the effect on the kinetic analysis on the assumption that a dimeric intermediate is formed. An expression is obtained for the exchange current, in terms of reactant species of the form:

$$i_0 = 2F(\underline{k}_0)^{1-\beta}(\underline{k}_0)^{\beta}K^{-\beta}[Cd^{2+}]^{2-\beta}[Cd]^{\beta} \tag{3.1.3}$$

where $K = [Cd][Cd^{2+}]/[Cd_2{}^{2+}]$, which differs from the usual form for first-order kinetics, and β is a symmetry factor:

$$i_0 = i = 2F\underline{k}[Cd^{2+}]^{1-\alpha}[Cd]^{\alpha} \tag{3.1.4}$$

A difficulty is encountered with the $Cd_2{}^{2+}$ mechanism in that a very high frequency factor is obtained and it therefore appears that for both cases the simple situation of the intermediate in solution is difficult to support with experimental evidence. Since there is no evidence for the existence of completely adsorbed intermediates, it seems that the most likely situation involves a transition state with a partially charged, adsorbed Cd^{2+} ion.

3.1.1.2. Solid Metal Electrodes. There has been considerably less study of the kinetics of the exchange reaction at solid metal electrodes. Table 3.1.2 shows the results of reported investigations.

The kinetics of deposition and dissolution of solid cadmium in neutral sulfate solutions has been extensively investigated by Lorenz [95, 96]. Lorenz made measurements in the Tafel region from which it was deduced that the rate-determining step is the charge transfer process between Cd^{2+} ions and the metal electrode. The charge transfer process was considered to occur as a one-step process with a charge transfer valence of two and a cathodic transfer coefficient of 0.45. Brodd [97], using complex plane analysis of impedance data, produced results in support of the conclusions of Lorenz [95, 96]. It has been emphasized from studies of the interphase that information can be obtained on electrolyte conditions for

TABLE 3.1.2. Kinetic Parameters of the Exchange Reaction at Solid Cadmium Electrodes in Aqueous Solution

Reaction	Conditions	Electrode	Technique	Rate constant	α	Remarks	Refs.
$Cd^{2+} + 2e \rightarrow Cd$	1 M $CdSO_4$	rd	i stp	$R_p = 2\text{–}6\ \Omega\text{-cm}^2$	0.5	$20 \pm 1°C$	[95]
	0.01 M $CdSO_4$ + 0.04 M K_2SO_4			$R_p = 20\text{–}50\ \Omega\text{-cm}^2$			
$Cd^{2+} + 2e \rightarrow Cd$	0.4–0.75 M Na_2SO_4 (5×10^{-3} M Cd^{2+})	stat dsk	i stp	$i_0 = 1.5 \times 10^{-3}$ A cm^{-2}	0.45 ± 0.03	$20 \pm 1°C$	[96]
$Cd^{2+} + 2e \rightarrow Cd$	25×10^{-3} M Cd^{2+}	stat dsk	i stp	-	0.5 ± 0.05	$20 \pm 1°C$	[96]
$Cd^{2+} + 2e \rightarrow Cd$	0.71 M $CdSO_4$	stat dsk	farad. imp	$i_0 = 1.4 \times 10^{-2}$ A cm^{-2}			[97]
$Cd^{2+} + 2e \rightarrow Cd$	1 M $NaClO_4$, pH 3, 0.028–0.452 M Cd^{2+}	stat dsk	i stp	$k = 3.8 \times 10^{-5}$ cm/sec	0.23	$23°C$, $\Delta H_D = 26.3$ kJ/mole	[98]
$Cd \rightarrow CdO$	6.25 M NaOH	stat dsk	farad. imp	$i_0 = 23 \times 10^{-2}$ A cm^{-2}	-	$80°C$	[99]
$Cd^{2+} + 2e \rightarrow Cd$	1 M $NaClO_4$, pH 3, 0.175 M Cd^{2+}	stat dsk	farad. imp	$i_0 = 5.8 \times 10^{-2}$ A cm^{-2}	0.35	$23°C$, $\Delta H_D = 29$ kJ/mole	[100]
$Cd(II) \rightarrow Cd(0)$	7 M KOH, 0.0099 M Cd^{2+}	stat dsk	farad. imp	$i_0 = 3.1 \times 10^{-2}$ A cm^{-2}	-	$23°C$	[101]
$Cd(II) \rightarrow Cd(0)$	7 M KOH, 0.0099 M Cd^{2+}	stat dsk	i stp, double pulse	$i_0 = 2.6 \times 10^{-2}$ A cm^{-2}	-	$23°C$	[101]

which the electrode interphase would be free of adsorption and the development of films. Several differential capacitance studies have been reported (see Section 3.2.). It is clear from a study of these that adsorption of the OH^- ion on the cadmium electrode is unimportant at potentials around the potential of zero charge, E_Z, and at more cathodic potentials. Adsorption of anions is, however, most marked on the anodic side of E_Z. The most reliable determinations of E_Z place it at ~-0.73 V (NHE) so that E°_{Cd} is ~0.4 V more positive than E_Z. Exchange measurements are therefore made with the electrode positively charged with respect to the electrolyte. It is likely that at these potentials the adsorption of OH^- is important and may lead to the development of films unless both potential and electrolyte conditions are carefully chosen. Thus, for example, if electrodes are electropolished at very anodic potentials, even in perchloric acid (10%), films may be produced.

Experiments have shown that for conditions of moderate anodic polarization (within ~150 mV of E°_{Cd}), the development of films may be avoided if the pH of the electrolyte is sufficiently low. Heusler and Gaiser [102] consider that a mechanism involving two consecutive transfer reactions cannot be excluded, namely:

$$Cd^{2+} + e \underset{k_1^+}{\overset{k_1^-}{\rightleftharpoons}} Cd_{ad}^+ \tag{3.1.5}$$

$$Cd_{ad}^+ + e \underset{k_2^+}{\overset{k_2^-}{\rightleftharpoons}} Cd \tag{3.1.6}$$

It was argued that an analogous mechanism with monovalent metal in the adsorbed state has been deduced from the kinetics of the solid zinc electrode in perchlorate solutions. Furthermore, measurements at cadmium amalgam electrodes yielded transfer coefficients close to $\alpha = 0.25$ assuming $Z = 2$ point toward the Mechanism (3.1.5)-(3.1.6). In addition, Lorenz [95, 96] found $\alpha = 0.25$ from measurements of transition times for the deposition of solid cadmium at high current densities. Biegler et al. [92] have adequately dealt with these arguments for the case of the amalgam electrode. However, Heusler and Gaiser [102] present some data which supports their hypothesis. This data is derived from measurements of current-potential curves with solid cadmium in solutions of cadmium perchlorate and barium perchlorate under stationary and transient conditions. The results were tested using the criteria established by Vetter [103] for the appearance of consecutive charge transfer reactions, and it was calculated that the first-order rate constants of Step (3.1.6) in both directions of the reaction are always larger than the rate constants of Step (3.1.5), even at the highest cadmium perchlorate concentration. Therefore, Step (3.1.5) may be considered to be rate determining. This does not mean that Step (3.1.6) is always in equilibrium. The respective reverse reactions of both steps do not influence the rate of the overall reaction at high overpotential. Although the rate constants of Step (3.1.6) are relatively high, the electronic interaction between the cadmium metal and the adsorbed monovalent cadmium is still

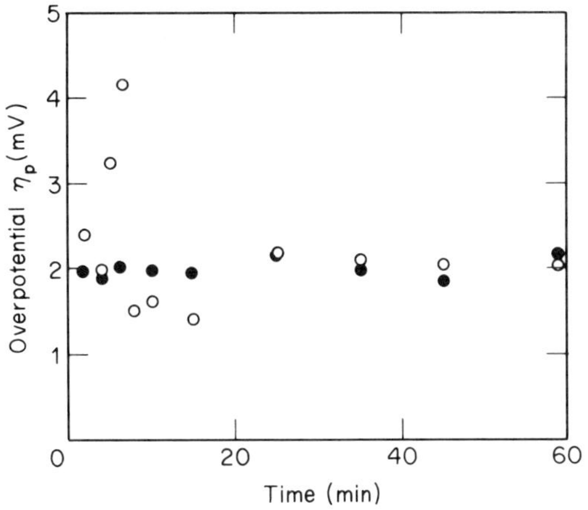

FIG. 3.1.2. The time-dependence of overpotential, 0.452 M Cd^{2+}; total [ClO_4^-], 2 M with $NaClO_4$; 23° C; faradaic current, 3.08 mA cm^{-2}; (•) electrochemically etched; (o) "normalized" perchloric acid (10%).

very weak. Transfer coefficients of α = 0.5 for both steps and a well-defined monovalent intermediate are necessary assumptions to explain the kinetics. If it is assumed that the monovalent cadmium is bound to the metal by a water bridge, then Step (3.1.6) is a metal/ion reaction [103] with the charge being transferred by the metal ion. Step (3.1.5) is a typical redox reaction with electrons carrying the charge across the double layer.

The comparatively low rate constants of Step (3.1.5) may be explained by significant differences in the solvation sheaths of cadmium ions in the solution and of adsorbed monovalent cadmium. Small rearrangements of the solvation sheets of the completely discharged metal and of the adsorbed monovalent metal are left for the further reaction of Step (3.1.6), which results in a relatively large rate. Conditions may be very different at a solid metal electrode than at the amalgam electrode, particularly at high overpotential, in the Tafel region, and this may account for the possible presence of Cd^+. For measurements in the region of the thermodynamic equilibrium potential work by Hampson, Latham, and Larkin [98] has shown that no evidence can be adduced for Cd^+. In this investigation, which was conducted at pH 3 to avoid the formation of films and using ultrapure conditions, a pronounced time dependence of overpotential was observed which depended upon the method of electrode pretreatment. This is shown in Fig. 3.1.2. Exchange currents are shown in Table 3.1.3 from which it can be calculated that the cathodic charge transfer coefficient is 0.23. The temperature variation is shown in Fig. 3.1.3. The slope of the line is consistent with a value of the enthalpy of activation of 26.3 kJ/mole. In a subsequent paper the authors present the results of an extension of this work to confirm earlier findings and to include measurements of the degree of metallurgical control exercised by the electrode on the exchange reaction. The method used was to study the faradaic impedance as a function of frequency of the applied

TABLE 3.1.3. Exchange Currents Calculated from Mean Slopes through the origin of η_D-i Plots

$[Cd^{2+}]$ (M)	i_0 (mA cm^{-2})	$[Cd^{2+}]$ (M)	i_0 (mA cm^{-2})
0.452	24.84	0.114	5.28
0.452	11.4	0.103	3.4
0.32	10.8	0.095	4.35
0.235	13.53	0.0617	3.74
0.196	6.69	0.0617	2.28
0.196	5.4	0.049	2.3
0.114	5.83	0.028	2.0

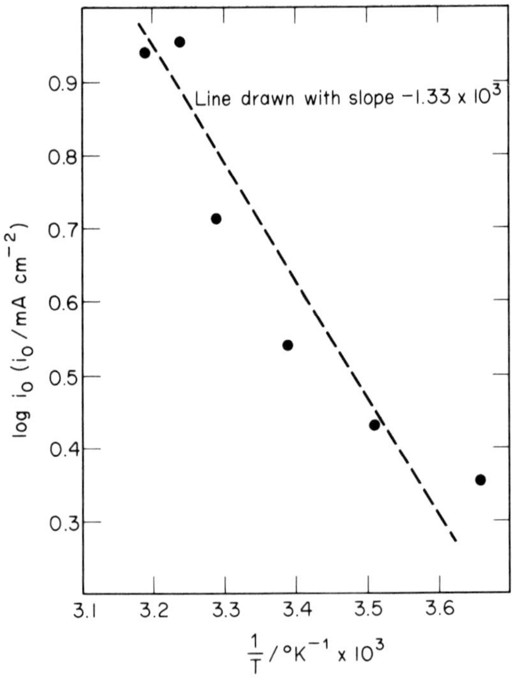

FIG. 3.1.3. Temperature-dependence of exchange current. 0.0617 M Cd^{2+}. Electrolytes maintained at 2 M $[ClO_4^-]$ with $NaClO_4$. (Data from Ref. 98, p. 216, by permission of Elsevier Ltd.)

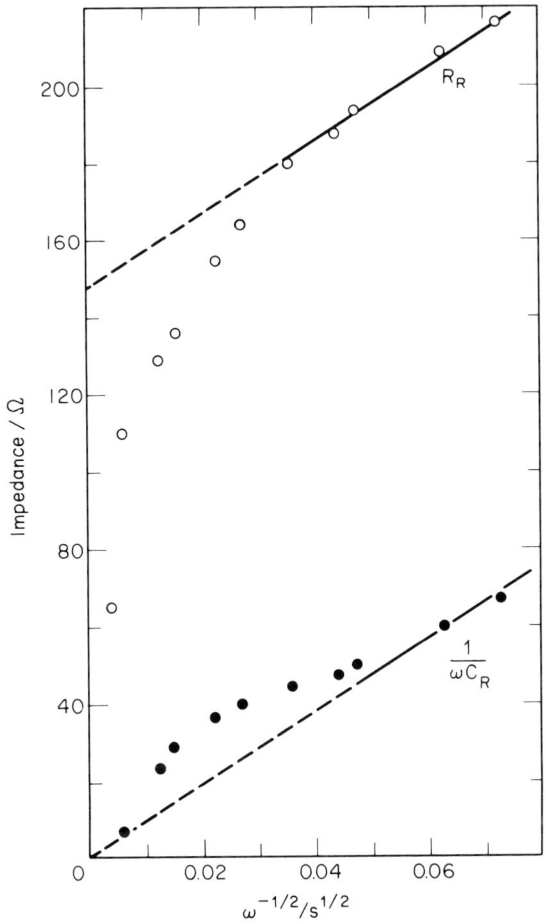

FIG. 3.1.4. Typical impedance spectrum. Electrode/electrolyte contact time 8 hr; 0.0253 mole/l Cd^{2+}; total $[ClO_4^-]$ 2 moles/l with $NaClO_4$, pH 3; 23°C; electrode area 3.46×10^{-2} cm^2. (Data from Ref. 100 by permission of Elsevier Ltd.)

ac and also as a function of temperature and concentration of electroactive species. Figure 3.1.4 shows a typical faradaic impedance spectrum.

From the intercepts at $\omega \rightarrow \infty$, the exchange current i_0 was calculated according to:

$$i_0 = \frac{RT}{ZF} \frac{1}{\Delta_{\omega \rightarrow \infty}} \qquad (3.1.7)$$

Figure 3.1.5 shows the variation of exchange current with $[Cd^{2+}]$. From the slope $\partial \log i_0 / \partial \log [Cd^{2+}]$, a value of 0.35 is obtained for the apparent charge transfer coefficient. The results are somewhat higher than those reported from galvanostatic measurements.

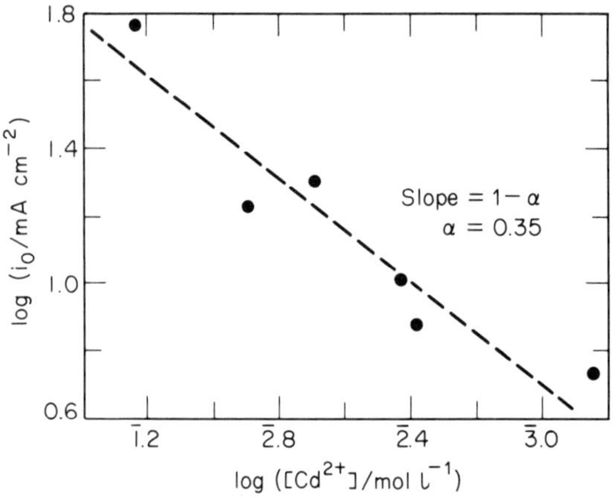

FIG. 3.1.5. Variation of exchange current with $[Cd^{2+}]$. Total $[ClO_4^-]$ 2 mole/l with $NaClO_4$; pH 3; 23°C. (Data from Ref. 100 by permission of Elsevier Ltd.)

TABLE 3.1.4. Comparison of Rates of Charge Transfer and Crystallization

$[Cd^{2+}]$ (mole/l)	i_0 (mA cm^{-2})	ZFV_g^0 (mA cm^{-2})	$10^{12}\,\Gamma_O$ (mole cm^{-2})
0.175	58.0	48.0	5
0.0755	18.6	7.45	2
0.0484	20.5	28.4	30
0.0253	10.2	5.1	16
0.024	7.6	6.5	21
0.007	5.5	1.9	22

In the range of $[Cd^{2+}]$ investigated, the relative magnitudes of the exchange current density and adatom flux indicate that the processes of charge transfer and crystallization have rates of the same order of magnitude. A comparison of the two parameters is given in Table 3.1.4.

The results presented in Table 3.1.4 show that Γ_O, the surface adatom concentration, is independent of $[Cd^{2+}]$. The relationship between exchange current and concentration of electroactive species:

$$i_0 = ZFkC_O^{1-\alpha} \tag{3.1.8}$$

is therefore justified and no adjustment of α is necessary to allow for changes in Γ_O. It is seen that the adatom flux is related to $[Cd^{2+}]$ in the same manner as the exchange current density. According to the mathematical treatments of Fleischmann et al., this indicates

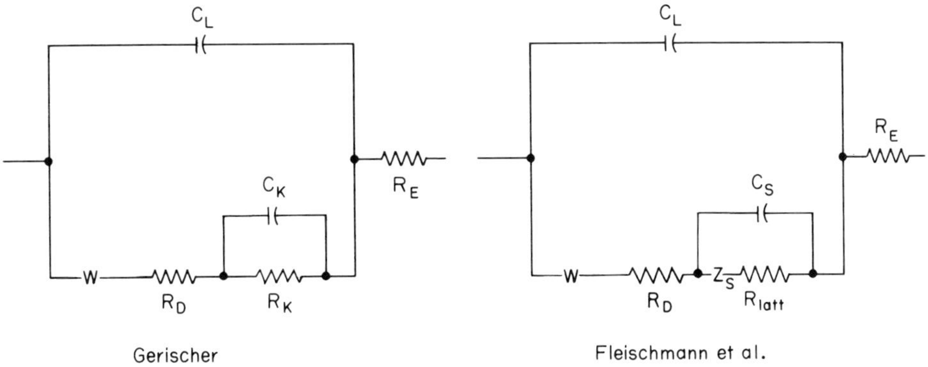

FIG. 3.1.6. Gerischer and Fleischmann et al. models.

that the process of diffusion of adsorbed species at the surface exerts some control over the rate of crystallization [104]. The significant additional controlling process has the effect of introducing an extra component in the electrode analog, and while it is clear that the incorporation of a further term in the electrode analog would complicate the frequency response, it has not been found possible to decompose the experimental data so that the component processes are resolvable. Nevertheless it is clear that the extra controlling process would further complicate the frequency response of the electrode interphase expected from a Gerischer-type model [105] (Fig. 3.1.6). The apparent value of α, obtained from the slope $\partial \log i_0 / \partial \log[Cd^{2+}]$, is 0.34 (cf. 0.23 obtained from galvanostatic measurements).

A value for the enthalpy of activation for the charge transfer process, ΔH_D, was obtained as ~29 kJ/mole. This value is slightly greater than that obtained for the galvanostatic measurements but is consistent with values for the charge transfer process in other solid metal exchanges. An inability to obtain results at higher temperatures was again observed, and the significant decrease in the high frequency impedance values was consistent with the formation of a film at the electrode surface.

The influence of surface-active substances on the kinetics of the electro-deposition of cadmium has been studied by Loshkarev et al. [106]. The dependence of the rate of the process upon the potential in the presence of most of the additives studied [tetrabutylammonium sulfate (TBAS), tribenzylmethylammonium sulfate (TBMAS), benzoylpiperidine, tribenzylamine (TBA), polyvinyl alcohol, and others], is N-shaped. The height of the maximum depends upon the nature of the additives, the temperature, the concentration of cadmium ions, and the anionic composition of the solution. The extreme values of the current are explained by the beginning of adsorption of the additive, followed by compacting of the adsorption layer. The current maximum (i_{max}) corresponds to the potential (E_{max}) at which a solid adsorption layer begins to form on the electrode surface. The formation of such a layer produces a sharp decrease in the current, which remains extremely low within

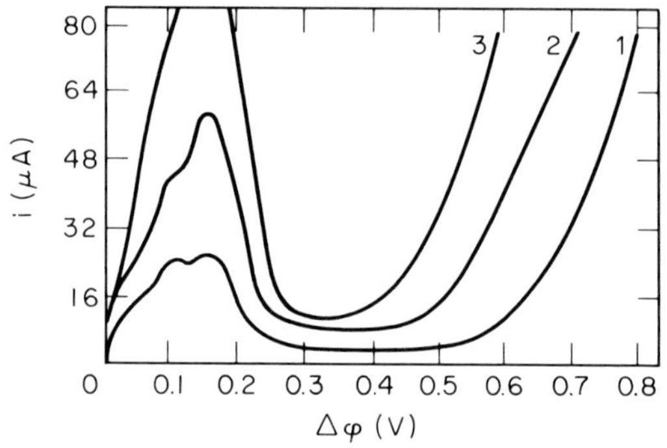

FIG. 3.1.7. Influence of temperature on polarization in the electrodeposition of cadmium: 0.25 M $CdSO_4$; 0.5 M H_2SO_4; 2 g/l OP-10. (1) 2°C, (2) 25°C, (3) 50°C; electrode area 2×10^{-3} cm^2. (Data from Ref. 106.)

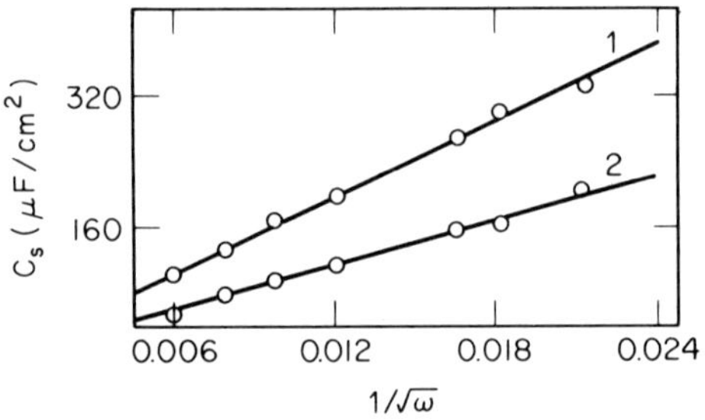

FIG. 3.1.8. Dependence of C_s on $w^{-1/2}$ at E°. (1) 1 M $CdSO_4$, 0.5 M H_2SO_4; (2) with addition of 2 g/l OP-10; 25°C.

a broad range of potentials in the presence of TBA, TBMAS, and certain other additives. This is evidence of a large value of the supplementary potential barrier. The highest polarization (above 0.9 V) is observed when TBA is introduced into the electrolyte. Among the tetrasubstituted ammonium compounds investigated, the greatest inhibiting effect was exhibited by TBMAS. This additive is almost as effective as TBA; however, in contrast to the latter it is readily soluble not only in acid, but also in neutral media.

Figure 3.1.7 illustrates the influence of temperature upon the process of electrocrystallization of cadmium from a solution with an addition of 2 g/l OP-10: increasing the temperature leads to a regular increase in the maximum values of the current.

The dependences of the polarization pseudocapacitance and the diffusion resistance upon the alternating current frequency in plots of C_s vs $\omega^{-1/2}$ and R_d vs $\omega^{-1/2}$ are straight lines (Fig. 3.1.8). The values of R_d are consistent, which is independent evidence of diffusion control of the process in the region of small polarization. The introduction of an additive

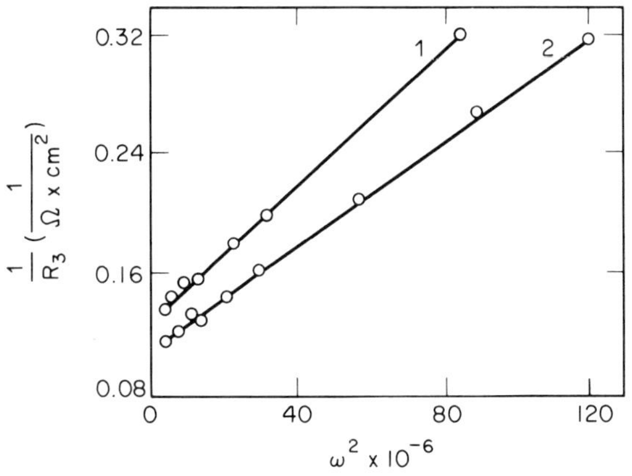

FIG. 3.1.9. Dependence of $1/R_R$ on ω^2. 0.25 M $CdSO_4$, 0.5 M H_2SO_4, 2 g/l OP-10; (1) $\Delta E = -0.4$ V; (2) $\Delta E = -0.7$ V; 25°C. (Data from Ref. 106.)

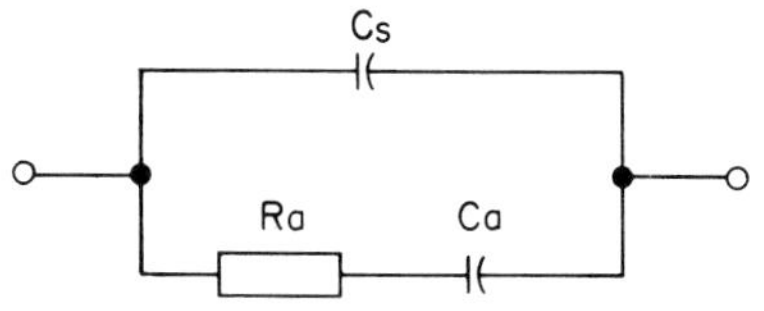

FIG. 3.1.10. Equivalent circuit.

leads to a decrease in C_s caused by a decrease in the effective surface of the electrode when it is partially covered with adsorbed molecules. The double-layer capacitance in this case drops from 36 to 15 $\mu F/cm^2$.

In the region of high polarization, when there is a solid adsorption film on the electrode surface, the limiting step of the electrode reaction becomes the penetration or discharging of ions through the layer of organic molecules. Under these conditions the active component of the conductivity is a linear function of ω^2 (Fig. 3.1.9). From the nature of the change in the components of the impedance with the alternating current frequency, it can be concluded that in the region of high polarizations the behavior of the electrode in the alternating current circuit is described by the equivalent circuit of Fig. 3.1.10. When there is an adsorption on the surface, the double-layer capacitance is reduced by more than fivefold compared with the pure electrolyte. The inhibition of the electrode process by an adsorption layer follows from the data in Figure 3.1.9 and agrees with the inhibiting action of additives. The results of a measurement of the effective activation energy in the region of high polarization (12 kcal/mole at $\eta = 0.4$ V and 7 kcal/mole at $\eta = -0.7$ V), as well as the low values of the transport coefficient (~0.1), lead to the same conclusion.

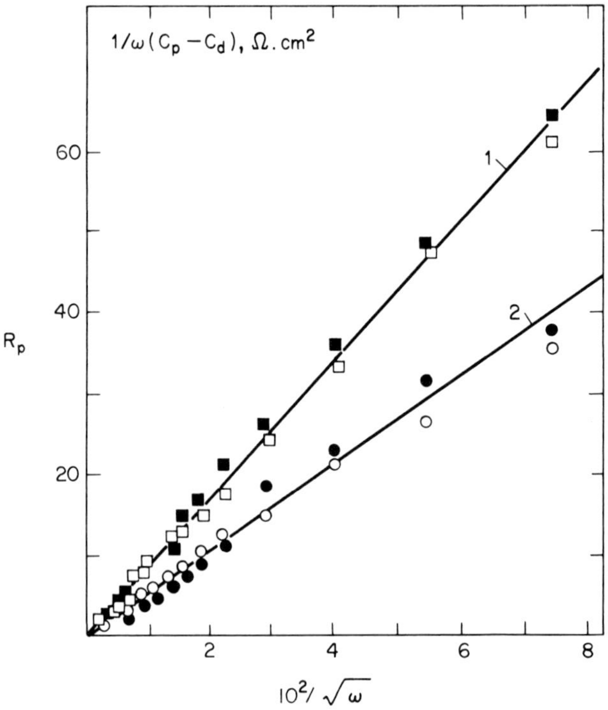

FIG. 3.1.11. Plots of R_p and $1/\omega(C_p - C_d)$ against $\omega^{-1/2}$ at the equilibrium potential for the system at 25°C; (1) 5 M KOH, (2) 10 M KOH. (Data from Ref. 107.)

The kinetics of exchange of the cadmium electrode in alkali is of some commercial importance. However, there is a difficulty in avoiding the intrusion of hydroxide films. It has been shown [107-109] that adsorption processes are important in the anodic oxidation of a cadmium electrode in alkaline solutions. L'vova et al. [99] have made impedance measurements at frequencies ranging from 30 Hz to 100 kHz at potentials close to the equilibrium potential of the system $Cd/Cd(OH)_2$ (-0.917 V); at a stationary potential (without current) of (-0.90) to (-0.93 V), at an anode polarization of -0.89 V; and at cathode polarizations of -0.95, -1.00, and -1.10 V. All potentials are given relative to a mercury-oxygen electrode in the test alkali solution.

The measurements [after subtraction of the electrolyte resistance, which was found by extrapolation to infinite frequency of the (R_R, $\omega^{-1/2}$) curve] were transformed according to a parallel equivalent circuit (C_p, R_p) (Fig. 3.1.11). The results are shown in Table 3.1.5.

This type of impedance frequency dependence is characteristic of diffusion, but in this case it cannot be attributed to diffusion of hydroxyl ions. At these solution concentrations the latter would give a diffusion conductivity four orders of magnitude greater than that observed experimentally. This diffusion impedance does not markedly alter when the potential

TABLE 3.1.5. Parameters of Equivalent Circuit of Cadmium Electrode in KOH Solution

T (°C)	$\underline{M}_{KOH}$	E_{HgO} (V)	C_d (μF/cm²)	R_a (Ω cm²)	C_a (μF/cm²)
25	2.5	-0.905	28 ± 7	200	-
25	5.0	-0.911	35 ± 15	200	-
25	10.0	-0.906	25 ± 3	200	-
25	1.5	-0.890	25 ± 5	200	-
25	5.0	-0.890	22 ± 3	200	-
25	10.0	-0.890	120 ± 15	200	-
40	5.0	-0.915	30 ± 10	90 ± 40	(1000)
60	5.0	-0.919	20 ± 4	16 ± 8	1300
80	5.0	-0.924	18 ± 3	4 ± 2	700
40	5.0	-0.890	19 ± 5	200	-
60	5.0	-0.890	25 ± 5	200	-
80	5.0	-0.890	35 ± 5	200	-
25	5.0	-0.950	24 ± 2	300	-
60	5.0	-0.950	16 ± 5	200 ± 100	-
80	5.0	-0.950	12 ± 4	110 ± 40	-
25	5.0	-1.000	18 ± 3	-	-
60	5.0	-1.000	10 ± 3	300 ± 100	-
80	5.0	-1.000	8 ± 2	140 ± 20	-

is shifted from -0.91 to -1.1 V, or when the alkali solution is saturated with cadmate ions, and therefore it cannot be due to diffusion of cadmate ions in the solution. It is suggested that it is due to electrochemical adsorption of hydroxyl ions: $OH^- \rightarrow OH_{ads} + e$. The movement of these species to a cadmium electrode in alkaline solutions has been demonstrated by other experimental methods [107, 108] and also by Breiter [109] by analysis of differential capacitance curves. The observed frequency dependence of the impedance of this reaction could be due to surface diffusion of hydroxyl radicals, or to surface inhomogeneity characterized by interaction of various areas with hydroxyl ions. At present there is no quantitative basis for either of these explanations.

The equivalent circuit found suggests the kinetic scheme of anodic oxidation of cadmium in alkaline solution:

$$OH^- \rightarrow OH_{ads} + e \qquad (3.1.9)$$

$$OH_{ads} + OH^- \rightarrow O_{ads}^- + H_2O \qquad (3.1.10)$$

$$Cd + O_{ads}^- \rightarrow CdO + e \qquad (3.1.11)$$

According to this scheme, the two electrochemical Stages (3.1.9) and (3.1.11) in ac polarization correspond to two faradaic circuit elements in parallel. At low temperatures Stage (3.1.11) is slow. When the temperature rises, the exchange current of the third stage increases rapidly. Shift of potential toward the cathode and accumulation of the products of anodic cadmium oxidation, as one would expect, greatly reduce the rate of the third stage.

Hampson and Latham [101] have investigated the exchange reaction using ultrapurified alkaline solutions. Experimental methods included faradaic impedance and double impulse galvanostatic techniques. It was reported that the reaction was fast, as shown by Table 3.1.6 (impedance data) and Table 3.1.7 (double impulse data).

TABLE 3.1.6. Ac Data for Solid Cadmium Electrodes in Alkaline Solution. 7 mole/l NaOH. Electrode Area = 3.46×10^{-2} cm^2

Cd^{2+} (mole/l)	Δ^a (Ω, 23°C)	$i_0 = \frac{RT}{ZF\Delta}$ (mA cm^{-2})	Slope of impedance plots (Ω-sec$^{-1/2}$, 23°C)
0.0099	12	31.0	3000
0.0053	20	18.5	5000
0.0002	9	41.0	3600
0.0001	12	31.0	4000

[a] Δ is the difference between the in-phase and out-of-phase components of the electrode impedance.

TABLE 3.1.7. Double Impulse Data for Solid Cadmium Electrodes in Alkaline Solution

Cd^{2+} (mole/l)	R_D (Ω-cm^2)	$i_0 = \frac{RT}{ZFR_D}$ (mA cm^{-2})
0.0099	0.49	26.0
0.0014	0.41	31.0
0.0001	0.40	32.0

An interesting observation reported by these authors was that in the range of temperature investigated, 0-50°C, no significant change of Δ with temperature was found. As with solutions at acid pH values, a variation of electrode characteristics with time was observed. The variation of the electrode capacitance C_0, is shown in Table 3.1.8. The variation of the reaction resistance with time is shown in Fig. 3.1.12. It was considered that characteristic data for the system were obtained after ~ 20-25 min electrode/electrolyte contact. It was emphasized that a certain uncontrolability is manifested with this system. At the equilibrium potential, or even at potentials slightly more negative, the development of a film of

TABLE 3.1.8. Variation of Electrode Capacitance, C_0, with Electrolyte Contact Time

Time	C_0 (μF cm^{-2})
0 min	120
15 min	84
20 min	87
45 min	93
60 min	81
135 min	64
220 min	58
450 min	43
22 hr	34

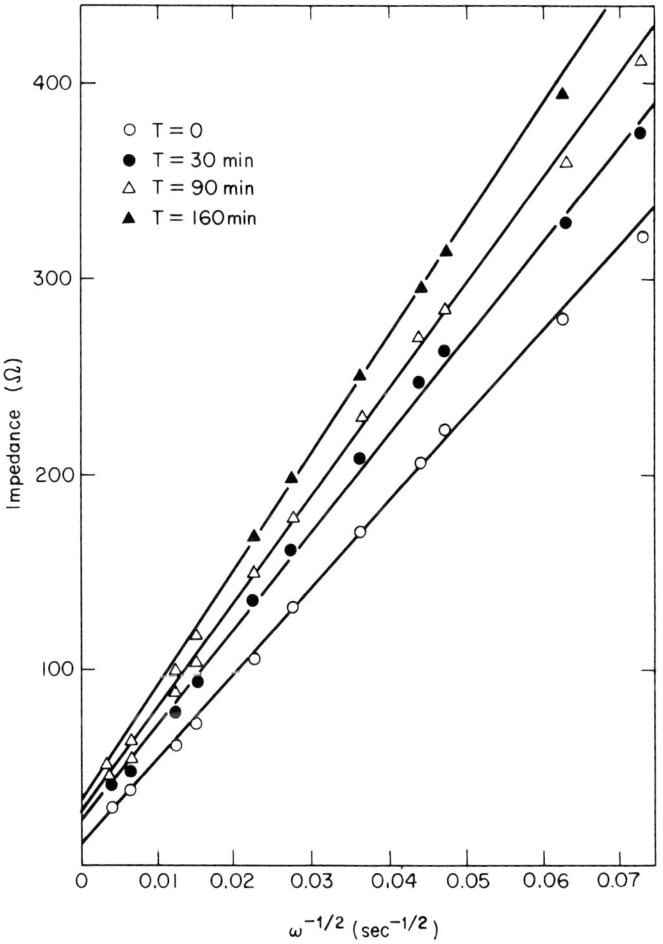

FIG. 3.1.12. Variation of R_R with time. 0.009 mole/l Cd^{2+}; total [OH^-] 7 mole/l with NaOH; supervicial electrode area 3.46 X 10^{-2} cm^2; 23°C. (Data from Ref. 101 by permission of Elsevier Ltd.)

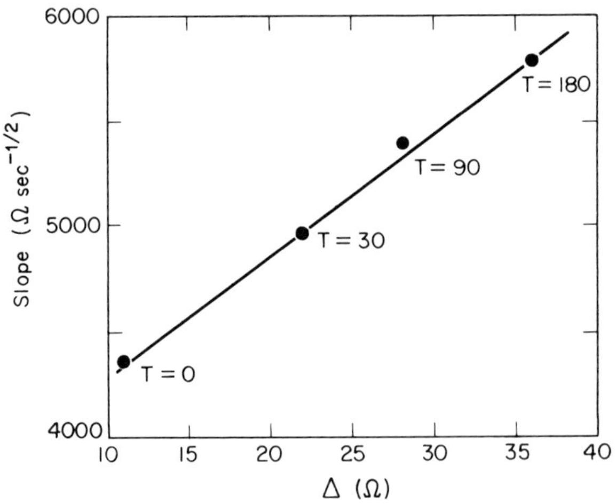

FIG. 3.1.13. Change in Δ with slope of impedance plots. 0.0053 mole/l Cd^{2+}; total [OH^-] 7 mole/l with NaOH; 23°C. T denotes time of electrode/electrolyte contact (min). (Data from Ref. 101 by permission of Elsevier Ltd.)

hydroxide occurred spontaneously. The effects of this film were reflected in changes of the electrode impedance, and from the magnitude of these it was clear that any conclusions drawn from such systems could only be qualitative.

The increase of electrode impedance with time supported the argument that the electrode reaction was continually being blocked. R_0, the electrolyte resistance, increased with time which suggested that the area of the electrode active for the reaction was continually diminishing. The slopes of the R_R vs $\omega^{-1/2}$ curves also indicated a progressive reduction in the area available for the diffusion process. Figure 3.1.13 shows that the slope of the impedance plots is proportional to the intercept. Thus the effective area of the surface is reduced with time of contact of the electrode with the electrolyte as the hydroxide (or oxide) film grows. The film growth is confirmed by the fact that the capacitance decreases with the time of contact. The value of the Warburg coefficient calculated theoretically, and based on the superficial electrode area, was found to be ~40 times less than the value obtained from the experimental data (from the first available measurements). Clearly the active area of the electrode was only a fraction of the total area available. In spite of these limitations the apparent charge transfer process was found to be considerably faster in alkali than in perchlorate electrolyte [98, 100].

The exchange process was shown to be independent of [Cd^{2+}] by both the impedance and the double impulse results. In the presence of the large excess concentration of supporting electrolyte, it is possible that a complexed or neutral species is transferred. The presence of cadmiate species in alkaline solutions has not been well established [110] (cf. zincates and plumbates). Gerischer [111, 112] has suggested that the species transferred most favorably are those with low electric charge and small coordination number. For example, in the overall reaction [44]

$$Zn(CN)_4^{2-} + 2e = Zn(Hg) + 4CN^- \tag{3.1.12}$$

the charge transfer process was found to be

$$Zn(OH)_2 + 2e = Zn(Hg) + 2OH^- \tag{3.1.13}$$

The transfer of a species such as $Cd(OH)_2$ explains why the process is much more rapid in alkaline solution.

The deviation from linearity on the impedance spectra observed at the lower end of the frequency range is interesting. The most likely cause is a frequency dispersion effect which results if different areas of the electrode have different capacitances [113]. Such a situation is enhanced in the presence of adsorption to form a hydroxide (oxide) film.

The lack of variation of the results with change in temperature is unusual. The control of the reaction by film growth obviously obscures any variation in the rate of the charge transfer process with temperature. The enthalpy of activation of the charge transfer process for polycrystalline zinc in alkaline solution was found to be ~12 kJ/mole. This is much smaller in magnitude than the values observed for systems free from strong adsorption. The lower value expected for ΔH_D indicates that the change of the rate of the charge transfer process with temperature would be minor and thus easily obscured.

3.1.2. Nonaqueous

3.1.2.1. Amalgams.

Biegler, Gonzalez, and Parsons [92] have investigated the Cd^{2+}/Cd(Hg) reaction in a series of solvent mixtures of almost constant permitivity (methanol + acetonitrile) and one of widely varying permitivity (N-methylformamide + N, N-dimethylformamide). The measured impedance was analyzed assuming that the double-layer capacity was unaffected by the reactants. The results are shown in Table 3.1.9 and are based on three or four determinations at different concentrations of cadmium in the range 0.3-0.8 m$\underline{M}$/1. The rate constants were sufficiently high to satisfy the conditions under which this method may be used, except for those obtained in the pure solvents methanol and propylene carbonate, and the acetonitrile-water mixture. This mixture appeared to give reliable results although the behavior in pure methanol solution may not be as satisfactory. The supporting electrolyte was 0.9 $\underline{M}$ $NaClO_4$ in each case. Temperature coefficients were measured at four temperatures in the range 15 to 50° C.

Biegler et al. [92] emphasize that the most significant outcome of studies in various pure solvents to date is the remarkable lack of dependence of kinetic parameters on the nature of the solvent. The rate constant for the system Cd^{2+}/Cd(Hg) has been measured in this work for six solvents and ranges from 0.01 to 0.45 cm/sec.

TABLE 3.1.9. Rate Parameters for the Reaction $Cd^{2+} + 2e \rightarrow Cd(Hg)$ in Various Solvents at 25°C; Base Electrolyte 0.9 mole/l $NaClO_4$.

Solvent	$\overleftrightarrow{k}$ (cm/sec)	α	$\Delta\overleftrightarrow{H}$ (kJ/mole)
H_2O	0.45	0.15	
0.5 mole H_2O + 0.5 mole CH_3CN	0.0145		
CH_3CN	0.19	0.15	22
0.8 mole CH_3CN + 0.2 mole CH_3OH	0.080	0.14	
0.5 mole CH_3CN + 0.5 mole CH_3OH	0.028	0.14	
CH_3OH	0.013	0.27	46
DMF	0.15	0.24	17
0.8 mole DMF + 0.2 mole NMF	0.081	0.14	
0.5 mole DMF + 0.5 mole NMF	0.089	0.20	22
NMF	0.12	0.40	30
Propylene carbonate	~0.01		

3.1.2.2. Solid Metal Electrodes. Plonski [114] has studied the kinetics of the Cd/Cd^{2+} (alcohol) electrode at equilibrium by means of radioactive ^{115}Cd. Indications are found that in absolute ethyl alcohol media the corrosion of cadmium is mostly avoided.

The equation for the isotopic exchange is

$$-\ln(1 - F) = \frac{A + B}{AB} Rt \tag{3.1.14}$$

in which F is the fraction exchanged, t is the exchange time, and R is the rate of appearance of ^{115}Cd in the form of metallic cadmium; that is, the rate of the isotopic exchange reaction. A and B are the quantities of reactants (in moles) involved in the exchange; in this case A refers to Cd^{2+} in solution and B refers to Cd^0 in the metal. Figure 3.1.14 shows ln (1 - F)/ time curves for 6×10^{-6} and 9.8×10^{-4} M $CdCl_2$ solutions. Table 3.1.10 shows the effect of cadmium ion concentration on the number of monolayers exchanged at equilibrium. The results also showed that

$$\frac{\partial \ln i_0}{\partial \ln a_{Cd^{2+}}} = \alpha \sim 0.6$$

It was calculated that for concentrations up to 1×10^{-5} M, the supply of the ions to the surface by diffusion through the liquid diffusion layer is the rate-determining process. Above 1×10^{-4} M recrystallization is thought to be responsible for the activity pick up by the metal.

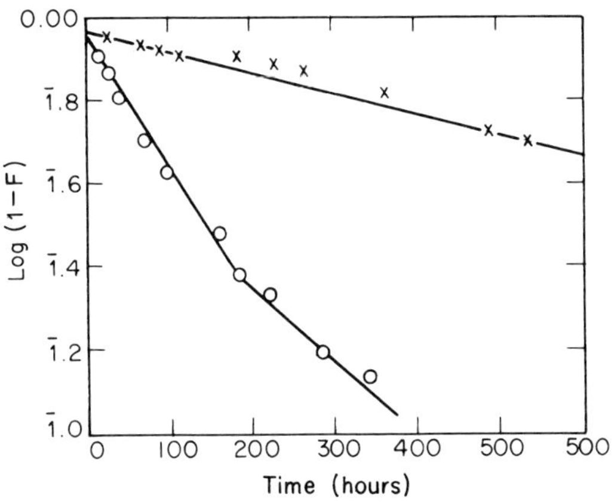

FIG. 3.1.14. Log (1 - F) against the time curves for isotopic exchange between Cd and $^{115}Cd^{+2}$ in alcoholic solution: o = 6×10^{-6}; x = 9.8×10^{-4} $CdCl_2$. (Data from Ref. 114, p. 67, by permission of Revue Roumaine de Chimie.)

TABLE 3.1.10. Effect of $[Cd^{2+}]$ on Extent of Monolayers Exchanged

Cd^{2+} concentration in solution (equiv g/l)	Monolayers exchanged at equilibrium
6×10^{-6}	2
2×10^{-4}	49
9.8×10^{-4}	132
2×10^{-3}	250
1×10^{-2}	932

3.1.3. Molten Salts

Laitinen and Osteryoung [115] made measurements at solid microelectrodes in a fused LiCl/KCl eutectic mixture at 450°C. A residual capacity (apparent double-layer capacity) of Pt showing a large frequency dispersion was reported. Laitinen and Gaur [116] extended these measurements, and experiments were made in the presence of Cd^{2+} ions in order to attempt a quantitative estimate of rate constants of electrode reactions from impedance data and to compare such values with those obtained from polarization data. Impedance and polarization measurements were made and repeated at various potentials, anodic as well as cathodic, to the equilibrium potential; the latter being the open circuit potential of the

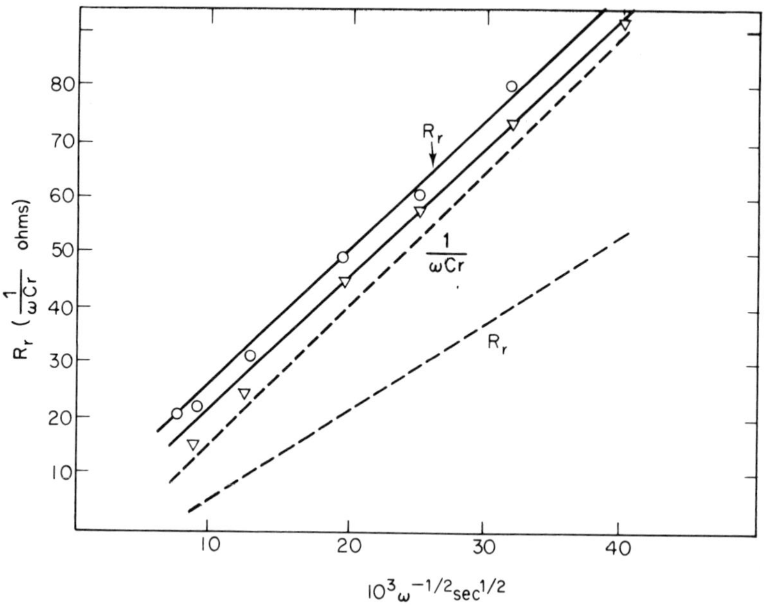

FIG. 3.1.15. The broken lines are the uncorrected reaction impedance plots for the reduction of $CdCl_2$ in KCl-LiCl at 450°C. The solid lines represent the plots obtained after correcting for the admittance due to the adsorbed ions. The Cd concentration was 5.95×10^{-5} mole cm^{-3}. (Data from Ref. 116, p. 734, by permission of the Journal of the Electrochemical Society.)

metal-coated electrode. Cadmium shows a tendency for predeposition on to the platinum. The faradaic admittance had an experimental phase angle for the faradaic path in excess of 45°. This was ascribed to the adsorption of metal ions at the electrode surface. Rate constants for the above processes were calculated from the faradaic admittance after an empirical correction for these effects. Figure 3.1.15 shows a typical impedance plot.

From the data the rate constants for the reduction of cadmium ion was found to be 1.27 cm/sec. This corresponded to an exchange current of 172 A cm^{-2} at unit concentration, 1 mole cm^{-3}. Current-potential curves showing the polarization behavior in the vicinity of equilibrium potential are described. By a curve fitting method the values of α and the exchange current were computed. These values of i_0 are significantly lower than those calculated by using the rate constants obtained from impedance measurements.

More reliable experimental data was probably obtained by the impedance method because of the appearance of unexplained inflections in several of the polarization curves.

Laitinen, Tischer, and Roe [117] have made measurements on the Cd^{2+}/Cd electrode in a KCl-LiCl melt at 450°C. The results were obtained using galvanostatic, double pulse, and ac impedance techniques, and values are shown in Table 3.1.11.

TABLE 3.1.11. Kinetic Parameters for Exchange at a Solid Cadmium Electrode in Melts

Reaction	Conditions	Technique	Rate constant	α	Remarks	Refs.
$Cd^{2+} + 2e = Cd$	KCl-LiCl 5.95×10^{-2} $\underline{M}$ Cd^{2+}	farad. imp	$j_0 = 7.7$ A cm^{-2}			[116]
$Cd^{2+} + 2e = Cd$	KCl-LiCl, 450°C	V stp	$j_0 = 210 \pm 50$ A cm^{-2} $k^\circ = 0.4$	0.13 ± 0.05	$\Delta H \approx 3$ kcal/mole	[117]
$Cd^{2+} + 2e = Cd$	KCl-LiCl, 450°C	i stp (double pulse)	$j_0 = 210 \pm 50$ A cm^{-2} $k^\circ = 0.4$	0.13 ± 0.05	$\Delta H \approx 3$ kcal/mole	[117]

3.2. DOUBLE-LAYER PROPERTIES

Of the various methods available for determining the potential of zero charge (p.z.c.) on solid metals, only capacitance and electrochemical methods appear to have been applied in the case of cadmium. The results of the various reported investigations are outlined in Table 3.2.1.

The form of the curves in flouride electrolyte resemble those of mercury in similar electrolytes. This is shown clearly in the data of Bartenev et al. [123]. Figure 3.2.1 shows capacitance curves which are shown in integrated form in Figure 3.2.2.

It is clear that in the case of fluoride electrolyte [119] there is no evidence for specific adsorption. However, with other electrolytes the position is by no means as well defined. It is reported by Russian workers and others that in the case of sulfate, chloride, bromide, and iodide some evidence for a shift in the diffuse layer minimum with change in concentration has been obtained, although in the case of sulfate electrolyte there is some dispute as to whether this is due to the asymmetry of the electrolyte or the operation of the Esin-Markov [127] effect. At sufficiently negative potentials the double-layer structure of the cadmium/aqueous solution interphase becomes independent of the anion.

It has been emphasized by several workers that the double-layer structure on cadmium is influenced greatly by the presence of adsorbable impurities which affect both capacitance data and kinetic parameters. In this connection, Russian workers [124] have emphasized the need to counteract the effects of atmospheric oxidation at the surface of the metal electrode and they have claimed that sulfate anion is not adsorbed on an "oxide free" surface whereas an

TABLE 3.2.1.

Electrode	Conditions	Technique	E_z	Refs.
stat sphere	5 X 10^{-3} 1.0 M KCl	farad. imp	-0.91 ± 0.025	[118]
-	-	Electroreduction of anions	-0.9	[119]
-	0.1 M Na_2SO_4/H_2SO_4	farad. imp	-0.7 -0.9	[120]
-	-	-	[-0.6 to -0.89] Work function ≈ 4.0 eV	[121]
stat dsk	0.001 1.0 M $NaClO_4$	farad. imp	-0.9 ± 0.025	[122]
stat dsk	0.001 0.1 M NaF 0.01 N Na_2SO_4, pH 2-10	farad. imp	-0.72 ± 0.02	[123]
Electrodeposited stat dsk	0.01 M NaF	farad. imp	-0.74	[124]
stat dsk	0.02 M NaF, pH 3.9, 23°C	farad. imp	-0.75 ± 0.02	[125]
-	KCl/NaCl melt at 700°C	farad. imp	-0.80	[126]
-	LiCl/KCl melt at 700°C	Electrocapillary max	-0.82	[126]

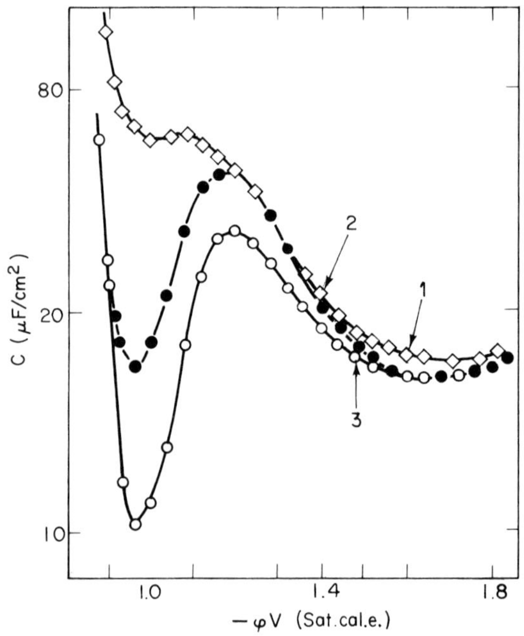

FIG. 3.2.1. Differential capacity curves on a cadmium electrode in NaF solutions: (1) 0.1; (2) 0.01; (3) 0.001 N. (Data from Ref. 123.)

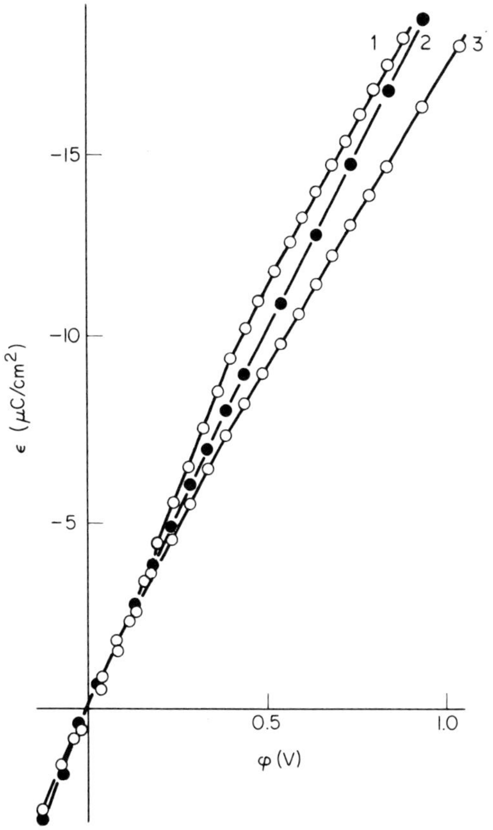

FIG. 3.2.2. Surface charge curves in 0.01 NaF: (1) on cadmium; (2) on lead; (3) on mercury. The potentials are given with respect to the zero-charge potential of mercury, lead, and cadmium electrodes. (Data from Ref. 123.)

insufficiently reduced cadmium electrode may well give rise to characteristic weakly developed minima in dilute solutions. It is noteworthy that other work [87, 128] has emphasized the need to purify electrode systems stringently before measurements are made. Prolonged circulation of electrolyte through purified, activated charcoal appears to be adequate. However, the nature of the impurity removed from the double layer is still uncertain. Hampson and Larkin [87] considered that there was some evidence that the borate ion, the presence of which arose from the degradation of borate glass, may be a possible contaminant. Certainly borate ions engender a serious reduction in double-layer capacitance. The importance of these effects on the electrochemical behavior of cadmium in general has been discussed by Bauer [88].

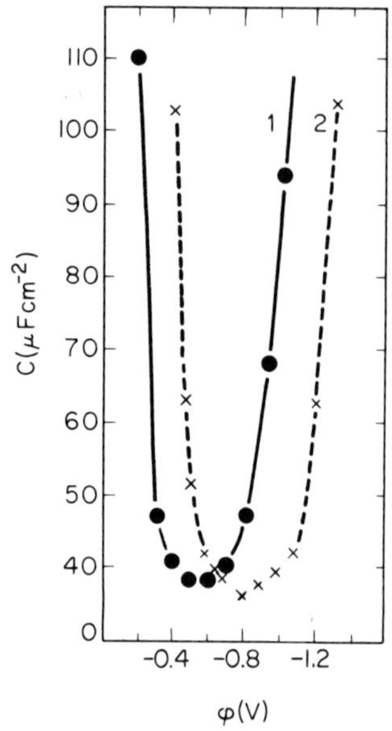

FIG. 3.2.3. Variation of double layer capacitance with potential in molten KCl-NaCl at 700°, f = 200 kHz: (1) Pb; (2) Cd. (Data from Ref. 126.)

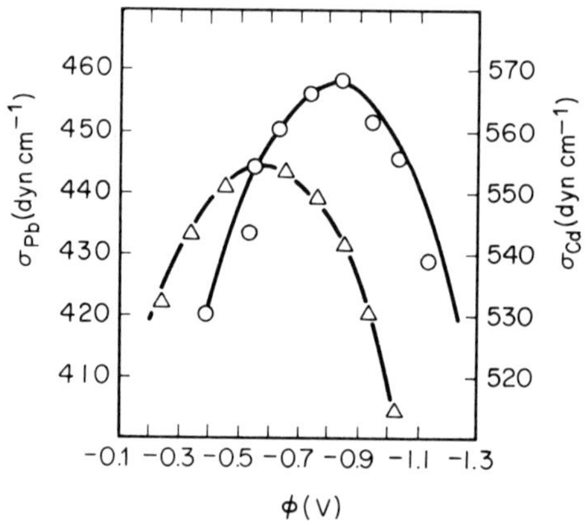

FIG. 3.2.4. Electrocapillary curves for Pb and Cd, obtained by integration of the C-φ curves in Fig. 3.2.3. The points show the experimental data. (Data from Ref. 126.)

The structure of the double layer in aqueous solution and its influence on electrode kinetics have been the subject of several investigations. However, there is considerably less information available on the properties of the double layer in molten salt systems. Ukshe et al. [126] have shown capacitance and electrocapillary curves (Figs. 3.2.3 and 3.2.4) for Cd in KCl-NaCl at 700°C over a frequency range 20-200 kHz. It is clear from the apparently high value of the capacitance at the minimum that a similar type of diffuse double layer envisaged for dilute aqueous solutions is not present. It is pointed out that such a structure at the interphase would in any case be incompatible with modern theories of the structure of ionic liquids such as molten alkali chlorides.

4. ELECTROCHEMICAL STUDIES

The anodic oxidation of cadmium metal and cathodic reduction of the oxidation product in strongly alkaline solutions is of interest in understanding the behavior of the cadmium electrode in nickel-cadmium batteries. Although there has been a large number of investigations, the reaction mechanism is still incompletely understood. Two different mechanisms have been proposed for the anodic reaction: the dissolution-precipitation mechanism [129-136] and the solid-state oxidation mechanism involving ionic transport through the $Cd(OH)_2$ film [137-139]. The formation of CdO has been the subject of considerable discussion. Croft [137], Falk [140], and Armstrong et al. [141] found no evidence of CdO formation.

Armstrong, Boult, Porter, and Thirsk [141] have investigated the structure of anodic films formed on cadmium single crystals in alkaline solutions. Anodic films were potentiostatically formed on electropolished polycrystalline cadmium and (0001) and (1010) single-crystals in 1 M NaOH solution. The films were examined by glancing incidence electron diffraction and electron microscopy. Glancing incidence electron-diffraction patterns from both the polycrystalline and single-crystal electropolished surfaces showed that there was no trace of any spots on the ring patterns, indicating the absence of surface films. Elongation of the diffraction spots in a direction normal to the shadow edge was observed; this effect is characteristic of a very smooth undulating surface. The observation of Kichuchi lines and bands for the single crystal faces demonstrated that the electropolishing procedure had successfully removed the deformed surface material. The anodic films were formed by holding the Cd electrodes at a potential ~200 mV negative to the reversible $Cd/Cd(OH)_2$ potential, -0.907 V (Hg/HgO), by means of a potentiostat for a few minutes and then applying an anodic pulse of potential for 1 min. In all cases, in addition to the patterns due to Cd for very thin films, electron diffraction revealed only ring or spot patterns which corresponded with those expected for $Cd(OH)_2$. The anodic phase was in parallel orientation with the single-crystal substrates at overpotentials of ~20 mV. For an overpotential of 20 mV these films are one-degree oriented.

Huber [130], Huber and Stucki [131], Breiter and Weininger [142], and Breiter and Vedder [143] have reported experimental results indicating the formation of CdO. Breiter and Vedder [143] investigated the nature of anodic films on cadmium. Smooth oxide-free cadmium electrodes were oxidized at constant potential between 0.05 and 1.0 V in various alkaline electrolytes. Film formation was followed by recording the current as a function of time. The nature of anodic films formed after different oxidation times at various potentials were investigated by electron microscopy, IR spectroscopy, and X-ray diffraction. The films appear to grow nonuniformly on the surface and the number of nuclei increases with oxidation potential. IR spectra showed the presence of β-$Cd(OH)_2$, γ-$Cd(OH)_2$, or a mixture of both types of hydroxide. This was supported by the X-ray diffraction results. The total amount of hydroxide decreases with potential and is, in general, less than expected for 100% efficiency of hydroxide formation. This is attributed to the simultaneous formation of amorphous CdO. During the reduction at constant cathodic potential, cadmium oxide and γ-$Cd(OH)_2$ disappear first, and then β-$Cd(OH)_2$. The conversion of anodically formed CdO into β-$Cd(OH)_2$ is a slow process.

Crystal growth appears to take place only at certain nuclei on the surface and, depending upon which of the two nucleates more easily, either β-$Cd(OH)_2$ or γ-$Cd(OH)_2$ is produced at 0.05 V. As expected, the density of nuclei differs on the various planes and it is concluded that the formation of β-$Cd(OH)_2$ or γ-$Cd(OH)_2$ at low potentials occurs by the precipitation of soluble cadmium-containing species which are produced anodically on the free metal surface. The evidence for high crystallinity of β-$Cd(OH)_2$ and γ-$Cd(OH)_2$ films apparently agreed with this conclusion. Provided that the crystals do not overlap, the anodic current rises with time. Overlapping causes the decrease of current, and film growth is determined by transport processes, probably in the pores of films of β-$Cd(OH)_2$, γ-$Cd(OH)_2$, or mixtures of both. At higher oxidation potentials a large number of small crystals is rapidly formed because of increased nucleation rate. If at higher voltages the formation of oxidation products continues to take place via a soluble intermediate, it would be difficult to see why cadmium oxide is the final product. Instead, the nature and growth of the amorphous CdO is controlled by a solid-state process through a film covering the entire metal surface. From the slow growth of the oxide layer on cadmium at room temperature in air, it cannot be assumed that solid-state transport occurs through all of the macroscopic thickness of the anodic oxide layer (1000 Å for 125 mC/cm^2). The continuous film through which diffusion takes place may be quite thin, transport through the remainder of the macroscopic layer occurring through film defects. It may well be that much of the cadmium hydroxide present in films formed at higher voltages has accumulated in these flaws.

Fleischmann, Rajagopolan, and Thirsk [144] investigated the formation of cadmium hydroxide on cadmium amalgam in alkaline solutions at constant potential. Approximately three monomolecular layers were deposited before passivation occurred with the basal plane of the hexagonal crystals parallel to the amalgam. It was shown that the slow stage of crystal growth is the formation of the lattice at the periphery of two-dimensional growth centers of monomolecular height, randomly nucleated on the surface. The variation of the crystal growth was found to be constant with potential and concentration and was interpreted such that the rate-determining step is probably a rearrangement of two adsorbed hydroxide ions into the lattice planes containing these ions. The most probable mechanism is therefore rearrangement of two "discharged" hydroxide ions into the two layer planes separating the cadmium ions. The results, however, do not show whether this rearrangement is due to a sequence $Cd^{2+}|OH^-|OH^-$ or to a sequence $OH^-|Cd^{2+}|OH^-$.

Okinaka [145] has investigated the oxidation-reduction processes of the $Cd/Cd(OH)_2$ electrode in concentrated alkaline solutions using a rotating ring-disk electrode in addition to a stationary electrode. The formation of a dissolved intermediate species during the anodic and cathodic process in concentrated alkaline solutions was indicated by the results of ring-disk electrode experiments. It is suggested that the dissolution-precipitation mechanism prevails in the active range of potential and the contribution of solid-state growth mechanism appears to be negligible in this potential range. The results of this investigation agree with previous workers that passivation is not caused by complete blocking of the cadmium metal surface by $Cd(OH)_2$ crystals but is possibly due to the formation of a thin, continuous film of CdO next to metallic cadmium. This passive film appears to be insoluble or only very slowly soluble in alkali, and in the passive range of potential the $Cd(OH)_2$ crystals grow on top of the passive layer by a solid-state ionic transport mechanism. The amount of CdO formed increases with increasing anodic potential, and in 6.9 M KOH it becomes detectable by cathodic reduction if the anodization potential is brought to a value more positive than +0.4 V vs the equilibrium potential of $Cd/Cd(OH)_2$. The cathodic reduction of CdO occurs at a potential about 40 mV less cathodic than that of $Cd(OH)_2$. The reduction of $Cd(OH)_2$ occurs both by a direct mechanism and through the solution phase. The contribution of the latter mechanism increases as the reduction progresses and as the area of exposed cadmium metal increases.

Armstrong and West [146] have made similar measurements on cadmium using both rotating ring-disk and disk electrodes. These authors find that cadmium exhibits a region in which dissolution of the metal occurs immediately before film formation. This film is formed by a solid-state mechanism. The film thickens according to an approximately parabolic law at short times of polarization. Dissolution of the film at the outer surface may be balanced by the formation of new film at lower current densities when a steady-state condition obtains at longer polarization times. The convective conditions at the electrode surface determine the steady current and the film thickness, emphasizing the reversibility of the film dissolution.

5. APPLIED ELECTROCHEMISTRY

5.1. ELECTROLYTIC RECOVERY

In electrolytic zinc plants the precipitate caused by the addition of zinc dust in the purification of the zinc sulfate liquor prior to electrolysis forms the raw material. This precipitate (12% cadmium with 30-40% zinc, plus lead, nickel, and cobalt) is then treated with electrolyte from the zinc cells (cadmium and zinc are dissolved, leaving copper and lead as residue). The cadmium sulfate solution is passed to a lead-lined tank in which are suspended bars of zinc metal. At 70°C the cadmium is precipitated in sponge form, floating on the surface. The sponge is collected, washed, stored to permit oxidation (unoxidized sponge is difficult to dissolve), and dissolved in return acid electrolyte from the cadmium cells.

Trouble has been experienced in cadmium deposition as a result of "treeing." This phenomenon consists of irregular outgrowths of metal on the cathodes, resulting in short circuits with consequent loss of current efficiency. The trouble is caused by gas bubbles adhering to the metal giving rise to a honeycomb structure, thus laying the foundation for irregular deposition. If the cathode is caused to rotate, smooth deposits of metal could be obtained. Such cells (having a semicircular cross-section) contain several aluminum circular cathodes mounted on a central horizontal shaft, the latter being slowly rotated (1-1/2 rpm) by a motor through reduction gearing.

Two submerged lead anodes serve each cathode. As the gas bubbles form on the cathode they are slowly carried to the surface of the electrolyte and thence dissipated. At stationary cathodes the use of a low current density (4 to 5 mA cm^{-2}) and more frequent stripping than that employed in the zinc cell causes "treeing" to be reduced to a minimum. Another factor found to influence treeing is temperature, low temperatures being conducive, so that cooling of the cells is unnecessary. It is necessary to maintain the optimum operating temperatures at 30°C. Bratt [147] has investigated impurity effects in the electowinning of cadmium. The following effects are observed:

1. Pb and Tl increase current efficiency,
2. Al, Fe, Hg, Ni, Be, Zn, and Co have no effect,
3. Ag, Au, Bi, Cu, Mo, Pd, Sb, Sn, and Ge decrease the current efficiency.

5.2. ELECTROPLATING AND ELECTRODEPOSITION

Cadmium plating provides a corrosion-protective coating with an attractive appearance on various base metals. The relatively high price of cadmium means that it is applied largely where thin coatings will suffice; usually on parts which are used indoors or in sheltered positions outdoors. Cadmium plating is seldom used as an undercoating for other metals. Thus, for example, thin silver coatings are adsorbed by the cadmium, causing the surface to turn grey. Since cadmium covers cast and malleable iron better than zinc from a cyanide bath, it is used as an undercoating, up to 0.1 mm thick, for zinc. The resistance of cadmium plate to chemicals is generally quite low. Cadmium plating is often used where there are dissimilar metals, such as steel and brass, to minimize corrosion. Cadmium plate is used on steel and other metals because it is easily soldered to and has low contact resistance. For maximum corrosion protection cadmium plating is preferred for applications subject to tropic atmospheres or to salt air. The corrosion products of cadmium, unlike those of zinc, do not form insoluble basic salts. Udy [148] in 1919 patented an electroplating process from the cyanide bath. The hydroxide bath contained sodium cadmicyanide and sodium hydroxide but no free sodium cyanide.

Commercial cadmium baths consist essentially of sodium cadmicyanide, sodium cyanide, and sodium hydroxide, plus organic and sometimes inorganic addition agents and brighteners [149, 150]. Cadmium oxide is dissolved in sodium cyanide solution, which automatically results in a bath with a sodium hydroxide content equivalent to the cadmium content. The composition of the cyanide complex is not known with certainty. Soderberg [150] concludes that for practical purposes it may be $Na_2Cd(CN)_4$ and considers the sodium cyanide used in excess of this formula to be free sodium cyanide, effective in dissolving the anodes. The following formulas cover the typical cadmium plating baths:

	Still plating (g/l)	Barrel plating (g/l)
CdO	23-25	17-23
Cd metal	20-30	15-20
NaCN	90-120	70-90

The cathode current density range for best results widens and moves toward higher current densities as the cadmium content increases. Much higher cadmium concentrations are used for deposition of heavy coatings, whereas lower concentrations are used in very low current density areas. The sodium cyanide content is important in that it provides conductivity and improves anode corrosion. The sodium hydroxide content is usually considered not very critical. Except where there is much drag-in of pickling acid through insufficient rinsing, it is maintained automatically by the difference in anode and cathode efficiencies. Its prime function is to provide conductivity. It should not, however, be added much in excess of what is formed originally, since an excess causes a narrowing of the bright range. The effect of pH of a low metal bath without brightener and a high metal bath with brightener on plating range has been determined. The organic brighteners used in cyanide cadmium plating solutions include complex nitrogen-containing condensation products of high molecular weight [151-155], piperonal with aluminum [156], and certain sulfonic acids [157]. Many other substances used as brighteners have been patented and the majority are effective in baths of relatively low concentration. As the bath concentration and cathode current density are increased, it becomes more difficult to obtain smooth bright deposits by the use of organic additives only; then small amounts of certain metallic salts are also added. Nickel and cobalt salts [150, 158], when employed in conjunction with suitable organic additives, permit the use of current densities considerably in excess of those obtainable with the organic additives alone. However, nickel tends to codeposit with cadmium to a slight extent when the bath composition becomes low in cadmium or sodium hydroxide. Cobalt is as effective as nickel, but it is required in larger amounts although it does not codeposit with cadmium under conditions observed in normal bath operation. The cathode current density may be as low as 5-7 mA cm^{-2} but is usually 15-25 mA cm^{-2} where no agitation is used. It may be increased to 30-50 mA cm^{-2} where agitation is provided. The usual anode current density is 20 mA cm^{-2}. The anodes should be of cadmium; anode dissolution is greater than would be expected from the current used; extra anode area is often obtained from case-hardened steel cages. Operating temperatures are in the range 24-29°C. At higher operating temperatures a higher free cyanide concentration is necessary for uniform plating. Barrel plating solutions are usually basically the same as those for still tanks. Occasionally higher concentrations are used to prevent excessive reduction of the metal content.

Cadmium deposited from acid baths tends to be coarsely crystalline, porous, and poorly adherent, and the electrodeposition baths have very poor throwing and covering powers. High-strength steels become embrittled [159] from hydrogen adsorption when plated in cadmium cyanide baths, and this difficulty is generally overcome by baking procedures [159]. It has been claimed that hydrogen embrittlement may be reduced by the use of electrolyte solutions of low pH. Of the sulfate, perchlorate, sulfamate, and fluoborate electroplating solutions, only the fluoborate has been used commercially. The generally accepted formulas for cadmium fluoborate plating baths fall within the following limits:

	Range (g/l)	Ideal (g/l)
Cadmium fluoborate, $Cd(BF_4)_2$	150-300	250
Ammonium fluoborate, NH_4BF_4	60-120	90

The operating conditions are generally:

	Range	Ideal
Temperature	10-40°C	25°C
pH	1-4	2.5-3

Cadmium fluoborate electrolytic plating systems have been described containing 0.2-5 g/l of an enzyme selected from the group consisting of pepsin and rennet [160]. Such systems are said to produce mechanically improved coatings.

5.2.1. Bright Dips

Cadmium coatings are generally used "as plated"; however, the most common method of finishing electrodeposited cadmium is bright dipping. The most commonly used bright dips are acid solutions of considerable oxidizing power in which the cadmium plate dissolves without hydrogen evolution, apparently in the same manner as a semipolarized cadmium anode, at a potential sufficiently high to prevent preferential etching of the crystal faces. Typical compositions are either 100 g/l chromic acid with 1 ml of concentrated sulfuric acid (sp gr 1.84), or 7% by volume of 30% (100-volume) hydrogen peroxide with 0.3% by volume of concentrated sulfuric acid (sp gr 1.84). The time of immersion is usually from 2 to 30 sec depending on concentration. Chromic acid bright dip is characterized by the high degree of passivity which it imparts to the cadmium. Nitric acid bright dip has the disadvantage that it tends to discolor a cadmium surface when the articles are subsequently stored in a small, confined space. The acidified hydrogen peroxide dip has the advantages of being free-rinsing and not producing any stains or other aftereffects. The anodic brightening of cadmium in a cyanide bath has been proposed for treatment of dull deposits.

Cadmium may be removed from steel, brass, or copper. The parts are immersed at room temperature in a solution of 1 liter of hydrochloric acid, antimony trioxide, and 62.5 ml of water; another immersion treatment at room temperature is a solution of 120 g/l of ammonium nitrate [161]. An electrolytic process solution to remove cadmium plate from steel, brass, zinc-base die casting, or aluminum is [162]:

Orthophosphoric acid	61-86% by weight
Sulfuric acid	5-12% by weight
Glacial acetic acid	3-6.5% by weight
Water	6-25% by weight

Copper cathodes are used, with the plated parts as anodes and a current density of 5.5-16.6 A/dm^2. The cathode area should be 4-6 times the anode area.

5.2.2. Cadmium-Zinc Alloys

Cadmium-zinc alloys are readily deposited from cyanide electrolytes containing free cyanide ion and/or free sodium hydroxide. The plating solutions are prepared by introducing the metals into the electrolyte as the cyanides or oxides.

One of the characteristics that the compositions of the various baths have in common is the relatively large molarity of the complexing agents. The molarity of the total cyanide and hydroxide ion in the baths is considerably larger than the molarity of the total metal content of the bath. The deposition of this alloy from the cyanide bath has been investigated in detail [163, 164] with reference to the metal concentration ratio and current density on alloy composition, cathode efficiency, and cathode polarization. The efficiency was 60-90%, and the limiting current density 30 mA cm^{-2}. When the ratio of zinc to cadmium in the solution reached about 30, further increase in the ratio had little effect on the deposit composition. All alloy deposits containing from 0 to 100% zinc were dense, bright, with good adhesion to the base metal. In the case of codeposition of zinc, cadmium, and hydrogen from the cyanide bath, changes in the kinetics of the process have been observed. These may be due to changes in the structure of the double layer in the presence of, and under the influence of, the codepositing ions. Studies on the structure indicated the existence of a supersaturated solid solution of zinc in cadmium which forms a heterogeneous mixture with pure zinc.

Ethylenediamine solutions have been investigated also, but satisfactory coatings containing more than 30% zinc could not be obtained. The structure of the deposit and its distribution on the surface were somewhat improved by increasing the ethylenediamine concentration and by adding up to 1 g/l glue. The effect of variation in the metal, sodium hydroxide or EDTA concentration of the solution, current density, and temperature on the deposit composition has been determined. The current distribution between separate anodes of zinc and cadmium was unequal. The solution had good throwing power, and produced dense, light-grey deposits which possessed good adhesion to iron [165]. Bright coatings of the alloy have been obtained with metal sulfate solutions containing polyethylene polyamine [166] at a cathode current density of 1-30 A/dm^2.

The sulfamate bath has been studied in detail [167] and the optimum conditions established for the deposition of alloys containing 1-78% zinc at a cathode efficiency of 70-100%, anode efficiency of 98-100% with alloy anodes, and current density in the range 5-80 mA cm^{-2}. High current densities are used to obtain high-zinc alloys. Reconstruction of single metal potential curves from alloy composition and potential indicated that the extent of overpotential for cadmium is much more than that for zinc during codeposition. X-ray studies indicated the existence of the alpha or a mixture of alpha and beta phases.

Deposition of this alloy has also been carried out from acetate solutions [168]. The limiting composition of the bath at which the alloy is deposited varies approximately linearly with the logarithm of the current density and linearly at higher zinc concentration. The discharge cathode potential becomes more negative as electrolysis continues, then falls more sharply to a constant value, the equilibrium potential. The partial current density for the zinc deposit falls linearly as the cadmium concentration of the bath is increased and rises linearly with the total current density. The partial current density for the cadmium deposit rises linearly with cadmium concentration in the bath, but little with the total current density. The data indicated a discharge ion mechanism in which zinc acetate acts as a background electrolyte for the cadmium ion discharge.

5.2.3. Silver-Cadmium Alloys

Although neither zinc- nor silver-cadmium alloy has achieved any practical importance, each has been of considerable academic interest, chiefly because of the numerous phases it forms. The structure has been examined and compared with that of the cast alloys. The structure of the silver-cadmium alloys probably has been investigated more than that of any other electrodeposited alloy.

The standard electrode potential of silver (0.799 V) and those of cadmium (-0.403) and zinc (-0.763) are too far apart to permit codeposition of silver with these metals from solutions of simple salts. The alloys are best deposited from cyanide plating baths, although the potentials are still not very close together. Deposition of the alloys from other complexes has been investigated to a slight extent but no feasible process has resulted.

Cadmium-silver electrodes for use in galvanic cells have been prepared in the same manner as for cadmium. A ternary alloy containing cadmium 20-30, silver 60-75, and lead 5-10%, with improved antifriction properties, has been obtained from cyanide solutions at room temperature and 10-15 A/cm^2 current density.

5.2.4. Cadmium-Iron Alloys

The sulfamate bath has been found to be satisfactory for the deposition of alloys containing 1-93% iron, the cathode efficiency being 47-97% and the limiting current density 250 mA/cm^2. There are two breaks in the alloy potential curves corresponding to the deposition of cadmium and iron. Accordingly almost pure cadmium is deposited at low current densities; the sharp change in the potential indicates the feasibility of codeposition of iron in the higher current density range. There is irreversibility at low and at high current densities for the deposition of iron in the presence of cadmium. The deposit is a mechanical mixture of the two metals with hcp and bcc structures.

5.2.5. Cadmium-Copper Alloys

Cadmium-copper alloys can be electrodeposited from cyanide plating baths similar to those used for depositing copper-zinc alloys, but the range of plating conditions that produces good deposits is more restricted, and the deposits are not easily obtained with a uniform appearance. The alloys are not commercially electrodeposited, although they have some potentialities. They have an interesting range of color and could therefore be used for decorative purposes. The alloys are harder and more corrosion resistant than brass, but more brittle. There is a difference of about 0.7 V between the standard electrodepotential of cadmium (-0.403 V) and that of copper (0.337 V) in the solution of their simple ions. This difference is too great to permit satisfactory codeposition of the two metals from their simple ions, and solutions containing the metals as complex ions must therefore be used. In cyanide baths the potentials of the two metals are only a few tenths of a volt apart. The main characteristics of the bath compositions are the large ratio of copper to cadmium and the low free-cyanide content. The metal percentage of copper in these baths is about 90%. The high ratio of copper to cadmium is necessary since copper is less noble than cadmium. The copper content of the deposits is lower than the metal percentage of copper in the bath; for example, a typical bath contains 87 metal-% of copper but the deposit contains only 46% of copper.

The preparation of the baths involves no difficulty as the cyanides of the metals readily go into solution to form the complexes. To keep the cadmium in solution as the complex, a small amount of free cyanide must be present because cadmium hydroxide precipitates in its absence. The content of free cyanide required to maintain a clear solution depends on the concentration of copper and cadmium in the bath. These relations were studied by Longhurst [169]. The precipitation of a hydroxide from a cyanide bath is unusual. It does not occur with the analogous brass-plating bath. A comparison has been made of the structures, as of copper-cadmium alloys in the range 14-99% cadmium, prepared by melting and casting and by electrodeposition from the cyanide bath. The existence of metastable alloys in the as-deposited state and their transformation on annealing to the equilibrium phases found in the melt-formed alloy have been discussed. The cathode potential during deposition decreases from 2.27 to 1.15 V, and the potential range for the different phases from alpha to beta has been determined [170].

Electrodeposition of these alloys from cyanide solutions has been studied [171] at frequencies of 18.5 and 38 kHz and at acoustic intensities equal to or less than 0.5 W/cm^2. Marked changes occurred in the composition of the deposited alloys. Some ultrasonic deposits were brighter, harder, more adherent, or of finer grain size than nonirradiated deposits, and the copper content was lowered. Cathode efficiency was increased by as much as 36%, and anode corrosion was doubled in some cases.

The effect of surface-active substances on the phase structure of alloys plated from sulfate solutions has been studied. X-ray analysis indicated the formation of a supersaturated solid solution of cadmium in copper, an alloy with liquid-like structure, and a compound CuCd with an increase in the cadmium content. Organic substances had little effect on the structure. The codeposition of copper was strongly inhibited with increase in the cadmium content of the alloy obtained from electrolytes containing sulfuric acid without and with surface-active substances. This inhibition is attributed to a change in the structure of the electrical double layer and increase in the adsorption of the additives [172]. The formation of supersaturated solid solutions on a copper base enhanced formation of compact, smooth deposits at potentials at which copper is deposited under limiting diffusion current conditions. Cadmium in the copper lattice hindered copper deposition. Inhibition of copper deposition is due to a change in the potential difference at the metal solution interface as a result of a shift in the zero charge point of the alloy with an increase in the cadmium content [173].

5.2.6. Cadmium-Nickel Alloys

Electrodeposited alloys of nickel with cadmium have potential uses in the decorative and protective finishing of metals. Attempts have been made to deposit a cadmium-nickel alloy from sulfate, chloride, chloride-ammonia, ammonia citrate, boron hydrofluoride, pyrophosphate, and cyanide electrolytes. However, in most cases only cadmium ion was deposited. Alloy deposition was, however, possible from pyrophosphate solutions containing cadmium oxide and nickel sulfate, with additions of ethylenediamine and glue. The nickel content of the deposit increased with an increase in the nickel concentration, pH of the solution, temperature, and current density, but decreased by agitation [174]. Cathode efficiency and polarization on the composition and temperature of the plating solution, current density, and other factors has been studied, together with some aspects of the kinetics of ion discharge.

5.2.7. Cadmium-Tin Alloys

Typical acid cadmium-tin-alloy plating baths contain addition agents since dull coarse-grained deposits are obtained without. Bennett [175] made a detailed study of more than 30 addition agents, both singly and in pairs. The appearance of the deposit was dependent on the type of addition agent used. Bright deposits were yielded by the combination of certain addition agents added to the bath as aqueous solutions. The addition agents codeposited with the alloy, as shown by the residue remaining after alloys were dissolved in acid. All the baths listed contained excess acid, the function of which was to prevent the hydrolysis of tin compounds. However, the baths should not be made too acid because higher acid concentrations lower the cathode current efficiency. The addition of ammonium fluoborate and boric acid may be made for the purpose of buffering. Additions of ammonium fluoride to baths for the purpose of refining the grain size of the deposit are suggested.

The cathode current efficiency of the acid baths was about 100% when operated at current densities below 20 mA cm^{-2}. At higher current densities the efficiency dropped, depending on the type and metal concentration of the bath, to as low as 70%. In still baths the current density may be as high as 40 mA cm^{-2} and in an agitated bath the current density may be 400 mA cm^{-2} [176]. The fluosilicate bath can operate at a higher current density than the others without producing dendrite deposits.

Fine-grained deposits of cadmium-tin have been obtained from chloride [177] solutions containing ammonium fluoride and additions of glue and phenol, and with good throwing power, at a pH of 2.5-4 and 10-20 A/cm^2. Polarization during hydrogen evolution on electrodeposited cadmium-tin alloys has been studied in sulfuric acid solutions without and with additives gelatin, thymol, β-naphthol, and diphenylamine.

5.3. NICKEL-CADMIUM CELLS (ALKALINE TYPE)

Nickel-cadmium cells were little known until after 1930. They are now used for emergency lighting, control and switch-gear operation, and auxiliary power systems. Cadmium is apparently less subject to self-discharge than iron and has considerable resistance to passivity at low temperatures. Nickel-cadmium cells have been constructed in two distinct forms. The older, the Jungner type of cell, was developed first in Sweden about 1900, and the other is an impregnated sintered-plate type which apparently originated in Germany during World War II. The sintered plate type at the present time generally finds more application than the Jungner type.

The negative plates are similar in construction for both types of cell, and consist of perforated pockets containing the active materials. The positive plates in some types of nickel-cadmium cells are of tubular construction. The pockets for both positive and negative plates are made from perforated steel ribbon which has been nickel-plated and annealed in hydrogen. Provision must be made for some expansion of the positive active materials within the pockets. Pockets of the negative plates are filled initially with cadmium oxide, CdO, or cadmium hydroxide, $Cd(OH)_2$, and reduced to metallic cadmium in spongy form on the first charge. Iron (5 to 30%) is added to the cadmium in order to obtain the required degree of fineness of the cadmium. The beneficial effect of the iron seems to be a more or less empirically determined fact, and the real function of the iron is not entirely clear. It has been thought that it forms an alloy with the cadmium; some have regarded it as an increase to the effective surface area of the electrode, whereas it may be that the sole function of the iron is that of a disperser to bring about a finely divided state of the electrolytically precipitated cadmium. Iron in the metallic state increases the conductivity of the active material; however, cadmium oxide itself is a good conductor. Whether iron oxidizes during discharge is also subject to differences of opinion. Crennell and Lea [178] indicated that it may do so and contribute a small part of the discharge capacity, but Hauel [179] considered that the iron takes no part in the current-producing process.

5.3.1. Mechanism of Discharge

Many studies on the discharge mechanism of the cadmium electrode in alkaline solution have been published, resulting in different proposed mechanisms due to the complexity of the system, and results have been referred to in the section describing electrochemical studies. Yoshizawa and Takehara [135] have made a more applied study on the discharge mechanism of cadmium in the alkaline battery. It was concluded that cadmium dissolves electrochemically, through cadmium oxide as an intermediate product, into the solution as $HCdO_2^-$; after saturation with $HCdO_2^-$ of the solution next to the electrode, cadmium oxide accumulates on the electrode surface. It then dissolves chemically in the solution as $HCdO_2^-$, which then becomes cadmium hydroxide.

The larger part of the overpotential depends on the ionic migration of Cd^{2+} in the cadmium oxide layer.

There are two possibilities of decreasing such an overpotential:

1. Diffusion resistance of Cd^{2+} in the cadmium oxide layer formed by the discharge must be decreased,
2. The decomposition rate of cadmium oxide must be increased.

Diffusion resistance for Cd^{2+} may be decreased by impurity additions, such as In, as shown by Croft [137] and Hosono and Matsui [180], or Tl and Ag, as shown by Hosono and Matsui [180]. The decomposition rate of cadmium oxide can be increased by the increase of its dissolution rate or by the increase of the decomposition rate of cadmium hydroxide.

REFERENCES

1. H. S. Harned and M. E. Fitzgerald, J. Amer. Chem. Soc., 58, 2624 (1936).
2. J. Kielland, ibid., 59, 1675 (1937).
3. W. B. Trueman and L. M. Ferris, ibid., 80, 5048 (1958).
4. W. G. Parks and V. K. La Mer, ibid., 56, 90 (1934).
5. A. J. de Bethune and N. A. S. Loud, Encyclopedia of Electrochemistry (C. A. Hampel, ed.), Reinhold, New York, 1964.
6. R. G. Bates, J. Amer. Chem. Soc., 61, 308 (1939).
7. Y. Okinaka, S. Toshima, and H. Okaniwa, Talanta, 11, 203 (1964).
8. J. L. Burnett and M. H. Zirin, J. Inorg. Nucl. Chem., 28, 902 (1966).
9. F. H. Getman, J. Phys. Chem., 35, 588 (1931).
10. A. W. Andrews, D. A. Armitage, R. W. C. Broadbank, K. W. Morcom, and B. L. Muju, Trans. Faraday Soc., 67, 128 (1971).
11. T. Paulopoulos and H. Strehlow, Z. Phys. Chem., 2, 89 (1954).
12. R. Lorenz and H. Velde, Z. Anorg. Chem., 183, 81 (1929).
13. M. F. Lantratov and A. F. Alabyshev, J. Appl. Chem. USSR, 26, 321 (1953).
14. I. A. Menzies, G. J. Hills, L. Young, and J. O'M. Bockris, Trans. Faraday Soc., 55 1580 (1954).
15. S. N. Flengas and T. R. Ingram, Can. J. Chem., 36, 780 (1958); J. Electrochem. Soc., 106, 714 (1959).
16. M. Fiorani, G. A. Sacchetto, and G. G. Bombi, Electrochim. Acta, 11, 717 (1966).
17. J. A. Plambeck, J. Chem. Eng. Data, 12, 77 (1967).
18. R. Marassi, V. Bartocci, P. Cescon, and M. Fiorani, J. Electroanal. Chem., 22, 215 (1969).
19. E. Deltombe, M. Pourbaix, and N. de Zoubov, in Atlas of Electrochemical Equilibrium in Aqueous Solution (M. Pourbaix, ed.), Pergamon, London, 1966.
20. G. Charlot, L'analyse qualitative et les reactions en solution, 4th ed., Masson, Paris, 1957.
21. A. Ya. Chatalov, Dokl. Acad. Nauk SSSR, 86, 775 (1952).
22. D. T. Hurd, An Introduction to the Chemistry of the Hydrides, Wiley, New York, 1952, p. 197.
23. J. Piater, Z. Anorg. Allgem. Chem., 174, 321 (1928).
24. W. Feitknecht and R. Reinmann, C. R. 3e Reunion du CITCE, Berne, 1951, 93; Manfredi, Milan, 1952.
25. K. Huber, C. R. 3e Reunion du CITCE, Berne, 1951, 117, Manfredi, Milan, 1952.
26. I. Kolthoff and P. I. Elving, Treatise on Analytical Chemistry, Part 2, Vol. 3, Interscience, New York, 1961.
27. A. M. Bond, J. Electroanal. Chem., 16, 21 (1968).
28. J. E. B. Randles, Trans. Faraday Soc., 44, 327 (1948).
29. T. F. Retajczyk and D. K. Roe, J. Electroanal. Chem., 16, 21 (1968).
30. J. Heyrovsky and J. Kuta, Principles of Polarography, Publ. House of Czech Acad. Sci., 1966.
31. L. Meites, Polarographic Techniques, Wiley (Interscience), New York, 1965.
32. N. Tanaka, R. Tamamushi, and M. Kodama, Z. Phys. Chem., 14, 141 (1958).

33. J. V. Cakenburghe, Bull. Soc. Chem. Belges, 60, 3 (1951).
34. R. Tamamushi and N. Tanaka, Z. Phys. Chem., [N. F.] 21, 89 (1959).
35. S. S. Mesaric and D. N. Hume, Inorg. Chem., 2, 1063 (1963).
36. C. Rulfs, J. Amer. Chem. Soc., 76, 2071 (1954).
37. C. Auerbach and D. K. McGuire, J. Inorg. Nucl. Chem., 28, 2659 (1968).
38. D. D. DeFord and D. N. Hume, J. Amer. Chem. Soc., 73, 5321 (1951).
39. L. G. Sillen, Acta Chem. Scand., 10, 186 (1956).
40. J. L. Jones and H. A. Fritsche, J. Electroanal. Chem., 12, 334 (1968).
41. R. T. Burrus, Ph.D. Thesis, University of Tennessee, 1962.
42. C. K. Mann and K. K. Barnes, Electrochemical Reactions in Nonaqueous Systems, Dekker, New York, 1970.
43. M. Francini, S. Martini, and H. Geiss, Electrochim. Metall., 3, 355 (1968).
44. D. Inman, D. G. Lovering, and R. Narayan, Trans. Faraday Soc., 63, 3017 (1967).
45. M. Steinberg and N. H. Nachtrieb, J. Amer. Chem. Soc., 72, 3558 (1950).
46. G. C. Barker and R. L. Faircloth, A.R.E.E., 1956, C/R 2032.
47. M. Francini and S. Martini, Euratom Report E.U.R., 1964, 1908e.
48. H. S. Swafford and C. L. Hollifield, Anal. Chem., 37, 1513 (1965).
49. M. Saito, Nippon Kagaku Zasshi, 83, 883 (1962).
50. R. M. de Fremont, in Polarography 1964 (G. J. Hills, ed.), Macmillan, London, 1966.
51. G. G. Bombi, Chem. Commun., 455 (1966).
52. E. L. Colichman, Anal. Chem., 27, 1559 (1965).
53. D. Inman, D. G. Lovering, and R. Narayan, Trans. Faraday Soc., 64, 2476 (1967).
54. J. H. Christie and R. A. Osteryoung, J. Amer. Chem. Soc., 82, 1841 (1960).
55. G. G. Bombi, G. A. Massocchin, and M. Fiorani, Ric. Sci., 36, 573 (1966).
56. J. O. Liljenzin, H. Reinhardt, H. Wirries, and R. Lindner, Z. Naturforsch., 18a, 840 (1963); Radiochim. Acta, 1, 161 (1963).
57. J. Braustein and A. S. Minano, Inorg. Chem., 3, 218 (1964).
58. M. G. Francini and S. Martini, in Polarography 1964 (G. J. Hills, ed.), Macmillan, London, 1966.
59. H. M. N. H. Irving, "International Conference on Coordination Chemistry," Chem. Soc. Spec. Publ. (London), 13 (1959).
60. P. Delahay, J. Phys. Chem., 70, 2373 (1966).
61. M. E. Berglund, Bull. Soc. Chim. Belges, 29, 422 (1876).
62. Y. Ito, Kogyo Kagaku Zasshi, 73, 1866 (1969).
63. V. Bartocci, R. Marrassi, and F. Pucciarelli, Chim. Ind., 52, 1201 (1970).
64. M. Francini, S. Martini, and C. Monfrini, Electrochem. Metall., 2, 3 (1969).
65. M. Francini and S. Martini, Electrochem. Metall., 3, 136 (1968).
66. M. Francini and S. Martini, in Polarography 1964 (G. J. Hills, ed.), Macmillan, London, 1966, pp. 1153-1164.
67. H. A. Laitinen and H. C. Gaur, Anal. Chim. Acta, 18, 1 (1958).
68. L. L. Naryshkin, V. P. Yurkinskii, and B. S. Yavich, Elektrokhimiya, 2, 807 (1966).
69. R. Kal'voda and N. Kh. Tumanova, Ukr. Khim. Zh., 33, 102 (1967).

70. H. A. Laitinen and W. Ferguson, Anal. Chem., 29, 4 (1957).

71. C. S. Thalmayer, S. Bruckenstein, and D. M. Gruen, J. Inorg. Nucl. Chem., 26, 347 (1964).

72. J. E. B. Randles and K. Somerton, Trans. Faraday Soc., 48, 951 (1952).

73. H. Gerischer, Z. Elektrochem., 57, 604 (1953).

74. T. Berzins and P. Delahay, J. Amer. Chem. Soc., 77, 6448 (1955).

75. W. Vielstich and P. Delahay, ibid., 79, 1874 (1957).

76. H. H. Bauer and P. J. Elving, Anal. Chem., 30, 334 (1958).

77. G. L. Barker, Anal. Chim. Acta, 18, 118 (1958).

78. H. H. Bauer and P. J. Elving, J. Amer. Chem. Soc., 82, 2091 (1960).

79. J. E. B. Randles, Electrode Processes Symposium (E. Yeager, ed.), Wiley, New York, 1961.

80. T. Kambara and J. Ishii, Rev. Polarog., 9, 30 (1961).

81. J. H. Sluyters, Rec. Trav. Chim., Pays-Bas, 82, 535 (1963).

82. Y. Okinaka, S. Toshima, and H. Okaniwa, Talanta, 11, 203 (1964).

83. T. Biegler and H. A. Laitinen, J. Electrochem. Soc., 113, 852 (1966).

84. T. Biegler and H. A. Laitinen, Anal. Chem., 37, 572 (1965).

85. A. Hamelin, C. R. Acad. Sci., Paris, Ser. C, 262, 520 (1966).

86. D. J. Kooijman and J. H. Sluyters, Electrochim. Acta, 12, 693 (1967).

87. N. A. Hampson and D. Larkin, J. Electroanal. Chem., 18, 401 (1968).

88. H. H. Bauer, ibid., 12, 64 (1966).

89. B. Lovrecek and A. Marincic, Electrochim. Acta, 11, 237 (1966).

90. A. R. Despic, D. R. Jovanovic, and S. P. Bingulac, ibid., 15, 459 (1970).

91. M. Anbar, Quart. Rev., 22, 578 (1968).

92. T. Biegler, E. R. Gonzalez, and R. Parsons, Collect. Czech. Chem. Commun., 36, 414 (1971).

93. J. D. Corbett, Inorg. Chem., 1, 700 (1962).

94. R. A. Potts, R. D. Barnes, and J. D. Corbett, ibid., 7, 2558 (1968).

95. W. Lorenz, Naturwissenschaften, 40, 578 (1953).

96. W. Lorenz, Z. Elektrochem., 58, 912 (1954).

97. R. J. Brodd, J. Res. Nat. Bur. Stand., A, 65, 275 (1961).

98. N. A. Hampson, R. J. Latham, and D. Larkin, J. Electroanal. Chem., 23, 211 (1969).

99. L. A. L'vova, D. K. Grachev, and V. A. Panin, Elektrokhimiya, 5, 627 (1969).

100. N. A. Hampson and R. J. Latham, J. Electroanal. Chem., 32, 175 (1971).

101. N. A. Hampson and R. J. Latham, ibid., 32, 337 (1971).

102. K. E. Heusler and L. Gaiser, J. Electrochem. Soc., 117, 762 (1970).

103. K. J. Vetter, Electrochemical Kinetics, Academic, New York, 1967.

104. M. Fleischmann, S. K. Rangarajan, and H. R. Thirsk, Trans. Faraday Soc., 63, 1256 (1967).

105. H. Gerischer, Z. Phys. Chem., 201, 55 (1952).

106. Yu. M. Loshkarev, L. P. Smetkova, V. N. Kovtun, and N. P. Tkalik, Elektrokhimiya, 4, 298 (1968).

107. L. A. L'vova, Work of Young Scientists Vyp. Khim., Izd-vo Sarat, Gos. Un-ta, 1965, p. 3 (in Russian).

108. L. A. L'vova and A. V. Fortunatov, Anodic Protection of Metals, Mashinostroenie, Moscow, 1964, p. 395 (in Russian).

109. M. W. Brieter and I. L. Weininger, J. Electrochem. Soc., 111, 707 (1964).

110. "Stability Constants," Chem. Soc. Spec. Publ. (London), 17 (1964).

111. H. Gerischer, Anal. Chem., 32, 33 (1969).

112. H. Gerischer, Z. Phys. Chem., 202, 302 (1953).

113. L. Ramaley and C. G. Enke, J. Electrochem. Soc., 112, 947 (1965).

114. I. H. Plonski, Rev. Roum. Chim., 13, 65 (1968).

115. H. A. Laitinen and R. A. Osteryoung, J. Electrochem. Soc., 102, 598 (1955).

116. H. A. Laitinen and H. L. Gaur, ibid., 104, 730 (1957).

117. H. A. Laitinen, R. Tischer, and D. Roe, ibid., 107, 546 (1960).

118. T. A. Borisova and B. V. Ershler, Zh. Fiz. Khim., 24, 337 (1950).

119. N. V. Nikolavea, N. S. Shapinocend, and A. N. Frumkin, Dokl. Akad. Nauk SSSR, 86, 851 (1952).

120. V. A. Kheifets and B. S. Krasikov, Dokl. Akad. Nauk SSSR, 109, 586 (1956).

121. L. I. Antropov, Ukr. Khim., 29, 555 (1963).

122. N. A. Hampson and D. Larkin, J. Electrochem. Soc., 114, 933 (1967).

123. V. Ya. Bartenev, E. S. Sevast'yanov, and D. I. Leikis, Elektrokhimiya, 4, 1502 (1969).

124. V. Ya. Bartenev, E. S. Sevast'yanov, and D. I. Leikis, ibid., 5, 1502 (1969).

125. N. A. Hampson and R. J. Latham, J. Electroanal. Chem., 34, 57 (1971).

126. E. A. Ukshe, N. G. Bukun, and D. Leikis, Zh. Fiz. Khim., 36, 2322 (1962).

127. O. A. Esin and B. F. Markov, Acta Physicochem. USSR, 10, 353 (1939).

128. N. A. Hampson and R. J. Latham, J. Electroanal. Chem., 31, 57 (1971).

129. S. A. Rozentsveig, B. V. Ershler, E. L. Strum, and M. M. Ostania, Tr. Soveshch. po Elektrokhim. Acad. Nauk SSSR, Otd. Khim. Nauk, 1950, 571 (1953).

130. K. Huber, J. Electrochem. Soc., 100, 376 (1953).

131. K. Huber and S. Stucki, Helv. Chim. Acta, 51, 1343 (1968).

132. P. E. Lake and E. J. Casey, J. Electrochem. Soc., 105, 52 (1958).

133. P. E. Lake and E. J. Casey, ibid., 106, 913 (1959).

134. I. Sanghi, S. Visvanathan, and S. Ananthanarayanan, Electrochim. Acta, 3, 65 (1960).

135. S. Yoshizawa and Z. Takehara, ibid., 5, 240 (1961).

136. M. A. V. Devanathan and S. Lakshmanan, ibid., 13, 667 (1968).

137. G. T. Croft, J. Electrochem. Soc., 106, 278 (1969).

138. G. T. Croft and D. Tuomi, J. Electrochem. Soc., 108, 915 (1961).

139. J. P. G. Farr and N. A. Hampson, Electrochem. Technol., 6, 10 (1969).

140. S. U. Falk, J. Electrochem. Soc., 107, 661 (1960).

141. R. D. Armstrong, E. H. Boult, D. F. Porter, and H. R. Thirsk, Electrochim. Acta, 12, 1245 (1967).

142. M. W. Breiter and J. L. Weininger, J. Electrochem. Soc., 113, 651 (1966).

143. M. W. Breiter and W. Vedder, Trans. Faraday Soc., 63, 1042 (1967).

144. M. Fleischmann, K. S. Rajagopolan, and H. R. Thirsk, ibid., 59, 741 (1963).

145. Y. Okinaka, J. Electrochem. Soc., 117, 194 (1970).

146. R. D. Armstrong and G. D. West, J. Electroanal. Chem., 30, 385 (1971).

147. G. C. Bratt, Electrochem. Technol., 2, 32 (1964).

148. P. C. Udy, U.S. Patent 1,383,174 (1919).

149. I. R. Westbrook, Trans. Amer. Electrochem. Soc., 55, 333 (1929).

150. K. G. Soderberg, J. Electrodep. Tech. Soc., 13(9), 1 (1937).

151. A. W. Young and M. E. Louth, U.S. Patent 1,692,240 (1928).

152. J. A. Kendricks, U.S. Patent 2,085,747-50 (1937).

153. J. V. Vaughen, U.S. Patent 2,085,776 (1937).

154. R. O. Hall, U.S. Patent 2,090,049 (1937).

155. J. A. Hendricks, U.S. Patent 2,350,163 (1944).

156. W. Burkart, Oberflachen Tech., 19, 81 (1942).

157. L. R. Westbrook, U.S. Patent 1,818,179; 1,826,159 (1931).

158. L. R. Westbrook. U.S. Patent 1,681,509 (1928).

159. J. R. Gustafson, ASTM Proc., 47, 782 (1947).

160. R. E. Alexander, U.S. Patent 2,703,311 (1955).

161. G. A. Miller, Plating, 49, 157 (1962).

162. B. F. Phillips, U.S. Patent 2,578,898 (1950).

163. G. Dehmel, East German Patent 30,406 (1964).

164. N. P. Fedat'er, P. M. Viacheslaoor, and G. P. Andreeva, Zh. Prikl. Khim., 1, 572 (1963).

165. V. S. Galinker and A. I. Saprykin, ibid., 37, 342 (1964).

166. N. R. Kokrec and B. A. Popov, USSR Patent 185,659 (1966).

167. T. H. Rama Char, Proceedings of the First Australian Conference on Electrochemistry (A. Friend and F. Gutmann, eds.), Pergamon, New York, 1964.

168. J. Ibarz and A. Estape, An. Real Soc. Espan. Fis. Quim, Ser. B, 63, 169 (1967).

169. E. E. Longhurst, J. Electrodep. Tech. Soc., 26, 71 (1950).

170. P. L. Ahuja and T. Banerjee, J. Inst. Metals, 94, 200 (1966).

171. C. B. Kenaham, D. Schlain, and E. Chin, U.S. Bur. Mines, Rep. Invest., 6938 (1967).

172. Yu. M. Polukarov and V. V. Grinina, Elektrokhimiya, 1, 350 (1965).

173. Yu. M. Polukarov and V. V. Grinina, Zh. Fiz. Khim., 39, 1176 (1965).

174. N. T. Kudryavstev, S. M. Firged, and N. N. Dakiua, Tr. Mosk. Khim. Tekhnol. Inst., 44, 91 (1963).

175. P. S. Bennett, J. Electrodep. Tech. Soc., 26, 91 (1950).

176. R. D. Gray and W. A. Paocht, U.S. Patent 2,609,338 (1952).

177. N. T. Dakry-avtsec, K. M. Tgutina, and A. M. I. Fatkh, Metal. Fin. Abstr., 9, 89 (1967).

178. J. T. Crennell and F. M. Lea, Alkaline Accumulators, Longmans Green, London, 1928.

179. A. P. Hauel, Trans. Electrochem. Soc., 76, 435 (1939).

180. T. Hosono and M. Matsui, Presented at the discussion meeting on batteries of the Chemical Society of Japan, November 1959.

Chapter I-5

LEAD

THOMAS F. SHARPE

Electrochemistry Department
Research Laboratories
General Motors Corporation
Warren, Michigan

1. STANDARD AND FORMAL POTENTIALS

1.1. AQUEOUS SOLUTIONS

Standard and formal potentials involving the lead electrode in aqueous solutions are given in Table 1.1.1. It is important to note the difference in potential between lead and lead amalgam electrodes, since the latter were used in many of the determinations. For numerous solvents a potential difference of 5.7-5.8 mV is detected [2, 15] as compared to the slightly lower value of 5.3 mV reported by Bates et al. [16].

1.1.1. The Pb/Pb^{2+} Couple

Carmody [2] measured the potential of lead amalgam electrodes against the Ag/AgCl reference in solutions containing $PbCl_2$. By extrapolating to infinite dilution according to the procedure of Randall [17], and correcting for the Pb(Hg)/Pb potential difference, a value of -0.1263 V was obtained, in excellent agreement with the value of -0.126 V calculated from thermodynamic data [1].

The stability of lead in acid solution is a result of the high hydrogen overvoltage on lead [18]. Passivation of the local anodes of the corrosion cells may also be a contributing factor, since the majority of lead salts are insoluble and frequently form protective films [19, 20]. In corrosion studies the potential of lead has been determined in the presence of a number of anions, and these values are included in Table 1.1.1. Values in these media are approximate because the potential can change depending on degree of aeration as well as condition and purity of the metal surface.

In the work of Fleischmann and Liler [9] involving the electrodeposition of PbO_2, a knowledge of the lead ion activity in various acetate buffer solutions was required. Reported potential values of Pb(Hg) electrodes in these solutions (Table 1.1.1) were used to calculate these data.

TABLE 1.1.1. Standard and Formal Potentials in Aqueous Solutions

Half-cell reaction	Potential (V)	Conditions	Refs.
$Pb^{2+} + 2e^- = Pb$	-0.126		[1], [2]
	-0.280	0.1-1.3 M NaCl	[3]
	-0.204	0.1 M $NaNO_3$	[4]
	-0.334	0.1 M KBr	[4]
	-0.472	0.1 M NaF	[4]
	-0.393	0.2 M Na_2SiO_3	[4]

(continued)

TABLE 1.1.1 (continued)

Half-cell reaction	Potential (V)	Conditions	Refs.
	-0.180	0.01 M $NaHCO_3$	[5]
	-0.230	0.2 M $NaHCO_3$	[4]
	-0.128	0.3 M H_3PO_4	[4]
	-0.259	0.1 M Na_2SO_4	[4]
	-0.356	0.1 M NaAc	[4]
	-0.177	H_2S (satd)	[6]
	-0.230	5×10^{-4} to 2 M H_3PO_4 in $HClO_4$, pH 1.05	[7]
	-0.280	Phthalate buffer, pH 5.0	[8]
$Pb^{2+} + Hg + 2e^- = Pb(Hg)$	-0.195	1 M $Pb(Ac)_2$, pH 5.6	[9]
	-0.239	1 M $Pb(Ac)_2$, 1 M HAc, pH 3.9	[9]
	-0.193	1 M $Pb(Ac)_2$, 1 M NaAc, 1 M HAc, pH 4.72	[9]
$PbCl_2 + 2e^- = Pb + 2Cl^-$	-0.268		[1]
$PbBr + 2e^- = Pb + 2Br^-$	-0.284		[1]
$PbI + 2e^- = Pb + 2I^-$	-0.365		[1]
$PbHPO_4 + 2e^- = Pb + HPO_4{}^{2-}$	-0.465		[7]
$Pb_3(PO_4)_2 + 6e^- = 3Pb + 2PO_4{}^{3-}$	-0.658		[7]
$PbSO_4 + 2e^- = Pb + SO_4{}^{-2}$	-0.353 to -0.359		[10], [1], [11]
$PbSO_4 + Hg + 2e^- = Pb(Hg) + SO_4$	-0.351		[1], [12]
$PbO_r + H_2O + 2e^- = Pb + 2OH^-$	-0.580		[1]
$PbO_2 + H_2O + 2e^- = PbO_r + 2OH^-$	-0.247		[1]
$PbO_2 + 4H^+ + 2e^- = Pb^{2+} + 2H_2O$	1.455		[1]
α-$PbO_2 + SO_4{}^{-2} + 4H^+ + 2e^- = PbSO_4 + 2H_2O$	1.697 to 1.698		[13], [14]
	1.709	4.4 M H_2SO_4, 31.8°	
β-$PbO_2 + SO_4 + 4H^+ + 2e^- = PbSO_4 + 2H_2O$	1.690		[13]
	1.692	4.4 M H_2SO_4, 31.8°	

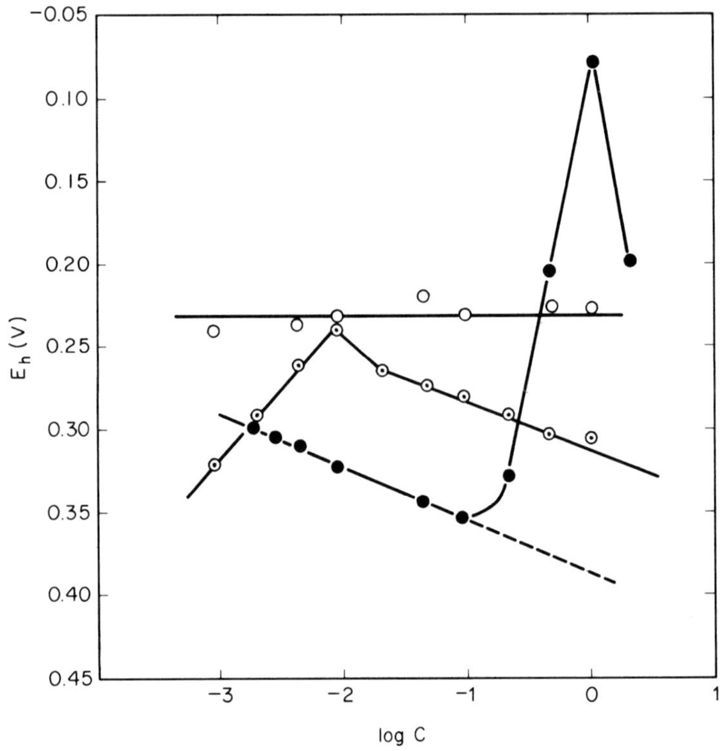

FIG. 1.1.1. Potential of lead in phosphate solutions of various concentrations. (o) H_3PO_4 in $HClO_4$, pH 1.05; (⊙) $NaH_2PO_4 + H_3PO_4$, pH 2.05; (•) NaH_2PO_4, pH 4.6 [7]. (Courtesy of Elsevier, Ltd.)

1.1.2. The Pb/Pb Phosphate Couple

The behavior of lead as a metal-metal phosphate electrode was recently studied by Awad et al. [7] and some results are presented in Fig. 1.1.1. At pH 1.05 the measured potential is independent of phosphate concentration, indicating the absence of phosphate films. At pH 2.05 and 4.5 a slope of -29 mV/log C over a range of phosphate concentrations is consistent with the presence of a secondary phosphate layer on the electrode surface, giving the electrode reaction

$$Pb + HPO_4^{2-} = PbHPO_4 + 2e^- \qquad (1.1.1)$$

This was further confirmed by favorable comparison of the standard potential for the Pb/$PbHPO_4$ couple calculated from the experimental results with that calculated from thermodynamic data. The standard potential given by Awad for Reaction (1.1.1) is -0.055 V.

As indicated in Fig. 1.1.1, deviations from the reversible Pb/$PbHPO_4$ reaction were observed at both high and low concentrations of phosphate, and they were ascribed to various adsorption effects. At pH higher than 8.7, evidence was presented that the electrode behavior became controlled by the reaction

$$3Pb + 2PO_4^{3-} \rightarrow Pb_3(PO_4)_2 + 6e \qquad (1.1.2)$$

within certain limits of phosphate concentration. The standard potential calculated by Awad for Reaction (1.1.2) is -0.658 V.

1.1.3. The Pb/Pb Halide Couple

The standard potential for the Pb/$PbCl_2$ electrode calculated from thermodynamic data is in excellent agreement with the work of Gerke [15], who measured the potential of the Pb(Hg)/$PbCl_2$ electrode in HCl solution against the Ag/AgCl reference. Using the accepted standard potential for the Ag/AgCl electrode (0.222 V) [1] and correcting for the Pb/Pb(Hg) potential gives a value of 0.268 V for the Pb/$PbCl_2$ half-cell. Potentials involving the other halides have been calculated from thermodynamic data and are included in Table 1.1.1.

1.1.4. The Pb/Pb Sulfate Couple

The half-cell has been widely studied using both pure lead and lead amalgam electrodes. Satisfactory preparation of amalgam electrodes for this purpose was effected by cathodizing Hg in $PbNO_3$ solution until the Hg contained about 6% lead. The amalgam was washed and then heated to insure homogeneity, and results were improved by covering the electrode with dry lead sulfate before immersing it in the oxygen-free test solution. Emf measurements with this electrode against the Hg/Hg_2SO_4 reference electrode in neutral sulfate solution have given highly reproducible results [11, 21-23].

At low pH it is necessary to take into account incomplete dissociation of the H_2SO_4. When measurements on the cell

$$Pb(Hg)\,|\,PbSO_4\,|\,H_2SO_4\,|\,H_2, Pt \qquad (1.1.3)$$

were extrapolated in a manner which allowed for incomplete dissociation, a value of 0.351 V was obtained by Shrawder and Cowperthwaite [12]. Correcting for the Pb/Pb(Hg) potential difference gives a value of 0.357 for the Pb/$PbSO_4$ half-cell. Results based on thermodynamic calculations range from 0.356 to 0.359 V (Table 1.1.1). A lower value of 0.353 V was reported from the work of Harned and Hamer [11].

Investigations have been made concerning the potential-time relationship for lead electrodes in H_2SO_4. A clean lead surface quickly assumes the Pb/$PbSO_4$ reversible potential when immersed in H_2SO_4, whereas one which has not been properly cleaned may exhibit a potential near 0 V for a substantial period of time before drifting to the Pb/$PbSO_4$ value [24]. The potential near 0 V has been ascribed to a film of oxide (or hydroxide) on the lead surface.

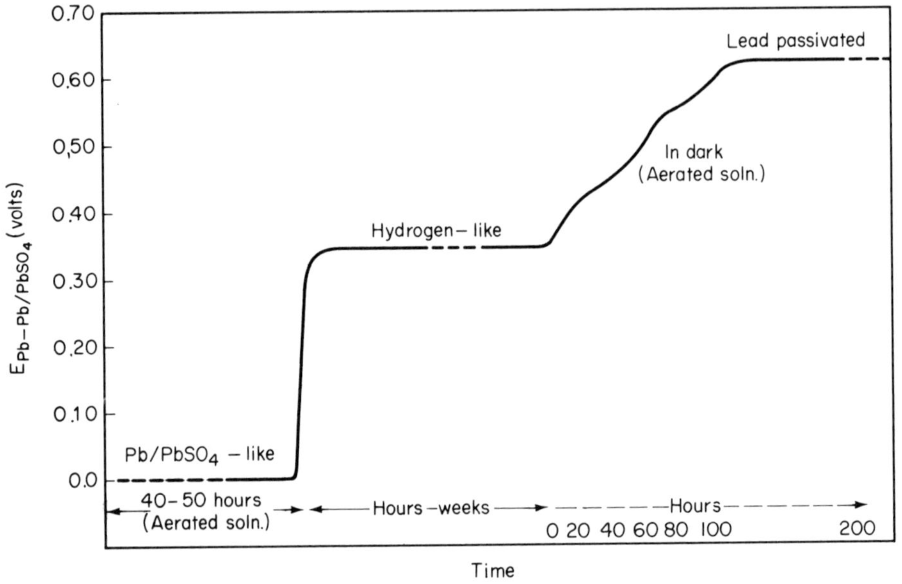

FIG. 1.1.2. Potential of lead in oxygen-saturated H_2SO_4 as a function of time [28]. (Courtesy of Electrochemical Society.)

A surface that is reportedly [24] free of oxides (at least to the extent that they are not potential-determining) is effected by treating the lead with a suitable etchant and then immersing the lead in the H_2SO_4 electrolyte while it is still wet from the final water rinse.

After the electrode remains at the Pb/$PbSO_4$ value for a period of time, the potential shifts in the anodic direction and ultimately reaches a poorly reproducible value near 300 mV, where the electrode is considered passivated. The early work of Haring and Thomas [25] indicated a direct shift from the Pb/$PbSO_4$ potential to the passivation potential with time, but others [26, 27] have observed arrests in the potential-time curve for this process.

The potential-time curve reported by Casey and Campney [28] is shown in Fig. 1.1.2. At the Pb/$PbSO_4$ plateau the corrosion is under cathodic control and the measured potential varies from -420 to -240 mV over a range of acid concentrations [26]. The length of time that the lead electrode maintains the Pb/$PbSO_4$ potential decreases with increasing H_2SO_4 concentration [26], which has been attributed to increased solubility of $PbSO_4$ in more concentrated solutions.

As indicated in Fig. 1.1.2, the electrode shifts from the Pb/$PbSO_4$ potential to a value near 0 mV (~350 mV vs Pb/$PbSO_4$), where the lead behaves as a hydrogen-like electrode. Here, the corrosion rate may be anodically controlled by a partially covering layer of $PbSO_4$. The hydrogen-like potential has been attributed by Ekler [26] to $PbO.PbSO_4$ which may be present in the pores of the $PbSO_4$ layer. Ohse [29] gives a value of 34 mV for the electrode potential of Pb, $PbSO_4$/$PbO.PbSO_4$ at pH = 0.

A number of explanations have been offered for the attainment of the passivation potential. Müller [30] described the passivation potential in terms of electrolyte iR drop thought to exist in the pores of the $PbSO_4$ film. Haring and Thomas [25] found that the presence of oxygen at the electrode/solution interface decreased the time required for passivation, and attributed the passivation potential to a mixed potential of hydrogen electrodes and an unspecified species of oxygen electrodes.

Shortcomings in these theories have been pointed out by Casey and Campney [28]. Photomicrographs obtained by these workers showed that the only apparent physical difference before and after passivation was the number and depth of the $PbSO_4$ crystals on the surface of the lead. Their studies indicated that the potential of the passivated lead was independent of the partial pressure of oxygen but that additions of H_2O_2 to the system substantially reduced the time required for passivation and eliminated the hydrogen-like step. Moreover, the potential value at passivation was dependent on both $[H_2O_2]$ and $[H_2SO_4]$. They also found that the passivated lead exhibited ac rectifying properties and proposed that the $PbSO_4$ on the surface becomes a defect-type semiconductor brought about by the possible sorption of oxygen atoms into the $PbSO_4$ lattice. Casey and Campney concluded that the passivation potential is based on an equilibrium of the system SO_4^{-2}-$S_2O_8^{-2}$-SO_5^{-2}-H_2O_2.

A decision on the probable mechanism by which lead becomes enobled in H_2SO_4 is difficult to make. Eckler [26] points out that the passivation potential corresponds to the value at which PbO and $Pb(OH)_2$ have been detected by X-ray diffraction. The static behavior of lead in H_2SO_4 is important in view of observations by Ekler that the time of immersion in the acid affects the results obtained during subsequent polarization experiments.

1.1.5. Potential-pH

A number of couples, along with their pH dependency, are found in the listing of Delahay et al. [10]. Of particular interest is the potential-pH diagram for lead in the presence of sulfate ion. The diagram originally given by Delahay [10] was later modified to take into account the formation of basic lead sulfates, and such a diagram for lead in sulfate ions of unit activity is given in Fig. 1.1.3 [31]. The equilibria between solid phases is represented by solid lines, whereas those between the soluble lead species are given by broken lines. Diagrams calculated for other values of sulfate activity [32] have the same pattern as Fig. 1.1.3 except, of course, that the areas of stability for the various phases are shifted with respect to the axes.

As indicated in Fig. 1.1.3, lead should rapidly decompose water in acid solution, but as previously stated, this may not be observed in practice because of the high hydrogen overvoltage and passivation effects. The end product of oxidation of all the phases is PbO_2; however, two modifications, α-PbO_2 and β-PbO_2, are known. The line showing the variation of the Pb^{2+}/PbO_2 potential was constructed for the β-modification. Additional thermodynamic

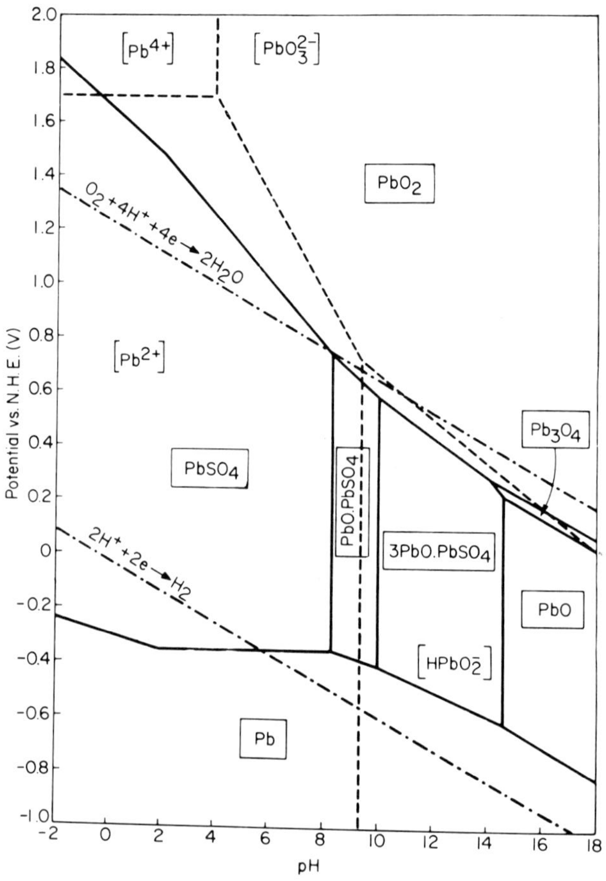

FIG. 1.1.3. Potential-pH diagram of lead in the presence of sulfate ion at unit activity and at 25°C [31]. (Courtesy of Pergamon Press.)

data (Fig. 1.1.4) [33] shows that in solutions of pH 9.4 and higher, β-PbO_2 would not be expected to form at any potential, but below this pH, α-PbO_2 is the stable phase over a range of potentials. Also, the Pb/PbO equilibrium in Fig. 1.1.3 was calculated for the stable (red) modification, although it was pointed out in the discussion [31] of the diagram that the Pb/PbO (yellow) and Pb/$Pb(OH)_2$ potentials would appear as parallel lines at potentials slightly positive to that of Pb/PbO (red).

Potential-pH diagrams for lead have been used extensively in discussions of the phase changes that occur during the manufacture of lead-acid battery plates [31, 34] as well as a supplement to kinetic data for the elucidation of corrosion mechanisms [10, 20, 32, 33]. Potential-pH relationships have also been determined from analysis of the products obtained by the potentiostatic oxidation of lead and lead alloys in buffered solutions [31, 35, 36].

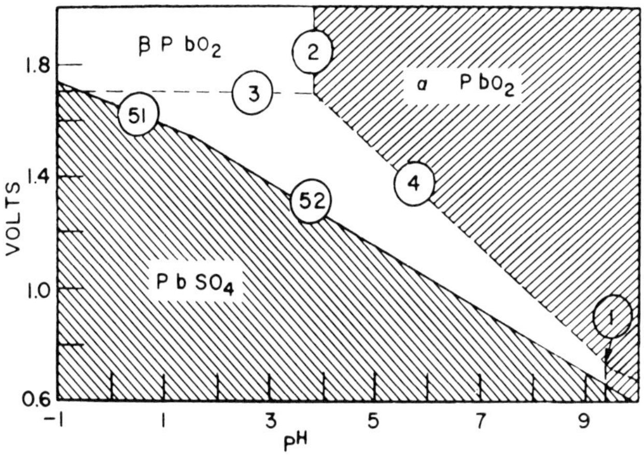

FIG. 1.1.4. Portion of the potential-pH diagram showing stability regions for α- and β-PbO_2 [33].

1. $HPbO_2^- + 3H^+ = Pb^{2+} + 2H_2O$
2. $PbO_3^{2-} + 6H^+ = Pb^{2+} + 3H_2O$
3. $Pb^{2+} + 2e^- = Pb^{2+}$
4. $PbO_2^{2-} + 6H^+ + 2e^- = Pb^{2+} + 3H_2O$

51. $PbO_2 + HSO^-_4 + 3H^+ + 2e^- = PbSO_4 + 2H_2O$
52. $PbO_2 + SO_4^{2-} + 4H^+ + 2e^- = PbSO_4 + 2H_2O$

(Courtesy of Electrochemical Society.)

1.1.6. Oxide Couples

Couples involving the oxides of lead have been calculated from thermodynamic data and a number of them are included in Table 1.1.1. The $PbO_2/PbSO_4$ couple has been widely studied in connection with lead-acid battery research and was the half-cell selected by Harned and Hamer for establishing thermodynamic data for sulfuric acid solutions [11]. Early determinations of the standard potential based on data from the cell

$$\mathrm{Pt, H_2 | H_2SO_4 | PbSO_4 | PbO_2} \tag{1.1.4}$$

have ranged from 1.681 to 1.686 V [11, 37, 38] as compared to the value of 1.682 V calculated from thermodynamic data [1]. A factor not recognized in the early work was that an orthorhombic (α) and tetragonal (β) modification of PbO_2 exists, for which different electrode potentials have subsequently been noted.

Wynne-Jones and co-workers [14] studied the electrode potential of pure α-PbO_2 prepared by carefully-controlled electrodeposition from plumbous salts. Against a hydrogen reference electrode, the α-PbO_2 registered constant potential values up to about 2 m H_2SO_4, but at higher acid concentrations, the potential fell to more negative values. The effect was attributed to a gradual conversion of α- to β-PbO_2 at the higher acid concentrations. It was also found that the exchange from the α- to β-modification could be rendered complete by raising the temperature above 100°C, provided sufficient time were allotted. Evidence for this exchange reaction is supported by the work of Hampson et al. in perchloric acid (see Section 3).

The measurements of Wynne-Jones et al. [14], when combined with the activity coefficient data of Stokes [39], yield the same standard potential for α-PbO_2 as that determined by Rüetschi and Cahan [40]. According to the latter investigators, electrodeposited α-PbO_2 gave a stable electrode potential even in 4.4 M H_2SO_4. But if the α-PbO_2 was permitted to stand for long periods in the acid to permit a buildup of a $PbSO_4$ layer, subsequent electrochemical oxidation of the lead sulfate gave an electrode containing a mixture of α- and β-PbO_2. Apparently the oxidation caused the electrode potential to be raised sufficiently to provide the extra nucleation energy required for the formation of β-PbO_2.

Electrodeposited β-PbO_2 prepared by Rüetschi and Cahan [40] exhibited a stable potential about 7 mV below that of α-PbO_2 in 4.4 M H_2SO_4, and the calculated [13] standard potentials for α- and β-PbO_2 based on these results are given in Table 1.1.1. β-PbO_2 electrodes prepared by pressing β-PbO_2 powder into perforated lead sheet also exhibited potentials lower than those of α-PbO_2 prepared in the same manner [40], although these electrodes exhibited limited stability because of self-discharge effects.

It has been pointed out [14] that the properties of α-PbO_2 suggest that its free energy of formation would be higher than that for β-PbO_2. Hence, a higher electrode potential would be expected for α-PbO_2, as supported by the values in Table 1.1.1. The earlier observation of Bode and Voss [41] that the α-PbO_2 potential is 30 mV lower than that for β-PbO_2 does not appear to be substantiated.

It is evident from the potential pH diagram (Fig. 1.1.3) that PbO_2 is thermodynamically unstable in H_2SO_4 and should oxidize water to oxygen, since its rest potential is about 0.5 V higher than the H_2O/O_2 potential. However, because of the high oxygen overvoltage on PbO_2 and protection of the PbO_2 surface by adsorbed sulfate ions or films [13], the stability of PbO_2 in H_2SO_4 is very high. In the absence of impurities the self-discharge reaction

$$PbO_2 + SO_4^{-2} + 2H^+ \rightarrow PbSO_4 + H_2O + \frac{1}{2}O_2 \qquad (1.1.5)$$

proceeds so slowly that ultrasensitive techniques are required to measure its rate.

Rüetschi and co-workers [13] measured the oxygen evolved, corresponding to the equation for pure α- and β-PbO_2, and found that, despite its lower surface area and lower electrode potential [40, 42], the β-PbO_2 had the greater self-discharge rate. The results could be correlated to steady-state overvoltage studies where oxygen was observed to have a greater overvoltage on α-PbO_2.

TABLE 1.2.1. Electrode Potentials in Nonaqueous Solvents

Solvent	Electrode	Potential	Reference electrode	Conditions (°C)	Refs.
Formamide	$Pb(Hg)PbCl_2$	-0.193 ± 0.012	$Cd/CdCl_2$	18	[44]
DMSO	$Pb(Hg)PbCl_2$	1.777	Li(Hg)LiCl (0.1 M) in DMSO		[45]
DMF	$Pb(Hg)PbCl_2$	1.845	Li(Hg)LiCl (0.1 M) in DMF		[45]
Acetonitrile	$Pb/Pb(ClO_4)_2$	-0.120	NHE		[43], [46]

1.2. NONAQUEOUS SOLVENTS

Interest has been meager in this area. The lead/lead chloride couple has been considered for a reference electrode of the second kind in nonaqueous solvents but other electrode systems are much more reproducible and reversible [43]. The $Pb(Hg)/PgCl_2$ behaves reversibly in formamide [44] and dimethylsulfoxide (DMSO) [45], and reported potentials in these solvents are given in Table 1.2.1. Less reversibility was evident for cells in dimethylformamide (DMF) [45] but this seemed to be related in part to reaction of a soluble Pb species at the Li/LiCl reference.

Pleskov [46] measured the potential of the lead electrode, along with other metals, in acetonitrile, to establish an emf series in this solvent. Using the cell

$$Pb \left| \begin{array}{c} Pb(ClO_4)_2\ (0.01\ \underline{N}) \\ \text{in } CH_3CN \end{array} \right| \begin{array}{c} AgNO_3 (0.01\ \underline{N}) \\ \text{in } CH_3CN \end{array} \Bigg| Ag \qquad (1.2.1)$$

a value of 0.3544 V was obtained. From a knowledge of the difference between the standard potentials of Ag and hydrogen electrodes in acetonitrile, Butler [43] recently reported the result on the hydrogen scale (Table 1.2.1).

1.3. FUSED SALTS

Electrode potentials of lead in molten salts have been obtained for the calculation of thermodynamic properties of fused salts as well as to establish the position of lead in the activity series in various solvents [47-50]. Commanding high interest has been the $Pb/PbCl_2$ electrode, which has been thoroughly studied in the cell

$$Pb | PbCl_2 | Cl_2, C \qquad (1.3.1)$$

TABLE 1.3.1. Electrode Potentials in Fused Salts

Fused salt	Electrode	Potential (V)	Reference electrode	Conditions (°C)	Remarks	Refs.
$PbCl_2$	Pb/Pb^{2+}	-1.27 (1.271)	Cl_2	500	$dV/dt = 5.31 \times 10^{-4}$ V/deg	[47], [51]
$PbCl_2$	Pb/Pb^{2+}	-1.17 (1.162)	Cl_2	700		[47]
NaCl	Pb/Pb^{2+}	2.27	Na	700		[47]
NaBr	Pb/Pb^{2+}	2.07	Na	700		[47]
PbI_2	Pb/Pb^{2+}	-0.616	I_2	550		[52]
NaI	Pb/Pb^{2+}	1.82	Na	700		[47]
NaCl-KCl 1:1	Pb/Pb^{2+}	-1.352	Cl_2	475		[49], [53]
LiCl-KCl eut	Pb/Pb^{2+}	-1.317	Cl_2	450		[50], [53]
$MgCl_2$-NaCl-KCl 50/30/20 mole %	Pb/Pb^{2+}	-1.247	Cl_2	475		[53], [54]
PbI_2-KI 1:1	Pb/Pb^{2+}	-0.678	I_2	550		[52]
KSCN	$Pb/PbCl_2$	-0.259	Ag	185		[48]
$KHSO_4$	Pb/Pb^{2+}	-0.830	NHE	260-280		[55]
LiAc-NaAc-KAc 20/30/50 mole %	$Pb(Hg)Pb^{2+}$	0.602	Zn	495	$dV/dt = 1.24 \times 10^{-4}$ V/deg	[56]

Values for the electrode potential against the chlorine reference are given in Table 1.3.1. Included in parentheses are potential values based on the free energy of formation of lead chloride calculated by Hamer et al. [57]. It is seen that experimental values over a wide temperature range agree well with the thermodynamic values.

Potentials lower than those cited in Table 1.3.1 for the $Pb/PbCl_2$ electrode have been reported, but this has been attributed [47, 57] to impurity effects. An important consideration is the correct preparation of the chlorine reference electrode. An effective method of chlorine electrode preparation, which led to the reliable experimental values reported by Lantratov et al. (data summarized in Ref. 47) was to heat a thin-walled graphite tube in a stream of dry chlorine at 700°C for 3-4 hr. Then, a vacuum and a chlorine atmosphere were alternately applied to the tube at 400-600° for 24 hr, which served to fill the pores with chlorine and promote removal of impurities from the graphite. Prior to making measurements, the tube was again exposed to a stream of dry chlorine for 30-60 min. This procedure resulted in a constant emf value for the cell over a long time interval at constant temperature as well as reproducible values as the cell temperature was raised or lowered.

The experimental setup employed by Lantratov et al. is shown in Fig. 1.3.1. The chlorine electrode compartment was separated from the bulk of the melt by a test tube having a capillary opening at the bottom.

Because of its stability and ease of preparation, the $Pb/PbCl_2$ electrode has been used extensively as a reference electrode in numerous fused salt studies [47]. The electrode has been constructed of fluoride-resistant materials for use in fluoride melts.

Thermodynamic formation functions calculated from potential values of the Pb/PbI_2 electrode against the iodine electrode are in good agreement with those calculated from thermochemical data [52].

Potentials of the lead electrode in mixed halide electrolytes have also been of interest. Potentials in KCl-NaCl melt were measured against the Ag/Ag^+ reference by Flengas et al. [49], and those in $MgCl_2$-NaCl-KCl melt were measured against a platinum reference electrode by Gaur and Behl [53]. Laitinen et al. [58] measured the potential of lead in LiCl-KCl eutectic also, using a platinum reference. In recent work by Gaur and Behl [53] the potential of the Ag and Pt electrode was measured against the chlorine reference, thereby permitting conversion of the electrode potentials for lead in the above melts to the chlorine reference scale. These values are reported in Table 1.3.1.

The potentials in the mixed halides are different than those obtained in the pure Pb halide melts because of composition changes as well as interactions between the components of the melt [47]. The formation of numerous possible chloro complexes in the mixed chloride melts has been suggested [49, 51]. For a system in which the pure PbI_2 is diluted with KI, observed deviations of the measured electrode potentials (activity coefficients) from ideality suggest that $2PbI_2 \cdot KI$ is the potential-determining species [52].

The $Pb/PbCl_2$ electrode in liquid KSCN has been studied by Bennett et al. [48], and a potential of -0.259 V (Table 1.3.1) was recorded against a Ag/Ag^+ reference. By constructing a concentration cell

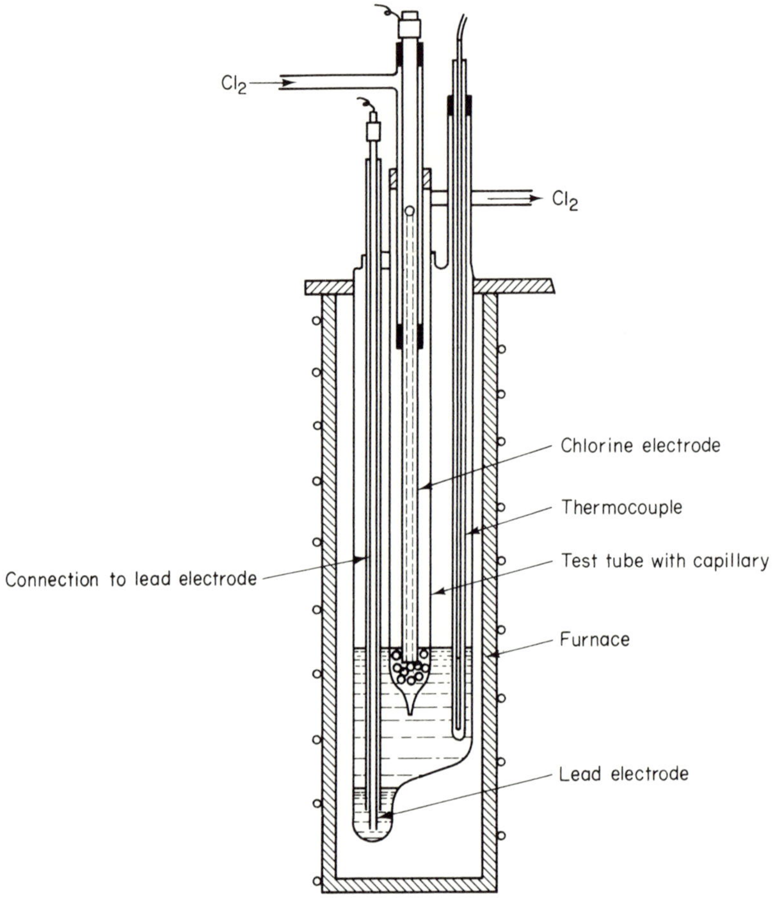

FIG. 1.3.1. Cell for measuring the $Pb/PbCl_2$ electrode potential against a chlorine reference [47]. (Courtesy of Sigma Press.)

$$Pb(solid) \left| \begin{matrix} PbSCN(C_1) \\ KSCN \end{matrix} \right| \begin{matrix} Glass \\ frit \end{matrix} \left| \begin{matrix} PbSCN(C_2) \\ KSCN \end{matrix} \right| Pb(solid) \qquad (1.3.2)$$

the interaction of Pb^{2+} with CN^- was studied by adding KCN to the working compartment of the cell. The corresponding change in potential was used to calculate the association constant for the $PbCN^+$ complex ($\log K_i = 3.23$). When KCl was added to the working compartment of the cell, no change in potential was observed, indicating that chloride does not complex with Pb^{2+} in this melt.

The initial potential/time behavior for the lead electrode in molten $KHSO_4$ is complicated, a factor attributed to the slow buildup of a $PbSO_4$ layer on the electrode surface [55]. After 1 hr the potential stabilizes at the value indicated in Table 1.3.1.

The lead electrode becomes oxidized in molten alkali acetate, but if the lead is amalgamated to prevent oxidation, a reproducible electrode potential is obtained [56]. In molten alkali nitrate, oxidation of the lead is also evident and the potential varies continuously, even for amalgamated electrodes [59].

2. VOLTAMMETRIC CHARACTERISTICS

2.1. POLAROGRAPHIC CHARACTERISTICS

2.1.1. Aqueous Solution

Polarographic characteristics for the reduction of lead ion in aqueous systems from a variety of supporting electrolytes are summarized in Table 2.1.1. The voluminous literature on this subject does not permit an exhaustive listing, and much of the information prior to 1964 was taken from the tabulation by Meites [72]. A section devoted to the polarography of lead is included in the text by Kolthoff and Lingane [63], and numerous references to the polarography of lead are contained in the volumes by Zuman and Kolthoff [97].

Complex formation is not indicated in perchlorate and dilute nitrate supporting electrolytes, and the half-wave potential exhibits values close to the standard electrode potential for lead in these systems [61]. At high nitrate concentrations a shift in half-wave potential ($E_{1/2}$) to more negative values is observed which may be accounted for partly from the increase in ionic strength of the solution, although the presence under these conditions of the weak complex $PbNO_3^+$ has been confirmed by potentiometric and spectrophotometric identification [60]. The excessive negative shift of $E^{1/2}$ in chloride medium is in agreement with the known value of the formation constant of the $PbCl^+$ ion [63].

The use of fluoride solution is especially useful for the determination of lead in mineralogical and metallurgical samples because of its ability to form strong complexes with many cations. Furthermore, HF can be used as both solvent and supporting electrolyte and, because of its strong buffering action, results do not depend markedly on the amount used in dissolving the samples [65, 66]. Etching of the conventional dropping mercury electrode (DME) has required the use of DME's constructed from Teflon or other fluoride-resistant material [66, 98], although Bond et al. [65] retained the advantages of the glass DME by using fast potential scans (facilitated by short Hg drop times of 0.16 sec) during which etching was minimized.

Shifts in $E^{1/2}$ and a slight lowering of the diffusion current with increasing fluoride concentration were interpreted by Mesaric and Hume [98] in terms of complex formation,

TABLE 2.1.1. Polarographic Characteristics in Aqueous Solvents

Substance	Product	Condition	$E_{1/2}$ (V vs SCE)	I	$E_{3/4}-E_{1/4}$ (mV)	Remarks	Refs.
Pb^{2+}	0	1 M $NaClO_4$ or $HClO_4$	-0.38		rev		[60]
Pb^{2+}	0	0.1 M KNO_3 or $NaNO_3$	-0.38		rev		[60-62]
Pb^{2+}	0	1 M HNO_3, KNO_3, or $NaNO_3$	-0.40	3.67	rev		[61], [63], [64]
Pb^{2+}	0	1-12 M HF	-0.37 to -0.44		-38		[65], [66]
Pb^{2+}	0	1 M NaF	-0.41	4.08		pH 1-3, 0.01% gelatin	[67]
Pb^{2+}	0	0.1 M KCl, LiCl, or $(CH_3)_4NCl$	-0.40	3.85		0.005% gelatin	[61], [63], [64], [68-72]
Pb^{2+}	0	0.1 M KCl, 0.1 M HCl		3.99			[73]
Pb^{2+}	0	1 M KCl or 1 M HCl	-0.44	3.86	-28	0.01% gelatin	[61], [63], [64], [70], [72], [74-77]
Pb^{2+}	0	8 M HCl	-0.62				[72]
Pb^{2+}	0	12 M HCl	-0.90				[72]
Pb^{2+}	0	5 M $CaCl_2$	-0.53				[78]
Pb^{2+}	0	1 M NH_4Cl, 1 M NH_3	-0.67fw			Low solubility	[72]
Pb^{2+}	0	0.1 M KSCN	-0.39	4.10	rev	Limited solubility	[79], [80]
Pb^{2+}	0	1 M KCN	-0.72w				[63], [81]
Pb^{2+}	0	1 M NaOH	-0.76	3.40	-28	0.005% gelatin	[64], [70], [72], [75], [76], [81-83]
Pb^{2+}	0	10 M NaOH	-0.83w		-28		[72]
Pb^{2+}	0	7.3 M H_3PO_4	-0.53		rev		[72]
Pb^{2+}	0	0.1 M $Na_4P_2O_7$	-0.69		-28		[72]
Pb^{2+}	0	1 M HAc, 1 M NH_4Ac	-0.50w	2.70	rev	0.01% gelatin	[84-86]

Pb^{2+}	0	0.1 M KH phthalate	-0.40		rev		[87]
Pb^{2+}	0	0.1-0.25 M $Na_2C_2O_4$	-0.50		rev		[72], [88]
Pb^{2+}	0	1 M $K_2C_2O_4$	-0.58w		-28	pH 7-10	[83], [89]
Pb^{2+}	0	0.1 M $(NH_4)_2C_2O_4$, 0.1 M NH_3	-0.53w		-25		[72]
Pb^{2+}	0	Satd malonic acid (H_2Mal)	-0.375w		-29		[72]
Pb^{2+}	0	1 M $(NH_4)_2Mal$	-0.457w		-30		[72]
Pb^{2+}	0	0.05 M Adipate ion	-0.39		rev	pH 3	[90]
Pb^{2+}	0	0.5 M Na_2tart	-0.58	2.40	rev	pH 9	[64], [70], [83], [91]
Pb^{2+}	0	0.25-0.5 M Na_2tart, 0.01-0.1 M NaOH	-0.70	2.40	rev		[64], [70], [72], [83]
Pb^{2+}	0	0.1 M $(NH_4)_2tart$, 0.1 M NH_3	-0.54	2.67	-26		[72]
Pb^{2+}	0	Satd citric acid (H_3cit)	-0.36		-28		[72]
Pb^{2+}	0	0.1 M $(NH_4)_3cit$, 0.1 M NH_3	-0.53			pH 8.5	[72]
Pb^{2+}	0	0.1 M $(NH_4)_3cit$	-0.54		-27	pH 6.1	[72]
Pb^{2+}	0	0.25 M $(NH_4)_3cit$	-0.77			pH 5	[64], [75], [92]
Pb^{2+}	0	1 M Na_3cit, 0.1 M NaOH	-0.78		rev		[72]
Pb^{2+}	0	0.5 M Ascorbic acid	-0.40		rev	pH 1.2	[93]
Pb^{2+}	0	Satd $N_2H_4 \cdot 2HCl$	-0.54w		-29		[72]
Pb^{2+}	0	0.5 M Ethylenediamine	-0.71				[72]
Pb^{2+}	0	0.05 M EDTA, 0.8 M HAc	-0.75w		-125	pH ~ 2	[72]
Pb^{2+}	0	0.05 M EDTA, 1 M HAc, 1 M NaAc	-1.1			pH 4	[94], [95]
Pb^{2+}	0	0.1 M EDTA, 0.8 M NaAc	-1.37fi			pH ~ 7	[72]
Pb^{2+}	0	0.1 M Pyridine	-0.40		-27		[72]
Pb^{2+}	0	0.1 M Pyridinium chloride	-0.31		-28		[72]
Pb^{2+}	-	0.3 M Triethanolamine, 0.1 M KOH	-0.88				[96]
Pb^{2+}	-	0.3 M Triethanolamine, 0.7 M NH_3, 1 M NH_4Cl	-0.56				[96]
Pb^{2+}	-	0.6 M Triethanolamine, 0.5 M NaOH	-1.02	2.82	-80		[72]

TABLE 2.1.2. Diffusion Current Constants as a Function of Lead Ion and Supporting Electrolytic Concentration [110]

Supporting electrolyte = 0.1 M KNO_3	
Pb ion concn (mM)	Diffusion current constant (I)
5.00	4.83
2.00	4.81
1.00	4.39
0.50	4.36

Pb ion concentration = 2 mM	
KNO_3 concn (M)	Diffusion current constant (I)
1.0	4.65
0.5	4.22
0.2	4.32
0.1	4.81

although complex formation is not clearly indicated from other investigations. Moderate irreversibility has been observed in this system from both a.c. and d.c. polarographic techniques [65].

Sodium hydroxide supporting electrolyte (1 M) has been useful in the determination of lead in the presence of tin, antimony, and arsenic, and tartrate supporting electrolyte is recommended for the simultaneous determination of Pb, Cu, Bi, and Cd [63]. Information on the application of other solvents to permit wave resolution for lead in mixtures of cations can be gained from a comparison of the half-wave potentials in Table 2.1.1 with those compiled for the other elements.

Data in complexing media have also been of interest in fundamental studies concerning complex formation. Formation constants for the complexes of lead with organic acids and their salts [72, 88, 90, 93, 99-107] as well as with pyrophosphate [108] and nitrate [109] have been calculated from half-wave potential data.

The polarographic diffusion current constants (I) vary with the nature of the supporting electrolyte, which may be attributed to variations in complex formation between lead and the anion of the supporting electrolyte. Kumar et al. [110] have shown that the I increases with an increase in lead ion concentration, and that the value of I falls and then rises with a decrease in concentration of supporting electrolyte (Table 2.1.2). Values of I as a function of pH (Fig. 2.1.1) [64] are related to the precipitation of oxide or hydroxide at intermediate

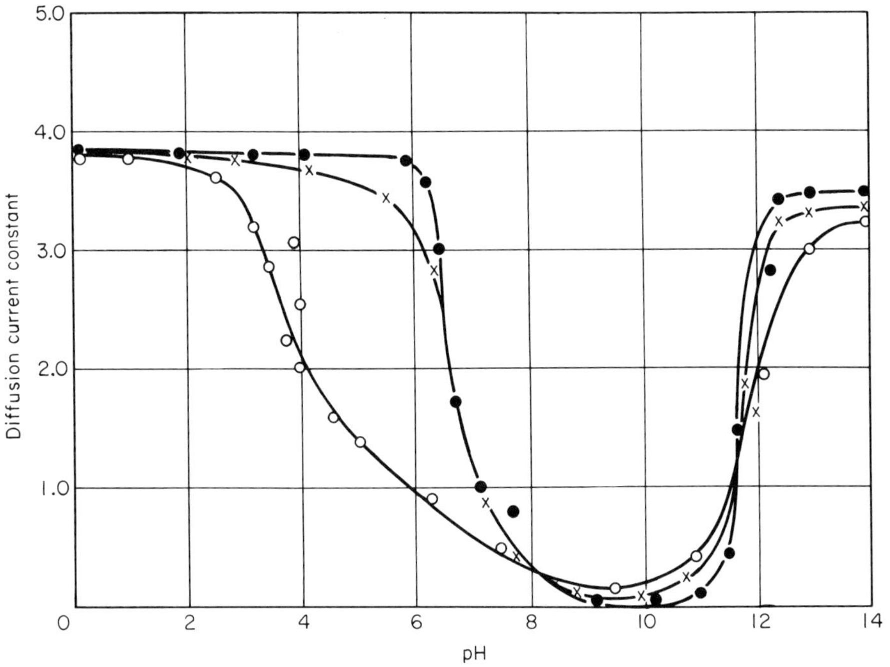

FIG. 2.1.1. Diffusion current constant for lead ion as a function of pH and maximum suppressor content. (•) 0.01% gelatin; (X) 0.1% gelatin; (o) 1.0% gelatin [64]. (Courtesy of American Chemical Society.)

pH and the formation of plumbite ion at high pH. The values are appreciably influenced in acid media by the content of maximum suppressor.

2.1.2. Nonaqueous Solution

From an analytical standpoint, polarography of Pb in nonaqueous media has been limited. These solvents are generally not suitable because considerable preliminary work is often required to obtain the lead species in a soluble form. In work by Headridge et al. [111] solution of Pb^{2+} in acetic acid-acetic anhydride for polarographic study was effected by adding an excess of acetic anhydride to an aqueous solution containing Pb^{2+}, and then refluxing until reaction between water and acetic anhydride was complete. This gave a solution containing $<2 \times 10^{-4}$ M Pb^{2+}. Diffusion currents measured in this system at the DME were directly proportional to $[Pb^{2+}]$ over the range 5×10^{-5} to 5×10^{-4} M.

The experimentally determined and calculated values for the half-wave potentials on the aqueous SCE scale (Table 2.1.3) have permitted some comparisons on the relative solvating properties of the various organic solvents compared to that of water. Both DMF and DMSO

TABLE 2.1.3. Polarographic Characteristics in Nonaqueous Solvents

Substance	Product	Condition	$E_{1/2}$*	$E_{3/4}-E_{1/4}$ (mV)	Remarks	Refs.
Pb^{2+}	Pb(Hg)	Acetic acid-acetic anhydride + 0.25 M $NaClO_4$ + 0.25 M LiCl	-0.16	34	0.002% Methyl cellulose	[111]
Pb^{2+}	Pb(Hg)	Acetic acid + 1.0 M $LiClO_4$	-0.22			[112], [113]
Pb^{2+}	Pb(Hg)	Acetic acid + 0.25 M $NH_4C_2H_3O_2$	-0.57			[112]
Pb^{2+}	Pb(Hg)	Propionic acid + 1.0 M $LiClO_4$	-0.20			[113]
Pb^{2+}	Pb(Hg)	Isobutyric acid + 1.0 M $LiClO_4$	-0.20			[113]
Pb^{2+}	Pb(Hg)	Acrylic acid + 1.0 M $LiClO_4$	-0.25			[113]
Pb^{2+}	Pb(Hg)	DMF + 0.1 M Et_4NClO_4	-0.03 vs Ag/AgCl/satd Et_4NClO_4	40		[114]
Pb(Hg)	Pb^{2+}	DMF + 0.1 M $NaClO_4$	(-0.03) vs Ag/AgCl/satd Et_4NClO_4	rev		[114]
Pb^{2+}	Pb(Hg)	DMF + 0.1 M $NaNO_3$	1.63 vs Na(Hg) $NaClO_4$			[115]
Pb^{2+}	Pb(Hg)	DMSO + 0.1 M $NaNO_3$	0.526 vs Zn			[115]
Pb^{2+}	Pb(Hg)	Ethylenediamine + 0.29 M LiCl	-0.33 vs Hg pool	rev	$D = 3.4 \times 10^{-6}$	[115]

appear to be similar to H_2O in their solvating abilities of Pb^{2+}, but ethylenediamine appears to have significantly greater solvating ability, as indicated by the significant shift in half-wave potential for this solvent [115].

As in aqueous polarography, the half-wave potentials in the nonaqueous solvents are dependent on the supporting electrolyte. In acetic acid containing 0.25 M ammonium acetate as supporting electrolyte, Bachman et al. [116] noted that Pb^{2+} ions are reduced at potentials ~0.25 V more positive than those in water and yield normal polarograms having wave heights 2/3 of those in water. On the other hand, Pb^{4+} ions gave a highly abnormal reduction curve containing unexplained "discontinuities," and could not be determined polarographically in this solvent.

2.1.3. Fused Salts

Polarographic studies for the reduction of Pb ions in fused salts are summarized in Table 2.1.4. Where discrepancies exist, they are not readily explained, although Inman et al. [118] have shown that the data obtained may depend markedly on the purity and dryness of the melt.

Much of the data in nitrate melts have been obtained as a preliminary step in the study of adsorption and complex ion formation involving Pb and halide ions in these melts. Formation constants for chloro- and bromo-lead complexes have been calculated from the shifts in half-wave potentials to negative values when these halide ions are added to the melt [126, 127]. The formation constants for $PbCl^+$ and $PbBr^+$ calculated by Inman et al. [127] from polarographic data agree well with values reported by Braunstein et al. [128] from emf measurements using the Ag/Ag^+ halide electrode.

In $NaNO_3$-KNO_3 melt when the $[Pb^{2+}]$ exceeded 10^{-3} M, a maximum or "peak" on the cathodic polarograms prior to reaching the limiting current plateau has been reported by Swofford et al. (Fig. 2.1.2) [117]. From current-time plots for individual mercury drops, it appeared that the reduction process was being inhibited by the formation of an insoluble reaction product. A reaction product could not be identified but it was suggested that an oxide film might have formed following reduction in the highly oxidizing nitrate melt.

Similar observations have been noted by others [126, 129] but in these cases the presence of bromide in the melt was a requisite for the formation of the prewave peak. According to Inman et al. [129], the peak current depended on the $[Pb^{2+}]$ and on the bromide to lead ion ratio. The peak current increased with mercury column height, indicative of an adsorption controlled process. The concentrations of the individual complexing species in the melt could be calculated from the aforementioned formation constants, and the peak current was found to be directly proportional to the concentration of $PbBr_2$. This suggests that $PbBr_2$ was the species directly adsorbed although, as pointed out by Inman et al., the bromide ions may be initially adsorbed to act as bridges for the electron transfer process.

TABLE 2.1.4. Polarographic Characteristics in Fused Salts

Substance	Product	Conditions	$E_{1/2}$	Ref. electrode	Rev	Remarks	Refs.
Pb^{2+}	0	KNO_3-$NaNO_3$ eut, 250°	-0.85	Ag/Ag^+	irr	Prewave at Pb^{2+} 3×10^3 M	[117]
Pb^{2+}	0	$LiNO_3$-$NaNO_3$-KNO_3 eut, 145°	-0.46	Ag/AgCl	rev	$D = 0.92 \times 10^{-6}$	[118]
Pb^{2+}	0	$LiNO_3$-$NaNO_3$-KNO_3 eut, 160°	-0.47	Hg/Hg_2Cl_2	rev	$D = 1.80 \times 10^{-6}$	[119]
Pb^{2+}	0	$LiNO_3$-$NaNO_3$-KNO_3 eut, 156°	-0.48	Hg/Hg_2Cl_2	q-rev	$D = 0.78 \times 10^{-6}$	[120]
Pb^{2+}	0	$LiNO_3$-$NaNO_3$-KNO_3 eut, 165°	-0.38	Ag/AgCl	rev	$D = 1.3 \times 10^{-6}$	[121]
Pb^{2+}	0	LiCl-KCl-$AlCl_3$, 160°	-0.35	Hg/Hg_2Cl_2	rev		[122]
Pb^{2+}	0	NaCl-KCl-$AlCl_3$, 140°	0.44	Al	rev		[123]
Pb^{2+}	0	LiAc-NaAc-KAc, 197°	-0.95	Ag/Ag^+	rev		[124]
Pb^{2+}	0	$NH_4 \cdot O \cdot CO \cdot H$, 125°	-0.21	Hg pool			[125]

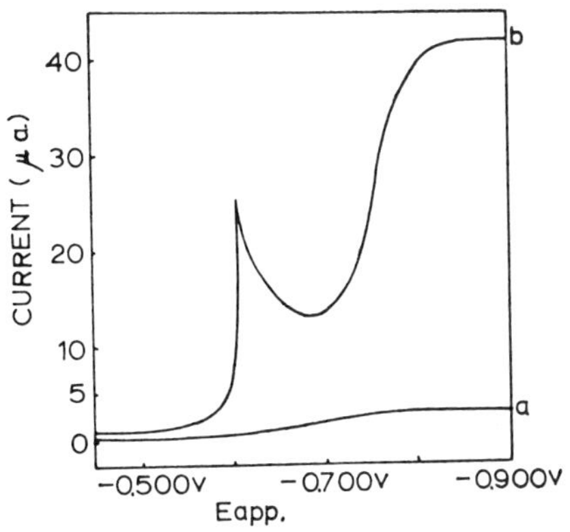

FIG. 2.1.2. Polarogram for the reduction of lead ion in $NaNO_3$-KNO_3 eutectic at 250° C. (a) 2.0×10^{-4} M Pb^{2+}; (b) 3.2×10^{-3} M Pb^{2+} [117]. (Courtesy of American Chemical Society.)

The prewave peaks are not observed in melts containing chloride ion [126, 127, 129]. This is not surprising since the chloro complexes are much weaker than the bromo complexes and Cl^- is less strongly adsorbed than Br^-.

The diffusion coefficients for Pb^{2+} in molten salts appear to be about five times greater than those in aqueous systems [119]. Francini et al. [121] reported good agreement between diffusion coefficients obtained from conventional polarography and those obtained from potential sweeps at low sweep rates (0.6 V/sec). They observed the relationship

$$D = A \exp(-E/RT)$$

where A is a constant and E is the activation energy for the diffusion process. In $LiNO_3$-$NaNO_3$-KNO_3 eutectic, E was determined as -8.4 kcal/mole.

2.2. VOLTAMMETRIC CHARACTERISTICS

Voltammetric techniques have been especially useful for the analysis of small quantities (10^{-2} to 10^{-9} M) of lead in aqueous media. A typical chronopotentiogram for the deposition of lead from 0.01 M lead nitrate onto a stationary mercury electrode is shown in Fig. 2.2.1 [130] and corresponding data for this system are summarized in Table 2.2.1. The quarter-wave potential is in excellent agreement with the polarographic half-wave potential for lead in comparable electrolytes. From the average value of the transition time constant ($i\tau^{1/2}/C$) and the geometric area of the electrode (2.57 cm^2), a diffusion coefficient of 1.07×10^{-5} cm^2/sec was calculated, in excellent agreement with the value based on conductance measurements. An overall precision of ±1% was estimated for the results over the lead concentration range studied (10^{-2} to 10^{-4} M).

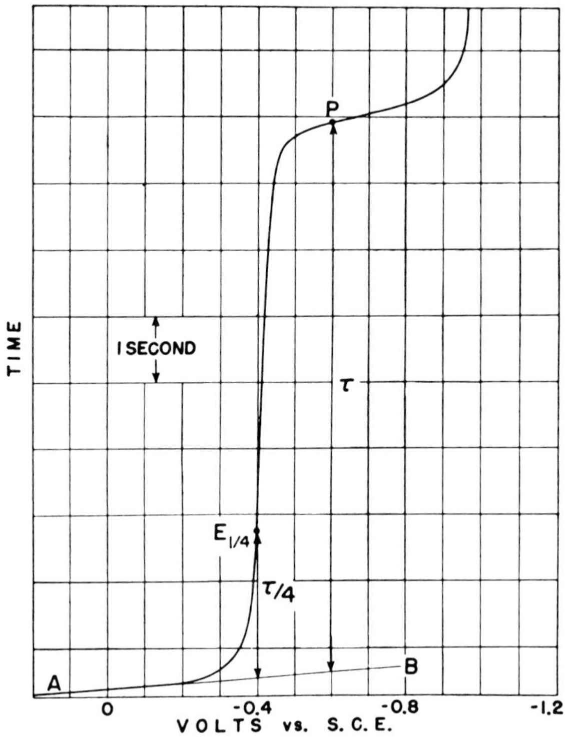

FIG. 2.2.1. Chronopotentiogram for 0.01 M $PbNO_3$ at 5 mA [130]. (Courtesy of American Chemical Society.)

TABLE 2.2.1. Chronopotentiometric Data on Lead Nitrate in 0.2 M Nitric Acid at 25°C [130]

C (moles/l)	i (mA)	τ_{av} (sec)	$E_{\tau/4}$ (V vs SCE)	$i\tau^{1/2}/C$ [A $sec^{1/2}$ X cm^3 $moles^{-1}$ (X10^{-3})]
1 X 10^{-4}	0.100	3.1	-0.47	1.8
2	0.150	4.0	-0.43	1.51
5	0.250	8.4	-0.42	1.45
1 X 10^{-3}	0.450	10.4	-0.40	1.45
2	1.200	5.48	-0.40	1.40
5	2.50	8.33	-0.41	1.44
5	3.00	5.67	-0.39	1.43
1 X 10^{-2}	5.00	8.04	-0.41	1.42
1	5.00	8.26	-0.40	1.44
2	10.00	8.21	-0.40	1.43
				av 1.44[a]

[a]Omitting 1.8 X 10^3.

TABLE 2.2.2. Voltammetric Characteristics

Substance	Product	Conditions	Electrode	Technique	Potential (V vs SCE)	Remarks	Refs.
Pb(Hg)	Pb^{2+}	0.5 M $NaClO_4$	Hg-coated Pt	ac stp	-0.42		[134]
Pb(Hg)	Pb^{2+}	1.0 M KNO_3	Pg(Hg) drp	E swp	-0.42		[135]
Pb(Hg)	Pb^{2+}	1.0 M NH_4SCN	Pb(Hg) drp	E swp	-0.43		[135]
Pb(Hg)	Pb^{2+}	High chloride medium (sea water)	Pb(Hg) drp	E swp	-0.51		[135]
Pb(Hg)	Pb^{2+}	2.0 M HAc, 2.0 M NH_4Ac	Pb(Hg) drp	E swp	-0.52		[135]
Pb(Hg)	Pb^{2+}	1.0 M Koxalate	Pb(Hg) drp	E swp	-0.58		[135]
Pb(Hg)	Pb^{2+}	0.1 M KNatart	Pb(Hg) drp	E swp	-0.58		[133]
Pb(Hg)	Pb^{2+}	0.5 M Nacit	Pb(Hg) drp	E swp	-0.56	pH 7	[135]
Pb(Hg)	Pb^{2+}	0.2 M H_2SO_4	Pb(Hg) drp	E swp	-0.07		[140]
Pb	Pb^{2+}	0.5 M $NaClO_4$	Bright Pt	ac stp	-0.53		[134]

The chronopotentiometric reduction of lead has also been studied at platinum electrodes [131], but lead is deposited in acid solution at a potential close to that for hydrogen ion reduction at platinum, and a clear wave is not obtained.

As an amalgam-forming metal, lead can be accumulated in stationary Hg electrodes during the deposition process and then analyzed by anodic stripping techniques. The anodic current obtained is influenced to some extent by the rate of diffusion of the lead through the amalgam [132] and to minimize this effect, small Hg drop electrodes (area = 0.05 cm^2) [133], as well as platinum electrodes that have been coated with a thin layer of Hg [134], have been devised.

Anodic stripping has been applied to the analysis of lead contaminants in water [135-138], spectrographically pure zinc [133], and in uranium salts [139]. Data in a number of supporting electrolytes are included in Table 2.2.2. When lead is anodically stripped from platinum electrodes, loss or retention of lead during the deposition step has been demonstrated [131] and the measured peak potential for ac stripping, compared to that for Hg electrodes, suggests some degree of irreversibility [134].

Both stationary and rotating platinum electrodes have been applied to the cathodic stripping voltammetry of lead [141]. Anodic and cathodic voltage sweeps for Pb^{2+} at the platinum electrode in acetate buffer are shown in Fig. 2.2.2. The anodic peak for PbO_2 formation is partially masked by oxygen evolution, but the cathodic peak is well-defined. The cathodic

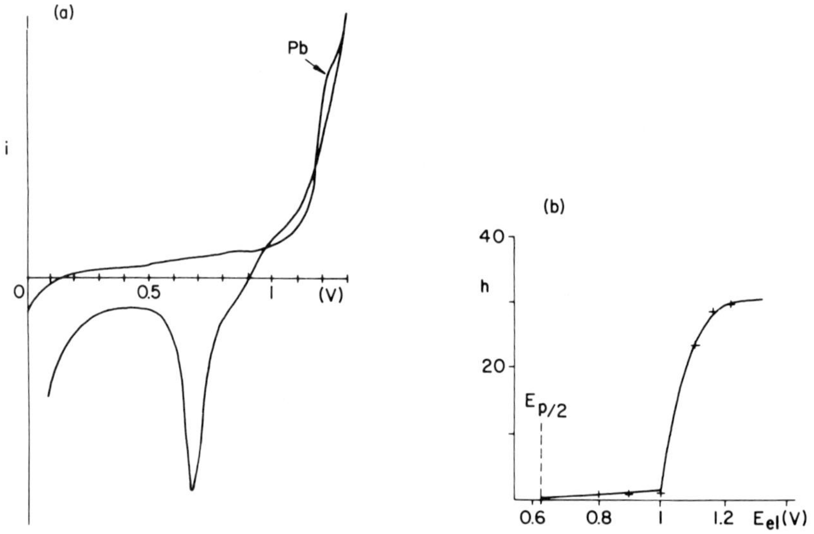

FIG. 2.2.2. (a) Polarization curve for lead (10^{-4} M) in acetate buffer solution, pH 4.9; sensitivity 3 μA; 80 mV/sec. (b) Dependence of peak height on electrolysis potential, 5×10^{-6} M Pb^{2+}; preelectrolysis 1 min [141]. (Courtesy of Elsevier, Ltd.)

peak height varies with the deposition potential, as also indicated in Fig. 2.2.2. Both the cathodic peak height and peak potential are dependent on pH, and the best defined peaks are observed at pH 5 (acetate or borate buffer). Lead has been determined under these conditions at concentrations as low as 5×10^{-9} M, and is easily distinguished from a number of other cations.

3. KINETIC PARAMETERS AND DOUBLE-LAYER PROPERTIES

3.1. KINETIC PARAMETERS

3.1.1. Solid Electrodes in Aqueous Media

The exchange mechanism between solid lead electrodes and lead ions in solution may be dependent upon metallurgical properties of the electrode as well as the extent to which active centers on the electrode are removed by adsorption effects or by the nucleation and growth of films. Fast and slow processes have been distinguished through the attainment

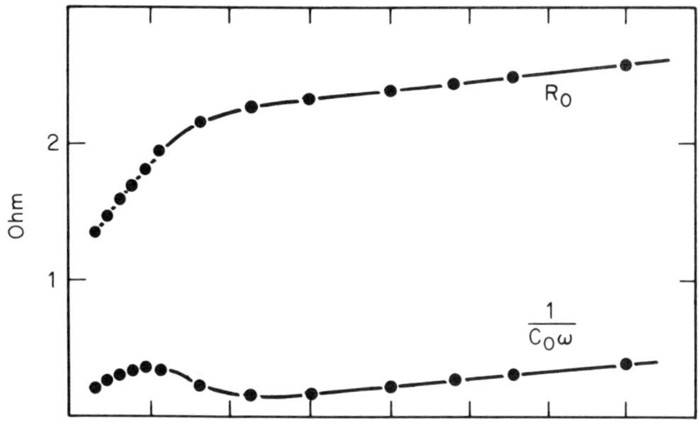

FIG. 3.1.1. Frequency response diagram for the Pb/Pb^{2+} exchange in 1 N $HClO_4$. $Pb^{2+} = 10^{-3}$ M, 16°C [142]. (Courtesy of Electrochemical Society of Japan.)

of faradaic impedance values over a wide range of ac frequencies. As the frequency of the ac is increased from low values, each component process successively becomes rate determining and then relaxes. Provided that their relaxation times are sufficiently different, it is possible to identify each process from frequency response diagrams.

3.1.1.1. Acid Solution. Haruyama [142] studied the exchange reaction at polycrystalline lead electrodes over the 30 Hz to 50 kHz range in perchloric acid containing a range of lead ion concentrations. Frequency response diagrams, typified by Fig. 3.1.1, were treated by dividing them into diagrams representing diffusion plus electron transfer (fig. 3.1.2a) and crystallization (Fig. 3.1.2b). For the charge transfer reaction, the excess of the in-phase component (R_0) over the out-of-phase component ($1/\omega C_0$), e.g.,

$$\Delta = R_0 - (1/\omega C_0) \tag{3.1.1}$$

is related to the exchange current density (j_0) for the charge transfer by

$$j_0 = RT/2F\Delta \tag{3.1.2}$$

Values for the j_0's at various lead ion concentrations are given in Table 3.1.1.

The crystallization process involves the diffusion of adatoms (or adions) across the electrode surface. The resistance due to adatom diffusion is given by the value of Δ from Fig. 3.1.2b at low frequencies. Adatom diffusion rates calculated from Δ's are also given in Table 3.1.1. It is seen that the rate of the adatom diffusion process was about one-tenth that of the charge transfer process. Both these processes, however, are so fast that the diffusion of lead ions in solution will be an important factor in controlling the overall reaction rate.

It was not readily apparent in the work of Haruyama why the crystallization process should be slower than the charge transfer process. Frequency scans obtained by Lorenz

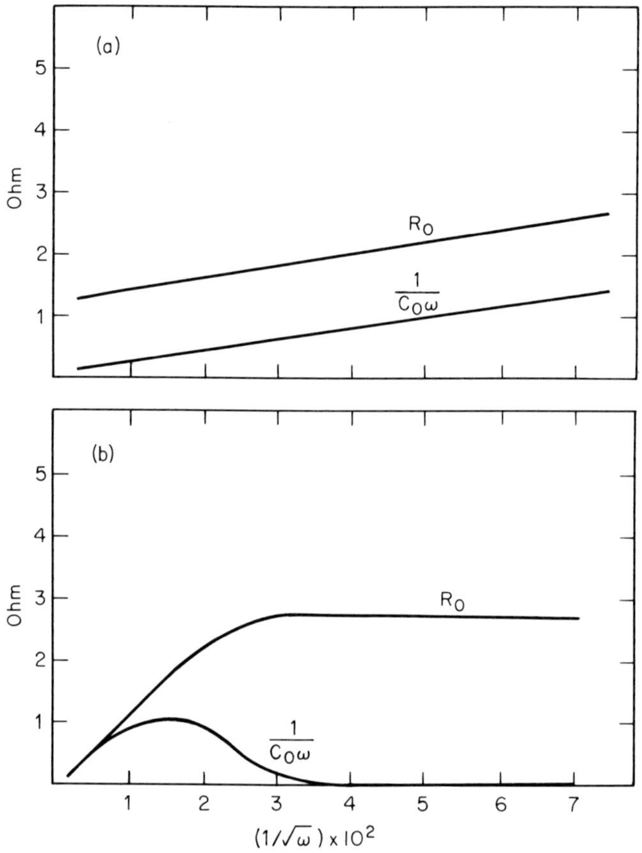

FIG. 3.1.2. Analysis of frequency response diagram for the Pb/Pb^{2+} exchange in 1 N $HClO_4$. Pb^{2+} = 10^{-3} M, 16°C. (a) Diffusion + electron transfer. (b) Surface reaction [142]. (Courtesy of Electrochemical Society of Japan.)

[143] and Hampson et al. [144] in nitrate electrolytes did not show effects due to adatom diffusion. Hampson [144] pointed out, however, that the traces of sulfate contained in Haruyama's electrolyte (where the Pb^{2+} was added as $PbSO_4$) would be adsorbed onto the electrode, thereby removing active centers and increasing the importance of crystallization as the rate-controlling step.

Impedance measurements obtained by Hampson et al. [144] for polycrystalline lead electrodes in nitrate electrolyte containing a range of lead ion concentrations are shown in Fig. 3.1.3. Current density (j) vs overpotential (η) curves from double pulse measurements were also obtained (Fig. 3.1.4) where the j_0's are related to the slopes of the curves by

$$-\left(\frac{\partial\eta}{\partial j}\right)_{\eta\to 0} = RT/2Fj_0 \qquad (3.1.3)$$

TABLE 3.1.1. Rates of Electron Transfer and Adatom Diffusion for Polycrystalline Lead Electrodes in 1 N $HClO_4$ at 16°C (Pb^{2+} added to electrolyte as $PbSO_4$) [142]

Pb^{2+} [a] (mole/l)	j_0 electron transfer (A/cm^2)	Adatom diffusion (A/cm^2)
10^{-3}	2×10^{-2}	3×10^{-3}
10^{-2}	6×10^{-2}	8×10^{-3}
5×10^{-2}	9×10^{-2}	1.5×10^{-2}
1	3×10^{-1} [b]	4×10^{-2} [b]

[a] Added as $PbSO_4$.

[b] Extrapolated to 1 M Pb^{2+}.

Table 3.1.2 gives values of j_0 at various Pb^{2+} concentrations calculated from the faradaic impedance and double pulse methods.

The R_R and $1/\omega C_R$ plots (Fig. 3.1.3) were linear throughout the entire frequency range (20 Hz to 10 kHz), thereby indicating that a single process (in addition to diffusion in solution) controlled the electrode reaction. Satisfactory agreement between j_0's from the impedance and double pulse techniques (Table 3.1.2) would indicate that this process was charge transfer, since the time interval for the double pulse (2 μsec) was equivalent to a frequency sufficiently high for adatom diffusion to have relaxed completely.

From the dependence of j_0 on the lead ion concentration, Hampson et al. [144] calculated a value of ~0.8 for α, the cathodic charge transfer coefficient. High values of α have been ascribed to adsorption of reactants at the electrode [145]. Adsorption of Pb^{2+} was indicated in the present case by an increase in the double-layer capacity (C_0) as the Pb^{2+} in solution was increased (Table 3.1.2).

In nitrate electrolyte containing lead ion concentrations of or below 3×10^{-3} M, faradaic impedance curves (Fig. 3.1.5) indicated that the exchange was no longer entirely controlled by charge transfer. Relaxation (convergence of the R_R and $1/\omega C_R$ plots) at high frequencies indicated that the process which was rate-controlling at low frequencies could not follow the ac at high frequencies. It was suggested that at the low Pb^{2+} concentrations the exchange became controlled by the process of lattice growth and dissolution.

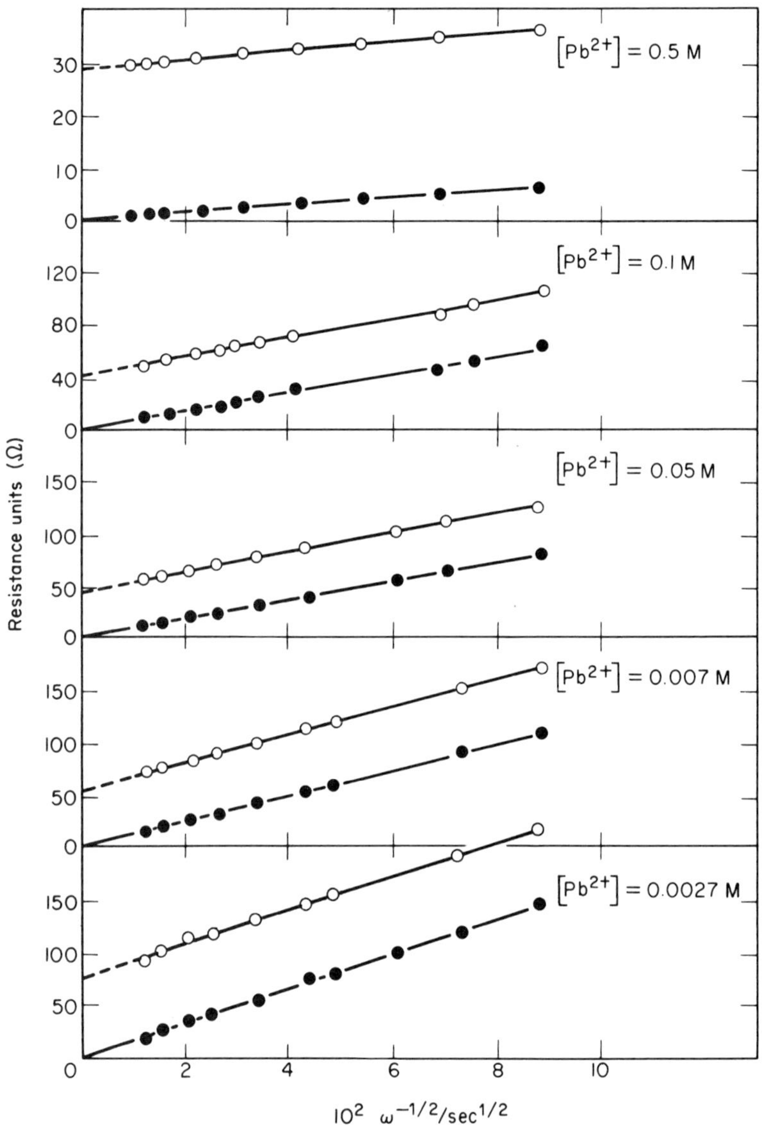

FIG. 3.1.3. Faradaic impedance curves for polycrystalline lead electrodes in 1.0 M nitrate at various Pb^{2+} concentrations, 23°C. (o) R_R; (•) $1/\omega C_R$ [144]. (Courtesy of Faraday Society.)

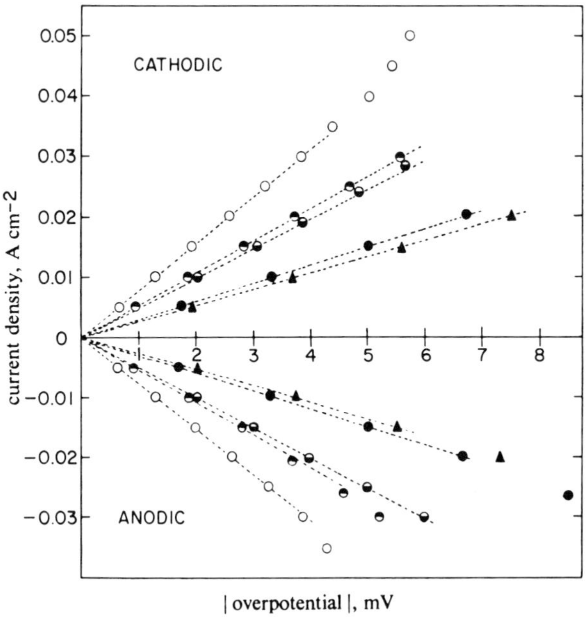

FIG. 3.1.4. Current density (j) vs overpotential (η) for polycrystalline lead electrodes in 1.0 M nitrate at 23°C. Pb^{2+}: (o) 0.5 M; (◒) 0.1 M; (◓) 0.05 M; (●) 0.007 M; (▲) 0.0027 M [144]. (Courtesy of Faraday Society.)

TABLE 3.1.2. Kinetic Data for Polycrystalline Lead Electrodes in Nitrate Electrolyte at 23°C [144]

	Faradaic impedance data				Double pulse data	
Pb^{2+} (mole/l)	Δ^a (Ω)	j_0 (mA/cm²)	C_0 (μF/cm²)	α	j_0 (mA/cm²)	α
0.5	29.3	71.4	67.3		100	
0.1	41.6	49.8	62.7		67	
0.05	45.3	46.8	49.1	0.85 ± 0.05	64	0.80 ± 0.05
0.007	59.4	34.9	34.5		39.5	
0.0027	79.1	26.2	32.3		35.6	

[a]Charge transfer resistance.

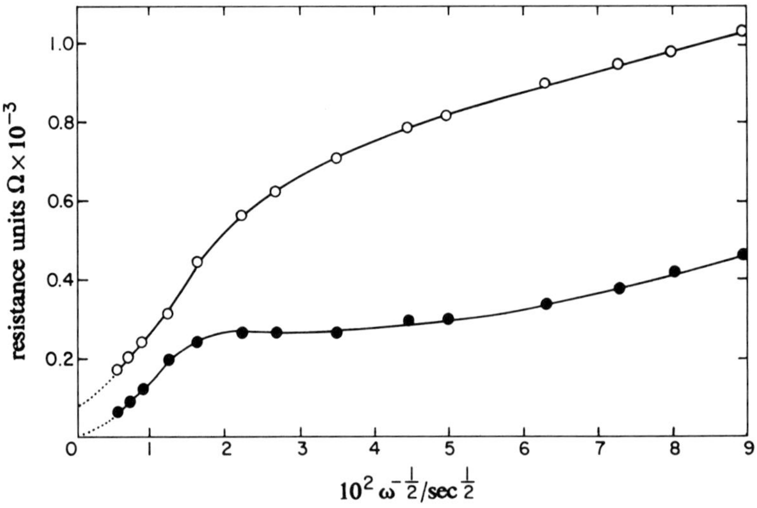

FIG. 3.1.5. Relaxation observed for polycrystalline lead electrodes in 1.0 M nitrate containing 0.0027 M Pb^{2+}, 23°C. (o) R_R; (•) $1/\omega C_R$ [144]. (Courtesy of Faraday Society.)

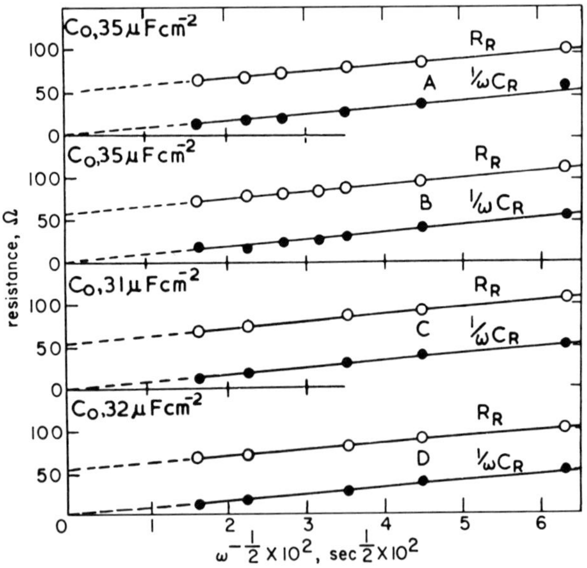

FIG. 3.1.6. Faradaic impedance curves for polycrystalline lead electrodes in 3 M NaOH at 20°C. (A) 0.09 M Pb^{2+}, 1 hr 0°C; (B) 0.09 M Pb^{2+}, 24 hr 0°C; (C) 0.04 M Pb^{2+}, 24 hr 0°C; (D) 0.04 M Pb^{2+}, 24 hr 0°C, $HClO_4$ etch, 7 hr further 0°C [146]. (Courtesy of Faraday Society.)

TABLE 3.1.3. Exchange Rates for Lead Electrodes in 3.0 M NaOH at 20°C [146]

	Single crystal[a]	Normal polycrystal	Etched polycrystal[b]
Δ (Ω-cm^2)	1.0	3.1×10^{-1}	4.4×10^{-2}
Adatom diffusion rate (mA/cm^2)	13	42	294

[a]100 plane.

[b]HAc-H_2O_2 etch.

3.1.1.2. Alkali Solution. Faradaic impedance curves for polycrystalline lead electrodes in alkaline solution are given in Fig. 3.1.6 [146]. Stable values of Δ, which were obtained after a 24-hr electrode/electrolyte contact, were independent of the [Pb^{2+}] over the range indicated. Additional experiments revealed that the stabilized values of Δ were dependent entirely on electrode surface characteristics, thereby indicating that the exchange was controlled by adatom diffusion. Values of Δ and corresponding exchange rates for normal and etched polycrystalline electrodes and single crystal electrodes are given in Table 3.1.3. The exchange rate increased with increase in lattice disorder, which is expected if dislocations provide the sites for lattice dissolution. The reaction sequence may be written

$$Pb_{kink} \rightleftharpoons Pb_{ad} \quad \text{(slow)} \tag{3.1.4}$$

$$Pb_{ad} + 2OH^- \rightleftharpoons PbO + H_2O + 2e^- \quad \text{(fast)} \tag{3.1.5}$$

In contrast to the nitrate system, active centers may be removed by the possible nucleation and growth of oxide films, so that the rate-determining step would directly involve the lattice. Furthermore, the species that crosses the double layer is electrically neutral PbO rather than a charged ion, so that the charge transfer reaction is expected to be fast.

As the reactant concentrations are decreased, film formation becomes more favorable [146, 147]. At low [Pb^{2+}], impedance values for polycrystalline electrodes increased with time, and C_0 fell to low values as indicated in Fig. 3.1.7. This was attributed to film formation, especially since a chemical etch (20% $HClO_4$) restored the electrode to its original activity. Relaxation occurred at progressively lower frequencies with time, indicating that the process of adatom diffusion could not completely follow the ac. Similar effects were observed when the [NaOH] was reduced to 0.5 M (Fig. 3.1.8). After 24 hr electrode/electrolyte contact, impedance curves conformed to single process control. This was suggested as charge transfer control, since reduction of the adatom diffusion rate would make transfer directly from kink sites energetically feasible.

Kinetic data from the various investigations are summarized in Table 3.1.4.

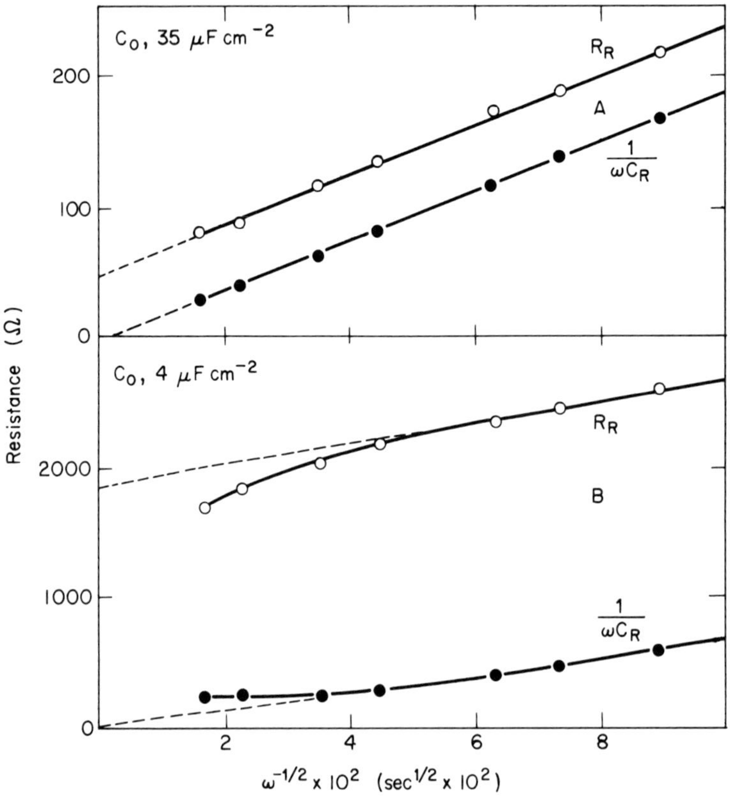

FIG. 3.1.7. Faradaic impedance curves for polycrystalline lead electrodes in 3 M NaOH at low lead ion concentration (Pb^{2+} = 0.001 M), 22°C. (A) 2 min, 0°C; (B) 60 min, 0°C [146]. (Courtesy of Faraday Society.)

3.1.2. Amalgam Electrodes in Aqueous Media

For the exchange at amalgam electrodes (data summarized in Table 3.1.5), charge transfer is most likely the rate-controlling step. In $HClO_4$ the impedance behaves normally [145] and the reactants do not appear to be adsorbed. For the exchange in KCl, KBr, and KI, strong specific adsorption of the reactants is indicated by high values for the charge transfer coefficient. Upon the addition of lead ion (5×10^{-4} M) to 1 M KCl, adsorption was indicated [151] from a marked increase in the differential capacitance of mercury in the potential range -0.15 to -0.35 V vs normal calomel electrode (NCE). From chronocoulometric studies by Barclay et al. [153], a potential dependence of the surface excess (Γ) of Pb^{2+} at the mercury electrode in chloride solution has been demonstrated (Table 3.1.6). The suggested mechanism was adsorption of the chloride ion, followed by adsorption of the $PbCl_3^-$ complex, e.g.,

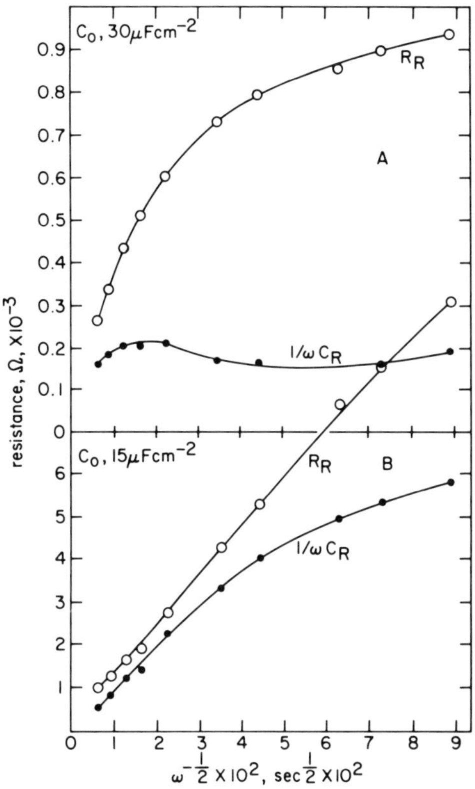

FIG. 3.1.8. Faradaic impedance curves for polycrystalline lead electrodes in dilute (0.5 M) alkali, 0.004 M Pb^{2+}, 22°C. (A) 1 hr, 0°C; (B) 24 hr, 0°C [146]. (Courtesy of Faraday Society.)

$$PbCl_3^- + Cl^- - Hg \rightarrow [Cl_3Pb - Cl]^{2-} - Hg \qquad (3.1.6)$$

Such a mechanism was indicated from the linearity and zero intercept of the plot $\Gamma_{Pb^{2+}}$ vs Γ_{Cl^-}, shown in Fig. 3.1.9.

3.1.3. PbO_2 Electrodes in Acid Media

Studies, notably by Hampson and co-workers [154-158] involving the exchange between PbO_2 electrodes and lead ions in solution, have shown that the exchange mechanism depends upon the magnitude of the overvoltage as well as the modification (α or β) of PbO_2. Results are different in acid than in alkali because of the different rational potentials involved. From the value of the point of zero charge of PbO_2 (~1.1 V), it follows that in acid medium

TABLE 3.1.4. Kinetic Parameters for the Pb^{2+}/Pb Exchange

Reaction	Condition	Electrode	Technique	Rate constant	α	Remarks	Refs.
$Pb^{2+} + 2e \rightleftarrows Pb$	0.05 M $Pb(NO_3)_2$ in $NaNO_3$ or KNO_3	Polycryst Pb	ac imp and double pulse	j_0 = 0.05–0.1 A/cm^2	0.80	ΔH = 12.6 kcal/mole	[143], [144]
$Pb^{2+} + 2e \rightleftarrows Pb$	1 M $HClO_4$, 10^{-3} M $PbSO_4$	Polycryst Pb sphere	ac imp	$j_0 = 2 \times 10^{-2}$ A/cm^2	0.63	ΔH = 1.2 kcal/mole	[142]
$Pb_{ad} \rightleftarrows Pb_{kink}$	1 M $HClO_4$, 10^{-3} M $PbSO_4$	Polycryst Pb sphere	ac imp	Adatom diffusion rate = 3×10^{-3} A/cm^2		ΔH = 2.9 kcal/mole	[142]
$Pb^{2+} + 2e \rightleftarrows Pb$	2 N H_2SO_4	Pb	ac imp	$j_0 = 3.2 \times 10^{-8}$ A/cm^2			[148]
$Pb^{2+} + 2e \rightleftarrows Pb$	10 N H_2SO_4	Pb	ac imp	$j_0 = 5 \times 10^{-6}$ A/cm^2			[149]
$Pb_{ad} \rightleftarrows Pb_{kink}$	0.03 M Pb^{2+}, 3 M NaOH, 22°C	Polycryst Pb	ac imp	Adatom diffusion rate = 0.04 A/cm^2		ΔH = 11.8 kcal/mole	[146]
		Polycryst Pb, (etched)	ac imp	Adatom diffusion rate = 0.3 A/cm^2			
		Single crystal, (100 plane)	ac imp	Adatom diffusion rate = 0.01 A/cm^2			

TABLE 3.1.5. Kinetic Parameters for the $Pb^{2+}/Pb(Hg)$ Exchange

Reaction	Condition	Electrode	Technique	Rate constant	α	Refs.
$Pb^{2+} + 2e^- \rightleftarrows Pb(Hg)$	1 M $HClO_4$	Hg drp	ac imp	k° = 2.0 cm/sec	0.50	[145]
$Pb^{2+} + 2e^- \rightleftarrows Pb(Hg)$	1 M $NaClO_4$, 2×10^{-5} M Pb^{2+}	Hg drp	farad. rect.	k° = 3.3 cm/sec	0.60	[150]
$Pb^{2+} + 2e^- \rightleftarrows Pb(Hg)$	1 M KF satd with PbF_2	Hg drp	farad. rect.	k° = 5.0 cm/sec	0.50	[151]
$Pb^{2+} + 2e^- \rightleftarrows Pb(Hg)$	1 M KCl, 2×10^{-5} M Pb^{2+}	Hg drp	farad. rect.	k° = 0.2 cm/sec	0.94	[152]
$Pb^{2+} + 2e^- \rightleftarrows Pb(Hg)$	1 M KBr, 2×10^{-5} M Pb^{2+}	Hg drp	farad. rect.	k° = 0.2 cm/sec	0.97	[151]
$Pb^{2+} + 2e^- \rightleftarrows Pb(Hg)$	1 M KI, 2×10^{-5} M Pb^{2+}, pH 2, 22°C	Hg drp	farad. rect.	k° = 0.2 cm/sec	1.0	[151]

TABLE 3.1.6. Potential Dependence of the Adsorption of Pb^{2+} from Chloride Solution. 0.7 m$\underline{M}$ $Pb(NO_3)_2$ in 1 $\underline{M}$ NaCl [153]

-E (V vs SCE)	$10^{10}\ \Gamma_{Pb^{2+}}$ (mole/cm^2)
0.050	1.14
0.100	0.84
0.150	0.66
0.200	0.56
0.250	0.49
0.300	0.47

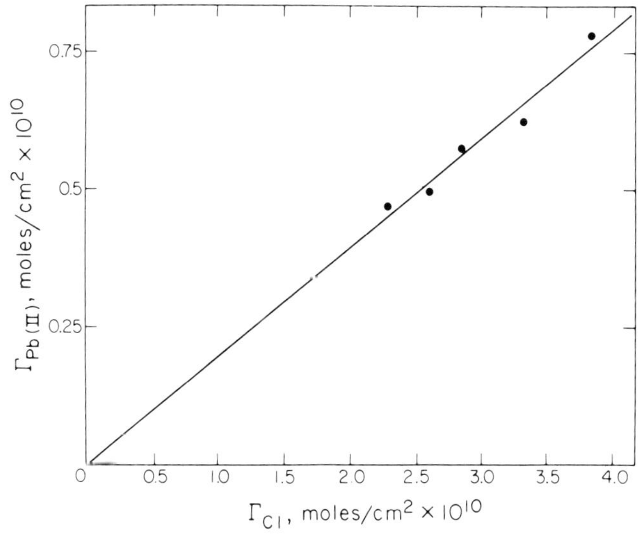

FIG. 3.1.9. Correlation between the adsorption of Pb^{2+} and free chloride anions on mercury. Electrolyte = 1 $\underline{M}$ NaCl + 0.7 m$\underline{M}$ Pb $(NO_3)_2$ [153]. (Courtesy of Elsevier, Ltd.)

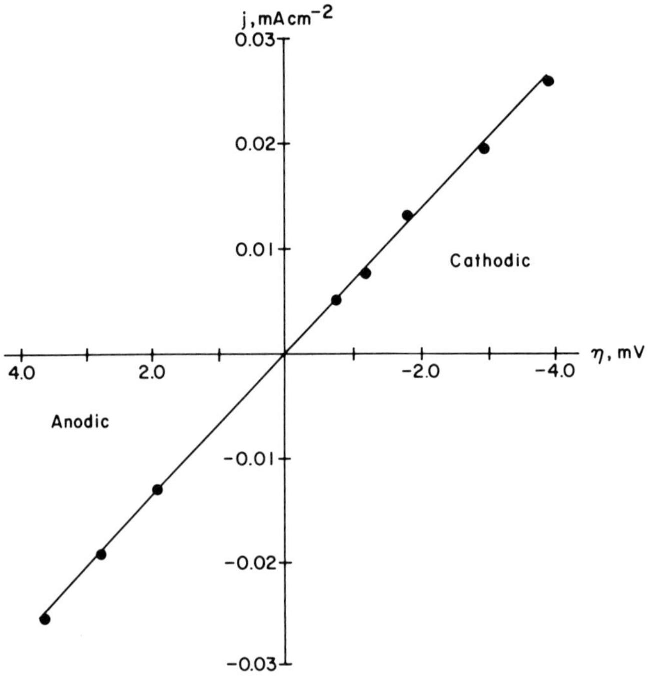

FIG. 3.1.10. Current density (j) vs overpotential (η) for β-PbO_2 electrode in perchlorate electrolyte at 23°C [156]. (Courtesy of National Research Council of Canada.)

($E^\circ \sim 1.5$ V) èxchange data are obtained at positive rational potentials, whereas in alkaline medium ($E^\circ \sim 0.36$ V) kinetic data are obtained at negative rational potentials.

3.1.3.1. Low Overvoltage Region. A current density (j) vs overpotential (η) curve for β-PbO_2 in perchloric acid is given in Fig. 3.1.10 [156]. The data for Fig. 3.1.10 were obtained from rapid (80 msec) galvanostatic pulses for which the η was limited to ±10 mV. Equality of anodic and cathodic current for a given η indicated that the rate-controlling reaction was the same for both anodic and cathodic processes. This was suggested as a simultaneous $2e^-$ exchange.

Exchange current densities (j_0's) calculated from Eq. (3.1.3), using slopes typified by the curve in Fig. 3.1.10, are given in Table 3.1.7. The j_0's are close to the order of magnitude for those observed for the $PbO_2 \rightleftarrows PbSO_4$ exchange ($j_0 = 10^{-4}$ to 10^{-5} A/cm^2), [148, 149]. The magnitude of the j_0's confirms that the exchange is indeed slow, about two orders of magnitude slower than the aforementioned charge transfer at solid lead electrodes in perchlorate solution.

TABLE 3.1.7. Exchange Current Densities and Charge Transfer Coefficients for PbO_2 Electrodes in Acid Lead Perchlorate Electrolyte[a] at 23°C

Concentration	j_0 (mA/cm^2)	j_0[b] (mA/cm^2)	α[b]
	β-PbO_2 [156]		
3.0 M H^+			
0.045 M Pb^{2+}	0.094	0.098	0.23
0.09 M Pb^{2+}	0.110	0.113	0.20
0.47 M Pb^{2+}	0.149	0.156	0.20
0.93 M Pb^{2+}	0.169	0.177	0.19
0.09 M Pb^{2+}			
6.65 M H^+	0.158	0.160	0.18
0.5 M H^+	0.086	0.093	0.22
0.05 M H^+	0.042	0.048	0.19
	α-PbO_2 [158]		
3.0 M H^+			
0.005 M Pb^{2+}	0.137	0.143	0.15
0.0098 M Pb^{2+}	0.156	0.163	0.17
0.063 M Pb^{2+}	0.194	0.197	0.20
0.50 M Pb^{2+}	0.266	0.276	0.20
0.1M Pb^{2+}			
0.24 M H^+	0.207	0.212	0.22
0.027 M H^+	0.214	0.210	0.17

[a]Total concentration of ClO_4^-, 6.85 M with $NaClO_4$.

[b]Determined iteratively.

Exchange current densities and cathodic charge transfer coefficients (α's), obtained by an iterative computer method which involved a statistical analysis of η vs j data, are also given in Table 3.1.7. The data were matched to the equation

$$j = j_0 \; e^{-\alpha zF\eta/RT} - e^{(1-\alpha)zF\eta/RT} \qquad (3.1.7)$$

and values of α chosen which made the mean deviation of j_0 a minimum. Values of j_0 calculated by the iterative method agreed with values from curves typified by Fig. 3.1.10. The

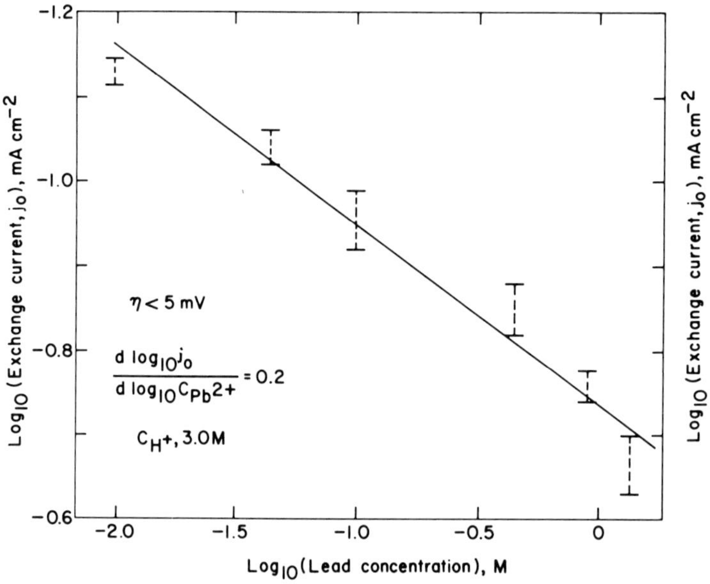

FIG. 3.1.11. Dependence of exchange current density on Pb^{2+} concentration for β-PbO_2 electrodes in perchloric acid at 23°C [156]. (Courtesy of National Research Council of Canada.)

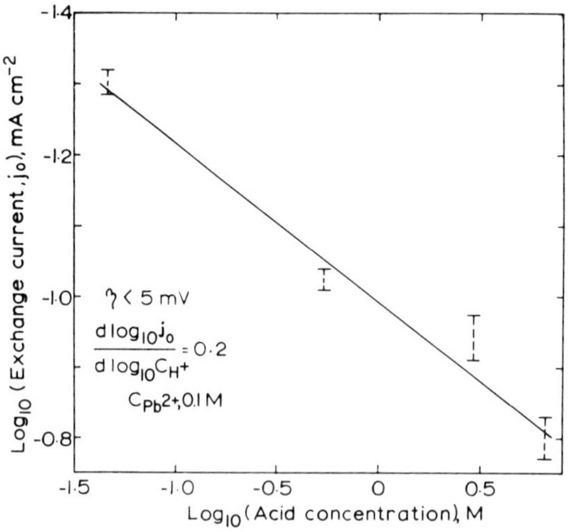

FIG. 3.1.12. Dependence of exchange current density on H^+ concentration for β-PbO_2 electrodes in perchloric acid at 23°C [156]. (Courtesy of National Research Council of Canada.)

α's derived from the iterative method agreed with the value calculated from the dependence of j_0 on $[Pb^{2+}]$ (Fig. 3.1.11), but not with the value calculated from the dependence of j_0 on $[H^+]$ (Fig. 3.1.12). From the value of 0.2 obtained for

$$\left(\partial \log j_0/\partial \log [H^+]\right)_{[Pb^{2+}]} \qquad (3.1.8)$$

and the rate equation for the charge transfer

$$j_0 = 2Fk[Pb^{2+}]^{\alpha}[H^+]^{(1-\alpha)}[PbO_2] \qquad (3.1.9)$$

where k is a rate constant and $[PbO_2]$ is taken as unity, an abnormally high value of 0.95 for α was indicated. This was ascribed to adsorption of hydrogen ions at the electrode.

Data corresponding to α-PbO_2 are also included in Table 3.1.7. Although data are similar for the two modifications of PbO_2, a rigorous comparison is not feasible. The attainment of precise kinetic data for α-PbO_2 appears to be limited by the suggested [158] formation of β-PbO_2 as a product of the exchange reaction. Kinetic data obtained with α-PbO_2 at high overvoltages in perchloric acid have been interpreted in terms of an α-PbO_2/β-PbO_2 equilibration process.

3.1.3.2. High Overvoltage Region. Tafel plots (η vs log j) characteristic of α-PbO_2 in perchloric acid are given in Figs. 3.1.13 and 3.1.14. Figure 3.1.13 refers to variable $[Pb^{2+}]$ at constant $[H^+]$, and the potentials ($E_{3\underline{M}}$) are against β-PbO_2 3.0 M H^+, 1.0 M Pb^{2+} as the reference. Figure 3.1.14 refers to variable $[H^+]$ at constant $[Pb^{2+}]$ and the potentials ($E'_{\underline{M}}$) are against the arbitrarily-selected reference β-PbO_2/0.1 M Pb^{2+}, 1.0 M H^+.

Data calculated from the Tafel plots are given in Table 3.1.8. The anodic and cathodic exchange current densities (j_{0a} and j_{0b}), obtained from extrapolations of the linear (Tafel) regions to $\eta = 0$, were not identical, and were sufficiently different from the values obtained in the low η region (Table 3.1.7) to indicate a change in mechanism from the simultaneous two-electron transfer at low η. A further indication of a change in mechanism was that charge transfer resistances (R_D) observed for $\eta < 10$ mV were different from those calculated from

$$R_D = \frac{RT}{4F}\left(\frac{1}{j_{0a}} + \frac{1}{j_{0b}}\right) \qquad (3.1.10)$$

at $\eta \sim 50$ mV. This comparison is given in Table 3.1.9.

On the basis of the kinetic data, Hampson et al. [157] proposed two successive single electron transfer steps at high η (>50 mV). The suggested mechanism was

$$PbO_2 + 2H^+ \rightleftharpoons PbO_2(H^+)_2 \text{ ads} \qquad (3.1.11)$$

$$PbO(H^+)_2 \text{ ads} + e^- \xrightleftharpoons{j_{0a},\ \alpha_a} [HO\cdot Pb\cdot OH]^+ \qquad (3.1.12)$$

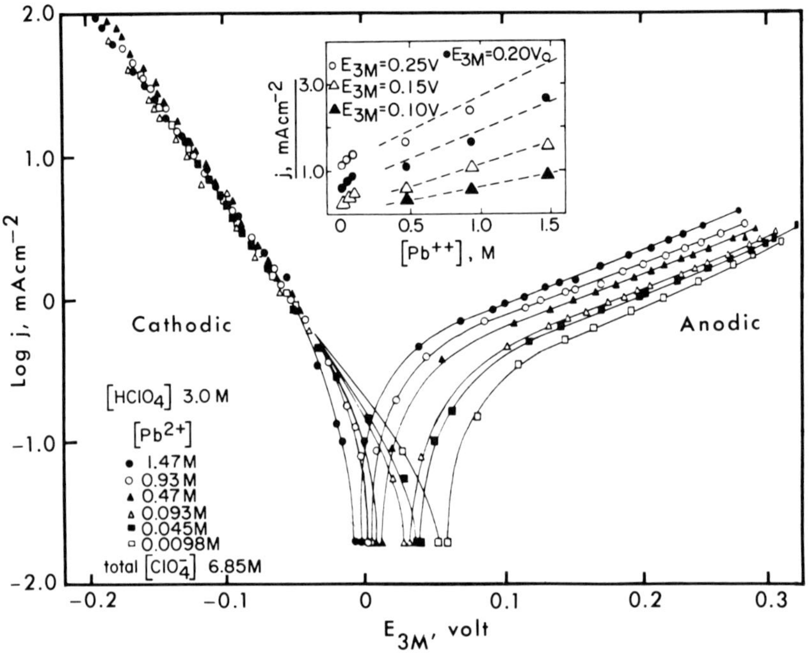

FIG. 3.1.13. Tafel plots for β-PbO_2 electrodes at various Pb^{2+} concentrations in 3.0 M $HClO_4$ at 23°C [157]. (Courtesy of National Research Council of Canada.)

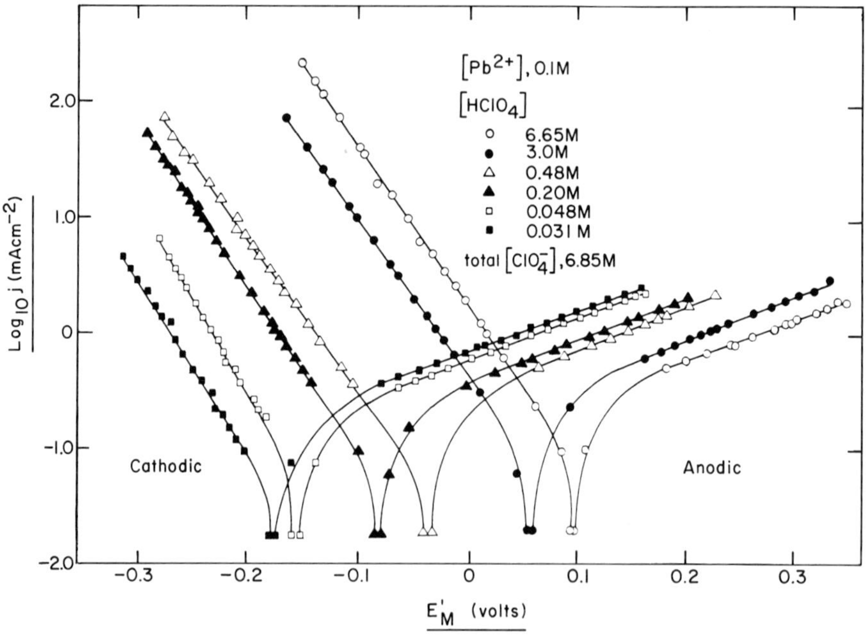

FIG. 3.1.14. Tafel plots for β-PbO_2 at various H^+ concentrations in perchloric acid; Pb^{2+} = 0.1 M; 23°C [157]. (Courtesy of National Research Council of Canada.)

TABLE 3.1.8. Characteristics of Tafel Plots. β-PbO_2 Electrode: Total $[ClO_4^-] = 6.85$ M (with added $NaClO_4$); 23°C [157]

Concentration (M)		Cathodic coeff., $-\frac{2.303RT}{F}\frac{\partial \log j_c}{\partial \eta}$	Anodic coeff., $\frac{2.303RT}{F}\frac{\partial \log j_A}{\partial \eta}$	Intercepts at $\eta = 0$			
Pb^{2+}	H^+			$2j_{0a}$ ($mA\ cm^{-2}$)	$2j_{0b}$ ($mA\ cm^{-2}$)	α_a	α_b
0.93	3.0	0.84	0.21	0.250	0.331		
0.47	3.0	0.88	0.22	0.118	0.285		
0.093	3.0	0.82	0.22	0.074	0.257	0.86 ± 0.1	0.78 ± 0.02
0.045	3.0	0.90	0.22	0.040	0.250		
0.089	0.48	0.84	0.22	0.050	0.220		
0.092	0.20	0.86	0.22	0.052	0.200		
0.093	0.0032	0.82	0.24	0.042	0.118		

TABLE 3.1.9. Charge Transfer Resistance for β-PbO_2 in Perchloric Acid [157]

Concentration (M)		Charge transfer resistance (ohm cm^2)	
Pb^{2+}	H^+	$\eta < 10$ mV	$\eta > 50$ mV
0.93	3.0	72	89
0.47	3.0	83	152
0.093	3.0	114	222
0.045	3.0	133	369
0.095	6.65	80	212
0.089	0.48	142	312

$$[HO\cdot Pb\cdot OH]^+ + e^- \underset{}{\overset{j_{0b},\ \alpha_b}{\rightleftarrows}} Pb(OH)_2 \qquad (3.1.13)$$

$$Pb(OH)_2 + 2H^+ \rightleftarrows Pb^{2+} + 2H_2O \qquad (3.1.14)$$

where each charge transfer reaction has its characteristic exchange current and transfer coefficient. For a two-electron transfer involving consecutive single-electron transfer steps where one step is rate determining, the Tafel slopes are related to the charge-transfer coefficient (α_i) for the rate-determining step by

$$\partial \log j_c/\partial\eta = \frac{-F}{2.3RT}(n_c^* - 1 + \alpha_i) \qquad (3.1.15)$$

and

$$\partial \log j_a/\partial\eta = \frac{F}{2.3RT}(n_a^* - \alpha_i) \qquad (3.1.16)$$

where n_c^* and n_a^* are, respectively, the ordinal number of the rate determining step in the cathodic and anodic direction. From the Tafel slopes indicated in Table 3.1.8, rational values of the charge transfer coefficients, α_a and α_b, were obtained only when n_c^* and n_a^* were each equal to 1. Therefore, the slow step in either the cathodic or the anodic direction was that leading to the Pb^{3+} intermediate (Eq. 3.1.12). Charge transfer coefficients calculated from the above relationships were $\alpha_a = 0.86$ and $\alpha_b = 0.78$ (Table 3.1.8).

In the proposed reaction mechanism, the species $PbO_2(H^+)_2$ ads (Eq. 3.1.11) was considered to arise through the adsorption of H^+ by lattice oxygen atoms at the surface of the PbO_2. A similar species, $Pb(OH)_2^{2+}$, was postulated by Fleischman and Liler [9] as an intermediate in the electrodeposition of PbO_2 from acetate solutions. Evidence for adsorption in the work of Hampson et al. [157] was reflected in the cathodic and anodic reaction

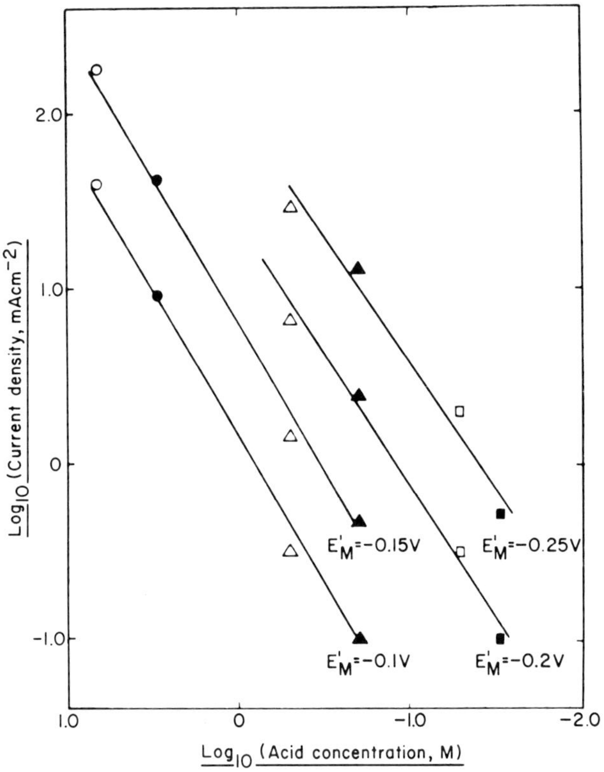

FIG. 3.1.15. Dependence of cathodic current density on H^+ concentration for β-PbO_2 electrodes in $HClO_4$ at 23°C [157]. (Courtesy of National Research Council of Canada.)

orders with respect to $[H^+]$ (Figs. 3.1.15 and 3.1.16). In the absence of adsorption, the rate equations for the cathodic and anodic reactions are, respectively,

$$j_c = k_c F[PbO_2][H^+]^2 \tag{3.1.17}$$

and

$$j_a = k_a F[Pb^{2+}][H^+]^{-2} \tag{3.1.18}$$

where the k's are appropriate rate constants. The effect of adsorption would be to reduce the exponent for the $[H^+]$ in Eq. (3.1.17), and to raise the exponent for the $[H^+]$ in Eq. (3.1.18), as was the case observed experimentally, e.g., from Fig. 3.1.15

$$\left(\frac{\partial \log j_c}{\partial \log [H^+]}\right)_{[Pb^{2+}]} = 1.5 \tag{3.1.19}$$

and from Fig. 3.1.16

$$\left(\frac{\partial \log j_a}{\partial \log [H^+]}\right)_{[Pb^{2+}]} = -0.4 \tag{3.1.20}$$

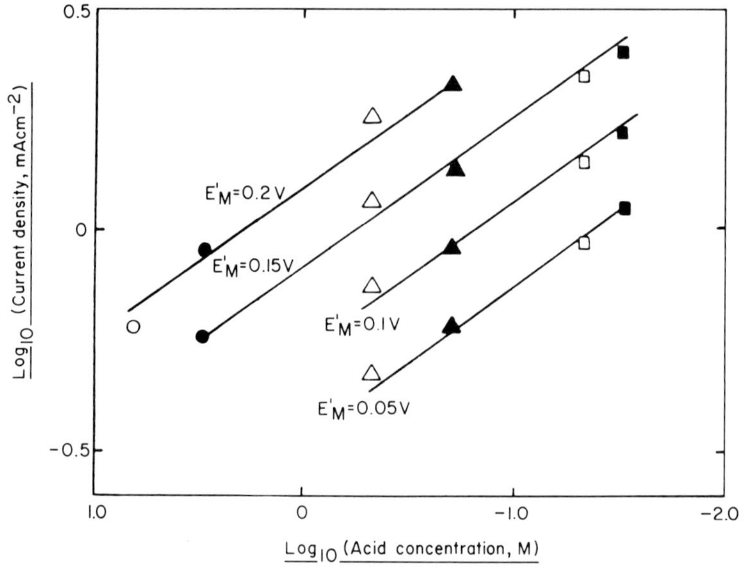

FIG. 3.1.16. Dependence of anodic current density on H^+ concentration for β-PbO_2 electrodes in $HClO_4$ at 23°C [157]. (Courtesy of National Research Council of Canada.)

For the dependence of current on $[Pb^{2+}]$ at constant $[H^+]$ (Fig. 3.1.13), experimental results fell on the same Tafel slope for the cathodic reaction, which is expected if the activity of PbO_2 remains constant. For the anodic plots the reaction order was fairly constant over a wide concentration range (Fig. 3.1.13, insert), which was consistent with the expected first-order dependence of anodic current on $[Pb^{2+}]$. However, at the lower concentrations, deviation from first-order dependence was indicated, possibly as a result of self-discharge which might be enhanced at the lower concentration region.

Comparable experiments for α-PbO_2 in perchloric acid are summarized in Figs. 3.1.17 and 3.1.18 [158]. A notable feature is that the cathodic reaction at α-PbO_2 did not follow the expected zero order with respect to $[Pb^{2+}]$. Participation of Pb^{2+} in the cathodic reaction would seem unlikely in view of the single cathodic Tafel line observed for β-PbO_2 over the range of $[Pb^{2+}]$. An explanation offered by Hampson et al. [158] was based on the equilibrium condition established during initial electrode/electrolyte contact (a 10-min interval preceding the start of the polarization). An increase in $[Pb^{2+}]$ would be expected to increase the rate of exchange between Pb^{2+} in solution and PbO_2 at the electrode surface. Since the formation of β-PbO_2 is favored in acid media, the effect of increasing the $[Pb^{2+}]$ would be to progressively alter the electrode surface from α-PbO_2 to β-PbO_2 at rates corresponding to the equilibrium exchange current densities (Table 3.1.7). As a result, the cathodic Tafel line for α-PbO_2 (Fig. 3.1.17) shifts toward that exhibited by β-PbO_2 as the $[Pb^{2+}]$ is increased.

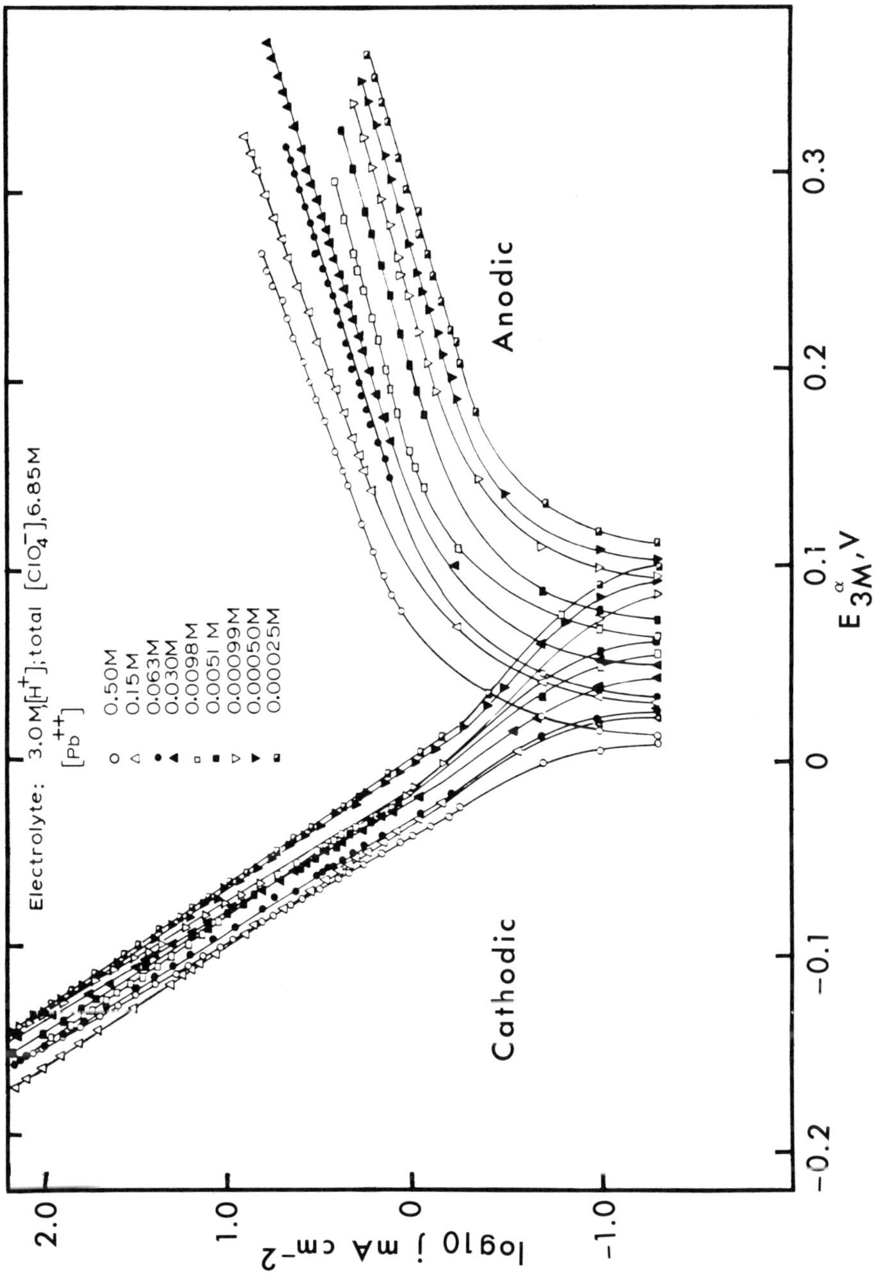

FIG. 3.1.17. Tafel plots for α-PbO_2 electrodes at various Pb^{2+} concentrations in $HClO_4$ at 23°C [158]. (Courtesy of National Research Council of Canada.)

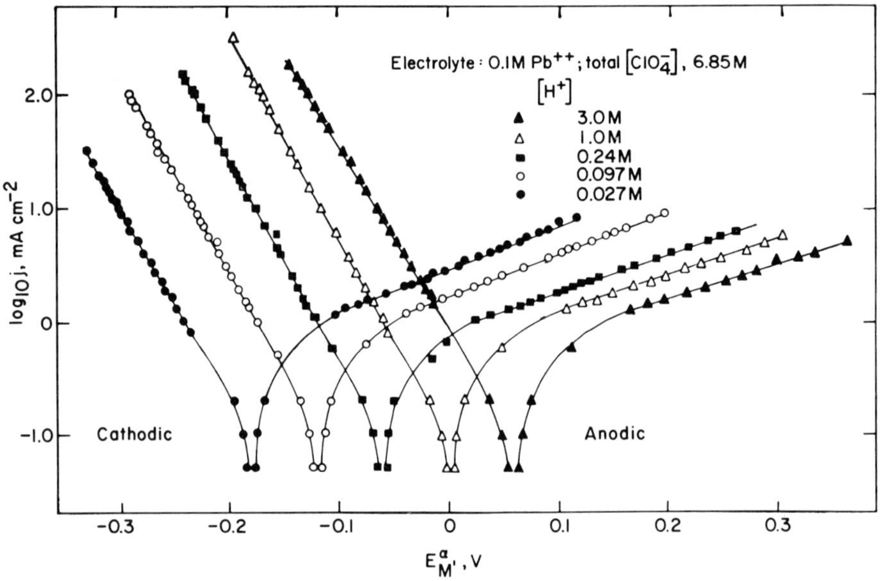

FIG. 3.1.18. Tafel plots for α-PbO_2 electrodes at various H^+ concentrations in $HClO_4$ at 23°C [158]. (Courtesy of National Research Council of Canada.)

A similar explanation was applied to the dependence of the cathodic current for α-PbO_2 on H^+ (Fig. 3.1.18), where the equilibrium quantity of β-PbO_2 was expected to increase in proportion to increasing $[H^+]$.

For α- and β-PbO_2 the anodic Tafel plots are essentially identical since the PbO_2 (whether α or β) is not involved as a reactant in the anodic process.

3.1.4. PbO_2 Electrodes in Alkali Media

3.1.4.1. Low Overvoltage Region. For PbO_2 electrodes in alkali, the relationships

$$\left(\frac{\partial E}{\partial \log[OH^-]}\right)_{[Pb^{2+}],\, j=0} = -29 \text{ mV} \tag{3.1.21}$$

and

$$\left(\frac{\partial E}{\partial \log[Pb^{2+}]}\right)_{[OH^-],\, j=0} = -29 \text{ mV} \tag{3.1.22}$$

are observed [154], which is consistent with the equilibrium electrode reaction

$$PbO_2 + 2H_2O + 2e^- \rightleftarrows Pb(OH)_3^- + (OH)^- \tag{3.1.23}$$

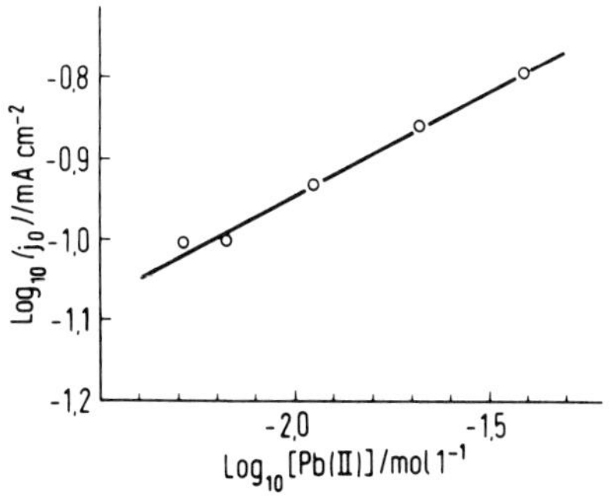

FIG. 3.1.19. Dependence of exchange current density on Pb^{2+} concentration for α-PbO_2 electrodes in alkali at 23°C. $[OH^-] = 0.68$ M [154]. (Courtesy of Berichte der Bunsen Gesellschaft.)

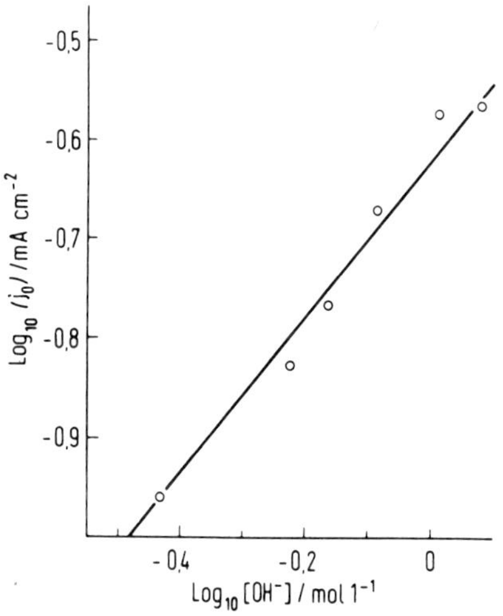

FIG. 3.1.20. Dependence of exchange current density on OH^- concentration for α-PbO_2 electrodes in alkali at 23°C. $Pb^{2+} = 0.0008$ M [154]. (Courtesy of Berichte der Bunsen Gesellschaft.)

Exchange current densities (j_0's) obtained from j-η transients (where $\eta < 6$ mV) for α-PbO_2 are given in Figs. 3.1.19 and 3.1.20. The j_0's are of about the same order of magnitude as those observed for PbO_2 in perchloric acid. Using the following expression derived from the appropriate rate equations

$$\left(\frac{\partial \log j_0}{\partial \log[Pb^{2+}]}\right)_{[OH^-]} + \left(\frac{\partial \log j_0}{\partial \log[OH^-]}\right)_{[Pb^{2+}]} = 2(1 - \alpha) \qquad (3.1.24)$$

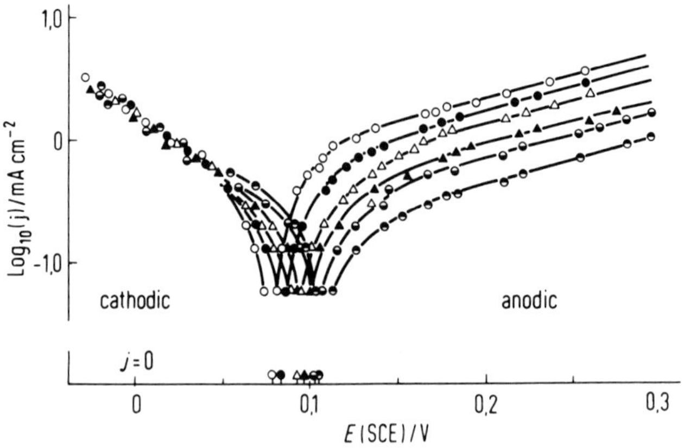

FIG. 3.1.21. Tafel plots for α-PbO_2 in alkali (OH^- = 0.68 M) at 23°C. Pb^{2+}: (○) 0.039 M; (▲) 0.021 M; (◒) 0.011 M; (△); 0.0067 M; (●) 0.0052 M [154]. (Courtesy of Berichte der Bunsen Gesellschaft.)

and the slopes obtained from Figs. 3.1.19 and 3.1.20, a value of 0.5 for α was calculated, indicating a symmetrical single step charge transfer reaction. The suggested [154] sequence for the electrode reaction was the charge transfer step

$$PbO_2 + H_2O + 2e^- \rightarrow PbO_{ads} + 2OH^- \qquad (3.1.25)$$

followed by

$$PbO_{ads} \rightleftharpoons PbO_{aq} \qquad (3.1.26)$$

and

$$PbO_{aq} + OH^- + H_2O \rightleftharpoons Pb(OH)_3^- \qquad (3.1.27)$$

According to Hampson et al. [154], it was not possible to obtain reliable kinetic data for β-PbO_2 in alkali. Although β-PbO_2 electrodes were stable in the absence of Pb^{2+}, addition of Pb^{2+} to the electrolyte caused mechanical deterioration of the electrodes. This was ascribed to the formation of α-PbO_2 (the stable polymorph in alkaline media) as a product of the exchange reaction between the β-PbO_2 and the Pb^{2+} in the electrolyte.

3.1.4.2. High Overvoltage Region. Tafel plots for α-PbO_2 are given in Figs. 3.1.21 and 3.1.22. In a manner comparable to that outlined for PbO_2 in perchloric acid, Hampson et al. determined from the kinetic data that at high η the exchange reaction at PbO_2 electrode in alkali changed to two consecutive electron steps. The consecutive step charge transfer reaction in the high η region was represented as

$$PbO_2 + H_2O + e^- \rightleftharpoons \left| PbO(OH)_2^- \right| \qquad (3.1.28)$$

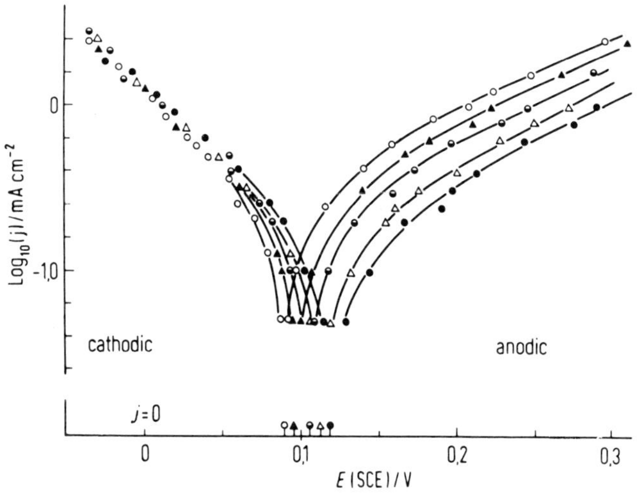

FIG. 3.1.22. Tafel plots for $\alpha-PbO_2$ in alkali for various OH^- concentrations at 23°C. OH^-: (○) 1.22 M; (●) 1.04 M; (△) 0.82 M; (▲) 0.64 M; (◒) 0.57 M; (◓) 0.37 M [154]. (Courtesy of Berichte der Bunsen Gesselschaft.)

$$\left\{PbO(OH)_2^-\right\} + e \rightleftharpoons PbO_{ads} + 2OH^- \tag{3.1.29}$$

and the slow step in either the anodic or the cathodic direction was that leading to the formation of the unstable intermediate $Pb(OH)_2^-$.

Since $\alpha-PbO_2$ appears to be the stable polymorph in alkali, the electrode surface is not likely to be modified by the electrolyte, in contrast to the case for $\alpha-PbO_2$ in perchloric acid. Hence, the cathodic Tafel plots in Figs. 3.1.21 and 3.1.22 indicate the expected zero-order dependence on the electrolyte constituents.

For the anodic reaction, the observed reaction order

$$\left(\frac{\partial \log j_a}{\partial \log[Pb^{2+}]}\right)_{[OH^-]} = 0.48 \tag{3.1.30}$$

and

$$\left(\frac{\partial \log j_a}{\partial \log[OH^-]}\right)_{[Pb^{2+}]} = 1.49 \tag{3.1.31}$$

were less than the expected values of 1.0 and 2.0, respectively. This has been attributed to adsorption of reactants at the electrode. Since the exchange occurs at negative rational potentials (negative charges on the electrode), a mechanism involving the adsorption of neutral PbO has been favored over that involving adsorption of the various negatively charged species involved in the reaction scheme. The observed reaction orders were consistent with calculations based on the use of a Freundlich isotherm to describe the adsorption of PbO.

Kinetic data for the exchange at PbO_2 electrodes are summarized in Table 3.1.10.

TABLE 3.1.10. Kinetic Data for the Exchange Reaction at PbO_2 Electrodes

Reaction	Condition	Electrode	Technique	Rate const	α	Remarks	Refs.
$PbO_2 \rightleftharpoons Pb^{2+} + 2e^-$	0.05–1 M Pb^{2+}, 3 M H^+, 6.5 M ClO_4^-	β-PbO_2	Galvanostatic pulse	$j_0 = 1 \times 10^{-4}$ to 2×10^{-4} A/cm^2	0.2	$\eta = \pm 10$ mV, $\Delta H = 7.4$ kcal/mole	[156]
$PbO_2 \rightleftharpoons Pb^{2+} + 2e^-$	0.005–0.5 M Pb^{2+}, 3 M H^+, 6.85 M ClO_4^-	α-PbO_2	Galvanostatic pulse	$j_0 = 1 \times 10^{-4}$ to 3×10^{-4} A/cm^2	0.22	$\eta = \pm 10$ mV, $\Delta H = 3$–10 kcal/mole	[158]
$PbO_2 \rightleftharpoons Pb^{2+} + 2e^-$	10^{-2} M Pb^{2+}, 0.68 M OH^-	α-PbO_2	Galvanostatic pulse	$j_0 = 10^{-4}$ A/cm^2	0.5	$\eta < 6$ mV, $\Delta H = 2$–10 kcal/mole	[154]
$PbO_2 \rightleftharpoons Pb^{2+} + 2e^-$	10 N H_2SO_4	50/50 α/β-PbO_2	ac imp	$j_0 = 3.2 \times 10^{-4}$ A/cm^2	-		[149]
$PbO_2 \rightleftharpoons Pb^{2+} + 2e^-$	2 N H_2SO_4	PbO_2	ac imp	$j_0 = 1.5 \times 10^{-5}$ A/cm^2	-		[148]

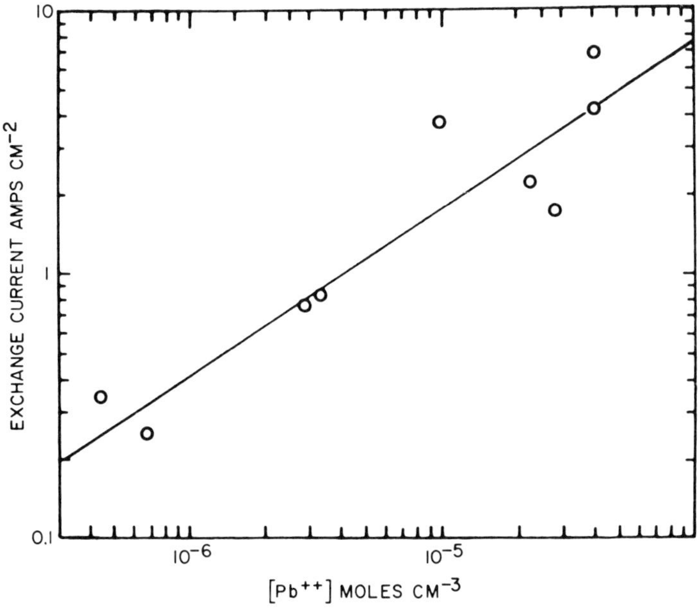

FIG. 3.1.23. Exchange currents for the lead electrode in LiCl-KCl eutectic at 450°C (data obtained with stationary drop electrodes) [159]. (Courtesy of Electrochemistry Society.)

3.1.5. Kinetic Parameters in Fused Salts

In much of the work concerning the kinetics of the Pb/Pb^{2+} system in fused salts, emphasis to date has been placed mainly on technique and reproducibility rather than on insight into the reaction mechanism. It seems likely, however, the mechanism of the exchange involves a complex species in the melt. In the early work of Laitinen et al. [159] the double pulse method of Gerischer and Krause [160] was applied to lead in LiCl-KCl melt. The electrode was a Pb drop contained in a tungsten cup. Data from replicate experiments (Fig. 3.1.23) showed a wide scatter which was attributed to uncertainty in surface area measurements of the drop.

Recent studies by Graves and Inman [161] involved the use of a syringe-type electrode (Fig. 3.1.24), which permitted exposure of fresh lead surface to the melt at the beginning of each experiment. In this case a conventional galvanostatic technique was applied, and the kinetic data could be analyzed by classical techniques provided that the overvoltage and time interval of the pulse were limited to 4 mV and 4–36 μ sec at 450°C. Under these conditions the impedance behaved as a linear circuit element.

Values of the exchange current density from these experiments, as a function of $[Pb^{2+}]$ in the melt, are given in Fig. 3.1.25.

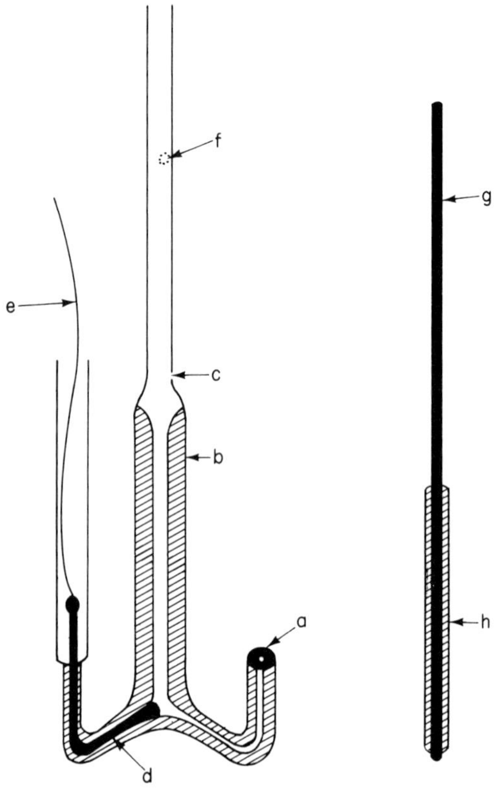

FIG. 3.1.24. Syringe electrode for liquid lead. Liquid lead is introduced into syringe via precision capillary a and controlled with plunger consisting of a tungsten rod g, coated with glass h, which has been ground to fit precision capillary b. Connection between liquid lead and pulse generator is made via tungsten contact d and conductor e. Connection between liquid lead and oscilloscope is made via tungsten rod g. Gas bubbles are removed from liquid lead with aid of holes c and f [161]. (Courtesy of Institute of Mining and Metallurgy, London.)

Alternating current methods have also been applied to the study of Pb/Pb^{2+} exchange in molten salts [150, 162]. For the Pb electrode in KCl-NaCl melt, Ukshe and Bukum [162] obtained good linear plots for the impedance components over a wide range of frequencies. They observed that below a critical concentration of lead ions (5×10^{-6} $\underline{M}$) the exchange current density began to rise. The behavior was attributed to corrosion of the lead electrode at the low lead ion concentrations.

The kinetic data obtained from these investigations are summarized in Table 3.1.11.

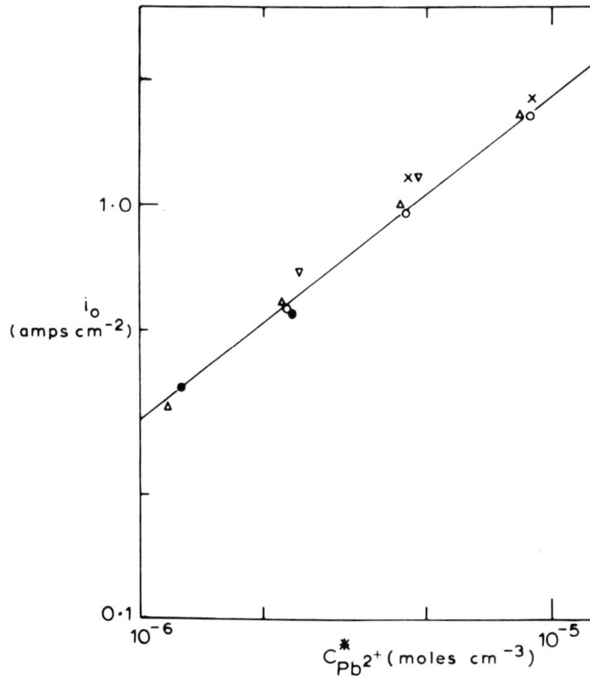

FIG. 3.1.25. Log j_0 vs log Pb^{2+} concentration for the Pb^{2+}/Pb exchange in LiCl-KCl eutectic at 450°C (data obtained with syringe type electrode shown in Fig. 3.1.24) [161]. (Courtesy of Institute of Mining and Metallurgy, London.)

TABLE 3.1.11. Kinetic Data for the Pb/Pb^{2+} Exchange in Fused Salts

Reaction	Condition	Elec-trode	Technique	Rate const	α	Refs.
$Pb^{2+} + 2e \rightleftarrows Pb$	LiCl-KCl eut, 450°C	stat Pb drp	Double pulse	$j_0 = 30 \pm 15$ A/cm^2 $K^\circ = 0.01$ cm/sec	0.38	[159]
$Pb^{2+} + 2e \rightleftarrows Pb$	LiCl KCl eut, 450°C	Syringe type	Pulse	$K^\circ = 0.07$ cm/sec	0.22	[161]
$Pb^{2+} + 2e \rightleftarrows Pg(Hg)$	$LiNO_3$-$NaNO_3$-KNO_3 eut + 0.02 mole %/KCl	DME	farad. rect	$K^\circ = 1.4$ cm/sec	0.17	[150]
$Pb^{2+} + 2e^- \rightleftarrows Pb$	NaCl-KCl, 720°C	Pb	ac imp	$j_0 = 2$-5 A/cm^2		[162]

TABLE 3.2.1. Point of Zero Charge (E_Z) for Pb in Aqueous Solution

Solution	E_Z	Method	Refs.
0.01-0.001 M NaF	-0.56	Capacity min	[163]
0.1-1 M NaCl	-0.62 to 0.63	Hardness max	[164], [165]
0.01 M KCl and 1 M NaCl	-0.75	Capacity min	[147]
KCl	-0.37	Hardness max	[165]
0.01 M KCl and 1 M NaCl	-0.60	Hardness max	[166]
5×10^{-4} M H_2SO_4	-0.67	Capacity min	[164]
0.002 M Na_2SO_4	-0.62 ± -0.01	Acoustical electrical method	[167]
0.02 M Na_2SO_4	-0.64	Capacity min	[165]
2 M H_2SO_4	-0.59	Hardness max	[165]
<0.02 M $S_2O_8^{-2}$	-0.70 to -0.94	Fall of current intensity	[168]
1 M NaOH	-0.58	Hardness max	[165]

3.2. DOUBLE-LAYER PROPERTIES

3.2.1. Lead in Aqueous Media

Table 3.2.1 shows a number of observed values for the point of zero charge (E_Z) on lead in contact with various solutions. According to the theory of Frumkin, the difference in the E_Z for a given metal and the electrocapillary maximum for mercury in the same solution should be equal to the metal/mercury contact potential. Although this relationship was observed in the early work of Borishova et al. [147, 169] for lead in sulfate and halide solutions, the results could not be confirmed by Rybalka and Leikis [170] who concluded that the effect of anions on the E_Z of mercury is greater than that on lead.

Typical capacitance-potential curves obtained by Rybalka et al. [163] for lead in fluoride solution are shown in Fig. 3.2.1. A reproducible lead surface for these experiments was effected by anodic polishing in acetate solution, followed by prolonged cathodic reduction in the measuring cell. The value of the capacitance at the minimum (C_{min}) compared favorably with that of mercury under the same conditions, indicating that the lead surface was

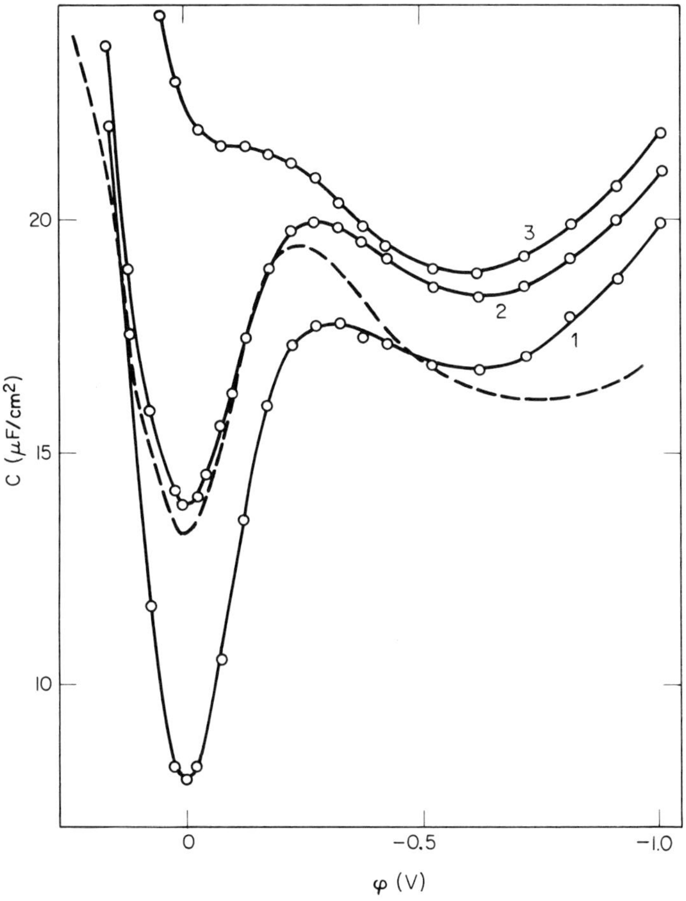

FIG. 3.2.1. Differential capacitance curves for a lead electrode at 25°C in NaF solutions (1) 0.001, (2) 0.01, (3) 0.1 N. Broken line shows differential capacitance for mercury in 0.01 N NaF [163]. (Courtesy of Consultants Bureau, New York.)

indeed smooth. Frequency dispersion for the lead electrode was negligible over the range 100 to 10,000 Hz, and the capacitance at a given potential remained constant over many hours.

The position of the C_{min} in Fig. 3.2.1 is independent of solution concentration, indicating that adsorption effects were absent. In the presence of the other halides [170], the C_{min} was progressively shifted to more negative values in the order Cl Br I, and the depth of the C_{min} also decreased in that same order, indicating that these ions are adsorbed. A shift in the C_{min} to more negative values was also exhibited in perchlorate and sulfate solutions. In the case of sulfate, the shift may partially result from asymmetry of the electrolyte, although sulfate ion adsorption must also be considered. In studies of hydrogen evolution on lead, sulfate adsorption has been well documented [18].

The effect of various cations on the double-layer structure of lead has also been determined [171]; however, these effects are marginal. Additions of tetraethylammonium ion to the electrolyte did not affect the C_{min}, but at sufficiently negative surface charges the capacitance was lowered, probably as a result of electrode surface coverage by the $(C_2H_5)_4N^+$ ions.

Variations of the E_Z values with the method of measurement (Table 3.2.1) are not readily explained. Where comparisons can be made, E_Z values from hardness measurements lie at potentials significantly negative to those values based on the C_{min}.

For a number of metals the potential registered by a freshly produced surface, effected by scraping the electrode in oxygen-free electrolytes, corresponds to the E_Z for the metal. The studies by Eyring et al. [172] indicate, however, that this is not the case for lead. The scrape potential for lead was largely independent of the type of anion involved and had significantly more positive E_Z values than those based on the C_{min}.

3.2.2. PbO_2

The zero charge potential (E_Z) reported for PbO_2 in various media are given in Table 3.2.2.

In nitrate solution, the position of the C_{min} (Fig. 3.2.2) is not dependent on solution concentration [173], an indication that nitrate ion is not adsorbed onto PbO_2. Based on the positions of C_{min}, Hampson and co-workers [173] reported E_Z values of 0.90 V vs SCE for β-PbO_2 and 0.82 V vs SCE for α-PbO_2. Integration of the capacitance curves were similar to those for Hg in a noninteracting electrolyte [178], after taking into account the roughness factor for the PbO_2. At constant charge density the measured potential of the electrode was a linear function of the chemical potential (Fig. 3.2.3), and the family of lines was symmetric about the potential value corresponding to the minimum in the capacitance curves, a further indication of the correct choices of E_Z.

In H_2SO_4 electrolyte a shift of E_Z to more negative values than 1.78 V has been observed [176, 177] with a decrease in H_2SO_4 concentration, indicating adsorption of sulfate ion. From ac impedance measurements on PbO_2, Carr and Hampson [179] noted a progressive rise in capacitance in sulfate electrolytes as the pH was decreased (Fig. 3.2.4). The observations were explained in terms of reversible redox reactions involving hydrogen ions and sulfate ions adsorbed on the electrode, e.g.,

$$H^+ + SO_4^{2-}\ ads = HSO_4^-\ ads \qquad (3.2.1)$$

$$4HSO_4^-\ ads + PbO_2 + 2e^- \rightarrow PbSO_4\ ads + 3SO_4^{2-} + 2H_2O \qquad (3.2.2)$$

A similar explanation was applied to phosphate electrolytes, where the magnitude of the capacitance was observed [178] to decrease progressively in the series H_3PO_4 KH_2PO_4 K_2HPO_4.

TABLE 3.2.2. Point of Zero Charge (E_Z) for PbO_2 in Various Media

Solution	E_Z	Method	Remarks	Refs.
KNO_3	0.81 ± 0.01 vs SCE	Capacitance min	α-PbO_2	[173]
KNO_3	0.91 ± 0.01 vs SCE	Capacitance min	β-PbO_2	[173]
0.05 M H_2SO_4	1.78	ac imp		[119], [174], [175]
0.2 M H_2SO_4	1.9	Hardness max		[176]
16 M H_2SO_4	1.7	Hardness max		[176]
0.02 M and 2 M H_2SO_4	1.80 ± 0.02	ac imp		[177]

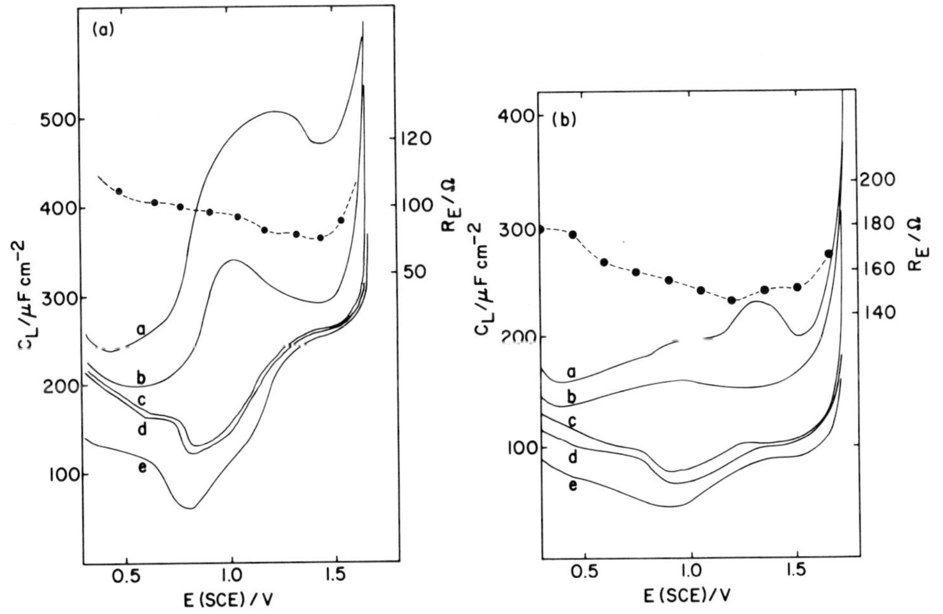

FIG. 3.2.2. Differential capacitance curves for electrodeposited α-PbO_2 (A) and β-PbO_2 (B) in KNO_3 electrolytes at 23°C, 120 Hz. (a) 0.307 M, (b) 0.0115 M, (c) 0.018 M, (d) 0.012 M, (e) 0.0043 M. ● = electrode resistance R_E [173]. (Courtesy of Elsevier, Ltd.)

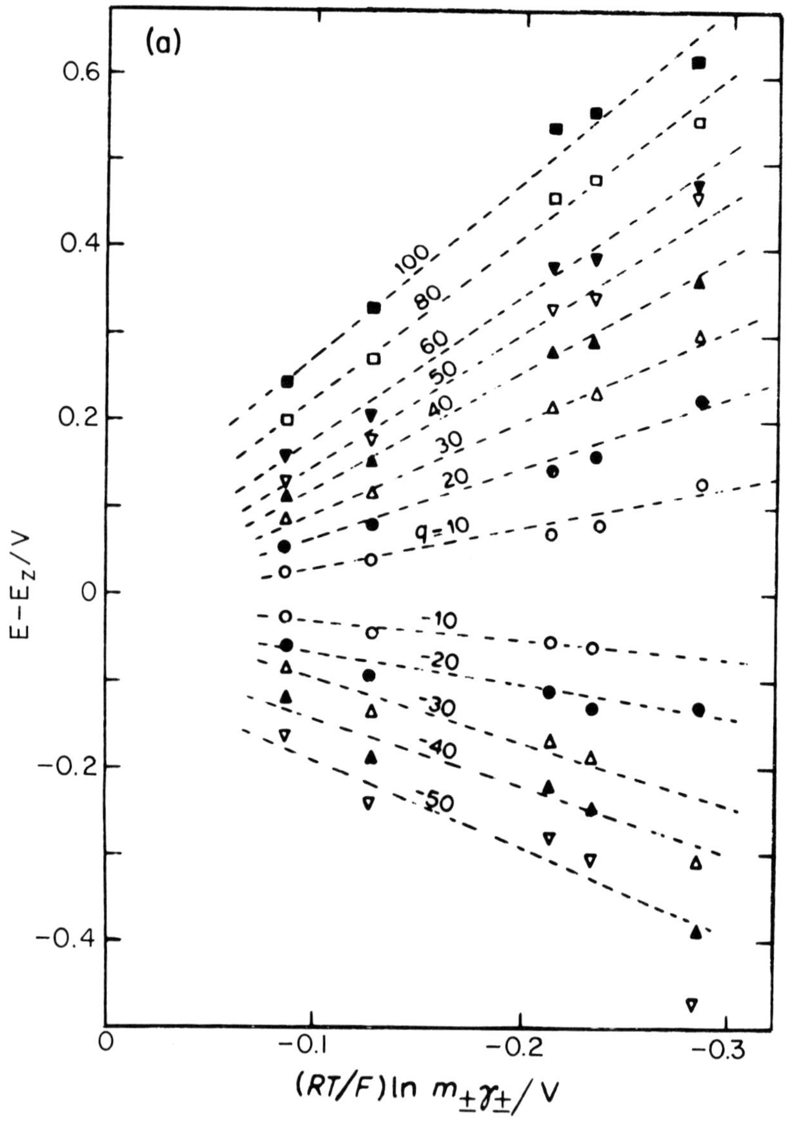

FIG. 3.2.3(a). Potential referred to E_Z as a function of activity, expressed as (RT/F) ln $a^{\pm}$ for electrodeposited α-PbO_2 [173]. (Courtesy of Elsevier, Ltd.)

At potentials negative to the ideally polarizable region, Farr and Hampson [180] noted a difference between the behavior of the PbO_2 in the sulfate and in the phosphate systems. The incipient formation of $PbSO_4$ at these potentials was accompanied by a pronounced reduction in capacitance, as expected from the flat capacitor formula. However, in H_3PO_4 electrolytes a significant reduction in capacitance was not observed at these potentials, indicating that an insoluble phosphate layer was apparently not formed on the electrode surface.

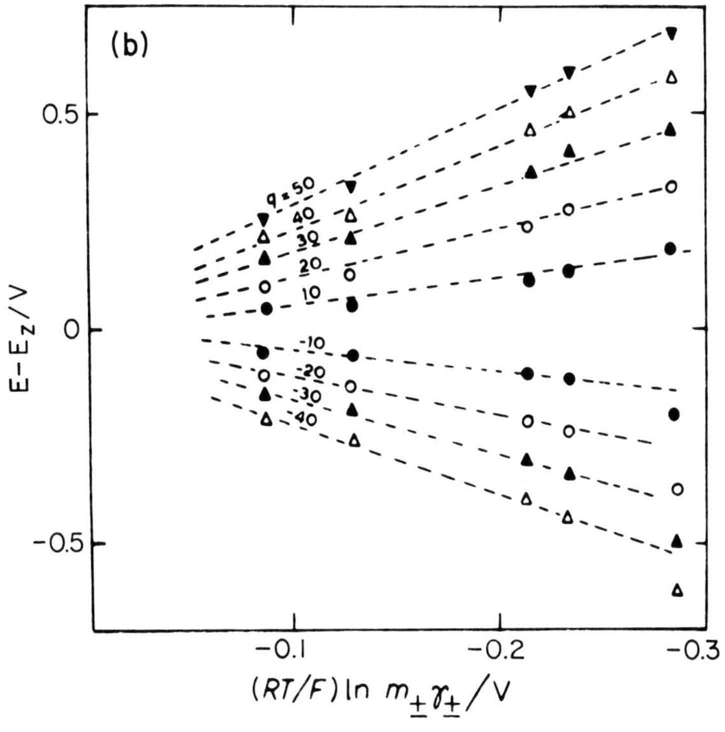

FIG. 3.2.3(b). Potential referred to E_Z as a function of activity, expressed as (RT/F) ln a ± for electrodeposited β-PbO_2 at constant charge [173]. (Courtesy of Elsevier, Ltd.)

In chlorate solutions an absence of a well-defined C_{min} developing with electrolyte dilution (Fig. 3.2.5) [181] indicates that ClO_4^- is adsorbed onto PbO_2. An increase in the capacitance of the PbO_2 with time, observed by Kiseleva et al. [174], was considered a result of ClO_4^- adsorption and lattice expansion of the PbO_2. Lattice expansion was also noted by Kokarev [175] who suggested the relationship

$$S_{true} = S_{limit} + (a + bt)t^{0.5} \qquad (3.2.3)$$

where a and b are constants, t is the time, and S is the electrode surface area.

3.2.3. Lead in Fused Salts

The shape of capacitance-potential curves for lead in fused salts and the point of zero charge are not affected by changing the anodic component of the melt [182-184] although the magnitude of the C_{min} increases in the order Cl Br I (Fig. 3.2.6), which has been associated with differences in polarizability of the anion. The influence of the anion in this

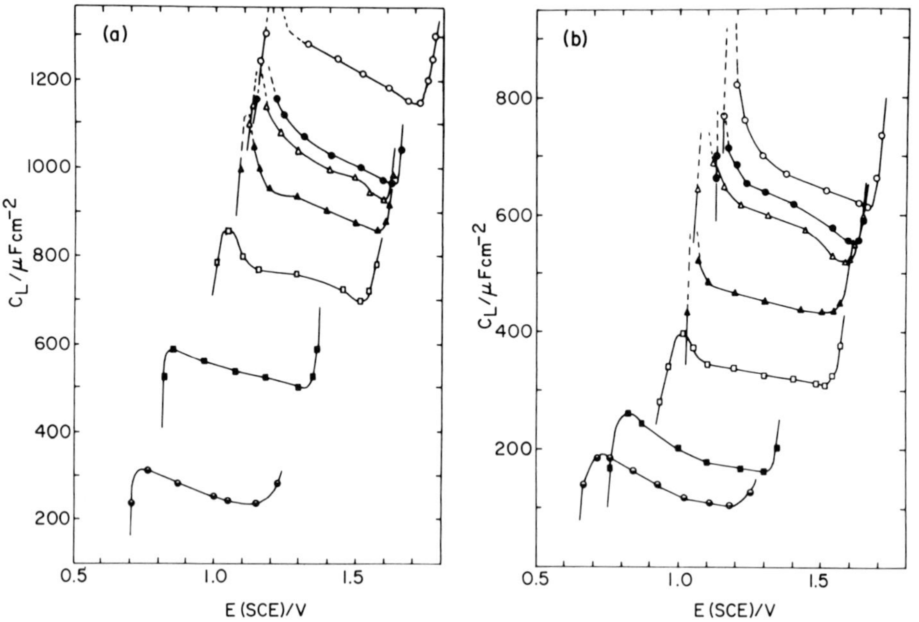

FIG. 3.2.4. Differential capacitance curves for electrodeposited α-PbO_2 (a) and β-PbO_2 (b) in K_2SO_4 (0.0344 M) at 23°C, 120 Hz. (◒) pH 12.0; (■) pH 6.0; (□) ph 3.8; (▲) pH 3.4; (△) pH 3.0; (●) pH 2.5; (○) pH 1.8 [179]. (Courtesy of Elsevier, Ltd.

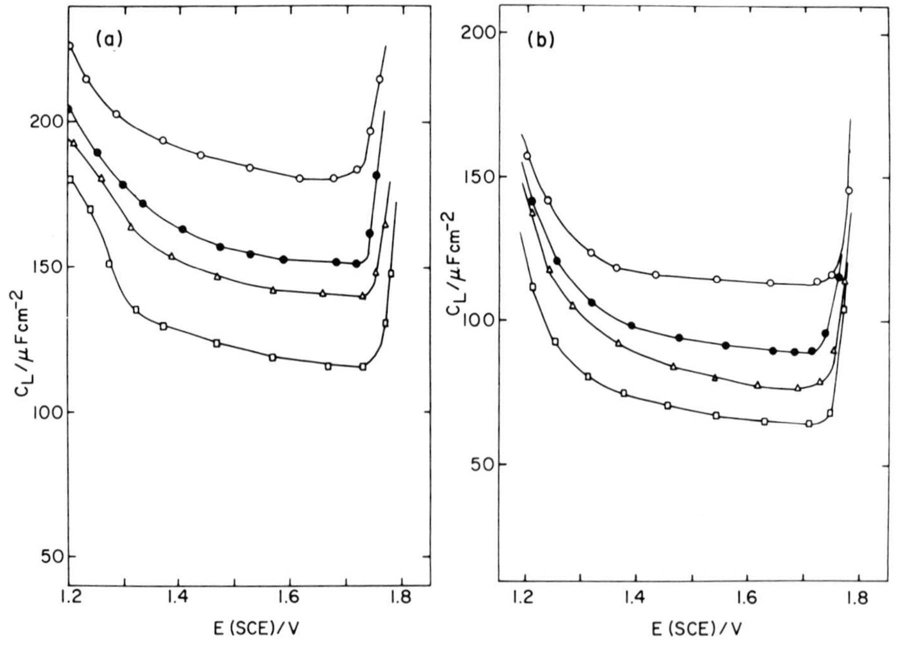

FIG. 3.2.5. Differential capacitance curves for electrodeposited α-PbO_2 (a) and β-PbO_2 (b) in $HClO_4$ at 23°C, 120 Hz. (○) 0.135 M; (●) 0.0645 M; (△) 0.0095 M; (□) 0.0003 M $HClO_4$ [181]. (Courtesy of Elsevier, Ltd.)

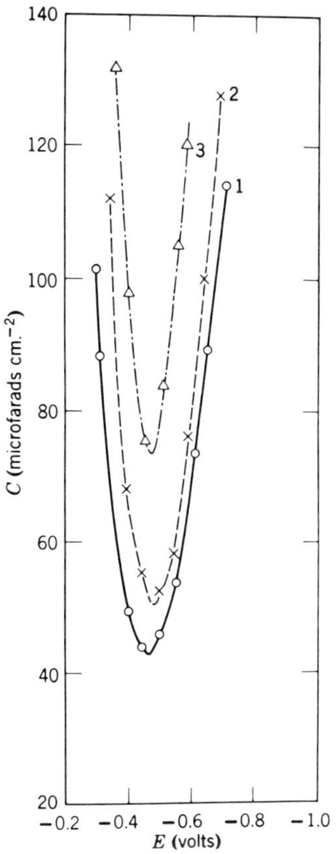

FIG. 3.2.6. Differential capacity vs potential for lead in sodium chloride at 820°C (Curve 1); sodium bromide at 800°C (Curve 2); and sodium iodide at 800°C (Curve 3). Potentials referred to a lead-acid chloride (10% by weight) electrode in the melt [184]. (Courtesy of Pergamon Press.)

respect is especially evident in Na or Li halide melts where the cation radii are small, but in K or Cs halide melts the anion/cation radius ratio is closer to unity. Hence the cationic contribution to the double-layer structure becomes more important and the influence of the anion on the magnitude of the C_{min} is correspondingly reduced.

The shapes of the capacitance-potential curves and the potential of zero charge are both affected by changing the cation component of the melt [183, 184] as exemplified in Fig. 3.2.7.

The effect of temperature in the capacitance-potential curves is illustrated in Fig. 3.2.8. According to Graves and Inman [161, 185], the effect of temperature on the shape of the curves can be understood by assuming that the ideally polarizable region diminishes as the temperature is increased. The increase in capacitance at anodic potentials is due to the charge transfer reaction

$$Pb^{2+} + 2e^- \rightleftarrows Pb \qquad (3.2.4)$$

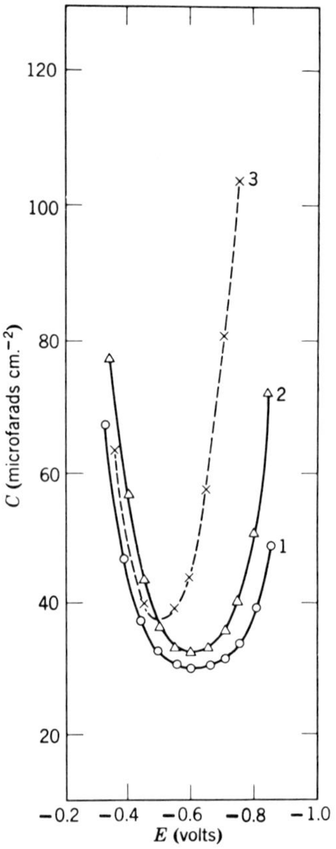

FIG. 3.2.7. Differential capacity vs potential for lead. Curve 1, potassium chloride; Curve 2, potassium iodide; and Curve 3, sodium chloride + potassium chloride (1:1). Temperature 800°C [184]. (Courtesy of Pergamon Press.)

which was shown by Graves and Inman to occur at potentials (designated by the arrows in Fig. 3.2.8) closer to the point of zero charge as the temperature is increased. Similarly, the sharp increase in capacitance in the cathodic region has been attributed to the onset of alkali metal deposition [186, 187] which is also expected to be facilitated at higher temperatures. The proposal by Graves and Inman is supported by the data of Heus et al. [188] who were unable to find any significant variation of capacitance with temperature in melts at low temperature (390–480°C) where the ideally polarizable region was expected to be large.

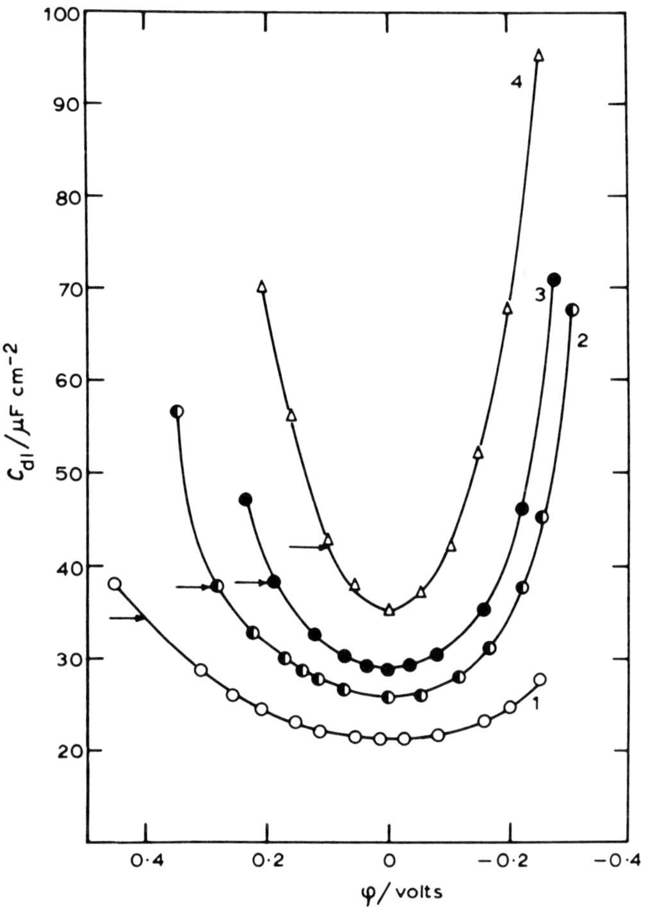

FIG. 3.2.8. Differential capacitance curves for the lead electrode in equimolar LiCl-KCl melt at (1) 450°C; (2) 600°C; (3) 700°C; (4) 800°C. Arrows indicate transitions from ideally polarizable to ideally reversible region [185]. (Courtesy of Elsevier, Ltd.)

4. ELECTROCHEMICAL STUDIES

4.1. ANODIC FILM STUDIES

Despite the numerous studies concerning the charging and discharging of lead electrodes in H_2SO_4, there are still many doubts regarding the exact mechanisms. Gladstone and Tribe [189] proposed the well-known "double-sulfate" process to describe the overall charge-discharge reaction

$$Pb + PbO_2 + 2H_2SO_4 \underset{charge}{\overset{discharge}{\rightleftharpoons}} 2PbSO_4 + 2H_2O \quad (4.1.1)$$

The reaction was confirmed by a number of methods [190] and supported by Beck and Wynne-Jones [191] as the only reaction consistent with the thermodynamics of the system. Later studies revealed PbO and basic lead sulfates as products of anodic oxidation, in contrast to earlier work where $PbSO_4$ and PbO_2 were the only products observed. Furthermore, two crystallographic modifications of PbO_2 have been noted.

Numerous studies, of which only a few are referenced [24, 35, 192-195], on film composition as a function of potential have been reported. The most definitive work has been that of Burbank [35], whose schematic of the structures formed on lead in H_2SO_4 appears in Fig. 4.1.1. A laminar structure results because of free energy gradients in the film, so that different compounds that are stable in contact with the solution may be different than those that are in contact with the metal. At all potentials, $PbSO_4$ is the first product detected [195], and growth of the film takes place under the protective $PbSO_4$ layer. It is thought that the superficial $PbSO_4$ coating is permeable to H^+ and H_2O, but not to SO_4^{-2} [24, 193-195]. As the potential is increased, the sulfate coating becomes progressively thinner, and the depicted porosity in the outer coating corresponds to regions of high corrosion rate, as determined by Lander [193].

4.1.1. Basic Lead Sulfates

The basic sulfates are reported from fusion products as well as from wet chemical methods of preparation. The monobasic, dibasic, and tetrabasic lead sulfates are the phases observed from the fusion of PbO and $PbSO_4$, although the dibasic form is unstable below about 640°C where it decomposes to the mono- and tetrabasic forms [196]. By boiling stoichiometric quantities of PbO and $PbSO_4$ in water, the mono- and tetrabasic forms are made, as well as a monohydrate of the tribasic form, $3PbO.PbSO_4.H_2O$ [197]. Because of discrepancies in reported data, the PbO-$PbSO_4$ system was recently reinvestigated, and IR adsorption, X-ray, and single crystal data for the basic lead sulfates have been reported [196]. The PbO-SO_3^{-}-H_2O phase diagram, as it relates to the manufacture of battery plates, has also been presented [198].

The basic lead sulfates form at potentials close to the calculated reversible potentials for these reactions, and both the monobasic and tribasic sulfates have been detected [194]. Although the tetrabasic form is not observed as a corrosion product, it is known to be present in lead-acid battery plates that have been cured at high temperatures [31]. The role of basic sulfates in the passivation of the lead electrode has been discussed extensively by Pavlov et al. [199].

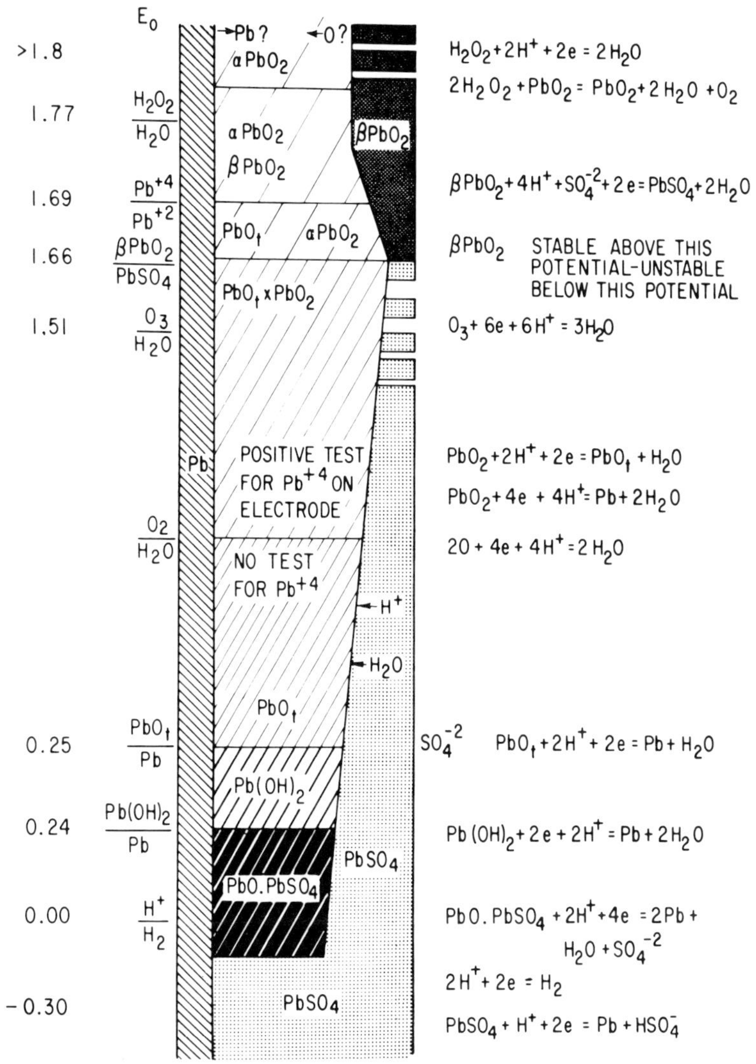

FIG. 4.1.1. Schematic drawing of the structures formed on lead in H_2SO_4 as a function of potential [35]. (Courtesy of Electrochemical Society.)

4.1.2. $Pb(OH)_2$ and PbO

The occurrence of $Pb(OH)_2$ as a corrosion product on Pb in H_2SO_4 was first observed by Burbank [24] who anodized a thin lead foil to translucency at about 0 V. The X-ray diffraction pattern showed primarily $Pb(OH)_2$ with a few strong lines for $PbSO_4$. Pavlov [200]

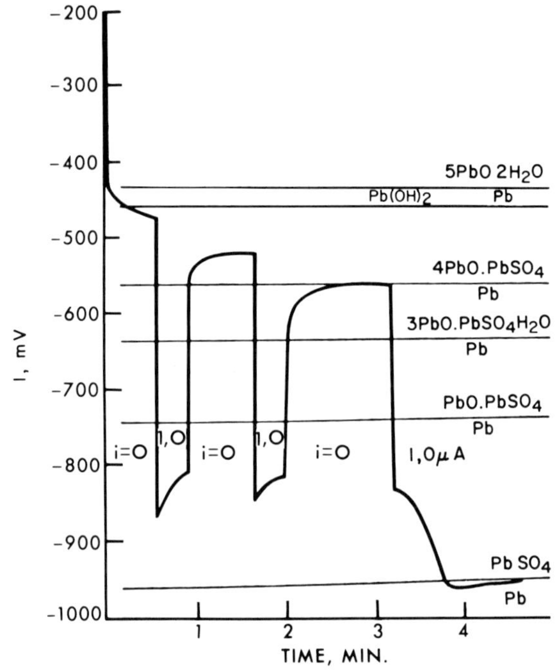

FIG. 4.1.2. Variation in potential upon galvanostatic reduction of the lead hydroxide electrode [200]. (Courtesy of Pergamon Press.)

applied cathodic pulses to a Pb/$Pb(OH)_2$ electrode that had been formed in 1 N H_2SO_4 during a 16-hr oxidation at 400 mV. Depending on the magnitude of the current, peaks were observed in the initial pulses and the potential shifted toward positive values despite the passage of a cathodic current (Fig. 4.1.2). The results were explained on the basis of an assumed increase in the rate of dissolution of the hydroxide brought about by the passage of the cathodic current.

On open circuit stand, a lead hydroxide electrode ultimately transforms into a Pb/$PbSO_4$ electrode with the intermittent formation of basic lead sulfates.

The material designated as PbO_t (Fig. 4.1.1) has been described by Burbank as having an orthorhombic-pseudotetragonal lattice, in contrast to the orthorhombic structure of litharge (PbO_y) or the tetragonal structure of masicot (red PbO). In alkaline medium the formation of PbO_t is associated with rapid corrosion, but in the presence of SO_4^{-2} the formation of a surface layer of $PbSO_4$ provides the metastable equilibrium. According to Burbank, the PbO_t exhibits a preferred orientation with the texture axis [110] parallel to the substrate, an observation that was recently confirmed by Pavlov et al. [194]. This causes the line d = 2.81 Å to be the strongest in the X-ray diffraction pattern. This line coincides with the strongest line for PbO_2 and seems to have led to some confusion regarding the structure of the corrosion film, as discussed later.

PbO_t is found to be the principal reaction product up to potentials approaching the $PbO_2/PbSO_4$ value. As the potential is increased in this interval, the time at which PbO_t is first observed markedly decreases [195], e.g., at 400 mV PbO_t is not observed until after 6 hr, but at 900 mV PbO_t is observed after the first minute of anodization. The amount of PbO_t increases with an increase in temperature and a decrease in H_2SO_4 concentration [193]. The latter effect was explained by showing a linear relation between the formation rate and the square of the activity of water, which supports the theory that PbO_t forms by the electrochemical reaction of Pb with H_2O.

4.1.3. $PbO.xPbO_2$

The area designated as $PbO.xPbO_2$ in Fig. 4.1.1 was determined by both X-ray diffraction and spot tests [35], and appears to result from various chemical combinations of PbO and PbO_2, e.g., Pb_2O_3, Pb_3O_4, Pb_5O_8, Pb_7O_{11}. The suggested mechanism is that, in this potential region, an intermediate oxidation product of water reacts with PbO_t to form the intermediate oxides.

4.1.4. PbO_2

Above the $PbO_2/PbSO_4$ potential, the $PbSO_4$ on the lead surface is converted to β-PbO_2 (the stable phase in acid solution) by a boundary nucleation and growth mechanism [33]. Water and/or hydroxyl ions are discharged beneath the layer of β-PbO_2, so that the potential and pH may be within the domain of α-PbO_2. The formation of the α-PbO_2 occurs as a solid phase oxidation of the base metal, and the rate of formation seems to be governed by the rate of penetration of oxygen through the β-PbO_2 layer. At very high potentials, according to Pavlov [195], the formation of β-PbO_2 becomes accelerated, and its quantity cannot be accounted for on the basis of oxidized $PbSO_4$ or PbO alone. Apparently, Pb^{4+} passes through the anodic layer at these potentials, which gives rise to an additional quantity of β-PbO_2 at the solution interface.

In the model of the anodic film proposed by Rüetschi and Cahan [32], α-PbO_2 is depicted in the film at potentials as low as 300 mV. Although this is possible from theoretical considerations, it does not confirm what others have observed. The discrepancy seems to have come about because many of the X-ray diffraction lines for PbO and PbO_2 coincide [194, 195]. However, when the presence of PbO_2 is confirmed by iodimetric titration [194], it is clear that PbO_2 does not form in the film at potentials below 900 mV.

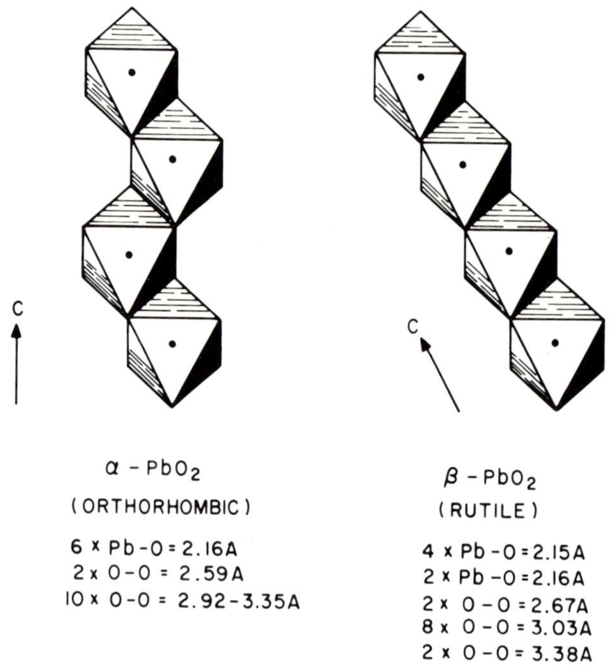

FIG. 4.2.1. Packing of octahedra in α- and β-PbO_2 [201]. (Courtesy of Electrochemical Society.)

4.2. PbO_2 PROPERTIES

4.2.1. Modifications

Two crystallographic modifications of PbO_2 exist: the commonly-occurring tetragonal or rutile form, called α-PbO_2, and the orthorhombic or columbite form, referred to as β-PbO_2. In both cases the Pb^{4+} ion is in the center of a distorted octahedron, but the essential difference is in the way in which the octahedra are packed (Fig. 4.2.1) [201]. Contrary to other oxide systems where the rutile structure is the high temperature form and the columbite type is generally the low temperature structure, α-PbO_2 is the high-temperature, high-pressure phase and β-PbO_2 is stable at normal pressures and temperatures [202]. The chemical preparation, analysis, and structure of each modification has been discussed in detail by Bagshaw and co-workers [203].

PbO_2 does not exist as a stoichiometric compound, and it is generally accepted that the composition lies within the limits $PbO_{1.80-1.98}(OH)_{0.11-0.26}$ [202]. Nonstoichiometry is observed regardless of the method of preparation [40, 203, 204].

4.2.2. Electrical Conductivity

The conductivity of PbO_2 is high (about $10^{-4}\ \Omega\ cm^{-1}$), and from negative values for the Hall coefficient it was established [205] that PbO_2 is an intrinsic semiconductor. Optical adsorption spectra indicate a band gap of 1.4-1.8 eV [201, 206, 207]. It has been suggested [208, 209] that the high concentration of free electrons in the conduction band may arise from an oxygen deficiency in the lattice, although Kittel found [210] that removal of oxygen by thermal decomposition leads to a decrease in conductivity. It is reported [211] that oxygen can be thermally removed from PbO_2 to a composition as low as $PbO_{1.66}$ without destroying the crystal lattice; however, this viewpoint is opposed by Burbank [202] and others. The conductivity data of Kittel might have resulted from a new insulating lattice created by the removal of oxygen.

Rüetschi and Cahan [40] suggested that the free electrons in PbO_2 may be due in part to OH groups substituting for oxygen in the lattice. Other explanations discussed recently [201] are based on the possible incorporation of hydrogen in the lattice as well as the presence of other impurities.

Electrical properties of α- and β-PbO_2 films have been recently determined by Mindt [201]. The higher conductivity of β-PbO_2 than of α-PbO_2 was also observed in earlier work [212], and seems to result from a lower electron mobility in the α-phase. The lower mobility compensates for the effect of a higher electron concentration in this phase. These results are shown in Fig. 4.2.2 [201].

4.2.3. Electrochemical Preparation

Pure forms of α- and β-PbO_2 can be obtained by electrodeposition, and detailed descriptions for the methods have been given [41, 42, 201, 204, 205, 213-216]. Pure β-PbO_2 has been electrodeposited from acidic solutions of lead nitrate, sulfamate, fluoborate, or perchlorate as well as by the anodic oxidation of lead in strong H_2SO_4. Pure α-PbO_2 has been deposited from neutral solutions of lead nitrate or acetate, or from lead perchlorate saturated with PbO, and by the anodic oxidation of lead in alkaline solution at high current densities.

The solution compositions would suggest that pH plays a significant role in the modification of the PbO_2 obtained. Results reported by Baker (Table 4.2.1) [213] indicated that the modification obtained during the initial anodic oxidation of lead as well as on subsequent cycling depends on the pH of the electrolyte. Kiseleva and Kabanov [217] observed that the formation of β-PbO_2 was retarded when lead was anodized in H_2SO_4 solutions containing Co^{2+}. They proposed that Co^{2+} displaces SO_4^{2-} which is chemisorbed on the oxide. Baker suggested, however, that the adsorption of protons was the controlling factor and that it must be the protons that are displaced by the Co^{2+}.

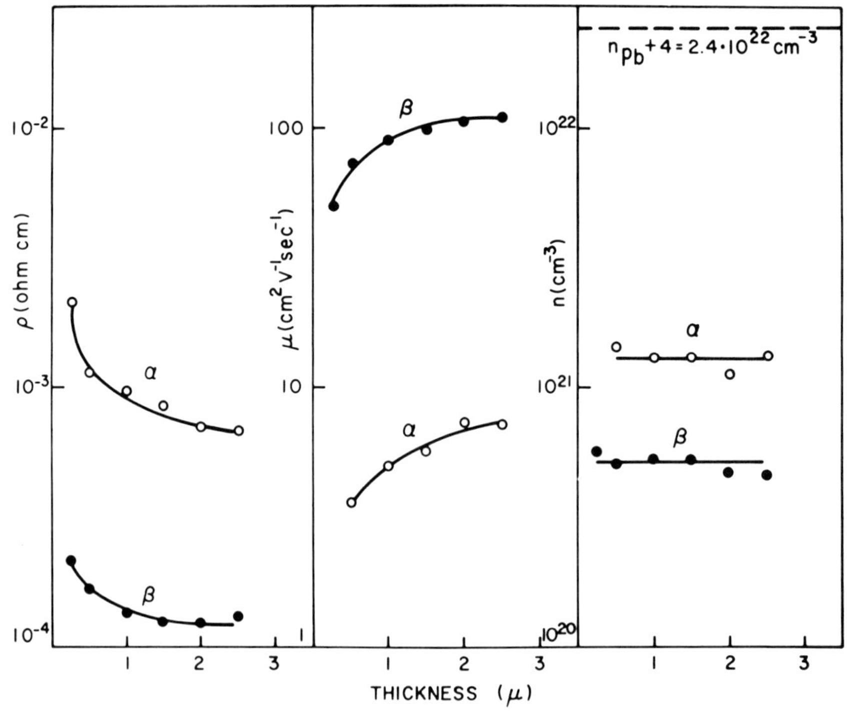

FIG. 4.2.2. Resistivity, electron mobility, and carrier density of α- and β-PbO_2 vs film thickness [201]. (Courtesy of Electrochemical Society.)

TABLE 4.2.1. PbO_2 Modification Obtained upon Anodization and Subsequent Cycling of Lead in Sulfate Electrolytes (anodic current density = 1.5 mA/cm^2; cathodic current density = 0.75 mA/cm^2) [213]

Electrolyte	pH	Electrode treatment	PbO_2 modification
2 M Na_2SO_4	7.6–8.0	Anodic only	100% α
		Anodic + 5 charge-discharge cycles	100% α
2 M $KHSO_4$	0.7	Anodic only	100% α
		Anodic + 5 charge-discharge cycles	30% α and 70% β to 60% α and 40% β
2 M H_2SO_4	<0	Anodic only	~50% α, ~50% β
		Anodic + 3 charge-discharge cycles	<10% α, >90% β

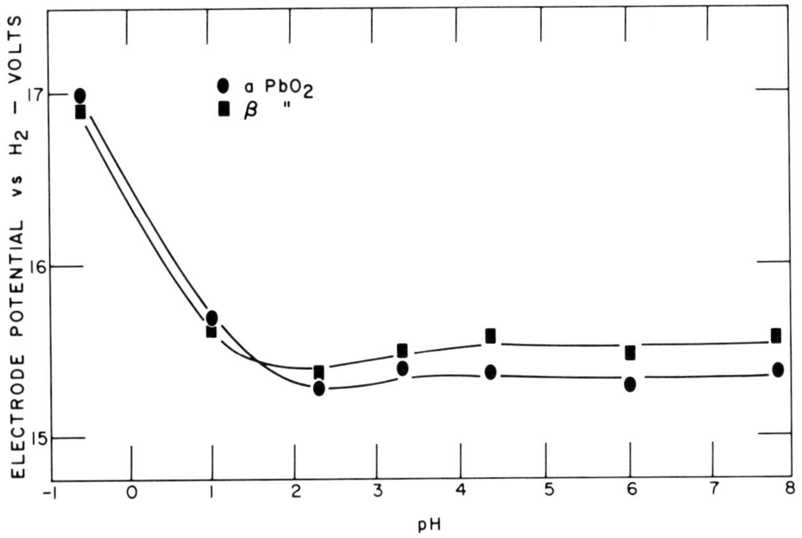

FIG. 4.2.3. Electrode potentials of α- and β-PbO_2 as a function of pH [42]. (Courtesy of Electrochemical Society.)

Rüetschi and co-workers [43] explained the pH effect based on the measured potentials of electrodeposited α- and β-PbO_2 (Fig. 4.2.3). In strong acid solution, β-PbO_2 exhibits the lower potential (hence a lower free energy) and would form preferentially. But at a pH above 1.5 the α-PbO_2 assumes the lower potential and would be the favored structure.

Bagshaw and co-workers [203] proposed a mechanism based on both pH and anion effects. At low pH, where the hydroxyl ion concentration is low, there is a greater probability that acid anions, rather than OH groups, would be involved in the coordination sphere around the Pb^{4+} ions. If the acid anions are large, steric considerations will force the chains of octahedra to nucleate in the β-structure. At higher pH, involvement of hydroxyl ions in the coordination sphere is favored, so that the α-PbO_2 would join in the more compact position.

As indicated above, the pure modifications of PbO_2 can be obtained from solutions containing a wide variety of anions. Therefore an influence of anions on the modification obtained would appear questionable. However, the recent observations by Mindt [201] are of interest and indicate that factors in addition to pH must be considered. He reported that a considerable amount of α-PbO_2 appears in electrodeposits obtained even from strong acid (pH < 0), and that pure β-PbO_2 is obtained from acid solution only within carefully controlled limits of current density and Pb^{2+} ion concentration. Similar observations have been reported by Duisman et al. [204].

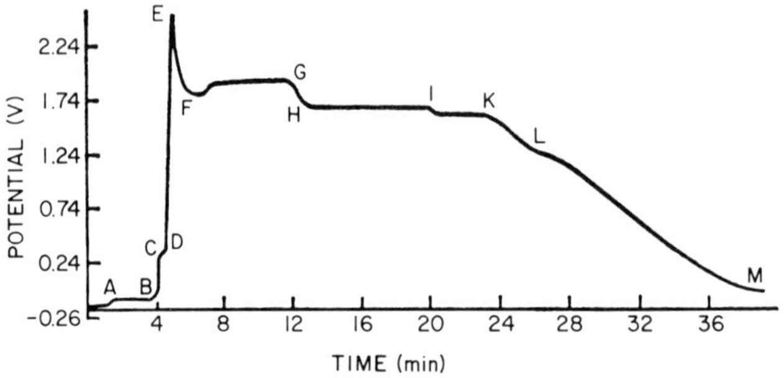

FIG. 4.3.1. Potential-time recording for constant current charge (A to G) and discharge (H to M) for lead in H_2SO_4 [218]. (Courtesy of National Research Council of Canada.)

4.3. CHARGING CURVES

4.3.1. H_2SO_4

The segment from A to G in Fig. 4.3.1 [218] represents a typical curve for the constant current (1 mA/cm^2) charging of lead in H_2SO_4. When the current is switched on, the potential rises to A and the metallic lead becomes converted to $PbSO_4$ along AB. At point B the electrode is passivated, leading to a rapid increase in potential. The time to passivation is longer in more dilute acids, and the longer time corresponds to progressively larger amounts of $PbSO_4$ on the electrode [219].

Metzler and Schwartz [220] found that the length of AB corresponded to 0.46 C/cm^2 in 1.28 g/ml H_2SO_4 and to 3.4 C/cm^2 in 1.12 g/ml H_2SO_4. The difference was attributed to the effect of acid concentration on the size of the $PbSO_4$ crystals produced. The onset of passivation has been indicated by a decrease in the double layer capacitance along AB, from initial values of 40-50 μf/cm^2 [221] down to 1-3 μf/cm^2 at B [199, 221, 222]. Rüetschi and Cahan [216] found that passivation occurred when a definite film thickness (about 640 Å) was reached. At passivation, film resistances of 0.1-1 Ω-cm have been measured [199, 222].

Pavlov suggested [199] that passivation occurs in the following manner. During growth of the lead sulfate crystals, the free lead surface area is decreased with a resultant increase in current density. Thus the electrochemical formation of Pb^{2+} ions exceeds the flux of SO_4^{2-} ions moving toward the lead surface. Diffusion and migration effects lead to an alkalization at the metal surface, thereby causing the precipitation of lead oxides and basic lead sulfates at the bases of the intercrystalline spaces. These compounds form a continuous layer with the $PbSO_4$ and the electrode passivates. The potential rises and the anodic deposit changes to a dense heterogeneous layer of $PbSO_4$ crystals held together by basic lead sulfates.

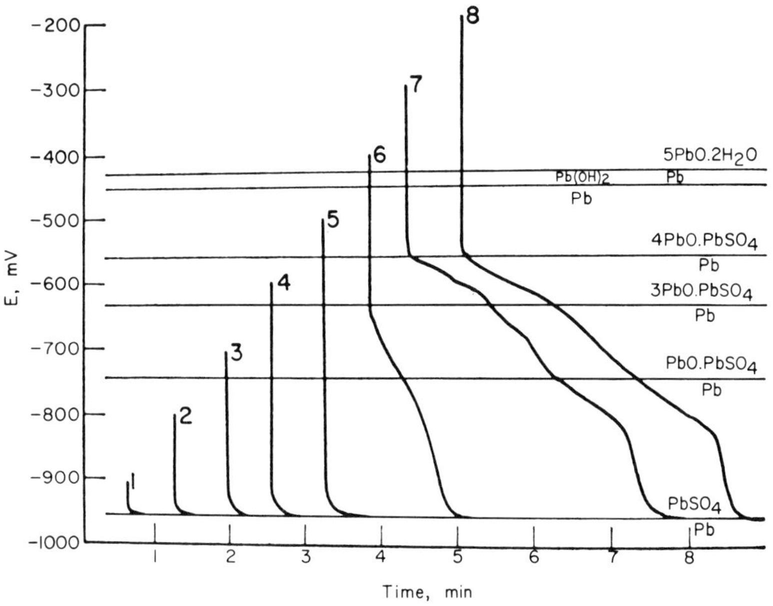

FIG. 4.3.2. Open circuit potential-time arrests for lead in H_2SO_4 after constant current polarizations to various potentials. Reference electrode was Hg/Hg_2SO_4 [200]. (Courtesy of Pergamon Press.)

Although Pavlov's passivation model suggests the presence of basic lead sulfates at high cathodic potentials, X-ray diffraction analysis [194, 195] as well as potential-time transients after polarizations to successively increasing potentials (Fig. 4.3.2) [200] indicate that $PbSO_4$ is the only material present in the anodic layer below 100 mV. However, the X-ray diffraction measurements would not be sensitive to minute quantities of basic sulfates that may be present in the corrosion film. Also, Pavlov's model was based, in part, on double-layer capacitance measurements which may be more sensitive to the presence of basic lead sulfates than the current-time transients.

In Fig. 4.3.2 the potential-time transients following polarizations to potentials between 200 and 400 mV have been attributed to basic lead sulfates in the anodic film.

An inflection CD was evident in the charging curve (Fig. 4.3.1) at current densities below about 0.5 mA/cm^2. The length of the inflection was dependent on H_2SO_4 concentration and appeared to reach a maximum in 10.5 N H_2SO_4. The inflection occurred at 300 $\pm$ 30 mV and seems to be related to the self-passivation potential (Fig. 1.1.2).

An inflection at close to 1 V has been observed in charging curves [216, 221], and it has been attributed [216] to the formation of PbO_2. Based on X-ray diffraction and chemical analysis [194, 195], the incipient formation of PbO_2 occurs at about this potential.

Following the overshoot at E (Fig. 4.3.1), the passivating film breaks down and the potential falls to the charging plateau FG. The overshoot has been discussed by Fleischmann and Thirsk [192] in terms of a nucleation overvoltage, and has also been attributed to the formation of nonconducting lower oxides of lead [223]. Breakdown of the passivating film is reflected in a rapid rise in the double-layer capacity after the overshoot is reached [221]. From an approximate calculation by Rüetschi and Cahan [216], the electric field in the passivating film at the breakdown point is 5×10^4 V/cm.

An inflection F is observed at the beginning of the charging plateau during which the $PbSO_4$ produced on charge is oxidized to PbO_2 [218, 219]. After the inflection the steady-state processes of oxygen evolution and oxidation of the metallic lead to PbO_2 are initiated. The corrosion products after anodization to the charging plateau have been identified by X-ray diffraction. The outer layer is primarily β-PbO_2, but the layer adjacent to the metal is primarily α-PbO_2 with traces of β-PbO_2 [224] and tetragonal PbO [195]. The relative amounts of these materials depend upon factors such as anodization time, current density, temperature, acid strength, and lead composition [195, 216, 224, 225].

Results similar to the above are reported [19] when lead is anodized in the presence of other anions that react to form highly passivating films. The passivating effect of borate and carbonate ion has recently been studied by Khairy et al. [226].

The anodic behavior of lead and its alloys in chloride media has been exhaustively reviewed by von Fraunhofer [227]. At sufficiently high current densities and within a range of chloride ion concentration of about 0.01 to 1.3 M [228], the electrode is passivated by a highly-resistive $PbCl_2$ film which is subsequently oxidized to PbO_2. Chlorine evolution on PbO_2 occurs simultaneously with oxygen evolution at high anodic potentials.

4.3.2. KOH

Charging curves for lead in KOH appear in Fig. 4.3.3 [229]. The first plateau corresponds to the formation of a layer composed exclusively of tetragonal PbO [230]. From ac impedance measurements Farr and Hampson [146] determined that the slow process for the overall reaction was removal of an atom from the lattice to form a mobile adatom

$$Pb_{kink} = Pb_{ad} \tag{4.3.1}$$

with subsequent diffusion before charge transfer

$$Pb_{ad} + 2OH^- \rightarrow PbO + H_2O \tag{4.3.2}$$

Contrary to the highly passivating $PbSO_4$ film, the PbO is loosely held to the surface and is appreciably soluble in the electrolyte. Conditions which favor the transport of PbO away from the electrode will therefore increase the time to passivation. This explains the longer time to passivation obtained with vertical rather than with horizontal electrodes

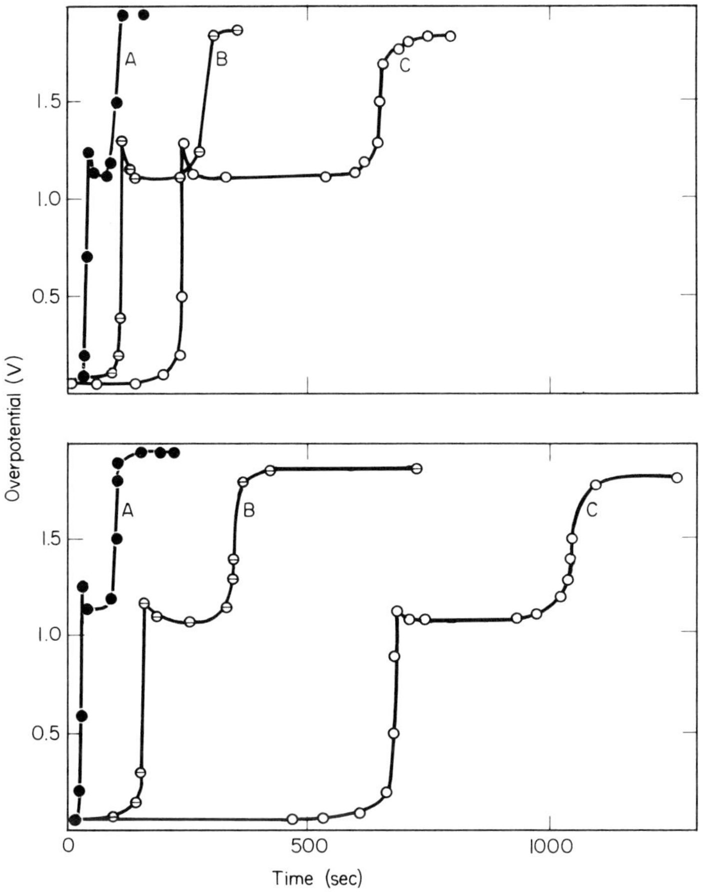

FIG. 4.3.3. Typical potential-time curves for lead anodes in 1.0 M KOH at 20°C. Upper curves, horizontal anodes; lower curves, vertical anodes. (A) 45 mA/cm^2; (B) 24 mA/cm^2; (C) 15 mA/cm^2 [229]. (Courtesy of Electrochemical Society.)

(Fig. 4.3.3). Also, the time to passivation reaches a maximum at concentrations of KOH between 2 and 4 M, where the solubility of PbO is also at a maximum.

According to the early results of Jones et al. [219], a current density of about 25 mA/cm^2 was required to passivate the lead in 1 M KOH at 25°C, and below this current density the lead simply dissolved as plumbite ion. Conversely, Farr and Hampson [229] found that the lead could be passivated at lower current densities provided that sufficient time is allotted. According to Popova et al. [231], the applied current (i), passivation time (t_p), and the critical current required for passivation (i_p) obeys the relationship

$$(i - i_p)(t_p)^{1/2} = \text{const} \qquad (4.3.3)$$

over a wide range of temperatures and concentrations.

At the second plateau in Fig. 4.3.3, the primary reaction is the oxidation of the PbO that has remained on the electrode to a film of α-PbO_2

$$PbO + 2OH^- \rightarrow \alpha\text{-}PbO_2 + H_2O + 2e^- \qquad (4.3.4)$$

along with some oxygen evolution. Orthorhombic and tetragonal PbO is also detected in the film [229]. Oxygen is evolved on the α-PbO_2 surface at the third plateau.

4.4. OXYGEN OVERVOLTAGE ON PbO_2

Oxygen evolution on PbO_2 has been explained in terms of a generalized theory of overvoltage [232]. In acid solution the rate-determining step appears to be the buildup of an adsorbed negatively charged oxygen species [42, 233], the exact nature of which is unknown. From kinetic data for PbO_2 in battery plates, Rüetschi and co-workers [234] suggested that more than one species may be involved, but they calculated that the average charge on the species is one electron per atom.

In H_2SO_4 the overvoltage obeys the Tafel equation except in the vicinity of the PbO_2/$PbSO_4$ potential, due to possible SO_4^{-2} adsorption effects [13]. Tafel slopes of 0.14 for β-PbO_2 and 0.07 for α-PbO_2 have been reported [42]. The different Tafel slopes for α- and β-PbO_2 indicate that the mechanism may be different for the two oxides. The slope for α-PbO_2 is typical of a large number of oxides in alkaline solution [232], whereas that for β-PbO_2 is exhibited by most electrodes in acid solution. The difference in the lattice spacing for the two oxides is offered [40] as an explanation for the difference in behavior.

An important factor in the procedure [42] for obtaining the Tafel slopes was to always maintain the electrodes above the PbO_2/$PbSO_4$ potential to avoid the formation of $PbSO_4$ on the electrode. Failure to do so would have changed the surface area upon subsequent anodization. This precaution was especially important in determining the Tafel slope for α-PbO_2, since the presence of $PbSO_4$ would lead to the formation of a surface layer of β-PbO_2.

Others [234] have reported Tafel slopes different from the above, but they seem not to have taken measures to avoid $PbSO_4$ formation.

The decay of the oxygen overvoltage is logarithmic with time, and the slope of the voltage-time plot is the negative of the Tafel slope [42]. When the overvoltage is forced down to the open circuit value by applying a reverse current, the voltage recovers to values substantially above the open circuit value when the current is interrupted. This effect was attributed to a strong adsorption of the oxygen species.

The Tafel equation is also obeyed in alkaline solution, except at low current densities where appreciable self-discharge takes place [235]. Adsorption of OH^- is probably the rate-determining step [229, 236].

4.5. PbO_2 DISCHARGE STUDIES

4.5.1. Self-Discharge

The discharge process will depend, of course, on the nature and amount of PbO_2 produced during charging. Concomitant with the electrochemical discharge is the process of self-discharge, which becomes a measurable part of the total reaction at low discharge rates. Anodic coatings of PbO_2 on lead undergo local action with the underlying lead according to

$$Pb + PbO_2 + 2H_2SO_4 \rightarrow 2PbSO_4 + 2H_2O \qquad (4.5.1)$$

Burbank [33] observed differences in the self-discharge behavior of the Pb/PbO_2 electrode for various lead alloys and correlated the results to microstructure and porosity effects. The rate of self-discharge also varies significantly for PbO_2 coatings on different crystal faces [216] and reaches a maximum in 2 N H_2SO_4, where the solubility of $PbSO_4$ also shows a maximum [218]. The self-discharge of PbO_2 in the lead-acid battery is markedly affected by impurities, as discussed by Vinal [190].

4.5.2. Discharge Curves in H_2SO_4

The segment from G to M in Fig. 4.3.1 represents a typical constant current discharge curve for the Pb/PbO_2 electrode in H_2SO_4. When the anodizing current is interrupted and the discharge is started, the potential drops from the oxygen overvoltage value G to the discharge plateau HK. Sublevels in the discharge plateau, represented by HI and IK in Fig. 4.3.1, have been noted. The length of the sublevels varies with alloy composition, and the material corresponding to HI has a significantly higher self-discharge rate, therefore appearing to be in greater contact with the underlying metal [40]. Rüetschi and Cahan suggested [40] that sublevels HI and IK corresponded, respectively, to the discharge of α- and β-PbO_2.

Differences in discharge overvoltage and coulombic capacity have been observed when studying electrodes of pure α- and β-PbO_2, as well as known mixtures thereof [33, 215, 237-242]. Continuous and interrupted discharge curves for electrodeposited α- and β-PbO_2 are given in Fig. 4.5.1 [241]. When the discharge current is interrupted, only the β-PbO_2 recovers to the original open circuit potential, but if the sulfate layer is chemically removed from the α-PbO_2 after partial discharge, the potential returns to the open circuit value [40].

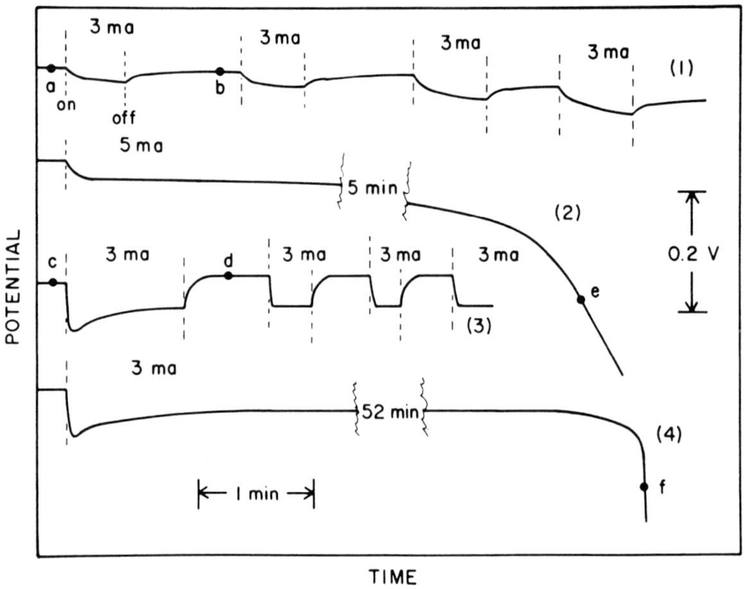

FIG. 4.5.1. Discharge characteristics of α-PbO_2 (Curves 1 and 2) and β-PbO_2 (Curves 3 and 4) in 0.1 M H_2SO_4. Apparent electrode area = 8 cm^2 [241]. (Courtesy of Electrochemical Society.)

The capacity delivered by the α-PbO_2 (Curve 2, Fig. 4.5.1) was less than 20% of the theoretical value. After discharge (Point e) a tenacious, uniform film of $PbSO_4$ was evident. By chemically removing the film, the electrode was rejuvenated and additional capacity could be obtained. Removal of the film after each successive discharge resulted in the attainment of 80-90% of the theoretical capacity.

When the α-PbO_2 was discharged in $HClO_4$, the theoretical capacity was obtained, a further indication that the discharge capacity of the α-PbO_2 in H_2SO_4 was limited by the film of $PbSO_4$.

The β-PbO_2 electrode represented by Curve 4, Fig. 4.5.1, delivered close to the theoretical capacity and the $PbSO_4$ produced on the discharge appeared as a dispersed crystalline deposit in contrast to the white film observed on the α-PbO_2.

The work of Mark [215, 241] suggests that the difference in discharge characteristics between the two modifications of PbO_2 are most pronounced in dilute (0.1 M) H_2SO_4. Others [243] have shown the effects of acid concentration and current density on the morphology of the $PbSO_4$ formed on the α- and β-PbO_2. In strong (4.4 M) acid no difference could be noted between the $PbSO_4$ formed on the two modifications.

For mixtures of α- and β-PbO_2, the α-PbO_2 is expected to discharge first at low current densities because of its higher open circuit potential in H_2SO_4 [13], but at high current densities the order of discharge could well be reversed because of electrolyte diffusion and surface area effects [13, 242]. In studies by Mark [241] of electrodes having a layer of β-PbO_2 on top of a layer of α-PbO_2 and vice versa, the outer layer reacted first.

The potential minimum exhibited by the discharge curve for β-PbO_2 (Fig. 4.5.1) before the steady-state discharge voltage is reached has also been noted in discharge curves for PbO_2 coatings on lead [40] as well as for lead-acid battery positive plates [243a]. In battery terminology, the effect is known as "Spannungssack" and was studied in detail by Simon [243-249]. His microscopic examinations showed that at the beginning of discharge there is an initial supersaturation preventing the formation of $PbSO_4$, thereby causing the voltage drop. When the supersaturation is relieved by the nucleation of $PbSO_4$, the potential rises to the equilibrium value. Berndt and Voss [250] found that the voltage minimum gradually disappeared when increasing amounts of $BaSO_4$ or $SrSO_4$ (e.g., materials that are isomorphous in $PbSO_4$ which may therefore provide nucleation sites for the $PbSO_4$) were added to the positive plate.

The observation that the voltage minimum is not present in the discharge curve for α-PbO_2 (Fig. 4.5.1) does not support the nucleation theory advanced by Simon. Mark [241, 251] proposed that the minimum is a result of a lattice expansion caused by the formation of lower oxides, and that such an expansion occurs only in the β-modification. Still another theory [40] is that the minimum is caused by oxygen desorption at the beginning of discharge.

4.5.3. Potential Arrests

For the discharge of the Pb/PbO_2 electrode in H_2SO_4, various potential arrests have been reported [24, 32, 33, 200] prior to the attainment of the $Pb/PbSO_4$ potential. Although Rüetschi and Angstadt [32] ascribed these arrests to the behavior of PbO_2 in various pH environments, the widely-accepted view is that the arrests are caused by mixed potentials between basic divalent compounds and metallic lead. Arrests reported by Burbank [24] were attributed to the presence of PbO, PbO-$PbSO_4$, and $Pb(OH)_2$. For PbO_2 coatings on Pb-Sb alloy, an arrest corresponding to the potential of the Sb electrode in H_2SO_4 was also noted.

4.5.4. Coulombic Capacity

The coulombic capacity for the discharge may be substantially less than the theoretical value, and substantial unreacted PbO_2, whose amount seems to be independent of the discharge rate [216, 218], is detected at the end of discharge. The end of discharge for the PbO_2 electrode on pure lead is associated with a covering over of the surface by $PbSO_4$, but for the discharge of PbO_2 on Pb-Sb alloy the mechanism appears to be different in that

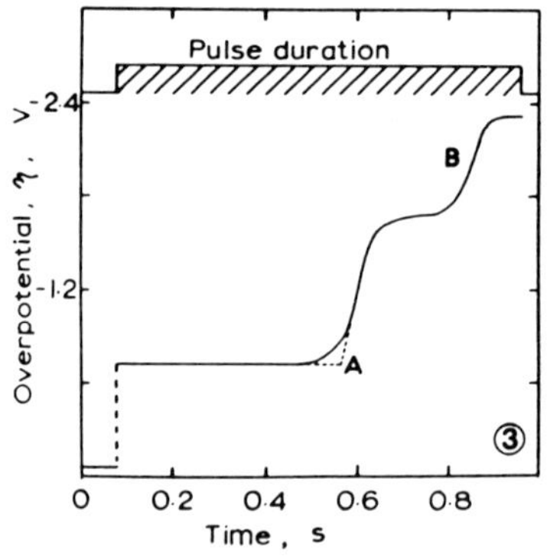

FIG. 4.5.2. Discharge of β-PbO_2 in $HClO_4$. j = 0.76 A/cm^2 [155]. (Courtesy of National Research Council of Canada.)

the $PbSO_4$ does not completely cover the PbO_2 at the end of discharge [252]. As suggested by Burbank [252], the termination of discharge in this case may result from an "exhaustion or covering over of the active centers," or the crystal growth rate itself may limit the discharge. Crystal growth rate as a limiting factor is supported by the observation that additional discharge capacity is obtained from electrodes on Pb-Sb alloy at reduced current densities.

4.5.5. Discharge in $HClO_4$

On cathodic polarization, PbO_2 dissolves freely in $HClO_4$:

$$PbO_2 + 4H^+ + 2e^- \rightarrow Pb^{2+} + 2H_2O \qquad (4.5.2)$$

until passivation occurs (Fig. 4.5.2, Point A) [155] and the potential falls rapidly, indicating transition to

$$Pb^{2+} + 2e^- \rightarrow Pb \qquad (4.5.3)$$

If the polarization is interrupted before passivation and the electrode allowed to remain on open circuit to allow cathodic products to leave the electrode surface, the subsequent passivation time is not reduced by the initial polarization. Hence the electrode remains free of substantial films before initial passivation sets in. For a given passivation time (t_p) and current density (j), (Fig. 4.5.3), the relationship

$$(j - j_i)t_p = k_p \qquad (4.5.4)$$

is obeyed, where j_i is the limiting current density below which passivation does not occur, and k_p is a constant which is characteristic of the system. Both j_i and k_p increase with increase in $[H^+]$.

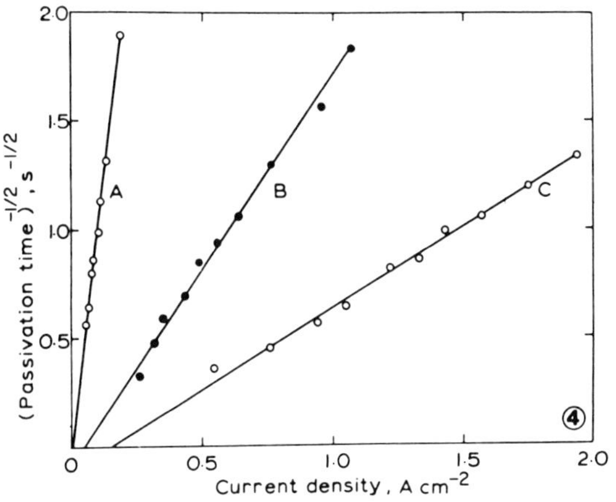

FIG. 4.5.3. Relationship between current density and passivation time for discharge of β-PbO_2 electrodes in $HClO_4$ [155]. (A) 0.09 M Pb^{2+} + 0.5 M H^+; (B) 1.4 M Pb^{2+} + 3.0 M H^+; (C) 0.1 M Pb^{2+} + 6.65 M H^+. (Courtesy of National Research Council of Canada.)

The above relationship was observed for both α- and β-PbO_2, even though the double-layer capacitance (obtained from the slope of the initial part of the transient) indicated a substantially greater true surface area for the α-PbO_2 [156, 158]. However, as pointed out by Hampson et al. [57], the capacitance is obtained at the beginning of the pulse when the diffusion layer is thin. But on the time scale of the passivation experiments (Fig. 4.5.2) the diffusion layer is thick and would not reflect the true surface area of the electrodes.

At the second transition (Fig. 4.5.2, Point B) a difference between the two polymorphs was noted [158]. Metallic lead was identified on the surface of the β-PbO_2, but on the α-PbO_2 no lead, and only Pb^{2+}, was identified. Absence of Pb in the latter case was attributed to the possibly greater oxidizing power of the α-PbO_2 promoting the reaction

$$Pb^{4+} + Pb \rightarrow 2Pb^{2+} \qquad (4.5.5)$$

and converting metallic lead to Pb^{2+} as it forms on the electrode.

4.6. FUNDAMENTAL CORROSION STUDIES

4.6.1. Anodic Corrosion

The effect of potential on the corrosion of pure lead in H_2SO_4 from weight loss measurements is shown in Fig. 4.6.1 [193]. Similar curves are obtained from measurements of the steady currents after 24 hr at the various constant potentials [32]. The corrosion increases with increasing temperature [193] and with decreases in acid concentration [32, 193]. Lander showed [193] that at potentials below the peak in Fig. 4.6.1, the corrosion occurs

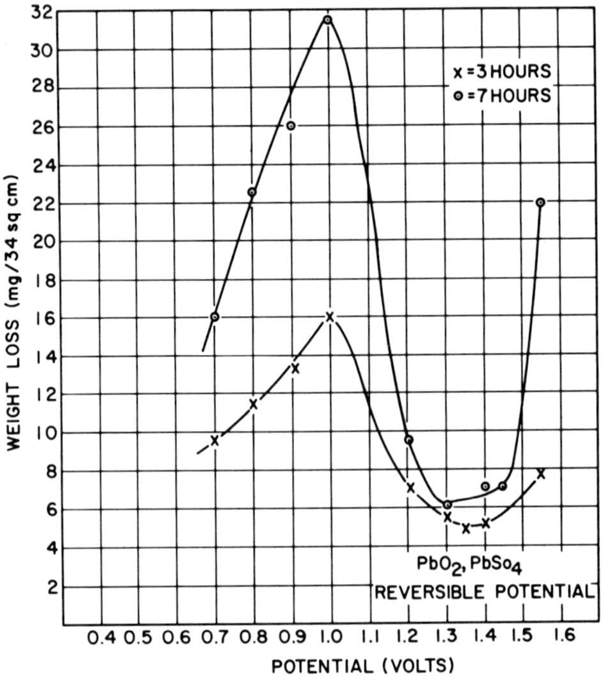

FIG. 4.6.1. Total attack of Pb anodes after 3 and 7 hr exposure in 40% H_2SO_4 at 30°C. Potential is referred to Hg/Hg_2SO_4 electrode in 40.8% H_2SO_4 at 20°C [193]. (Courtesy of Electrochemical Society.)

by a divalent mechanism, but that above the peak a tetravalent corrosion mechanism becomes predominant. Just above the $PbO_2/PbSO_4$ potential the corrosion rates fall off very rapidly because a protective film of β-PbO_2 is built up. At more elevated potentials where oxygen is evolved, the corrosion again increases by the conversion of metal to α-PbO_2 underneath the β-PbO_2 layer. As indicated by Burbank [33], the rate of corrosion of the underlying lead will depend on the rate of penetration of an oxygen species through the β-PbO_2 lattice and therefore depends on the effective oxygen concentration at the surface. In this potential region the rate of corrosion has been correlated to oxygen overvoltage measurements [40, 216].

The anodic corrosion of Pb is related to its microstructure and usually proceeds by grain boundary attack, or a combination of grain boundary and interdendritic penetration [230]. Generally, the finer the grain size, the lower the rate of corrosion [253, 254].

The corrosion of Pb-Sb alloys, as well as dilute Pb-Ca alloys, is characterized by deep penetration in the intergranular and interdendritic areas [33]. At constant current, corrosion of the Pb-Sb alloy passes through a minimum at 1% Sb and then increases slightly as the Sb concentration is increased [40]. At constant voltage, however, the destructive effect of

Sb on the corrosion rate is more noticeable. Samples containing 11% Sb corrode at a potential about 40 mV less positive than the potential of pure lead at the same current density and after the same time period.

For ternary alloys (e.g., Pb-Ca-Sn) the anodic attack is primarily in the areas of segregated ternary eutectic material [33]. In general the Sn binary and ternary alloys have lower corrosion rates than the alloys without Sn. The beneficial effect of Sn in this respect has long been recognized [255, 256].

During the constant current corrosion of Pb-Ag and Pb-Co binary alloys, dissolution of the alloying element occurs [257]. At the PbO_2 surface the resultant Co^{2+} or Ag^{+} ions are then oxidized to Co^{3+} or Ag^{2+}, which in turn oxidize H_2O to H_2O_2. The H_2O_2 is then catalytically decomposed in the acid solution to form O_2 and H_2O. The effect, then, is that the Co or Ag provides an alternate path for oxygen evolution, thereby reducing the oxygen overvoltage. For this reason these additives reduce the corrosion rate but are effective only at very high anodic potentials.

The corrosion of Pb-Li alloys has been of recent interest [258]. Chemically, Li resembles Ca in the free state, whereas it resembles antimony metallurgically. Small additions of Li substantially reduce the anodic corrosion rate, but for concentrations higher than 0.03% the corrosion is accelerated by pronounced intergranular attack. Examination of the corrosion layer for a 0.03% Li alloy indicated a predominance of α-PbO_2 with minimum quantities of free lead and β-PbO_2. The lattice parameter for the β-PbO_2 was different than that found on pure Pb, thereby suggesting that selective partitioning of Li occurs between the α- and β-PbO_2.

4.6.2. Static Corrosion

Under static conditions, the corrosion rate of lead in aqueous media will depend markedly on the solubility of the relevant lead salt and the physical nature of the corresponding protective film (Section 5.4).

4.7. STUDIES IN ORGANIC ELECTROLYTES

Significant contributions in this area have been confined to the paper by Rao [259] who was concerned with the operation of the lead electrode in rechargeable cells. During charge in propylene carbonate containing 1 M $LiAlCl_4$, the lead electrode displays periodic oscillation followed by passivation, as shown in Fig. 4.7.1. At passivation a layer of $PbCl_2$ was identified on the electrode surface, and the voltage fluctuations were explained in terms of breakdown and re-formation of the $PbCl_2$ film. Despite the voltage oscillations, the faradaic

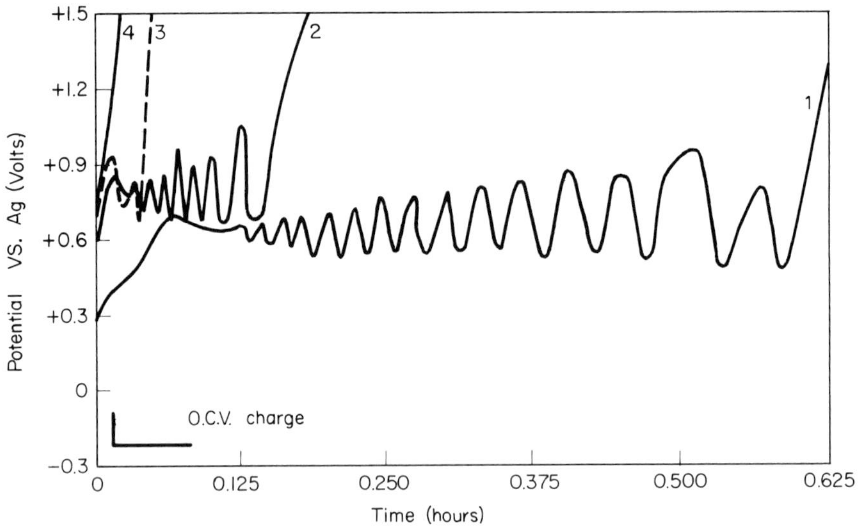

FIG. 4.7.1. Anodic charging of chemically etched lead in 1 M $LiAlCl_4$ in propylene carbonate. Curves 1, 2, 3 and 4 are for 2.5, 5, 7.5, and 10 mA/cm^2, respectively [259]. (Courtesy of Electrochemical Society.)

efficiency of subsequent discharge was 100% up to a current density of 2.5 mA/cm^2. It was also noted that the high iR drop observed during anodization was not evident during the discharge, probably because of the formation of a conducting lead matrix throughout the $PbCl_2$ layer.

The high discharge efficiency in this electrolyte could also be demonstrated with electrodes made from a paste of $PbCl_2$. These electrodes delivered 90% discharge efficiency based on the weight of $PbCl_2$ in the electrode.

The oscillations observed on charge were of concern from a practical standpoint, but it was found that the problem could be alleviated with the use of amalgamated lead electrodes. These electrodes neither oscillated nor passivated on charge, and the discharge efficiency was greater than 90%.

Also studied was α-butyrolactone containing 0.5 M LiCl, but the suitability of lead electrodes could not be demonstrated in this electrolyte. Although the lead electrode neither oscillated nor passivated up to anodic currents of 10 mA/cm^2, faradaic efficiency on subsequent discharge was less than 10%. The behavior was attributed to the dissolution of the divalent lead formed on anodization through the formation of a soluble $PbCl_4^{2-}$ complex.

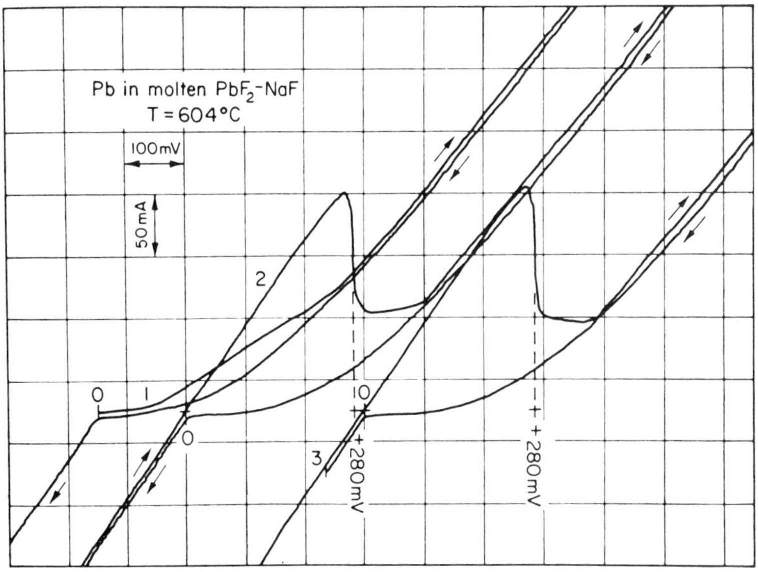

FIG. 4.8.1. Current-voltage plots for the potentiostatic anodic polarization of lead electrodes in PbF_2NaF (33/67 in moles) at 604°C. Bath contaminated with lead oxides. Electrode area = 0.85 cm^2. Sweep rate = 100 mV/min [262]. (Courtesy of Pergamon Press.)

4.8. STUDIES IN FUSED SALTS

Overvoltages for the lead electrode in various halide melts have been reported [151, 260, 261]. In the pure systems both the anodic and the cathodic overvoltages appear to be simply ohmic, but deviations from this behavior have been observed when various constituents have been added to the melt. In potentiostatic studies of Pb electrodes in PbF_2-NaF melts, Pizzini et al. [262] observed that traces of lead oxide added to the melt passivated the electrode during anodic polarization. At sufficiently noble potentials the curves exhibited a typical active-passive transition, as shown in Curves 2 and 3 of Fig. 4.8.1. The passive condition was attributed to a covering of the electrode surface with an oxide layer, and at the passive potentials the electrode was regarded as a $Pb/PbO/Pb_3O_4/O^{2-}$ half cell. Assuming the passivated potential to be controlled by

$$3PbO + O^{2-} = 2e^- + Pb_3O_4 \tag{4.8.1}$$

and the potential of the employed Pb/Pb ion reference electrode to be determined by

$$PbO + 2e^- = Pb + O^{2-} \tag{4.8.2}$$

Pizzini et al. proposed the net reaction

$$4PbO = Pb + Pb_3O_4 \tag{4.8.3}$$

The calculated standard potential for this reaction (0.181 V at 600°C) agreed well with the value of 0.176 V observed after correction for the ohmic potential drop.

Kitazawa et al. [260] studied the effect of additions of Pb^{2+} ion on the behavior of the lead electrode in a LiCl-KCl eutectic melt. Anodically, the lead dissolved as Pb^{2+} with an activation energy of 10-14 kcal/mole. The experimental results could be correlated with diffusion theory except at high Pb^{2+} ion concentrations where a large ohmic drop obscured precise measurements of the concentration overvoltage. When the electrode was polarized cathodically, a maximum current, proportional to the concentration of Pb^{2+} in the melt, appeared. The results could not be correlated to electrokinetic theory and the authors offered no explanation for the anomolous behavior.

5. APPLIED ELECTROCHEMISTRY

5.1. ELECTROREFINING

Electrorefining of lead is not done extensively today because of improved metallurgical refining techniques. Briefly, the process [263] consists of making lead bullion the anode in a bath containing lead fluosilicate and fluosilicic acid and using electrolytic lead in sheet form as the cathode. During electrolysis, a slime or mud forms on the anode which contains most of the impurities. The lead deposit on the cathode is melted, along with the starting sheet, and cast into bars.

Proper control of the process results in the production of spectrographically pure lead. Data on refining operations have been compiled by Mantell [263].

Lead anodes have been used in the electrowinning of numerous metals from H_2SO_4 leach liquors, notably copper, cadmium, cobalt, chromium, manganese, antimony, and zinc [263]. The PbO_2 formed on the anode is instrumental in regenerating H_2SO_4 during the electrolysis.

5.2. ELECTRODEPOSITION

5.2.1. Electroplating of Lead

Electroplating of lead is not extensively used because it is readily applied by hot dipping. It has been pointed out [264], however, that plated coatings are superior to dipped coatings since the deposit is less porous, contains a minimum of stress, and the plated object can be readily stamped and machined without damaging the coating. Plated lead and lead alloys have been useful in coating battery parts, bearing surfaces, strip steel, and in the manufacture

TABLE 5.2.1. Typical Lead Plating Baths [264]

Fluoborate bath		Fluosilicate bath		Sulfamate bath	
Lead (g/l)	120-240	Lead (g/l)	75-180	Lead (g/l)	54-140
Free fluoboric acid (g/l)	30- 60	Total fluosilicate (g/l)	140-150	Free sulfamic acid (g/l)	50-100
Excess boric acid (g/l)	13.3-26.6				
				pH	1.5
Temp (°C)	25- 40	Temp (°C)	35- 40	Temp (°C)	24- 50
Current density (A/ft^2)	5- 70	Current density (A/ft^2)	5- 80	Current density (A/ft^2)	5- 40

of printed circuits [264]. Electroformed lead and lead alloys has been of recent interest [265-268].

5.2.1.1. Baths. Lead has been plated from numerous baths [264, 269], but it is evident from early work [269] that deposits obtained from simple salts, e.g., nitrates or acetates, are structurally unsound. Deposition from complex salts containing various organic addition agents is required for smooth, coherent coatings. The baths commonly in use today are fluosilicate, fluoborate, and sulfamate, and typical compositions of each are given in Table 5.2.1. Operation and maintenance of these baths has been discussed by Du Rose and Blum [264].

The fluoborate bath has been widely used for obtaining electroforms of pure lead [265] as well as lead alloys [264, 266-268]. The excess fluoboric acid promotes a finer grained deposit with less tendency toward dendrite formation [264], and a dilute bath is especially advantageous for obtaining deposits of low hydrogen embrittlement [270]. Lead-tin alloy can be obtained by adding Sn as the fluoborate and/or by use of soluble Sn anodes. An additional quantity of H_3BO_3 is recommended to prevent oxidation and hydrolysis of stannous tin [271]. The amount of Sn in the deposit is, of course, proportional to the amount in the bath, and deposits ranging from 0.1 to 63% Sn have been obtained [267]. Lead-antimony deposits are similarly obtained, but for this process a pure lead anode is preferred. For good deposit control, precautions must be taken to minimize immersion deposition of antimony on the anodes.

Although the fluoborate bath is more costly than the fluosilicate bath, the former bath is somewhat more stable, gives finer grained, denser deposits, and can be used for plating lead directly onto steel [264]. The sulfamate bath has been favored to the fluoborate bath for barrel plating operations [272]. The low solubility of metallic impurities in the sulfamate bath permits the attainment of high purity deposits.

5.2.1.2. Additives. Electroplated lead is not suitable for decorative purposes, hence the use of organic additives in the bath is directed toward the attainment of smooth, adherent, nondendritic deposits rather than mirror bright finishes. A list of organic additives that have shown varying degrees of success would be extensive [264, 273], but recommendations have been made pertaining to specific applications. Frey and co-workers [267, 268] found that a mixture of lignin sulfonic acid and coumarin added to the fluoborate bath was instrumental in obtaining satisfactory Pb and Pb-Sn deposits up to 3.2 mm thick. Dini and Helms recommended a mixture of peptone, resorcinal, and gelatin for obtaining pure lead electroforms [265], and β-naphthol for obtaining Pb-Sb deposits [266]. A combination of glue and β-naphthol has been selected as an additive for the sulfamate bath [272].

5.2.1.3. Mechanical Properties. The tensile strength of electroformed lead is 2100 lb/in.^{2} [265] which compares favorably with the figure for chemical lead. A lead deposit containing 0.1% Sn has a yield strength of 5000 lb/in.^{2} and a tensile strength of 6000 lb/in.^{2}. Since lead deforms under constant load at ambient conditions, the creep strength is the most important design parameter for structural applications. Frey and co-workers [267] found that dispersion strengthening of electroforms could be effected by codeposition of TiO_2 particles. A deposit consisting of Pb-0.1% Sn-0.5% TiO_2 exhibited a steady state creep rate of 2.0×10^{-4}/hr at 300 lb/in.^{2}. Strengthening of lead electroforms has also been correlated to a buildup of carbon in the electroform, resulting from breakdown of the organic additives.

5.2.2. Lead Anodes in Electroplating

Lead has been used as an insoluble anode for the electroplating of other metals, e.g., cadmium and manganese. Perhaps the greatest use is in chromium plating [274], where the lead is frequently alloyed with antimony or tin. The PbO_2 coating on the anode provides the means for regenerating hexavalent chromium from the trivalent chromium produced at the cathode. A problem encountered with these anodes is that a resistive film of lead chromate is formed on the anode surface during idle periods. This necessitates removing the anodes when the bath is not in use, or "dummying" the bath to restore the anode to a satisfactory condition. Despite a number of studies [275-277], the mechanism controlling the film formation is not well understood, but the severity of the problem seems to depend entirely on the type of catalyst used in the plating bath [275].

5.3. ELECTROSYNTHESIS

Because of its relatively high conductivity and high oxygen overvoltage, PbO_2 has served as anode material for a number of electrosyntheses. Anode preparation for this purpose involves the electrodeposition of PbO_2, usually onto Ta, C, or Ni cores. PbO_2 may be deposited from numerous solutions, but a nitrate bath of composition $Pb(NO_3)_2$ (250-300 g/l) and $Cu(NO_3)_2 \cdot H_2O$ (1.5-4 g/l) has been favored [278-280] for commercial applications. The copper salt suppresses lead deposition on the cathode, which is usually graphite or copper. At current densities of 15-30 A/ft^2, strong, dense deposits up to a thickness of 2.5 cm or more have been obtained. Various addition agents for the bath have been recommended [280] to improve deposit strength.

The PbO_2 anode has been proposed as an insoluble anode for the electrolysis of aqueous solutions containing such anions as Cl^-, Br^-, I^-, F^-, ClO_3^-, ClO_4^-, SO_4^{2-}, NO_3^-, CO_3^{2-}, and $C_2H_3O_2^-$ [281]. A current efficiency of 100% is obtained for the oxidation of iodic to periodic acid at the PbO_2 anode [281]. The anode has been used as replacement for costly platinum anodes in the electrolytic production of perchlorates [278, 282, 283], where cumulative anode current efficiencies of 90% are reported [283]. The PbO_2 anode has been used in preference to graphite anodes for electrolytic production of bromates [279], where anode losses from corrosion were on the order of only 2% of their original weight after 1 year's service. PbO_2 anodes exhibit good stability in the electrolytic production of chlorine [284], where the low overvoltage of chlorine on PbO_2 is an advantage.

The high hydrogen overvoltage on lead makes it a desirable cathode for the reduction of a wide variety of organic compounds [285]. A recent advancement has been the use of a "wicking" lead cathode [286] to improve mass transfer of low-conducting reactants to reaction sites. Such cathodes are effective in obtaining large yields of phosphine by the electrolytic reduction of phosphorus suspended in aqueous electrolyte.

Lead cathodes have been suggested for the commercial reduction of nitrobenzene [287-289] to aniline and other reduction products, as well as for the electrochemical grafting of methyl methacrylate onto rayon fibers [290].

5.4. CORROSION

The anodic corrosion of lead, especially as it applies to the lead-acid battery, has been treated in other sections. This section is concerned with a brief discussion of the use of lead in cathodic protection systems, as well as the corrosion characteristics of lead in the absence of an externally applied current, e.g., the resistance of lead to various chemicals. Also, the use of lead as a protective sheath for underground cables warrants a brief note on underground corrosion.

5.4.1. Lead Anodes for Cathodic Protection

Lead anodes have been used extensively for the cathodic protection of numerous ocean-going vessels and offshore structures. The effectiveness of these anodes depends to a marked extent on the stability of the PbO_2 coating. At high current densities (overvoltages) the PbO_2 thickens markedly in seawater [228] and the volume difference between Pb and PbO_2 causes considerable expansive stress. This leads to blisters which rupture, giving corrosion of exposed lead and subsequent formation of $PbCl_2$. The nonconductive $PbCl_2$ causes a further increase in the overvoltage, leading to further rupture of the PbO_2 layer.

It is well-known, primarily from the work of Shrier et al. [228, 291-296], that insertion of platinum or other noble metal microelectrodes into the lead before polarization greatly improves the stability and electrochemical performance of the operating anode. Although the design and construction of such bielectrodes will vary with intended use, a typical bielectrode [296] might consist of a 1-1.5 in. diameter extruded bar of lead or lead alloy into which the microelectrodes (small platinum wires containing a small percentage of alloying element to insure hardness) are inserted at 6-12 in. intervals. The coating of PbO_2 is electrochemically formed after the bielectrode has been incorporated into a suitably designed cathodic protection scheme [297]. The nature of the lead is critical in controlling the growth and spalling of the PbO_2. Lead containing 0.1 or 1% Ag is beneficial in this respect, but alloys based on Sb, Bi, and Sn additions, as well as dispersion-strengthened lead, are detrimental [296].

Despite numerous basic studies [292, 294, 295, 298, 299] on these bielectrodes, the complete mechanism associated with their operation is not entirely clear, although it has been shown that the Pt microelectrode is instrumental during the initial electrochemical reactions involving passivation of the lead surface by $PbCl_2$ [298] and promotes conversion of the $PbCl_2$ to PbO_2 [294].

The Pb-Pt bielectrode can be used at current densities as high as 100 A/ft^2 for prolonged periods in flowing seawater [293, 295, 300]. Although lead alloy anodes that do not contain noble metal microelectrodes have been used in cathodic protection systems, they are generally restricted to current densities of 10-15 A/ft^2 for prolonged operations [295, 300]. The use of a Pb-Ag alloy has been reported [301, 302], although the PbO_2 is soft and easily removed [301], leading to high anode consumption rates. The PbO_2 formed on Pb-Sb alloy is hard and adherent, but unusually high current densities (~20 A/ft^2) are required to form the PbO_2 [302]. Perhaps the best compromise is a ternary alloy Pb/1-4% Ag/6% Sb, which is reported to have good corrosion resistance [303] and can be operated from 0.25 to 25 A/ft^2 depending mainly on seawater resistivity [110, 303]. This alloy has behaved favorably compared to carbon anodes in cathodic protection installations [302, 304].

Seawater resistivity and composition can markedly affect the formation and operation of the anodes. Generally, a higher current density is required to form and maintain the

PbO_2 as the electrolyte resistivity increases [303]. The presence of sulfate ions in seawater is beneficial to anode performance [295, 298, 303].

Further details concerning the use of lead anodes for cathodic protection are available in the excellent review by von Fraunhofer [227].

5.4.2. Resistance to Chemicals

In general, lead is very stable in media where insoluble lead salts can form. This is exemplified in the use of lead for the storage and handling of H_2SO_4 [20, 269, 305] having a wide range of temperatures and concentrations. The low solubility of $PbSO_4$ leads to the formation of protective film, and corrosion resistance is generally improved by any method that will quicken film formation, e.g., by inclusions that are cathodic to lead [266] or by impressing external current [306]. Dissolved oxygen in the acid may also be beneficial in this respect [266], and it is of interest to note that the corrosion magnitude, in addition to showing proportionality to $PbSO_4$ solubility, has also been directly correlated to the solubility of oxygen in the H_2SO_4 [269]. At substantial flow rates, corrosion is enhanced [307], most likely due to film erosion. Static and dynamic stresses may crack the $PbSO_4$ film, leading to increased corrosion [308, 309].

In boiling sulfuric acid the good corrosion resistance of commercial lead is unquestioned below concentrations of 30% H_2SO_4, but erratic results have been noted at higher concentrations. According to Hohlstein [310], a 50% solution is more corrosive than 30 or 70% acid. The corrosion rate of lead in boiling H_2SO_4 has been correlated to the measured electrode potential [306]. At -0.36 V the weight loss amounted to 5,000-15,000 g/m^2 per day, whereas at potentials between -0.2 and +0.2 V the loss dropped to about 1% of this value. In general, impurities in the lead play a smaller part at room temperature than at higher temperature [269]. Perhaps the worst offender is bismuth, which causes poor corrosion resistance even at moderate temperatures.

When structural durability is a primary consideration, Pb-Sb alloy is often used to handle H_2SO_4, and considerable work has been done to compare the corrosion resistance of this alloy with that of other alloys and pure lead [269, 311, 312]. In some studies [269], Pb-Sb has fared worse than pure lead, especially at high temperatures, although Weissbach [312] has observed good corrosion resistance of Pb 3-6% Sb alloy in 20-60% H_2SO_4 at temperatures up to boiling.

Other mineral acids to which lead is highly, but not completely, resistant are sulfurous, phosphoric, and hydrofluoric (up to 65% at room temperature) [269]. Lead is used in handling chromic acid and as a liner for chromium plating tanks. The behavior of lead toward hydrochloric acid is uncertain but, in general, storage of HCl by lead is restricted to low concentrations and temperatures. Corrosion in this medium is accelerated by noble inclusions in the lead and by excess oxygen.

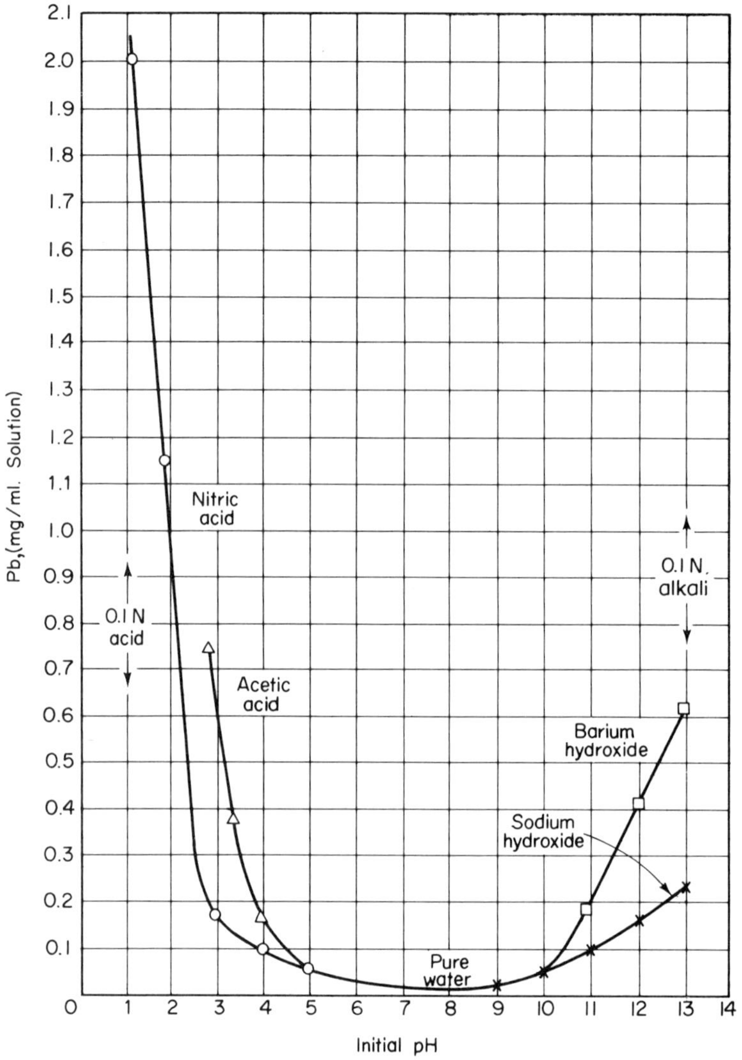

FIG. 5.4.1. Corrosion of lead as a function of pH in various solutions [20]. (Courtesy of John Wiley & Sons.)

As shown in Fig. 5.4.1, lead is attacked by nitric acid and dilute acetic acid, although attack in glacial acetic acid is slight [269]. Corrosion of lead by organic acids has been correlated to the solubility of the corresponding lead salts and to oxygen availability [20, 269]. Corrosion is related to the formation of soluble plumbates at high pH (Fig. 5.4.1).

5.4.3. Underground Corrosion

Numerous long-term corrosion tests [27, 313-317] have indicated that soil conditions affect corrosion to a greater extent than does the nature of the lead, although slightly greater attack on antimonial lead has been observed. Clay is generally more harmful than sandy soils [20, 314] and a combination of wet clay and cinders is disastrous [314]. The carbon in the cinders becomes cathodic to the portion of the pipe in the clay and the lead suffers severe anodic corrosion. Stray current corrosion resulting from nearby electrical installations can also be a problem, and serious corrosion is exhibited by lead pipes used to ground alternating currents [314]. Contact with other metals, such as support straps, can also accelerate corrosion [4].

When corrosion of lead is under anodic control, the principal corrosion product is tetragonal PbO [4] interspersed with numerous other divalent lead compounds [314]. The corrosion may also be under cathodic control, which is favored in alkaline soil [4] and in the presence of certain organic decomposition products [315]. Here, the principal corrosion product is Pb_3O_4. Pitting corrosion has also been noted [318], but this is not a serious problem.

In general, the constituents of soil that promote corrosion of lead are nitrate and chloride ions, alkalies and organic acids, whereas those that inhibit corrosion are silicates, carbonates, certain colloids, and certain organic compounds [314]. An effective inhibitor is sulfate ion, and the presence of sulfate-reducing bacteria promotes the doubly destructive effect of breaking down the sulfate layer and forming corrosive sulfides [319].

With increased aeration of the soil, corrosion may diminish through the formation of complex films, but on the other hand, the oxygen may accelerate corrosion by depolarizing cathodic sites or creating oxygen concentration cells [318].

5.5. THE LEAD-ACID BATTERY

A recent survey [320] indicates that the lead-acid battery accounts for almost 40% of the yearly consumption of lead in the United States. The lead-acid system is over 100 years old, and historical details have been adequately treated by Vinal [190]. An extensive review of the lead-acid cell has recently been compiled by Burbank, Simon, and Willihnganz [252].

Lead-acid batteries are marketed in a wide variety of sizes and capacities, ranging from small packages up to units of several tons [263]. Basically, however, the main types of secondary batteries can be categorized into the following: (a) starting, lighting, and ignition (SLI) batteries, designed for the cranking of internal combustion engines; (b) industrial batteries, which are generally of thick plate construction for heavy duty applications; and (c) reserve, or standby power batteries, which are maintained in a fully charged state by passing through them a small charging or "float" current until ready for use. Performance and design data for batteries of the first two categories are given by Mantell [263]. A current issue of the Bell System Technical Journal [321] is devoted to details concerning reserve batteries.

5.5.1. Plates

The electrodes, or plates, for the lead-acid battery may be of two types: Planté plates, which are made by electrochemically charging and discharging Pb sheet in H_2SO_4 electrolyte to build up coulombic capacity, and Faure plates (so-called "pasted plates"), which are made by electrochemically converting a paste of divalent lead compounds to active material. The latter are almost exclusively used today because of their substantially greater ampere-hour capacity. Both methods of preparation, however, ultimately give PbO_2 as active material for the positive plate and sponge lead for the negative plate, and the electrochemical process controlling the charge and discharge (Table 5.5.1) is the double sulfate reaction proposed by Gladstone and Tribe [189].

5.5.2. Paste

The paste is often comprised of litharge containing about 25% free lead and dilute sulfuric acid of a quantity to achieve suitable consistency. In addition, the paste for the negative plates contains minute quantities of materials known as expanders, whose function is discussed later. The paste is applied to Pb or Pb alloy grids that provide support for the material as well as electrical conductivity. Following the pasting operation, no additional treatment is required for the negative plates; however, the positive plates are subjected to a "curing" treatment, during which chemical reactions occur that are instrumental in the production of structurally sound plates. The chemistry involved in the pasting and curing of plates is extremely involved and has been discussed by Pierson [323], Barnes et al. [31], as well as Armstrong et al. [34]. Further details concerning paste composition are also available in Vinal [190].

The plates, along with microporous separators, are then assembled into cell elements and undergo the process known as formation, whereby the plates that are to become the negatives are cathodized to sponge lead and those that are destined for the positives are anodized to PbO_2. The plates are usually formed in dilute H_2SO_4, and during the early stages of formation the interior of the plate remains alkaline. Therefore, much of the paste is converted to active material below the potentials required for the decomposition of water, a factor which obviously contributes to the efficiency of the formation process [31]. The formation of both negatives and positives is initiated at the grid wires, and conversion of Pb in the negative plate takes place by dissolution of lead sulfate followed

TABLE 5.5.1. Reactions and Thermodynamic Data for the Lead-Acid Battery [322]

Negative electrode	$PbSO_4 + H^+ + 2e^-$	$\rightleftharpoons$	$Pb + HSO_4^-$
Positive electrode	$PbSO_4 + 2H_2O$	$\rightleftharpoons$	$PbO_2 + 3H^+ + HSO_4^- + 2e^-$
Cell reaction	$2PbSO_4 + 2H_2O$	$\rightleftharpoons$	$Pb + PbO_2 + 2H^+ + 2HSO_4^-$
Free energy of formation	ΔG (kcal) (T = 25°C) 2 X (-193.89) 2 X (-56.69)		0 -52.34 0 2 X (-179.94)
Entropy (standard)	S (cal/deg mole) 2 X 35.2 2 X 16.716		15.51 18.3 0 2 X 30.32
Change of free energy	$\Delta\Delta G$ = 88.84 kcal		Standard conditions (α_{HSO_4} = 1 mole/l)
Change of entropy	ΔS = -9.382 cal/deg mole		

$$\text{Standard cell voltage } E_0 = \frac{\Delta\Delta G}{nF} = \frac{88.84 \times 10^3 \times 4.186}{2 \times 96520} = 1.928 \text{ V}$$

$$\text{Temperature coefficient } \frac{dE}{dT} = -\frac{\Delta S}{nF} = \frac{9.382 \times 4.186}{2 \times 96520} = 0.203 \text{ mV/deg}$$

by reduction to lead metal [244]. Conversion to PbO_2 in the positive plate appears to take place by a solid-state reaction, and the kinetics of the oxidation may vary for different compounds and crystals in the paste [244, 324].

The positive plate contains α- and β-PbO_2 in varying amounts depending on the materials and manufacture. Positive plates containing α-PbO_2 in amounts ranging from traces to 70% have been reported [325], but the origin of the α-PbO_2 is not clearly defined. Studies by Dodson [325] indicate that PbO in the paste is responsible for the major portion of the α-PbO_2. Other suggested precursors to the formation of α-PbO_2 are the presence of metallic lead and tetrabasic $PbSO_4$ in the paste, as well as the presence of antimony in the grid metal [202]. The formation of α-PbO_2 in the plate is favored by a dense paste and when the temperature and specific gravity of the forming acid are high [325]. The high pH prevailing in the interior of the plate throughout much of the formation time creates a condition conducive to the formation of α-PbO_2. This is in agreement with the observation [326] that the α-PbO_2 is confined to the interior of the plate.

5.5.3. Grids

Grid composition for pasted plates may vary depending on the intended use of the battery, but grids are cast of an alloy of Pb and Sb, containing from 4-12% antimony, for many applications. The purpose of the antimony is to improve the castability and stiffness of the grid since pure Pb is difficult to cast and is easily deformed. Antimony is detrimental, however, in that it lowers the hydrogen overvoltage, thereby reducing the rechargeability and increasing the self-discharge of the negative plate. The harmful effects of antimony have been known for years [25] and the mechanism of antimony transfer in Pb-acid cells was recently studied in detail by Dawson et al. [327]. Pentavalent antimony ions are formed by the dissolution of antimony from the positive grid. The pentavalent antimony migrates through and around the separator where it becomes deposited onto the negative as Sb metal via the formation of a trivalent antimony species. Although corrosion of the negative grid during overcharging also releases trivalent antimony to the electrolyte by the formation of stibine, Herrmann et al. [328] showed that, even after extensive overcharging, at least 90% of the antimony deposited onto the negative plate originated from the positive grid.

The adverse effects of antimony have led to the use of various nonantimonial alloys, notably lead-calcium alloys. These alloys have proved to be an important replacement for lead-antimony alloys in batteries used for emergency telephone and submarine applications. However, their use for other applications has been limited because of their poor cycling characteristics.

5.5.4. Voltage

The open circuit voltage of the lead acid cell depends upon the acid concentration and temperature, as shown in Ref. 190, pp. 192 and 194. A fully-charged cell usually contains 1.26 sp gr (35%) H_2SO_4, and the corresponding voltage is about 2.10 V at 25°C. Reactions and thermodynamic data are summarized in Table 5.5.1.

Voltages observed for the positive and negative plates during typical constant current discharges at 23°C are given in Fig. 5.5.1 [329]. The rapid change of plate potential at the end of discharge is due to the formation of nonconducting $PbSO_4$, as well as changes in concentration of the acid at the electrode surface. As a result of the decrease in H_2SO_4 concentration during the discharge, a measurement of specific gravity of the electrolyte provides a convenient method for determining the level of charge of the system. A potential of -0.25 V for the negative plate and 1.60 V for the positive plate is considered as the end of useful discharge at all currents. The discharge duration of a cell will be limited, of course, by the plate having the lower capacity.

On recharge the voltage rises, usually within minutes, to about 2.1 V, then increases slightly until the charge is about three-quarters complete. Then there is a sharp rise in

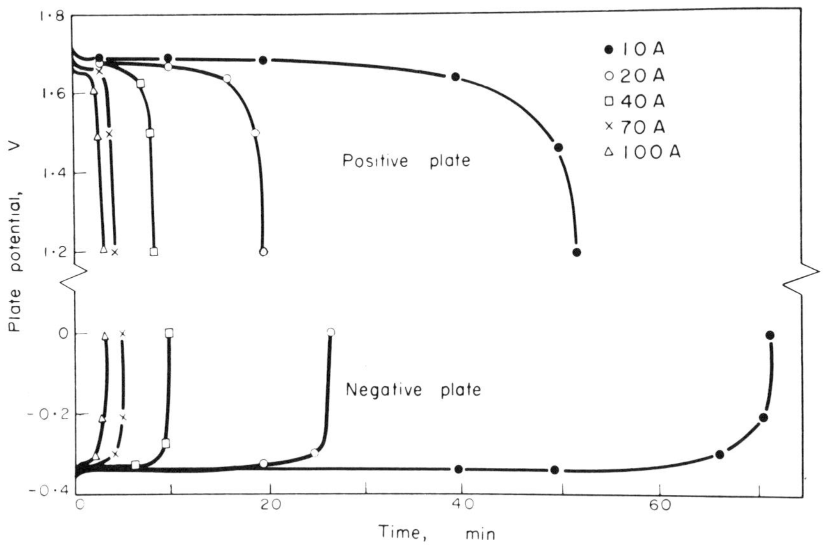

FIG. 5.5.1. Constant current discharge curves for lead-acid battery plates [329]. (Courtesy of Pergamon Press.)

voltage to a maximum of 2.6 V. Charge acceptance of the lead-acid cell at various charging rates and temperatures is given by Peters et al. [330].

5.5.5. Capacity and Cycle Life

Cell capacity will be determined by numerous factors, such as properties of the plate, amount and concentration of H_2SO_4, cell temperature, and discharge rate [190, 322]. The capacity of the cell generally increases with initial cycling up to a value that remains essentially constant throughout the life of the battery. A falloff to 80% of the original capacity usually signifies the end of battery life.

5.5.5.1. Negative Plate. Figure 5.1.1 depicts conditions where the positive plate limits the discharge, but it is well known that at low temperatures and at high discharge rates the capacity is severely limited by the negative plate. Loss in capacity has been related to encapsulation of the lead by sulfate layers of considerable thickness. Incorporation of materials known commercially as expanders into the plate substantially improves the low temperature performance [331, 332], although the ampere-hour capacity of the negative plate at low temperature is still only about 10% of the theoretical value. The expanders commonly used are a mixture of carbon black, $BaSO_4$, and various lignin derivatives.

Pierson and co-workers [333] showed that lignin modifies the lead crystal structure during the forming operation as well as the structure of the $PbSO_4$ crystals produced on discharge. During charge, lignin suppresses hydrogen evolution and promotes the transfer of Pb^{2+} through the crystal lattice of $PbSO_4$ [334]. Although $BaSO_4$ does not alter the microstructure of the formed plate [333], it does affect the size of the $PbSO_4$ produced on discharge [335]. It is thought that the $BaSO_4$ reduces passivation of the lead during discharge by providing alternate sites for the nucleation of $PbSO_4$.

There seems to be little doubt that the lignin acts by some sort of adsorption mechanism. Ultraviolet spectroscopy has been used to measure the adsorption of lignosulfonate onto lead surfaces [336]. The effectiveness of organic materials as expanders has been correlated to their effect on double-layer capacitance and cyclic voltammetric curves for smooth lead electrodes [337].

5.5.5.2. Positive Plate. The discharge capacity of the positive plate is usually 10-50% of the theoretical value of 0.224 A-hr/g PbO_2 [190]. Capacity limitations have been related to such factors as crystallite size and surface, area of PbO_2 [242, 338], modification of PbO_2 (α or β) [239, 242, 339], acid starvation effects [190, 242], pore size distribution [242], grid alloy [190, 340], and passivation by $PbSO_4$ [242, 247]. With repeated cycling the positive plate may fail from a combination of grid corrosion and shedding of active material.

Rüetschi and co-workers [40, 216] used cathodic stripping techniques to measure the amount of anodic corrosion of Pb and Pb alloys under conditions of oxygen evolution and reported [40] good agreement between the data obtained after 20-hr tests and the overcharge life of batteries containing positive plate grids of the various alloys. Nevertheless, it should be noted that short-term anodic corrosion tests may not always give the information required for the judicious selection of grid material [341]. For many alloys, corrosion begins as a uniform process [246] and the aforementioned (Section 4.6) grain boundary or other preferential attack may not be significant until the grids have been in service for a number of years. Local attack often manifests itself in grid growth, whereby the dimensions of the grid become enlarged beyond the normal size. An additional contributing factor may be the stresses induced by the PbO_2 corrosion product as a result of its substantially greater specific volume than that of the Pb from which it was formed [246]. The destructive effect of grid growth is demonstrated in Fig. 5.5.2 [342]. The plate has increased 10% in width, causing the PbO_2 pellets to crack and to lose contact with the grid members.

Cannone et al. [341] demonstrated that for batteries in float service the growth of both pure Pb and Pb-Ca grids are quadratically time-dependent, but because of its larger grain size (fewer corrosion sites per unit surface area), the former grow at a slower rate. The growth rate of Pb-Ca alloys critically depends on calcium content and increases markedly at calcium concentrations above 0.1% [343]. Excess calcium exists as large, local aggregates of Pb_3Ca, which seem to provide local sites for anodic attack [341, 343].

FIG. 5.5.2. Positive plate at end of life [342]. (Courtesy of Electrochemical Society.)

During overcharge, the growth of Pb-Sb grids is linearly time dependent, and is therefore substantially greater than nonantimonial grids [341]. The growth of Pb-Sb grids is undoubtedly enhanced by numerous surface imperfections exhibited by the alloy. As demonstrated by Simon [246], segregated antimony clusters become dislodged from the surface even before the paste begins its conversion to PbO_2, resulting in voids in the protective layer. Loss of antimony from the grid also influences the specific volume difference between the PbO_2 corrosion product and the grid metal.

It has been suggested that stresses leading to grid growth may develop from the weight of active material on the grid as well as from the volume change encountered by the PbO_2 active material during cycling. However, detailed microscopic examinations by Simon [246], as well as earlier studies by Thomas et al. [343], indicate that this is not the case.

Integrity of the PbO_2 has been related to its microstructure which, in turn, depend on factors such as formation conditions, and the composition and morphology exhibited by the original paste mix [244, 245, 324, 325, 340, 344-348]. As indicated by Burbank and Ritchie [340], the particle size of the PbO_2 increases with cycling, apparently as a result of grain growth. Particle size increase is reflected in a decrease in the bulk density of the PbO_2 [349] which serves to weaken the active material mass. Studies by Bode et al. [350] show that after prolonged cycling the large particles of PbO_2 become electrically isolated and are no longer discharged.

The influence of α- and β-PbO_2 content on positive plate life is not well understood, although studies by Simon [245] have indicated that α-PbO_2 having a "reticular" network may improve the durability of the active material. Other work [344] has shown that positive plate life may be enhanced by the presence of elongated prisms of β-PbO_2 and that a precursor to these prisms is the presence of $PbSO_4.4PbO$ in the paste.

It is well known that the use of antimonial grids delays the degradation of positive plate active material. As indicated by radioactive tracer studies [328] and by spectrographic data [340], a substantial portion of the antimony that is leached from the grid becomes associated with the PbO_2 in the positive plate. During cycling, the rate of density loss of the active material is less for antimonial grids [340], apparently because antimony retards the crystal growth rate. It has been suggested [340] that antimony enters the PbO_2 lattice by substitutional solid solution, thereby increasing the number of nucleation centers on the surface of the PbO_2. An alternate mechanism is that antimony is adsorbed preferentially onto the fastest growing crystal faces to act as a growth inhibitor.

5.5.6. Other Systems

Modifications of the secondary lead-acid battery have been cells based on perchloric and fluoboric acid electrolytes [351-354] for fuse applications. Lead dioxide has been used as a cathode in conjunction with zinc and cadmium anodes as a basis for high voltage primary systems [355]. Bipolar designs for the lead-acid system have recently been implemented [356-359].

REFERENCES

1. A. J. De Bethune and N. S. Loud, "Table of Standard Aqueous Electrode Potentials and Temperature Coefficients," in Encyclopedia of Electrochemistry (C. A. Hampel, ed.), Reinhold, New York, 1964.

2. W. R. Carmody, J. Amer. Chem. Soc., 51, 2905 (1929).
3. H. Helber and E. L. Littauer, Corros. Sci., 10, 413 (1970).
4. E. F. Wolf and C. F. Bonilla, Trans. Electrochem. Soc., 79, 307 (1941).
5. G. Butler, P. E. Francis, and A. S. McKie, Corros. Sci., 9, 715 (1969).
6. I. S. Shklovskii, Tr. Inst. Fiz. Khim, Akad. Nauk, SSSR, 1, 276 (1951); Chem. Abstr., 47, 6278 (1953).
7. S. A. Awad and Z. A. Elhady, J. Electroanal. Chem., 20, 79 (1969).
8. J. Elze and G. Oelsner, Metalloberflaeche, 12, 129 (1958).
9. M. Fleischmann and M. Liler, Trans. Faraday Soc., 54, 1370 (1958).
10. P. Delahay, M. Pourbaix, and P. Van Rysselberghe, J. Electrochem. Soc., 98, 57 (1951).
11. H. S. Harned and W. J. Hamer, J. Amer. Chem. Soc., 59, 33 (1935).
12. J. Shrawder and I. Cowperthwaite, ibid., 56, 2340 (1934).
13. P. Rüetschi, J. Sklarchuk, and R. T. Angstadt, Electrochim. Acta, 8, 333 (1963).
14. S. J. Bone, K. P. Singh, and W. F. K. Wynne-Jones, ibid., 4, 292 (1961).
15. R. H. Gerke, J. Amer. Chem. Soc., 44, 1684 (1922).
16. R. G. Bates, M. Edelstein, and S. F. Airee, J. Res. Nat. Bur. Stand., 36, 159 (1946); Chem. Abstr., 40, 3987 (1946).
17. M. Randall, Trans. Faraday Soc., 23, 498 (1927).
18. D. J. G. Ives and F. R. Smith, ibid., 63, 217 (1967), and references contained therein.
19. E. G. Yampol'skaya, M. I. Ershova, I. I. Astakhov, and B. N. Kabanov, Sov. Electrochem., 2, 1211 (1966).
20. E. L. Littauer and H. C. Wesson, in Corrosion, Vol. 1 (L. L. Shrier, ed.), Wiley, New York, 1963, p. 459.
21. W. E. Henderson and G. Stegman, J. Amer. Chem. Soc., 40, 84 (1918).
22. V. K. La Ner and E. L. Carpenter, J. Phys. Chem., 40, 287 (1936).
23. M. G. Melon and W. E. Henderson, J. Amer. Chem. Soc., 42, 676 (1920).
24. J. Burbank, J. Electrochem. Soc., 103, 87 (1956).
25. H. E. Haring and U. B. Thomas, Trans. Electrochem. Soc., 48, 293 (1935).
26. K. Ekler, Can. J. Chem., 42, 1355 (1964).
27. K. H. Logan and R. H. Taylor, J. Res. Nat. Bur. Stand., 12, 119 (1934).
28. E. J. Casey and K. N. Campney, J. Electrochem. Soc., 102, 219 (1955).
29. R. Ohse, Werkst. Korros., 11, 220 (1960).
30. W. Müller and W. Mach, Korros. Metallschutz, 16, 187 (1940); Chem. Abstr., 35, 3538 (1941).
31. S. C. Barnes and R. T. Mathieson, in Batteries 2 (D. H. Collins, ed.), Pergamon, New York, 1965, p. 41.
32. P. Rüetschi and R. T. Angstadt, J. Electrochem. Soc., 111, 1323 (1964).
33. J. Burbank, ibid., 104, 693 (1957).
34. J. Armstrong, I. Dugdale, and W. J. McCusker, Power Sources 1966 (D. H. Collins, ed.), Pergamon, New York, 1967, p. 163.
35. J. Burbank, J. Electrochem. Soc., 106, 369 (1959).
36. E. M. Khairy, A. A. Abdul Azim, and K. M. El Sobki, J. Electroanal. Chem., 12, 27 (1966).

37. W. J. Hamer, J. Amer. Chem. Soc., 57, 9 (1935).

38. W. C. Vosburgh and D. N. Craig, ibid., 51, 2009 (1929).

39. R. Stoker, Trans. Faraday Soc., 44, 298 (1948).

40. P. Rüetschi and B. D. Cahan, J. Electrochem. Soc., 105, 369 (1958).

41. H. Bode and E. Voss, Z. Elektrochem., 60, 1053 (1956).

42. P. Rüetschi, R. T. Angstadt, and B. D. Cahan, J. Electrochem. Soc., 106, 547 (1959).

43. J. N. Butler, in Advances in Electrochemistry and Electrochemical Engineering, Vol. 7 (P. Delahay, ed.), Wiley, 1970, p. 77.

44. T. Pavlopoulos and H. Strehlow, Z. Phys. Chem., 2, 89 (1954); Chem. Abstr., 49, 77c (1955).

45. J. C. Synnott and J. N. Butler, Anal. Chem., 41, 1890 (1969).

46. V. A. Pleskov, Zh. Fiz. Khim., 22, 351 (1948); Chem. Abstr., 42, 6249h (1948).

47. A. F. Alabyshev, M. F. Lantratov, and A. G. Morachevski, Reference Electrodes for Fused Salts, Sigma Press, Washington, D.C., 1965.

48. W. E. Bennett and W. P. Jensen, J. Inorg. Nucl. Chem., 28, 1829 (1966).

49. S. N. Flengas and T. R. Ingraham, J. Electrochem. Soc., 106, 714 (1959).

50. H. A. Laitinen and C. H. Liu, J. Amer. Chem. Soc., 80, 1015 (1958).

51. J. H. Hildebrand and G. C. Ruhle, ibid., 49, 722 (1927).

52. I. G. Murgulescu, S. Sternberg, and M. Teryi, Electrochim. Acta, 12, 1121 (1967).

53. H. C. Gaur and H. L. Jindal, ibid., 13, 837 (1968).

54. H. C. Gaur and W. K. Behl, Proceedings of the 1st Australian Conference on Electrochemistry, Pergamon, London, 1964.

55. A. J. Ariva and H. A. Videla, Electrochim. Acta, 11, 537 (1966).

56. R. Marassi, V. Bartocci, Paolo-Cesion, and M. Fiorani, J. Electroanal. Chem., 22, 215 (1969).

57. W. J. Hamer, M. S. Malmberg, and B. Rubin, J. Electrochem. Soc., 103, 8 (1956).

58. H. A. Laitinen, R. Tischer, and D. K. Roe, in Transactions of the Symposium on Electrochemical Processes, Philadelphia, 1949 (E. B. Yeager, ed.), Wiley, New York, 1961.

59. G. A. Mazzochim. G. G. Bombi, and M. Fiorani, Ric. Sci., 36, 338 (1966); Chem. Abstr., 65, 11763e (1966).

60. H. M. Hershenson, M. E. Smith, and D. N. Hume, J. Amer. Chem. Soc., 75, 507 (1953).

61. J. J. Lingane, ibid., 61, 2099 (1939).

62. C. A. Streuli and W. D. Cooke, Anal. Chem., 25, 1691 (1953).

63. I. M. Kolthoff and J. J. Lingane, Polarography, Interscience, New York, 1952.

64. J. K. Taylor and R. E. Smith, Anal. Chem., 22, 495 (1950).

65. A. M. Bond and T. A. O'Donnell, ibid., 41, 1801 (1969).

66. H. P. Raaen, ibid., 37, 1355 (1965).

67. P. W. West, J. F. Dean, and E. J. Breda, Collect. Czech. Chem. Commun., 13, 1 (1948).

68. A. Frisque, V. W. Meloche, and I. Shain, Anal. Chem., 26, 471 (1954).

69. J. J. Lingane and I. M. Kolthoff, J. Amer. Chem. Soc., 61, 825 (1939).

70. J. J. Lingane, Ind. Eng. Chem., 15, 583 (1943).
71. L. Meites and T. Meites, J. Amer. Chem. Soc., 73, 177 (1951).
72. L. Meites, Polarographic Techniques, Wiley, New York, 1965.
73. L. Meites, J. Amer. Chem. Soc., 73, 3724 (1951).
74. W. L. Belew and H. P. Raaen, J. Electroanal. Chem., 8, 475 (1964).
75. J. Heyrovsky and M. Kalousek, Collect. Czech. Chem. Commun., 11, 464 (1939).
76. J. J. Lingane, J. Amer. Chem. Soc., 67, 919 (1945).
77. J. Tomes, Collect. Czech. Chem. Commun., 9, 12 (1937).
78. G. F. Reynolds, H. I. Shalgosky, and T. J. Webber, Anal. Chim. Acta, 8, 564 (1953).
79. H. Chandra and A. Kumar, J. Inst. Chem. (India), 41, 245 (1969).
80. J. O. Hibbits and S. S. Cooper, Anal. Chem., 26, 1119 (1954).
81. J. Heyrovsky and D. Ilkovic, Collect. Czech. Chem. Commun., 7, 198 (1935).
82. J. J. Lingane, Chem. Rev., 29, 1 (1941).
83. T. Meites and L. Meites, J. Amer. Chem. Soc., 73, 1161 (1951).
84. M. A. De Sesa, D. N. Hume, A. C. Glamm, Jr., and D. D. Deford, Anal. Chem., 25, 983 (1953).
85. R. Pribil, Z. Roubal, and E. Svatek, Collect. Czech. Chem. Commun., 18, 43 (1953).
86. D. A. Skoog, Anal. Chem., 25, 1922 (1953).
87. G. B. Jones, Anal. Chim. Acta, 7, 578 (1952).
88. D. S. Jain, A. Kumar, and J. N. Gaur, J. Electroanal. Chem., 17, 201 (1968).
89. I. M. Kolthoff and J. I. Watters, Anal. Chem., 22, 1422 (1950).
90. J. K. Gupta and C. M. Gupta, J. Prakt. Chem., 311, 636 (1969).
91. D. Cozzi, Anal. Chim. Acta, 4, 204 (1950).
92. L. Meites, Anal. Chem., 24, 1374 (1952).
93. J. Musil, J. Dolezal, and J. Vorlicek, J. Electroanal. Chem., 24, 447 (1970).
94. R. Pribil and Z. Zabranski, Collect. Czech. Chem. Commun., 16, 554 (1951-2).
95. P. Souchay and J. Fancherre, Anal. Chim. Acta, 3, 252 (1949).
96. G. Jessop, Nature, 158, 59 (1946).
97. P. Zuman and I. M. Kolthoff, Progress in Polarography, Vols. 1 and 2, Wiley, New York, 1962.
98. S. S. Mesaric and D. N. Hume, Inorg. Chem., 3, 791 (1963).
99. J. N. Gaur, R. C. Mehrotra, M. Palrecha, D. S. Jain, A. Kumar, and C. Gupta, J. Polarog. Soc., 14, 122 (1968).
100. J. N. Gaur and M. Palrecha, J. Inst. Chem. (India), 41, 25 (1969).
101. H. M. Hershenson, R. Thompson, and M. E. Murphy, J. Amer. Chem. Soc., 79, 2046 (1957).
102. A. Iwase, J. Chem. Soc. Japan, 78, 141 (1953).
103. D. S. Jain and J. N. Gaur, Indian J. Chem., 12, 503 (1964).
104. D. S. Jain and J. N. Gaur, J. Indian Chem. Soc., 43, 425 (1966).
105. D. S. Jain and J. N. Gaur, J. Polarogr. Soc., 12, 49 (1966).
106. T. T. Lai and C. C. Hseih, J. Electrochem. Soc., 112, 218 (1965).

107. T. T. Lai and M. Chao, Talanta, 16, 544 (1969).

108. L. N. Popova and A. G. S. Tromberg, Sov. Electrochem., 4, 1033 (1968).

109. D. S. Jain and J. N. Gaur, Acta Acad. Sci. Hung., 51, 165 (1967).

110. G. T. Meoller, J. T. Patrick, and J. W. Caldwell, Mat. Prot., 1, 46 (1962).

111. J. B. Headridge and D. Pletcher, J. Chem. Soc., A, 1966, 757.

112. P. D. Coulter and R. T. Iwamoto, J. Electroanal. Chem., 13, 21 (1967).

113. P. D. Coulter and R. T. Iwamoto, ibid., 13, 28 (1967).

114. J. B. Headridge, M. Ashraf, and H. L. H. Dodds, ibid., 16, 114 (1968).

115. D. L. McMasters, R. B. Dunlap, J. R. Keumpel, L. W. Kreider, and T. R. Shearer, Anal. Chem., 39, 103 (1967).

116. G. B. Bachman and M. J. Astle, J. Amer. Chem. Soc., 64, 1303 (1942).

117. H. S. Swofford, Jr. and C. L. Holifield, Anal. Chem., 37, 1509 (1965).

118. D. Inman, D. G. Lovering, and R. Narayan, Trans. Faraday Soc., 63, 3017 (1967).

119. M. Steinberg and N. H. Nachtrieb, J. Amer. Chem. Soc., 72, 3558 (1950).

120. G. C. Barker and R. L. Faircloth, J. Polarogr. Soc., 1958, 11.

121. M. Francini and S. Martini, Z. Naturforsch., 23, 795 (1968).

122. M. Saito, Nippon Kagaku Zasshi, 83, 883 (1962).

123. R. M. de Fremont et al., in Polarography 1964 (G. J. Hills, ed.), Macmillan, London, 1966.

124. G. G. Bombi et al., Collect. Czech. Chem. Commun., 1966, 455.

125. E. L. Colichman, Anal. Chem., 27, 1559 (1955).

126. J. H. Christie and R. A. Osteryoring, J. Amer. Chem. Soc., 82, 1841 (1960).

127. D. Inman, D. G. Lovering, and R. Narayan, Trans. Faraday Soc., 64, 2476 (1968).

128. J. Braunstein and A. S. Minano, Inorg. Chem., 3, 218 (1964).

129. D. Inman, D. G. Lovering, and R. Narayan, Trans. Faraday Soc., 64, 2487 (1968).

130. M. M. Nicholson and J. H. Karchmer, Anal. Chem., 27, 1095 (1955).

131. A. R. Nisbet and A. J. Bard, J. Electroanal. Chem., 6, 331 (1963).

132. A. G. Stromberg and E. A. Zakharova, Sov. Electrochem., 1, 922 (1965).

133. J. G. Nikelly and W. D. Cooke, Anal. Chem., 29, 933 (1957).

134. M. Kodama and T. Noda, Bull. Chem. Soc. Japan, 42, 2699 (1969).

135. M. Ariel and U. Eisner, J. Electroanal. Chem., 5, 362 (1963).

136. Z. Kublik, Acta Chim. Hung., 271, 79 (1961); Chem. Abstr., 55, 22668f (1961).

137. I. Sinko and J. Dolezal, J. Electroanal. Chem., 25, 299 (1970).

138. G. C. Whitnack, ibid., 2, 110 (1961).

139. W. Kemula, E. Rakowska, and Z. Kublik, ibid., 1, 205 (1959).

140. E. Ladanyi, U. D. N. Radulescu, and M. Gavin, ibid., 24, 91 (1970).

141. E. Hrabankova, J. Dolezal, and P. Beran, ibid., 22, 203 (1969).

142. S. Haruyama, J. Electrochem. Soc. Japan, 35, 62 (1967).

143. W. Lorenz, Z. Phys. Chem., 19, 377 (1959).

144. N. A. Hampson and D. Larkin, Trans. Faraday Soc., 65, 1 (1969).

145. J. E. B. Randles, in Transactions of the Symposium on Electrode Processes, Philadelphia (E. B. Yeager, ed.), Wiley, New York, 1961.

146. J. P. G. Farr and N. A. Hampson, Trans. Faraday Soc., 62, 3502 (1966).

147. T. I. Borishova and B. V. Ershler, Zh. Fiz. Khim., 24, 337 (1950).

148. Y. D. Dunaev, G. Z. Kir'yakov, and Z. N. Cherny-sheva, Tr. In-ta, Khim. Nauk AN Kay SSR, 9, 18 (1962).

149. I. A. Aguf, Sov. Electrochem., 4, 1022 (1968).

150. G. C. Barker, R. L. Faircloth, and A. W. Gardner, Nature, 181, 247 (1958).

151. G. C. Barker, Transactions of the Symposium on Electrode Processes (E. B. Yeager, ed.), Wiley, New York, 1961, p. 325.

152. G. C. Barker, Anal. Chim. Acta, 18, 118 (1958).

153. D. J. Barclay, J. Electroanal. Chem., 28, 71 (1970).

154. J. P. Carr, N. A. Hampson, and R. Taylor, Ber. Bunsenges. Phys. Chem., 74, 557 (1970).

155. N. A. Hampson, P. C. Jones, and R. F. Phillips, Can. J. Chem., 45, 2039 (1967).

156. N. A. Hampson, P. C. Jones, and R. F. Phillips, ibid., 45, 2045 (1967).

157. N. A. Hampson, P. C. Jones, and R. F. Phillips, ibid., 46, 1325 (1968).

158. N. A. Hampson, P. C. Jones, and R. F. Phillips, ibid., 47, 2171 (1969).

159. H. A. Laitinen, R. P. Tisher, and D. K. Roe, J. Electrochem. Soc., 107, 546 (1960).

160. H. Gerischer and M. O. Krause, Z. Phys. Chem., 10, 264 (1947); 14, 184 (1958).

161. A. D. Graves and D. Inman, in E.m.f. Measurements in High Temperature Systems (C. B. Alcock, ed.), Institute of Mining and Metallurgy, London, 1968, p. 183.

162. E. A. Ukshe and N. G. Bukum, Zh. Fiz. Khim., 35, 2689 (1961).

163. K. V. Rybalka and D. I. Leikis, Sov. Electrochem., 3, 332 (1967).

164. A. N. Frumkin, Z. Electrochem., 59, 807 (1955).

165. P. A. Rebinder and E. K. Venstrem, Acta Physiochim. USSR, 19, 36 (1944).

166. P. A. Rebinder and E. K. Venstrem, Dokl. Akad. Nauk SSSR, 68, 329 (1949).

167. F. I. Kukoz, S. A. Semenchenko, and V. I. Bogdanov, Issled. Obl. Khim., 1955, 201.

168. N. V. Nikolaeva, N. S. Shapiro, and A. N. Frumkin, Dokl. Akad. Nauk SSSR, 86, 581 (1952).

169. T. I. Borishova, B. V. Ershler, and A. N. Frumkin, Zh. Fiz. Khim., 22, 925 (1948).

170. K. V. Rybalka and D. I. Leikis, Sov. Electrochem., 3, 1013 (1967).

171. K. V. Rybalka, ibid., 4, 1223 (1968).

172. T. N. Anderson, J. L. Anderson, and H. Eyring, J. Phys. Chem., 73, 3562 (1969).

173. J. P. Carr, N. A. Hampson, and R. Taylor, J. Electroanal. Chem., 27, 109 (1970).

174. I. G. Kiseleva, B. N. Kabanov, and D. Leikis, Dokl. Akad. Nauk SSSR, 49, 805 (1954).

175. G. A. Kokarev, N. C. Bakhchsaraits'yan, and V. V. Panteleeva, Tr. Mosk. Khim. Tekhnol. Inst., 54, 161 (1967).

176. D. I. Leikis and E. K. Venstrem, Proc. Acad. Sci. USSR, Phys. Chem. Sect. Engl. Transl., 112, 7 (1957).

177. B. N. Kabanov, I. G. Kiseleva, and D. I. Leikis, Dokl. Akad. Nauk SSSR, 99, 805 (1954).

178. P. Delahay, Double Layer and Electrode Kinetics, Wiley (Interscience), New York, 1965, Chap. 3.

179. J. P. Carr, N. A. Hampson, and R. Taylor, J. Electroanal. Chem., 27, 201 (1970).

180. J. P. Carr and N. A. Hampson, ibid., 28, 65 (1970).

181. J. P. Carr, N. A. Hampson, and R. Taylor, ibid., 27, 466 (1970).

182. Yu. K. Delimarskii and V. S. Kikhno, Ukr. Khim. Zh., 35, 468 (1969).

183. E. A. Ukshe, N. G. Bukum, and D. I. Leikis, Bull. Acad. Sci. USSR, 25, 1 (1963).

184. E. A. Ukshe, N. G. Bukum, D. I. Leikis, and A. N. Frumkin, Electrochim. Acta, 9, 431 (1964).

185. A. D. Graves and D. Inman, J. Electroanal. Chem., 25, 357 (1970).

186. R. Yu. Bek and A. S. Lifshits, Siberian Chem. J., 6, 713 (1967).

187. A. S. Lifshits and R. Yu. Bek, Izv. Sibirsk, Ser. Khim, Nauk SSSR, 4, 161 (1969).

188. R. H. Heus, T. Tidwell, and J. J. Egan, in Molten Salts: Characterization and Analysis (G. Mamantov, ed.), Dekker, New York, 1969.

189. J. H. Gladstone and A. Tribe, Nature, 27, 583 (1883).

190. G. Vinal, Storage Batteries, 4th ed., Wiley, New York, 1955.

191. W. H. Beck and W. F. K. Wynne-Jones, Trans. Faraday Soc., 50, 136 (1954).

192. M. Fleischmann and H. R. Thirsk, ibid., 51, 71 (1955).

193. J. J. Lander, J. Electrochem. Soc., 98, 213 (1951).

194. D. Pavlov, C. N. Poulieff, E. Klaja, and N. Iordanov, ibid., 116, 316 (1969).

195. D. Pavlov and N. Iordanov, ibid., 117, 1103 (1970).

196. H. W. Billhardt, ibid., 117, 690 (1970).

197. J. J. Lander, Trans. Electrochem. Soc., 95, 174 (1949).

198. ILZRO Research Digest No. 19, International Lead Zinc Research Organization, New York, April 1967.

199. D. Pavlov and E. Popova, Electrochim. Acta, 15, 1483 (1970).

200. D. Pavlov, ibid., 13, 2051 (1968).

201. W. Mindt, J. Electrochem. Soc., 116, 1076 (1968).

202. J. Burbank, N.R.L. Report 6859, Naval Research Laboratory, Washington, D.C.

203. N. E. Bagshaw, R. L. Clarke, and B. Halliwell, J. Appl. Chem. (London), 16, 180 (1966).

204. J. A. Duisman and W. F. Giauque, J. Phys. Chem., 72, 562 (1968).

205. U. B. Thomas, J. Electrochem. Soc., 94, 42 (1948).

206. F. Lappe, J. Phys. Chem. Solids, 23, 1563 (1962).

207. I. P. Shapiro, Opt. Spectrosk., 4, 256 (1958).

208. D. A. Frey and H. E. Weaver, J. Electrochem. Soc., 107, 930 (1960).

209. B. N. Kabanov, E. S. Weisberg, I. Romanova, and E. V. Krivolapova, Electrochim. Acta, 9, 1197 (1964).

210. A. Kittel, Dissertation, Tecknische Hockschule, Prague, Czechoslovakia, 1944. (Data summarized in Ref. 120.)

211. T. Katz and R. LeFarvre, Bull. Soc. Chim. Fr., 16, D124 (1949).

212. I. A. Aguf, A. I. Rusin, and M. A. Dasoyan, Zash. Metal, 1965, 328.

213. R. A. Baker, J. Electrochem. Soc., 109, 337 (1962).

214. J. Giner, A. B. Gancy, and A. C. Makrides, "Preparation and Characterization of Lead Dioxide Electrodes for Reserve Batteries," Report No. 265, Harry Diamond Laboratories, 1967.

215. H. B. Mark and W. C. Vosburgh, J. Electrochem. Soc., 108, 615 (1961).

216. P. Rüetschi and B. D. Cahan, ibid., 104, 406 (1957).

217. I. G. Kiseleva and B. N. Kabanov, Dokl. Akad. Nauk SSSR, 108, 864 (1956); 122, 1042 (1958).

218. E. Tarter and K. Ekler, Can. J. Chem., 47, 2191 (1969).

219. P. Jones, R. Lind, and W. F. K. Wynne-Jones, Trans. Faraday Soc., 50, 972 (1954).

220. H. Metzler and W. Schwarz, Electrochim. Acta, 11, 111 (1966).

221. B. D. Cahan and P. Rüetschi, J. Electrochem. Soc., 106, 543 (1959).

222. B. N. Kabanov and D. Leikis, Z. Electrochem., 62, 660 (1958).

223. S. Ikari and S. Yoshizawa, Denki Kagaku., 28, 675 (1960); Chem. Abstr., 62, 2504 (1965).

224. J. Burbank, Power Sources 1970, Preprint Paper 2.

225. L. M. Levinzon, I. A. Aguf, and M. A. Dasoyan, J. Appl. Chem. USSR, 39, 556 (1966).

226. E. M. Khairy, A. A. Abdul Azim, and K. M. El-Sobki, J. Electroanal. Chem., 11, 282 (1966).

227. J. A. von Fraunhofer, Anti-Corrosion, 15, 4, 9 (1968).

228. E. L. Littauer and L. L. Shrier, Electrochim. Acta, 11, 527 (1966).

229. J. P. G. Farr and N. A. Hampson, Electrochem. Technol., 6, 10 (1968).

230. J. Burbank and A. C. Simon, J. Electrochem. Soc., 100, 11 (1953).

231. S. S. Popova and A. V. Fortunatov, Sov. Electrochem., 2, 413 (1966).

232. P. Rüetschi and P. Delahay, J. Chem. Phys., 32, 556 (1955).

233. P. Rüetschi, J. B. Ockerman, and R. Amlie, J. Electrochem. Soc., 107, 325 (1960).

234. J. E. Puzey and R. Taylor, in Batteries 2 (D. H. Collins, ed.), Pergamon, New York, 1965, p. 29.

235. P. Jones, H. R. Thirsk, and W. F. K. Wynne-Jones, Trans. Faraday Soc., 52, 1003 (1956).

236. S. S. Popova and A. V. Fortunatov, Sov. Electrochem., 4, 444 (1968).

237. W. H. Beck, P. Jones, and W. F. K. Wynne-Jones, Trans. Faraday Soc., 50, 1249 (1950).

238. W. H. Beck, R. Lind, and W. F. K. Wynne-Jones, ibid., 50, 147 (1950).

239. V. H. Dodson, J. Electrochem. Soc., 108, 406 (1961).

240. S. Ikari, S. Yoshizawa, and S. Okada, J. Electrochem. Soc. Japan, 27E, 223 (1959).

241. H. B. Mark, J. Electrochem. Soc., 109, 634 (1962).

242. E. Voss and J. Freundlich, in Batteries (D. H. Collins, ed.), Pergamon, New York, 1963, p. 73.

243. A. I. Rusin, M. A. Dasoyan, and N. V. Merylikina, Issled. Obl. Khim. Istochnikov Toka, 1966, 176; Chem. Abstr., 68, 18034c (1968).

243a. A. C. Simon, Electrochem. Technol., 3, 307 (1965).

244. A. C. Simon and E. L. Jones, J. Electrochem. Soc., 109, 760 (1962).

245. A. C. Simon, in Batteries 2 (D. H. Collins, ed.), Pergamon, New York, 1965, p. 63.

246. A. C. Simon, J. Electrochem. Soc., 114, 1 (1967).

247. A. C. Simon, C. P. Wales, and S. M. Caulder, ibid., 117, 987 (1970).

248. A. C. Simon in Power Sources 2 1968 (D. H. Collins, ed.), Pergamon, New York, 1970.

249. W. Simon, Bosch. Techn. Ber., 1, 234 (1966).

250. D. Berndt and E. Voss, in Batteries 2 (D. H. Collins, ed.), Pergamon, New York, 1965, p. 17.

251. H. B. Mark, J. Electrochem. Soc., 110, 945 (1963).

252. J. Burbank, A. C. Simon, and E. Willihnganz, in Advances in Electrochemistry and Electrochemical Engineering, Vol. 8 (C. W. Tobias, ed.), Wiley, New York, 1971.

253. M. A. Dasoyan, Dokl. Akad. Nauk SSSR, 107, 863 (1956).

254. S. Felin, L. Galan, and J. A. Gongaley, "Influence of Cast Structure on Lead Electrochemical Corrosion," Final Report, Project No. LE 130 for ILZRO, 1970.

255. C. G. Fink and A. J. Dornblatt, Trans. Electrochem. Soc., 79, 269 (1931).

256. J. J. Lander, J. Electrochem. Soc., 98, 467 (1952).

257. J. J. Lander, ibid., 105, 289 (1958).

258. G. W. Mao, T. L. Wilson, and J. G. Larson, ibid., 117, 1323 (1970).

259. M. L. B. Rao, ibid., 114, 665 (1967).

260. K. Kitazawa, S. Asakura, K. Fulki, and T. Mukaibo, J. Electrochem. Soc. Japan, 37, 1 (1969).

261. R. Piontelli and G. Stermheim, J. Chem. Phys., 23, 1358, 1971 (1955).

262. S. Pizzini and L. Agace, Corros. Sci., 5, 193 (1965).

263. C. L. Mantell, Batteries and Energy Systems, McGraw-Hill, New York, 1970, Chap. 13.

264. A. H. Du Rose and W. Blum, in Modern Electroplating (F. A. Lowenheim, ed.), Wiley, New York, 1967, Chap. 11.

265. J. W. Dini and J. R. Helms, Metal Finishing, 67, 53 (August 1969).

266. J. W. Dini and J. R. Helms, J. Electrochem. Soc., 117, 269 (1970).

267. R. R. Vanderwoort, E. L. Raymond, H. J. Wiesner, and W. P. Frey, Plating, 57, 362 (1969).

268. H. J. Wiesner, W. P. Frey, R. R. Vanderwoort, and E. L. Raymond, ibid., 57, 358 (1970).

269. W. Hoffman, Lead and Lead Alloys, Springer, New York, 1970.

270. C. Zapffe, Plating, 37, 610 (1950).

271. J. Culbertson, J. Electrodepositors' Tech. Soc., 26, 99 (1950).

272. B. A. Shenoi, R. Subramanian, and K. S. Indria, Electroplating Metal Finishing, 21, 399 (1968).

273. A. K. Graham and H. L. Pinkerton, Proc. Amer. Electroplaters' Soc., 50, 135 (1963).

274. G. Dubpernell, in Modern Electroplating (F. A. Lowenheim, ed.), Wiley, New York, 1967, Chap. 5.

275. D. W. Hardesty, Plating, 56, 705 (1969).

276. H. Rickert and G. Holzaepfel, Ber. Bunsenges. Phys. Chem., 70, 171 (1966); Chem. Abstr., 64, 12183 (1966).

277. T. M. H. Saber, A. M. El Din, and A. M. Shams, Electrochim. Acta, 13, 937 (1968).

278. J. C. Grigger, H. C. Miller, and F. D. Loomis, J. Electrochem. Soc., 105, 100 (1958).

279. T. Osuga and K. Sugino, ibid., 104, 448 (1957).

280. Y. Shibasaki, ibid., 105, 624 (1958).

281. J. C. Grigger, in The Encyclopedia of Electrochemistry (C. A. Hampel, ed.), Reinhold, New York, 1964, p. 762.

282. K. C. Narasimham, S. Sundararajan, and H. V. K. Udupa, J. Electrochem. Soc., 108, 798 (1961).

283. J. C. Schumacher, D. R. Stern, and P. R. Graham, ibid., 105, 151 (1958).

284. A. J. Giuffrida, U.S. Patents 3,454,472 and 3,294,667.

285. M. J. Allen, Organic Electrode Processes, Van Nostrand-Reinhold, New York, 1958.

286. G. T. Miller, U.S. Patent 3,361,656 (January 2, 1968).

287. G. T. Miller, British Patent 1,092,294.

288. K. Sugino, Japanese Patent 4962 (1959); Chem. Abstr., 54, 5298 (1960).

289. C. E. Van Winckel, R. V. Smith, and E. H. Hutz, U.S. Patent 2,925,371.

290. T. Shwarski, Polinery, 13, 407 (1968).

291. E. L. Littauer and L. L. Shrier, Proceedings of the 1st International Congress on Metal Corrosion, Butterworths, London, 1961, p. 374.

292. E. L. Littauer and L. L. Shrier, Electrochim. Acta, 12, 465 (1967).

293. D. B. Peplow and L. L. Shrier, Corros. Technol., 4, 16 (1964).

294. L. L. Shrier and I. Weinraub, Chem. Ind., 36, 1326 (1958).

295. L. L. Shrier, Corrosion, 17, 90 (1961).

296. L. L. Shrier, Platinum Metals Rev., 12, 42 (1968).

297. J. A. Lehmann, J. Metals, 22(3), 56 (1970).

298. H. Helber and E. L. Littauer, Corros. Sci., 10, 411 (1970).

299. E. Sato, Bull. Chem. Soc., Japan, 39, 1592 (1966).

300. G. W. Walkiden, Corros. Technol., 9, 38 (1962).

301. K. N. Barnard, G. L. Christie, and D. G. Gage, Corrosion, 15, 581 (1958).

302. J. H. Morgan, Corros. Technol., 5, 347 (1958).

303. S. Tudor and A. Ticker, Mater. Prot., 3, 52 (1964).

304. A. E. Hiller and D. A. Lipps, ibid., 4, 36 (1965).

305. N. D. Tomashov, Theory of Corrosion and Protection of Metals, Macmillan, New York, 1966, p. 628.

306. G. Hohlstein and E. Pelzel, Metall., 14, 765 (1960). (Data summarized in Ref. 243.)

307. K. H. Roll, Corrosion, 7, 454 (1951).

308. H. J. Gough, J. Inst. Metals, 49, 17 (1932).

309. D. J. Mack, Amer. Soc. Test. Mat., 45, 629 (1945).

310. G. Hohlstein and E. Pelziel, Metall, 16, 764 (1962).

311. A. Turner, J. R. Wellington, and L. Williams, Can. J. Chem. Eng., 37, 55 (1959).

312. H. Weissbach, Werkst. Korros., 15, 555 (1964).

313. F. O. Anderegg and R. V. Achatz, J. Inst. Metals, 33, 372 (1925).

314. R. M. Burns, Bell System Tech. J., 15, 603 (1936).

315. I. A. Denison, Trans. Electrochem. Soc., 81, 435 (1942).

316. K. H. Logan, J. Res. Nat. Bur. Stand., 17, 781 (1936).

317. K. H. Logan and S. P. Ewing, ibid., 18, 361 (1937).

318. I. A. Denison and M. Romanoff, ibid., 44, 259 (1950).

319. B. B. Reinitz, Corrosion, 9, 425 (1953).

320. J. H. B. George, The Battery Industry, U.S. and International, Arthur D. Little, Cambridge, Massachusetts, May 1970.

321. Bell System Tech. J., 49 (September 1970).

322. D. Berndt, in Power Sources 2 1968 (D. H. Collins, ed.), Pergamon, New York, 1970.

323. J. R. Pierson, in Power Sources 2 1968 (D. H. Collins, ed.), Pergamon, New York, 1970.

324. J. Burbank, Power Sources 1966 (D. H. Collins, ed.), Pergamon, New York, 1967, p. 147.

325. V. H. Dodson, J. Electrochem. Soc., 108, 401 (1961).

326. I. Dugdale, Discussion of Paper by R. G. Acton in Power Sources 1966 (D. H. Collins, ed.), Pergamon, New York, 1967, p. 142.

327. J. L. Dawson, M. I. Gillibrand, and J. Wilkinson, Power Sources '70, Preprint Paper 1.

328. W. Herrmann and G. Z. Proepstl, Electrochimie, 61, 1154 (1957). [Data summarized in M. Fleischmann and H. R. Thirsk, in Advances in Electrochemistry and Electrochemical Engineering, Vol. 3 (P. Delahay, ed.), Wiley, New York, 1963, Chap. 3.]

329. M. I. Gillibrand and G. R. Lomax, Electrochem. Acta, 8, 693 (1963).

330. K. Peters, A. I. Harrison, and W. H. Durant, in Power Sources 2 1968 (D. H. Collins, ed.), Pergamon, New York, 1970.

331. A. P. Hauel, Trans. Electrochem. Soc., 78, 231 (1940).

332. E. Willihnganz, ibid., 92, 281 (1947).

333. J. R. Pierson, P. Gurbisky, A. C. Simon, and S. M. Caulder, J. Electrochem. Soc., 117, 1463 (1970).

334. S. Sekido and H. Ichimura, Denki Kagaku, 37, 168 (1969); Chem. Abstr., 71, 35391w (August 25, 1969).

335. A. C. Zachlin, J. Electrochem. Soc., 98, 325 (1951).

336. T. F. Sharpe, Electrochim. Acta, 14, 635 (1969).

337. T. F. Sharpe, J. Electrochem. Soc., 116, 1639 (1969).

338. D. Kordes, Chem. Ing. Tech., 38, 638 (1966).

339. I. I. Astachov, I. G. Kiseleva, and B. N. Kabanov, Doklady Akad Nauk SSSR, 126, 1041 (1959).

340. E. J. Ritchie and J. Burbank, J. Electrochem. Soc., 117, 299 (1970).

341. A. C. Cannone, D. O. Feder, and R. V. Biagetti, Bell System Tech. J., 49, 1279 (1970).

342. E. Willihnganz, Electrochem. Technol., 6, 338 (1968).

343. U. B. Thomas, F. T. Forster, and H. E. Haring, Trans. Electrochem. Soc., 92, 313 (1947).

344. R. V. Biagetti and M. C. Weeks, Bell System Tech. J., 49, 1305 (1970).

345. J. Burbank, in Batteries (D. H. Collins, ed.), Pergamon, New York, 1963.

346. J. Burbank, J. Electrochem. Soc., 111, 1112 (1964).

347. J. Burbank and E. J. Ritchie, ibid., 116, 125 (1969).

348. J. R. Pierson, Electrochem. Technol., 5, 323 (1967).

349. J. F. Dittmann and J. F. Sams, J. Electrochem. Soc., 105, 553 (1958).

350. H. Bode et al., Electrochim. Acta, 11, 1211, 1221, 1231 (1966).

351. M. A. Barron, in Proceedings of the 23rd Annual Power Sources Conference, PSC Publications Committee, Atlantic City, New Jersey, 1969, p. 134

352. T. J. Kilduff and E. F. Horsey, in Proceedings of the 24th Annual Power Sources Conference, PSC Publications Committee, Atlantic City, New Jersey, 1970, p. 30.

353. F. G. Turrill, in Proceedings of the 24th Annual Power Sources Conference, PSC Publications Committee, Atlantic City, New Jersey, 1970, p. 36.

354. J. C. White, W. H. Power, R. L. McMurtrie, and R. T. Pierce, Jr., Trans. Electrochem. Soc., 91, 73 (1947).

355. G. W. Heise, in The Primary Battery (G. W. Heise and N. C. Cahoon, eds.), Wiley, New York, 1971, p. 27.

356. R. E. Bidick and R. D. Nelson, Intersociety Energy Conversion Engineering Conference, IEEE, New York, 1968, p. 147.

357. A. M. Biggar, U.S. Patent 3,507,707 (April 21, 1970).

358. W. G. Darland, U.S. Patent 3,033,908 (May 8, 1962).

359. M. Shoeld, U.S. Patent 3,494,800 (February 10, 1970).

Chapter I-6

MANGANESE

C. C. LIANG

Laboratory for Physical Science
P. R. Mallory & Co., Inc.
Burlington, Massachusetts

1. STANDARD AND FORMAL POTENTIALS

The thermodynamic properties of manganese and its compounds have been extensively studied and well documented. Latimer's [1] work on the standard electrode potentials and other thermodynamic data for reactions and compounds containing manganese has been a standard reference for many years. Mah [2] examined the thermodynamic data for manganese and its compounds and compiled them into a single publication for convenient reference. Recently, Morgan [3] investigated the chemical equilibria of manganese and natural waters. Zordan and Hepler [4] published a comprehensive review on the thermochemistry and the oxidation potentials of manganese and its compounds. The most recent data on the thermodynamics of manganese and its compounds appeared in the National Bureau of Standards (NBS) Technical Note 270-4 [5], updating data published in the NBS Circular 500 [6].

Under a variety of conditions manganese(II) is the most stable oxidation state of the element. The stability of Mn^{2+} may be attributed to its electronic structure which contains a half filled d-band ($3d^5$). Manganese(II) salts of strong acids are quite soluble; the hydroxide and salts of weak acids are relatively insoluble. Manganese(II) forms many complexes and chelates. However, these complexes are generally less stable than the corresponding complexes of cobalt(II) or iron(II).

Manganese(III) is unstable in aqueous solution because it is strongly oxidizing and undergoes disproportion to manganese(II) and manganese dioxide. Stable manganese(III) species are either solids of low solubility, such as manganic fluoride and manganic oxide, or weakly dissociated complexes, such as hexacyanomanganate(III).

Manganese dioxide is one of the most important compounds of manganese(IV) because of its practical value as the active cathode material in batteries and as an oxidizing agent. Its electrochemical characteristics from both the thermodynamic and the kinetic points of view will be discussed in some detail throughout this chapter.

The hypomanganate ion (MnO_4^{3-}) containing manganese(V) has been observed [4, 7, 8], and it is moderately stable in strongly alkaline solutions. Manganese(VI) generally exists as the manganate ion MnO_4^{2-}, and it is quite stable in alkaline solutions. The permanganate ion (MnO_4^-) is the only important species containing manganese(VII). It is a strong oxidizing agent and is slowly reduced by water.

1.1. AQUEOUS SOLUTIONS

1.1.1. Mn^{2+}-Mn

On the basis of direct electrode potential measurements of an electrolytic manganese electrode in solutions containing various amounts of $(NH_4)_2SO_4$ and $MnSO_4$, Garkavi [9] reported the standard potential of the Mn^{2+}(aq)/Mn electrode as -1.06 V at 25°C. Zordan and Hepler [4], using -111.3 kcal/mole as the standard Gibbs free energy of formation (ΔG_f°) for manganese dioxide and -1.239 V as the standard potential (E°) for the MnO_2/Mn^{2+}(aq) electrode [10], calculated the ΔG_f° for Mn^{2+}(aq) to be -55.1 kcal/mole and the E° for the Mn^{2+}(aq)/Mn electrode to be -1.20 V at 25°C. Jangg [11] measured the electrode potential of the amalgamated manganese electrodes in $MnSO_4$ solutions. The results of these measurements led to the value of -1.168 ± 0.005 V as the standard potential for the Mn^{2+}(aq)/Mn electrode at 25°C. Based on the revised National Bureau of Standards value of

TABLE 1.1.1. Standard and Formal Potentials in Aqueous Solution

Half-reaction	Standard or Formal Potential (V)	Conditions	Refs.
$Mn^{2+}(aq) + 2e = Mn$	-1.18		[1], [5]
$Mn(OH)_2(c) + 2e = Mn + 2OH^-(aq)$	-1.56		[5]
$Mn^{3+}(aq) + e = Mn^{2+}(aq)$	+1.5	15 N H_2SO_4	[12], [13]
$Mn_2O_3(c) + 3H_2O(l) + 2e = 2Mn(OH)_2(c) + 2OH^-(aq)$	-0.25		[5]
$Mn_2O_3(c) + 6H^+(aq) + 2e = 2Mn^{2+}(aq) + 3H_2O(l)$	+1.48		[5]
$MnO_2 + 4H^+(aq) + 2e = Mn^{2+}(aq) + 2H_2O(l)$	+1.233 to +1.241		[10]
$MnO_2(c) + 2H_2O(l) + 2e = Mn(OH)_2(c) + 2OH^-(aq)$	-0.1		[5]
$2MnO_2(c) + 2H^+(aq) + 2e = Mn_2O_3(c) + H_2O(l)$	+0.98		[5]
$MnO_4^-(aq) + 2H_2O(l) + 3e = MnO_2(c) + 4OH^-(aq)$	+0.588		[48]
$MnO_4^-(aq) + 4H^+(aq) + 3e = MnO_2(c) + 2H_2O(l)$	+1.69		[23], [48]
$MnO_4^-(aq) + 8H^+(aq) + 5e = Mn^{2+}(aq) + 4H_2O(l)$	+1.51		[1], [5], [48]
$MnO_4^-(aq) + 4H_2O(l) + 5e = Mn(OH)_2(c) + 6OH^-(aq)$	+0.34		[5], [48]
$MnO_4^-(aq) + e = MnO_4^{2-}(aq)$	+0.558		[49]
$MnO_4^{2-}(aq) + 2H_2O(l) + 2e = MnO_2(c) + 4OH^-(aq)$	+0.603		[5], [49]
$MnO_4^{2-}(aq) + e = MnO_4^{3-}(aq)$	+0.3	6-12 N KOH	[8]

ΔG_f° for $Mn^{2+}(aq)$ (-54.5 kcal/mole) and the work of Latimer, we have selected -1.18 V as the standard potential (Table 1.1.1) for the electrode:

$$Mn^{2+}(aq) + 2e = Mn \qquad E^\circ = -1.18\ V \tag{1.1.1}$$

The standard potential of the $Mn(OH)_2(c)/Mn(c)$ electrode may be calculated from the NBS values of the Gibbs free energies of formation [ΔG_f° ($Mn(OH)_2$) = -147 kcal/mole, ΔG_f° (OH^-) = -37.594 kcal/mole]:

$$Mn(OH)_2(c) + 2e = Mn(c) + 2OH^-(aq) \qquad E^\circ = -1.56\ V \tag{1.1.2}$$

1.1.2. Mn^{3+}-Mn^{2+}

Investigations of the $Mn^{3+}(aq)/Mn^{2+}(aq)$ electrode were conducted in 15 N H_2SO_4 by Vetter and Manecke [12, 13]. From the ΔG_f° of $Mn^{2+}(aq)$ and the formal potential of the $Mn^{3+}(aq)/Mn^{2+}(aq)$ electrode:

$$Mn^{3+}(aq) + e = Mn^{2+}(aq) \qquad E^\circ = +1.5 \text{ V} \qquad (1.1.3)$$

the standard Gibbs free energy of formation for $Mn^{3+}(aq)$ is estimated to be about -190 kcal/mole.

In alkaline solution the Mn^{3+}/Mn^{2+} couple may be written as follows, with the standard potential calculated from the Gibbs free energies of formation:

$$Mn_2O_3(c) + 3H_2O(l) + 2e = 2Mn(OH)_2(c) + 2OH^-(aq) \qquad E^\circ = -0.25 \text{ V} \qquad (1.1.4)$$

Both manganic and manganous oxides are quite soluble in highly concentrated alkaline solution. The solubilities of manganic oxide and manganous oxide in 9 M KOH at room temperature, determined by Kozawa et al. [14, 15], are approximately 4×10^{-3} and 0.5×10^{-3} M, respectively. Presumably, the Mn^{3+} and Mn^{2+} ions form complexes with OH^- such as $Mn(OH)_4^-$ and $Mn(OH)_4^{2-}$ [14, 15]. From the results of a study on these complexes, Kozawa et al. estimated the formal potential of the $Mn(OH)_4^-/Mn(OH)_4^{2-}$ electrode in 9 M KOH to be -0.4 V vs the Hg/HgO reference electrode.

The electrode reaction of the Mn_2O_3/Mn^{2+} couple is written as follows, with the standard potential of 1.48 V calculated from the standard Gibbs free energies of formation:

$$Mn_2O_3(c) + 6H^+(aq) + 2e = 2\,Mn^{2+}(aq) + 3H_2O(l) \qquad (1.1.5)$$

1.1.3. MnO_2-Mn^{2+}

The importance of manganese dioxide as the cathode material in LeClanché dry cells has led to many investigations of the standard potential of the $MnO_2/Mn^{2+}(aq)$ electrode [1, 10, 16-25]. However, many difficulties have been encountered in obtaining steady and reproducible values of the electrode potential, especially when γ-MnO_2 electrodes were used [1, 10, 19, 21, 22, 25].

Difficulties in determining the standard potential for the $MnO_2/Mn^{2+}(aq)$ electrode may be attributed to: 1) The cation exchange capacity of manganese dioxide resulting in a change of pH at the vicinity of the electrode, and 2) the capacity of manganese dioxide to form homogeneous solid solutions with lower manganese oxides resulting in an average oxidation state of manganese lower than +4.

The ionic exchange property of manganese dioxide was reported over 40 years ago [26]. Johnson and Vosburgh [27] proposed the formula of $H_2MnO_3 \cdot xMnO_2$ for the hydrated manganese dioxide. They observed that H^+ in the hydrated manganese dioxide can be replaced by cations such as NH_4^+ and Na^+ from the solution resulting in a change in pH. Kozawa [28] demonstrated the ion exchange capacity of various types of manganese dioxide with respect to divalent cations such as Cu^{2+}, Ni^{2+}, Co^{2+}, Zn^{2+}, and Mg^{2+}. He also found that the ion exchange capacity of γ-MnO_2, prepared by electrodeposition from the $MnSO_4$ solution at 97°C, is much higher than that of β-MnO_2 prepared by thermal decomposition of $Mn(NO_3)_2$. The ion exchange reaction will change the pH in the vicinity of the manganese dioxide electrode. Furthermore, in Mn^{2+}-containing solutions, the absorption of the Mn^{2+} ions by the manganese dioxide electrode can be expected. Therefore, in the presence of the manganese dioxide electrode the concentrations of both H^+ and Mn^{2+} ions in the solution can change, resulting in an erratic and irreproducible potential for the $MnO_2/Mn^{2+}(aq)$ electrode. Since the ionic exchange capacity of the β-MnO_2 is much less than that of the γ-MnO_2, it is expected that the β-$MnO_2/Mn^{2+}(aq)$ electrode potential should be more reproducible than the γ-$MnO_2/Mn^{2+}(aq)$ electrode. Covington, Cressey, Lever, and Thirsk [10] investigated the standard potential of the β-$MnO_2/Mn^{2+}(aq)$ electrode and found that careful pretreatment of the oxide with the electrolyte in which the measurements are to be made is necessary to bring the exchange to equilibrium before commencing the cell potential measurements. Electrode potentials were measured in acetate buffered solutions. A standard potential range of 1.233–1.241 V was obtained at 25°C.

$$\beta\text{-}MnO_2 + 4H^+(aq) + 2e = Mn^{2+}(aq) + 2H_2O(l) \qquad (1.1.6)$$

It is interesting that despite a similar pretreatment of the oxide in the acetate buffered solutions, Covington, Talukdar, and Thirsk [25] could not obtain satisfactory results in determining the standard potential for the γ-$MnO_2/Mn^{2+}(aq)$ electrode.

The potential of the γ-$MnO_2/Mn^{2+}(aq)$ electrode may be examined according to Vetter's general thermodynamic theory for metal oxide electrodes [29, 30]. The presence of manganese in lower oxidation states, hydroxyl ions, and water in the crystal structure of γ-MnO_2 was postulated by Brenet [19, 31]. Vetter and Jaeger [24, 32] proposed the general formula, $MnO_{2n-2}(OH)_{4-2n}$ [or $MnO_n \cdot (2 - n)H_2O$] to signify the presence of OH^- and/or H_2O in the crystal structure and the wide range of homogeneity from n = 2 to 1.5 [33, 34]. The degree of oxidation n = 2 represents MnO_2 whereas n = 1.5 signifies MnO(OH).

According to Vetter, the electrode potential of such a metal oxide depends upon 1) the stoichiometry of the oxide (the average degree of oxidation, n), 2) the activity of the metal ion in the solution (e.g., Mn^{2+}), and 3) the pH of the solution. Only two of these three variables are independent. In agreement with Vetter's theory, Vetter and Jaeger [24, 32] established that a solution with a given activity of Mn^{2+} and H^+ ions can exist in equilibrium with only one manganese oxide of a fixed degree of oxidation. When the initial degree of oxidation of the oxide is different from that of equilibrium value, transfer of Mn^{2+} and OH^- between the oxide and the solution occurs until the equilibrium degree of oxidation is reached. The resulting manganese oxide is a homogeneous solid solution of manganese dioxide and lower oxidation state manganese oxides. This behavior is shown by Fig. 1.1.1 and it is noted that the potential also approaches the same final value from both directions

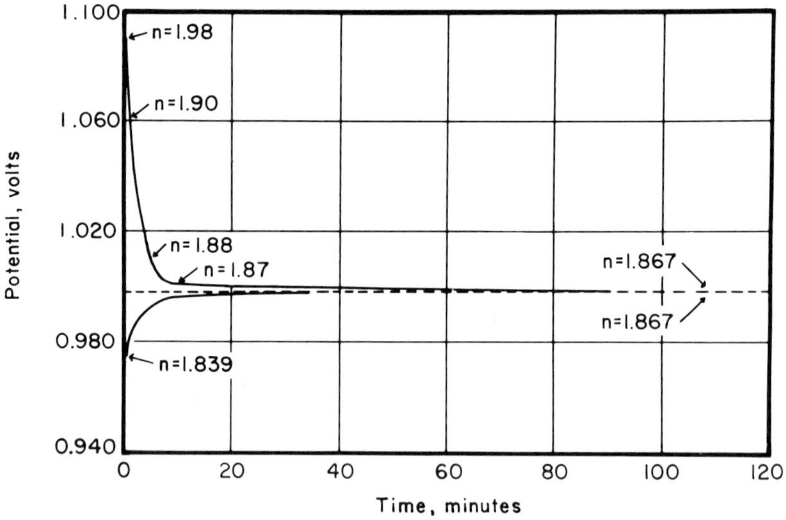

FIG. 1.1.1. Establishment of the equilibrium potential of a γ-$MnO_n \cdot (2 - n)H_2O$ electrode. Solution: 5×10^{-3} M Mn^{2+}, 1×10^{-1} M K_2SO_4. pH: 3.54. Equilibrium degree of oxidation: 1.867 [32].

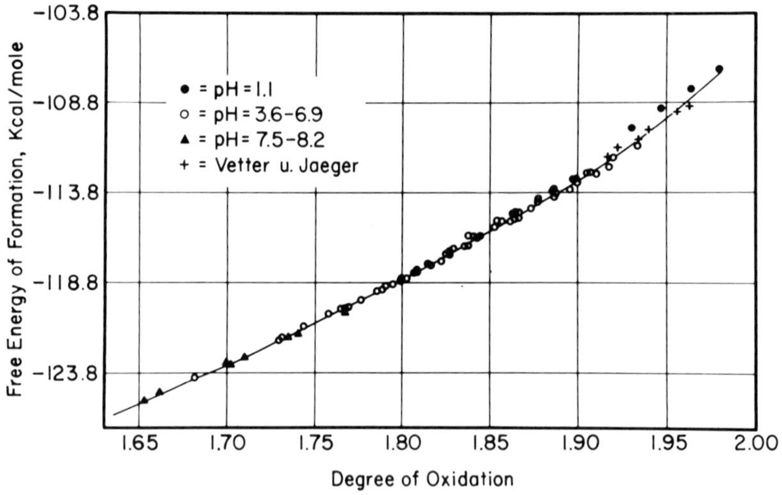

FIG. 1.1.2. Free energy of formation of γ-$MnO_n \cdot (2 - n)H_2O$ at 20° C as a function of degree of oxidation, n [32].

where the manganese oxide is in equilibrium with the solution. From the equilibrium potentials of the γ-$MnO_n \cdot (2 - n)H_2O$ electrodes in solutions of different activities of Mn^{2+} and H^+, Vetter and Jaeger calculated the Gibbs free energy of formation for γ-$MnO_n \cdot (2 - n)H_2O$ (or γ-$MnO_{2n-2}(OH)_{4-2n}$) as a function of the degree of oxidation, n. The results are shown in Fig. 1.1.2 and Table 1.1.2.

TABLE 1.1.2. Free Energy of Formation of γ-$MnO_n \cdot (2 - n)H_2O$ Determined by Electrochemical Measurement at 20°C [32]

Electrolyte		Degree of oxidation of Mn in "manganese dioxide" (n in MnO_n)	Free energy of formation of $MnO_n \cdot (2 - n)H_2O$ (kcal/mole)	Equilibrium MnO_n/Mn^{2+} electrode (V vs NHE)
pH	$-\log C_{Mn^{2+}}$			
1.1	2	1.931	-110.2	1.295
	3	1.947	-109.1	1.328
	4	1.964	-108.0	1.358
	5	1.980	-106.9	1.389
3.66	1	1.845	-106.3	0.928
	2	1.878	-114.2	0.971
	3	1.898	-113.1	1.003
	4	1.920	-111.8	1.034
	5	1.934	-111.2	1.061
4.14	1	1.825	-117.2	0.870
	2	1.854	-115.4	0.912
	3	1.878	-114.2	0.943
	4	1.899	-113.1	0.975
	5	1.918	-112.3	1.000
4.60	1	1.809	-118.1	0.807
	2	1.841	-116.2	0.849
	3	1.867	-114.9	0.883
	4	1.888	-113.8	0.914
	5	1.911	-112.7	0.942
5.08	1	1.795	-118.9	0.742
	2	1.830	-116.9	0.785
	3	1.853	-115.8	0.819
	4	1.874	-114.8	0.851
	5	1.896	-114.0	0.882
5.56	1	1.768	-120.3	0.676
	2	1.807	-118.3	0.716
	3	1.836	-116.8	0.755
	4	1.862	-115.5	0.789
6.10	1	1.730	-122.0	0.601
	2	1.786	-119.3	0.644
	3	1.817	-117.9	0.681
	4	1.838	-116.8	0.716
6.58	1	1.682	-124.0	0.535
	2	1.758	-120.5	0.582
	3	1.799	-118.7	0.618
	4	1.823	-117.7	0.652
6.90	2	1.744	-121.3	0.536
	3	1.789	119.2	0.575
7.52	3	1.662	-124.9	0.464
	4	1.735	-121.9	0.503
7.88	4	1.710	-123.0	0.453
8.24	4	1.653	-125.4	0.401

It is noted from the above discussion that difficulties in obtaining a reproducible and steady value for the standard potential of the γ-MnO_2/Mn^{2+}(aq) electrode may be attributed to the ionic exchange property and the wide range of homogeneity of γ-manganese dioxide. More often than not, an equilibrium potential, rather than a standard potential, can be obtained (Table 1.1.2). The fact that a reliable value for the standard potential of the β-MnO_2/Mn^{2+}(aq) can be obtained [10, 25] is due to the limited ionic exchange capacity of β-manganese dioxide [28]. Furthermore, β-MnO_2 does not readily form solid solutions with lower manganese oxides, and the range of homogeneity is extremely limited [32]. As a result, the degree of oxidation, which is one of the two independent variables in Vetter's thermodynamic theory, is fixed. Consequently the standard potential of the β-MnO_2/Mn^{2+}(aq) can be obtained.

In more alkaline solutions the electrode reaction of the MnO_2-Mn^{2+} couple may be expressed by Eq. (1.1.7), with the standard potential calculated from the Gibbs free energies of formation; ΔG_f° (MnO_2) = -111.18 kcal/mole; ΔG_f° ($Mn(OH)_2$) = -54.5 kcal/mole:

$$MnO_2(c) + 2H_2O(l) + 2e = Mn(OH)_2(c) + 2OH^-(aq) \qquad E^\circ = -0.10 \text{ V} \qquad (1.1.7)$$

1.1.4. MnO_2-Mn_2O_3

Manganese(III) oxide was found by various investigators [35-42] to be the final discharge product of the manganese dioxide electrode in electrolytes with relatively low pH values. For the standard potential of the electrode reaction:

$$2MnO_2(c) + 2H^+(aq) + 2e = Mn_2O_3(c) + H_2O(l) \qquad (1.1.8)$$

Moussard et al. [43] and Ionin and Zvorykina [44] calculated a value of 1.01 V at 25°C. Neumann and Roda [45] recently investigated the equilibrium potential of the γ-manganese dioxide electrode in NH_4Cl solutions. In agreement with earlier work, these authors found that the discharge product [manganese(III) oxide] forms solid solutions with γ-manganese dioxide. As a result, the equilibrium potential of the γ-manganese dioxide/manganese(III) oxide electrode is a function of not only the pH value of the electrolyte but also of the activities of the constituents of the solid solution. The standard potential of the electrode determined at the electrode composition, where the activity of manganese dioxide is equal to that of manganese(III) oxide, is 1.04 V at 22 ± 2°C.

Based on the NBS values of standard Gibbs free energies of formation [5], the calculated value of the standard potential for the $MnO_2(c)/Mn_2O_3(c)$ electrode is 0.98 V at 25°C. The difference between this value and that due to Neumann and Roda may result from the following:

1. In calculating the standard potential, the standard Gibbs free energy of formation of β-MnO_2 was used whereas the electrode used by Neumann and Roda was γ-MnO_2. It is

conceivable that the standard Gibbs free energy for β-MnO_2 is more negative than that of γ-MnO_2. (See Fig. 1.1.2.) It must be noted, however, that the results by Vetter and Jaeger were obtained at 20°C rather than 25°C. Nevertheless, the difference in the Gibbs free energy of formation between the two temperatures is estimated to be less than 1 kcal/mole.

2. The manganese(III) oxide involved in the $MnO_2/MnO_{1.5}$ electrode may be MnO(OH) rather than Mn_2O_3, the ΔG_f° of which was used in the calculation of the standard potential. Bode et al. [34, 46] found that the reduction of γ-MnO_2 produces α-MnO(OH), and Vosburgh [35] and Scott [40] proposed the H^+ diffusion mechanism for the reduction process of MnO_2. Therefore, it may be suggested that the electrode reaction in the investigation of Neumann and Roda is

$$\gamma\text{-}MnO_2 + H^+ + e = \alpha\text{-}MnO(OH) \tag{1.1.9}$$

with a standard potential slightly different from that of the β-MnO_2/Mn_2O_3(c) electrode.

It has been observed [14, 47] that the electroreduction of γ-MnO_2 in alkaline solution also produces α-MnO(OH) through the diffusion of protons and electrons in the crystal lattice of γ-MnO_2. As a result, a solid solution of γ-MnO_2 and α-MnO(OH) is formed during the electroreduction process. For the standard potential of the electrode reaction

$$\gamma\text{-}MnO_2 + H_2O(l) + e = \alpha\text{-}MnO(OH) + OH^-(aq) \tag{1.1.10}$$

Kozawa and Powers [47] found a value of 0.103 V at 23°C. In their calculation of the standard potential they assumed unit activity coefficients for the constituents of the solid solution, γ-MnO_2 and α-MnO(OH), therefore the open circuit potential of the electrode in 1 M KOH when the concentration of γ-MnO_2 is equal to that of α-MnO(OH) in the solid solution was taken to be the standard potential of the electrode. However, according to Neumann and Roda [45] the activity coefficients of MnO_2 and MnO(OH) are not unity in the solid solution, and equal activity of the two forms of manganese oxide occurs when the mole fraction of γ-MnO_2 is 0.67. In this case the standard potential of the γ-MnO_2/α-MnO(OH) electrode in alkaline solution should be more positive than 0.103 V.

1.1.5. MnO_4^--MnO_2

Andrews and Brown [48] measured the potential of the $MnO_4^-(aq)/MnO_2(c)$ electrode in alkaline solution and obtained a value of 0.588 ± 0.001 V for the standard potential.

$$MnO_4^-(aq) + 2H_2O(l) + 3e = MnO_2(c) + 4OH^-(aq) \qquad E^\circ = +0.588\ V \tag{1.1.11}$$

Gabano and Brenet [23] found that the equilibrium potentials of the $MnO_4^-(aq)/\beta$-MnO_2 electrode could be obtained only in solutions with a pH of 9.6 to 10.6. The standard potential of 1.679 V was obtained from these measurements. Taking the results from both of these investigations into consideration, we conclude that the standard Gibbs free energy of formation for $MnO_4^-(aq)$ is at least 0.5 to 0.6 kcal/mole more negative than the revised

NBS value. Therefore, we shall adopt the value of -107.5 kcal/mole as the ΔG_f° for MnO_4^-(aq) and write:

$$MnO_4^-(aq) + 4H^+(aq) + 3e = MnO_2(c) + 2H_2O(l) \qquad E^\circ = +1.69\ V \qquad (1.1.12)$$

1.1.6. MnO_4^--Mn^{2+}

Using the standard Gibbs free energies of formation for MnO_4(aq), Mn^{2+}(aq), and H_2O(l), one calculates the standard potential for the MnO_4^-(aq)/Mn^{2+}(aq) electrode:

$$MnO_4^-(aq) + 8H^+(aq) + 5e = Mn^{2+}(aq) + 4H_2O(l) \qquad E^\circ = +1.51\ V \qquad (1.1.13)$$

The standard potential for the following electrode reaction is also calculated from the Gibbs free energies of formation:

$$MnO_4^-(aq) + 4H_2O(l) + 5e = Mn(OH)_2(c) + 6OH^-(aq) \qquad E^\circ = +0.34\ V \qquad (1.1.14)$$

1.1.7. MnO_4^--MnO_4^{2-}, MnO_4^{2-}-MnO_2, and MnO_4^{2-}-MnO_4^{3-}

From the investigation by Carrington and Symons [49] we obtain:

$$MnO_4^-(aq) + e = MnO_4^{2-}(aq) \qquad E^\circ = +0.558\ V \qquad (1.1.15)$$

and a ΔG_f° for MnO_4^{2-}(aq) of -120.4 kcal/mole, which is 0.7 kcal/mole more negative than the NBS value.

From the ΔG_f° for MnO_4^{2-}(aq) and the free energies of formations of MnO_2(c), H_2O(l), and OH^-(aq), the following standard potentials are calculated:

$$MnO_4^{2-}(aq) + 2H_2O(l) + 2e = MnO_2(c) + 4OH^-(aq) \qquad E^\circ = +0.603\ V \qquad (1.1.16)$$

The standard potential for the MnO_4^{2-}(aq)/MnO_4^{3-}(aq) electrode was estimated [8] in strongly alkaline solutions. From an approximate value of 0.3 V for the E°, the ΔG_f° for MnO_4^{3-}(aq) is estimated to be -127 kcal/mole.

1.1.8. Summary

The standard or formal potentials of various electrodes in aqueous solutions are summarized in Table 1.1.1. For the purpose of reference, the standard Gibbs free energies of formation for manganese species are shown in Table 1.1.3.

Based on the electrode reactions discussed in this section, a potential-pH diagram, as shown by Fig. 1.1.3, is constructed according to the pattern set by Pourbaix, Brenet et al. [50, 51].

TABLE 1.1.3. Standard Free Energies of Formation of Manganese Species at 25°C

Substance	Free energy of formation ΔG_f° (kcal/mole)	Refs.
α-Mn	0	[5]
γ-Mn	0.34	[5]
Mn^{2+}(aq), standard state, m = 1	-54.5	[5]
MnO(c)	-86.74	[5]
$Mn(OH)_2$	-147.0	[5]
Mn^{3+}(aq)	~-190	[4], [12], [13]
Mn_2O_3(c)	-210.6	[5]
Mn_3O_4(c)	-306.7	[5]
MnO_2(c)	-111.18	[5]
MnO_4^{3-}(aq)	~-127	[4], [8]
MnO_4^{2-}(aq), standard state, m = 1	-120.4	[49]
MnO_4^-(aq), standard state, m = 1	-170.5	[23], [48]
$Mn(OH)_3^-$(aq), standard state, m = 1	-177.9	[5]
$MnSO_4$(c)	-228.83	[5]
$MnSO_4$(aq), standard state, m = 1	-232.5	[5]
$MnCl_2$(c)	-105.29	[5]
$MnCl_2$(aq), standard state, m = 1	-117.3	[5]
$Mn(NO_3)_2$(aq), standard state, m = 1	-107.8	[5]
$MnCO_3$(c)	-195.2	[5]

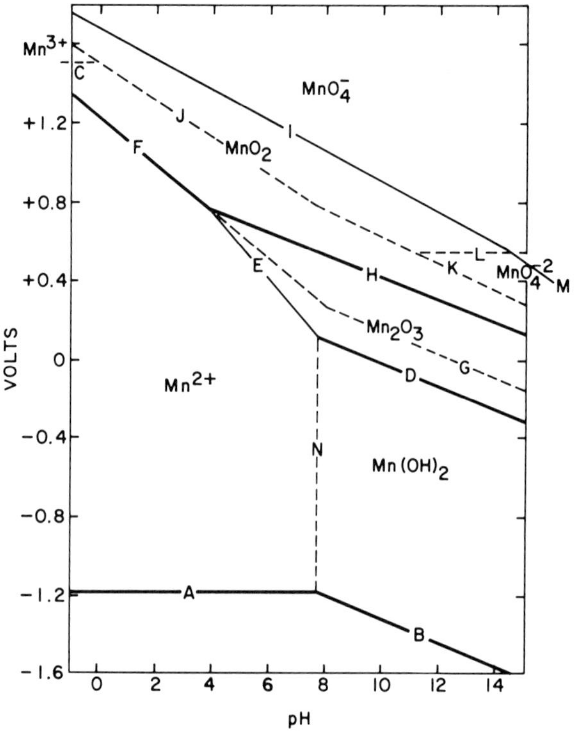

FIG. 1.1.3. pH-potential diagram of manganese system.

A. $Mn^{2+}(aq) + 2e = Mn \qquad (m_{Mn^{2+}} = 1)$

B. $Mn(OH)_2(c) + 2e = Mn + 2OH^-$

C. $Mn^{3+} + e = Mn^{2+}$

D. $Mn_2O_3(c) + 3H_2O(l) + 2e = 2\ Mn(OH)_2 + 2OH^-(aq)$

E. $Mn_2O_3(c) + 6H^+(aq) + 2e = 2\ Mn^{2+}(aq) + 3H_2O \qquad (m_{Mn^{2+}} = 1)$

F. $\beta\text{-}MnO_2 + 4H^+(aq) + 2e = Mn^{2+}(aq) + 2H_2O(l) \qquad (m_{Mn^{2+}} = 1)$

G. $MnO_2(c) + 2H_2O(l) + 2e = Mn(OH)_2(c) + 2OH^-(aq)$

H. $2MnO_2(c) + 2H^+(aq) + 2e = Mn_2O_3(c) + H_2O(l)$

I. $MnO_4^-(aq) + 4H^+(aq) + 3e = MnO_2(c) + 2H_2O(l) \qquad (m_{MnO_4^-} = 1)$

J. $MnO_4^-(aq) + 8H^+(aq) + 5e = Mn^{2+}(aq) + 4H_2O(l) \qquad (m_{MnO_4^-} = 1)$

K. $MnO_4^-(aq) + 4H_2O(aq) + 5e = Mn(OH)_2(c) + 6OH^- \qquad (m_{MnO_4^-} = 1)$

L. $MnO_4^-(aq) + e = MnO_4^{2-}(aq)$

M. $MnO_4^{2-}(aq) + 2H_2O(l) + 2e = MnO_2(c) + 4OH^-(aq) \qquad (m_{MnO_4^{2-}} = 1)$

1.2. FUSED ELECTROLYTES

The theoretical decomposition potentials of manganese oxides and halides at various temperatures were calculated by Hamer et al. [52-54] from the Gibbs free energies of formation. These values are shown in Table 1.2.1.

Formal potentials of the Mn^{2+}(l)/Mn(c) electrode were determined in fused chloride electrolytes by various investigators [55-58]. Delimarsky [59] calculated the manganese electrode potential in various complex electrolytes in reference to the sodium reference electrode. These results are shown in Tables 1.2.2 and 1.2.3, respectively.

Eluard and Tremillon [60] studied the thermodynamic properties of manganese in the NaOH-KOH eutectic melt containing various amounts of H_2O, which determines the acidity of the molten electrolytes. One of the interesting findings by Eluard and Tremillon is that the hypomanganate ion, MnO_4^{3-}, while unstable in aqueous media, is stable in the NaOH-KOH eutectic.

TABLE 1.2.1. Theoretical Decomposition Potentials of Solid or Fused Manganese Halides and Manganese Oxides at Various Temperatures [52-54][a]

Temperature (°C)	Theoretical emf of manganese-halogen (oxygen) cell or decomposition potential of manganese halide (oxide) (V)							
	MnF_2	MnF_3	$MnCl_2$	$MnBr_2$	MnI_2	MnO	Mn_2O_3	MnO_2
25	3.9	3.21	2.28	1.91	1.39	1.882	1.53	1.208
100	3.828	3.152	2.235	1.854	1.333	1.854	1.500	1.172
200	3.760	3.079	2.166	1.786	1.263	1.816	1.456	1.124
300	3.695	3.009	2.098	1.719	1.196	1.779	1.412	1.077
350	3.663	2.975	2.065	1.687	1.163	1.760	1.390	1.053
400	3.632	2.941	2.032	1.654	1.130	1.742	1.368	1.030
450	3.601	2.908	1.999	1.623	1.098	1.723	1.346	1.006
500	3.570	2.875	1.967	1.591	1.066	1.705	1.324	0.983
550	3.540	2.842	1.935	1.560	1.034	1.686	1.302	0.960
600	3.510	2.810	1.902	1.529	1.002	1.668	1.281	0.937
			......	......	......			
800	3.391	2.682	1.807	1.422	0.903	1.591	1.197	0.828
1000	3.289	2.557	1.725	1.334	0.890 (827B)	1.515	1.111	0.806 (874D)
		......						
1500	3.044	2.383	1.649 (1190B)	1.322 (1027B)		1.305	0.802 (1347D)	
1750						1.190		
						......		
2000						1.103		

[a]D = decomposition temperature. B = boiling point. Solid above the dotted line, fused salt below the dotted line.

TABLE 1.2.2. Formal Potential of the Mn^{2+}/Mn Electrode in Fused Electrolytes

Electrolyte	Formal potential, V vs Ag/AgCl	Temperature (°C)	Refs.
1 :1 KCl-NaCl	-1.230	450	[55]
	-1.206	700	
	-1.190	800	
	-1.172	900	
KCl-LiCl eut.	-1.212	450	[56]
$MgCl_2$(32.5 m/o)-KCl	-1.2078	475	[57]
$MgCl_2$-KCl-NaCl eut.	-1.794 vs Pt reference	475	[58]

TABLE 1.2.3. Electrode Potential of Mn^{2+}/Mn in Complex Electrolyte at 700°C, Calculated Against the Na/Na^+ Reference [59]

Electrolyte	NaCl-KCl-$SrCl_2$	NaCl-$AlCl_3$	NaBr-KBr	NaBr-$AlBr_3$	NaI	NaF
Potential (V)	1.41	1.77	1.26	1.40	0.98	0.67

2. VOLTAMMETRIC CHARACTERISTICS

2.1. POLAROGRAPHIC CHARACTERISTICS

Table 2.1.1 shows the polarographic characteristics of manganese in both aqueous and nonaqueous solutions containing various supporting electrolytes. In the absence of a complexing agent, the reduction of manganous ion produces manganese at the dropping mercury electrode. However, in the presence of complexing agents, the polarographic characteristics of the manganous ion are more complicated. Moros and Meites [66, 71] found that in neutral NaCN solution the reduction of manganese(II) [$Mn(CN)_6^{4-}$ and $Mn(H_2O)(CN)_5^{3-}$] proceeds reversibly to manganese(I) rather than to manganese(0) at the dropping mercury electrode. The addition of NaOH to the solution of Mn^{2+} in 1 M NaCN results in an additional wave with a half-wave potential of -1.72 V which is attributed to an irreversible two electron reduction of $Mn(OH)(CN)_5^{4-}$ to Mn^0. In the presence of complexing agents for manganese(III), anodic waves for the oxidation of manganese(II) have been observed (Table 2.1.1).

TABLE 2.1.1. Polarographic Characteristics

Substance	Product	Conditions	E (V vs SCE)	I	$E_{3/4} - E_{1/4}$[a] (mV)	Remarks	Refs.
			Aqueous				
Mn(II)	0	1 M KCl	-1.51w		sl irr	Trace gelatin	[61], [62]
		5 M $CaCl_2$	-1.44				[63]
		1 M NaF, pH 2.5-7	-1.55	3.93		0.01% Gelatin	[64]
		1 M NaCl, NaBr, or NaI, 30°C	-1.48		q-rev	0.005% Gelatin	[65]
		1 M $NaClO_4$	-1.472	3.85	-30		[66]
		0.1 M Na_2SO_4, $NaClO_4$ ionic strength = 1.5	-1.472		-33	0.005% Gelatin	[67]
		1 M Na_2SO_4	-1.533	2.94	-35		[66]
		1 M NH_4SCN	-1.544	3.96	-48		[66]
		1 M KSCN	-1.54fw		-36		[68]
		1 M HCOONa	-1.565	3.96	-70	0.005% Gelatin	[69]
		1 M HCOONa, NaClO ionic strength = 2	-1.516		-42	0.005% Gelatin	[67]
		1 M NH_3, 1 M NH_4Cl	-1.65			0.005% Gelatin	[61]
		1 M NH_3, 1 M NH_4Cl, 1 M N_2H_4	-1.116fi		-31	Acute max with or without Triton	[68]
		0.1 M $(NH_4)_3Cit$, pH 6.1	-1.615w		-52		[70]
		0.1 M $(NH_4)_3Cit$, 0.1 M NH_3 pH 8.5	-1.62i		-71		[70]
		0.1 M NH_3, 0.1 M $(NH_4)_2Ox$	-1.599fw		-37		[70]

(continued)

TABLE 2.1.1 (continued)

Substance	Product	Conditions	E (V vs SCE)	I	$E_{3/4} - E_{1/4}$ [a] (mV)	Remarks	Refs.
		0.1 M NH_3, 0.1 M $(NH_4)_2$Mal	-1.535w		-35		[70]
		0.1 M NH_3, 0.1 M $(NH_4)_2$Tart	-1.528w	2.1	-35		[70]
Mn(II) $[Mn(CN_6{}^{-4}]$ and $[Mn(H_2O)(CN)_5{}^{3-}]$	I	1 M NaCN	-1.364	1.6	-56		[66], [71]
Mn(II) $[Mn(OH)(CN_5{}^{-4}]$	0	0.88 M NaCN, 0.2 M NaOH	-1.72		-34.4	Wave at -1.36 V still present, i_d smaller	[66], [71]
Mn(II)	- 0	1 M NaOH, 1 M C_5H_5N	-1.53 -1.72i			Small irr prewave	[68]
Mn(II)	- - 0	1 M NaOH, std $Na_2P_2O_7$	-0.77i -0.96w -1.71fw		 rev	Two irr prewaves of equal height possibly due to Mn(III) formed	[68]
Mn(II) $[Mn(OH)_4{}^{-2}]$	III $[Mn(OH)_4{}^{-}]$	9 M KOH	(-0.29) vs Hg/HgO				[14]
Mn(II)	III 0	1 M NaOH, 1 M Na_2Mal	(-0.43)i -1.72i		irr sl irr	Extensive ppt from 1 mM Mn^{2+}	[73]
Mn(II)	III 0	1 M NaOH, 1 M Na_2Lact	(-0.46) -1.76fw		irr Nearly rev	Abnormally small	[73]
Mn(II)	III	0.1 M NaOH, 0.3 M $(C_2H_4OH)_3N$	(-0.5) -1.61		+56		[64], [72]
Mn(II)	III 0	2 M NaOH, 0.25 M Na_2Tart	(-0.46)w -1.82i	-0.95	+46 -40		[74]

Mn(III)	II	2 M NaOH, 0.25 M Na_2Tart	-0.88fi		-80		[74]
Mn(III) [$Mn(OH)_4^-$]	II [$Mn(OH)_4^{2-}$]	9 M KOH	-0.5 vs Hg/HgO				[14]
Mn(III)	II	0.4 M $K_2P_2O_7$, pH 2.3	>0	1.17		0.02%	[64]
		1 M K_2O_x, 0.4 M HOAc, 0.2 M NH_4OAc	>0	1.40			[64]
		1 M NaCN	>0				[66], [71]
Mn(III) (MnF_4^-)	II	0.08 M NaF	>0		irr	Triton x-100 or Na-DDBS	[75]
Mn(VII) (MnO_4^-)	II	$BaCl_2$	>0				[61]
			Nonaqueous				
Mn(II)	0	0.1 M $NaClO_4$ in acetonitrile	-1.12	4.66	-27		[76]
		0.1 M Et_4NClO_4 in benzonitrile	-0.98		-31		[77], [78]
		0.1 M Et_4NClO_4 in phenolacetonitrile	-1.02		-38		[78]
		0.1 M Et_4NClO_4 in acrylonitrile	-1.05		-32		[78]
		0.1 M Et_4NClO_4 in propionitrile	-1.10		-36		[78]
		0.2 M $NaClO_4$ in formamide	-1.57	1.78	-33		[79]
		0.1 M $NaClO_4$ in dimethylformamide	-1.12 (Ag/AgCl, satd Et_4NCl in DMF)		-32		[80]

[a]Some values of $E_{3/4} - E_{1/4}$ were calculated from the reported values of slopes of E vs log[i/i_d - i)] plots.

The reduction of manganese(II) at the dropping mercury electrode occurs at a more positive potential in various nitrile solutions than in aqueous solution. Larson and Iwamoto [77, 78] suggested that the relative ease of the reduction of Mn^{2+} in the nitrile solutions is due to the solvation sphere, surrounding the manganous ion, serving as a bridge for electron transfer. On the other hand the reduction of manganese(II) in the formamide and acetamide solutions occurs at a more negative potential than in aqueous solutions. The shift of the half-wave potential and the diffusion current with different solvents was discussed by Gaur and Goswami [81] in terms of the dielectric constant of the solvent, the viscosity of the solution, the effective ionic radius of the solvated manganese(II) ion, and the chemical nature of the solvent.

The reduction of most of the manganese(III) complexes to manganese(II) at the dropping mercury electrode begins at such a positive potential that it is impossible to determine the half-wave potential. However, in a basic tartarate solution and in 9 to 10 M potassium or sodium hydroxide solution, the manganese(III) complexes are relatively stable and the reduction of manganese(III) to manganese(II) can be determined (Table 2.1.1).

2.2. VOLTAMMETRIC CHARACTERISTICS

The polarographic characteristics of such higher valent manganese species as permanganate and manganate are not very well defined at the dropping mercury electrode. The reduction of these ions begins at such an anodic potential that oxidation of the dropping mercury electrode occurs. Although it is possible to measure the diffusion current for the reduction of permanganate to manganese(II) [61] or to manganese dioxide [82] under various conditions, the determination of half-wave potentials is virtually impossible at the dropping mercury electrode.

The reduction of permanganate in acidic solution at a bubbling platinum electrode [83] was studied by Desideri [84, 85] using the potential sweep technique. He found that the half-wave potential for the reduction of permanganate to manganese(II) in 1 M H_2SO_4 was 0.943 V vs the saturated calomel electrode and, with increasing acid concentration, shifted to more anodic values with a concurrent decrease in the diffusion current. Under controlled potential (0.4 V vs SCE) electrolyses, approximately 24 C were used in the reduction process of 0.05 mmole of MnO_4^- to Mn^{2+}; only about 11 C were used for the reduction process in 11 M or more H_2SO_4 solutions. These results indicate a partial decomposition of permanganate with evolution of oxygen and formation of intermediate ionic species between permanganate and manganous ions in concentrated H_2SO_4 solutions. Furthermore, it was found that in concentrated acidic solutions, MnO_4^- is reduced in the presence of Mn^{2+} according to:

$$MnO_4^- + 3Mn^{2+} + 6H^+ \rightarrow 3Mn^{3+} + MnO^{2+} + 3H_2O \qquad (2.2.1)$$

The Mn^{3+} ions are stabilized by forming manganese(III)-sulfate complexes and can be reduced at the bubbling platinum electrode with a half-wave potential of +0.65 V vs the saturated calomel electrode.

TABLE 2.2.1. Voltammetric Characteristics

Substance	Product	Condition	Electrode	Technique	Potential (V vs SCE)	Rev	Remarks	Refs.
MnO_4^-	Mn(II)	1 M H_2SO_4	bub Pt	E swp	0.943	irr	$E_{1/2}$ changes with H_2SO_4 concentration	[84], [85]
			Graphite	E swp	0.94-1.0	irr		[88]
MnO_4^-	MnO_4^{2-}	5 M NaOH or 0.3 M NaOH + 0.7 M $NaClO_4$	rot Pt dsk	E swp	0.584 (vs NHE)	rev		[86]
MnO_4^{2-}	Mn(V)	5 M NaOH	rot Pt dsk	E swp	0.282 (vs NHE)	rev	$E_{1/2}$ more anodic in lower conc NaOH	[86]
MnO_4^{2-}	Mn(IV)	1-8 M NaOH	bub Pt	E swp	0.15-0.17	irr		[84]
MnO_4^{2-}	Mn(VII)	2 M NaOH or 0.8 M KOH	rot Pt Graphite	E swp	(0.35)			[82], [87]
Mn(IV)	III	15 N H_2SO_4	Pt	E swp	~0.5 (vs $Hg/HgSO_4$)			[89]
Mn(III)	II	6 N H_2SO_4	bub Pt	E swp	0.65			[85]
		1 M KOH + 1% mannitol	DME	sq wave pol	-0.35			[90]
		0.5 M KOH + 0.2 M triethanolamine	DME	sq wave pol	-0.5			[91]

The polarographic reduction of permanganate to manganate and manganese(V) in alkaline solutions at the rotating platinum disk electrode was studied by Schurig and Heusler [86]. The reduction processes proceed reversibly as shown by the fact that the slopes of the E vs log $[i/(i_d - i)]$ plots are -59 mV for both the permanganate to manganate and the manganate to manganese(V) processes. The half-wave potential of the MnO_4^{2-}-Mn^{5+} wave was found to be influenced by the activity of OH^- in the electrolyte which was attributed to the hydrolysis of the MnO_4^{3-} ion:

$$MnO_4^{3-} + H_2O \rightleftharpoons HMnO_4^{2-} + OH^- \qquad (2.2.2)$$

The change in the half-wave potential with respect to the OH^- activity can be expressed as

$$E_{1/2} = E^\circ_{MnO_4^{2-}/MnO_4^{3-}} + \frac{RT}{F} \ln\left(1 + \frac{a_{H_2O}}{Ka_{OH^-}}\right) \qquad (2.2.3)$$

where $E_{1/2}$ is the half-wave potential of the MnO_4^{2-}-Mn^{5+} wave, $E^\circ_{MnO_4^{2-}/MnO_4^{3-}}$ the standard potential, a_{H_2O} and a_{OH^-} are the activities of H_2O and OH^-, respectively, and K the equilibrium constant for the hydrolysis Eq. (2.2.2). From the $E_{1/2}$ in electrolytes with various a_{OH^-}, K was calculated according to Eq. (2.2.3) to be 0.3 ± 0.1 at 25°C.

Landsberg, Hendal, and Muller [87] found that the oxidation of manganate to permanganate in potassium hydroxide solutions at both platinum and graphite electrodes proceeds reversibly with a half-wave potential of 0.345 ± 0.005 V vs the saturated calomel electrode. This is in good agreement with the value of 0.35 V obtained by den Boef and Poeder [82] at the rotating electrode in sodium hydroxide solutions.

The half-wave potential of permanganate to manganese dioxide at the bubbling platinum electrode in sodium hydroxide solution changes slightly with the sodium hydroxide concentration. In 1 M NaOH the half-wave potential is 0.15 V and it shifts to 0.17 V in 10 M NaOH [84].

The voltammetric characteristics of manganese species are shown in Table 2.2.1.

3. KINETIC PARAMETERS

The kinetics of the reduction of manganese(II) at the dropping mercury electrode have been investigated in a variety of supporting electrolytes [64-66]. From the results of their polarographic studies on manganese(II) in perchlorate and halide solutions, Gaur and Goswami [65] concluded that the standard rate constant at the standard potential for the reduction of manganese(II) at the dropping mercury electrode in these electrolytes is in the order of 5×10^{-4} cm/sec at 30°C and 10^{-3} cm/sec at 40°C as shown in Table 3.1. The rate constant for the reduction of manganese(II) at the dropping mercury electrode is slightly higher in the halide solutions than in the perchlorate solution. According to the authors this may be attributed to the formation of an ion-pair bridge between the halide ions adsorbed on the electrode and the manganese ions which facilitates the approach of Mn^{2+} to the electrode.

From a study on the dissolution and deposition of manganese at an electrolytic manganese electrode, Hurlen and Valand [92] found that in a solution of 0.01 M H_2SO_4, 0.39 M K_2SO_4, and 0.1 M $MnSO_4$, the Tafel lines for the anodic and the cathodic processes intercept at a potential of -1.232 V vs the normal hydrogen electrode and a current density of 1.9×10^{-4} A/cm^2. According to the crystal growth theory for the deposition and dissolution of metal [93], this current is half of the actual exchange current density. Accepting the value of -1.18 V as the standard potential of the manganese(II)-manganese electrode, the reversible potential -1.232 V gives a value of 0.017 M for the activity of $MnSO_4$. Therefore, according to Hurlen and Valand, the standard exchange current density of the manganese(II)-manganese electrode is 0.022 A/cm^2 at 22-23°C. Furthermore, these authors [94] also found that the standard exchange current density j_0 varies with the absolute temperature T according to

$$j_0 = 7.1 \times 10^6 T \times \exp\left(-\frac{15{,}100}{RT}\right) \text{ A/cm}^2 \tag{3.1}$$

Heusler and Bergmann [95] studied the deposition of manganese on a platinum electrode and the dissolution of the electrolytic manganese in $MnSO_4$ and $MnCl_2$ solutions containing $(NH_4)_2SO_4$ and NH_4Cl, respectively. The solutions were pre-electrolyzed by cathodically polarizing manganese-covered platinum with a surface area of 0.5 cm^2 for 2 to 3 days at a current of 25 mA. Both transient and steady-state measurements were made at 25°C. It was found that the anodic and cathodic transient polarization curves were Tafel lines with slopes close to 58 mV while the Tafel line for the deposition of manganese under steady-state conditions gave a slope of 29 mV. From the intercept of the anodic and the cathodic Tafel lines and their slopes, Heusler and Bergmann found the standard exchange current density to be 0.006 ± 0.002 A/cm^2 at the standard potential, with a transfer coefficient for the cathodic process of 0.50 ± 0.04 for the manganese(II)-manganese electrode in the $MnCl_2$-NH_4Cl solutions. The difference between the values of standard exchange current density obtained by Heusler and Bergmann and by Hurlen and Valand is because Hurlen and Valand considered the experimental value to be one-half of the exchange current density [92].

The Mn^{3+}(aq) ion is stable only in concentrated H_2SO_4 solutions or solutions containing other complexing anions. Issa, El Sammahy, and Ghoneim [75] studied the reduction of MnF_4^- in 0.08 M NaF solutions at the dropping mercury electrode and found that the kinetic parameters for the manganese(III) to manganese(II) process were affected by the concentration and the type of surfactant as shown in Table 3.1.

Thiele and Landsberg [8] studied the kinetics of the manganate-hypomanganate couple at a rotating platinum electrode in concentrated KOH solution. From the galvanostatically measured initial current-voltage curves the standard exchange current density and the

TABLE 3.1. Kinetic Parameters

Reaction	Conditions	Electrode	Technique	Rate[a]	α	Remarks	Refs.
$Mn^{2+} + 2e \rightarrow Mn(Hg)$	1 M $NaClO_4$	DME	dc pol	$k = 4.5 \times 10^{-5}$			[66]
	1 M Na_2SO_4	DME	dc pol	$k = 1.2 \times 10^{-5}$			
	1 M NH_4SCN	DME	dc pol	$k = 2.7 \times 10^{-5}$			
	1 M $NaClO_4$	DME	dc pol	$k = 5.4 \times 10^{-4}$	0.66	30°C	[65]
	1 M NaCl	DME	dc pol	$k = 5.9 \times 10^{-4}$	0.66	30°C	
				$k = 1.8 \times 10^{-3}$	0.45	40°C	
	1 M NaBr	DME	dc pol	$k = 6.5 \times 10^{-4}$	0.66	30°C	
				$k = 1.8 \times 10^{-3}$	0.40	40°C	
	1 M NaI	DME	dc pol	$k = 6.8 \times 10^{-4}$	0.66	30°C	
				$k = 1.8 \times 10^{-3}$	0.54	40°C	
$Mn^{2+} + 2e \rightarrow Mn$	0.1 M $MnSO_4$, 0.3 M K_2SO_4, 0.01 M H_2SO_4	Mn	i stp	$j_0 = 1.9 \times 10^{-4}$ at $[Mn^{2+}] = 0.39$ M	0.5	Experimental value	[92]
	1 M NH_4Cl, 2 M $MnCl_2$	Pt	i stp	$j_0 = (6 \pm 2) \times 10^{-3}$	0.5 ± 0.04	Standard exchange current density	[95]

$Mn^{3+} + e \rightarrow Mn^{2+}$	15 <u>N</u> H_2SO_4	Pt			0.28		[12]
$MnF_4^- + 2H^+ + e \rightarrow Mn^{2+} + 2HF_2^-$	0.08 <u>M</u> NaF	DME	dc pol	$k_0 = 3.8 \times 10^{-5}$	0.27	4.6×10^{-3}% Triton x-100	[75]
				$k_0 = 4.6 \times 10^{-6}$	0.32	1.8×10^{-2}% Triton x-100	
				$k_0 = 3.7 \times 10^{-6}$	0.32	5.2×10^{-2}% Triton x-100	
				$k_0 = 2 \times 10^{-3}$	0.14	3.6×10^{-3}% $DDBSO_3Na$	
				$k_0 = 7.5 \times 10^{-4}$	0.16	1.1×10^{-2}% $DDBSO_3Na$	
				$k_0 = 3.8 \times 10^{-4}$	0.16	6.8×10^{-2}% $DDBSO_3Na$	
$MnO_4^{2-} + e \rightarrow MnO_4^{3-}$	10.8 <u>M</u> KOH	Pt rot dsk	i stp	$j_0 = 0.6 \pm 0.1$	0.15 ± 0.05	$10^\circ C$, standard exchange current density	[8]
$MnO_4^- + e \rightarrow MnO_4^{2-}$	1 <u>M</u> NaOH	Pt	E stp	$j_0 = (12.3 \pm 0.5) \times 10^{-3}$ at $[MnO_4^-] = [MnO_4^{2-}] = 5 \times 10^{-3}$ <u>M</u>			[96]
		Au	E stp	$j_0 = 10.3 \times 10^{-3}$ at $[MnO_4^-] = [MnO_4^{2-}] = 5 \times 10^{-3}$ <u>M</u>			

[a] k = rate constant at $E = E^\circ$, cm/sec. k_0 = rate constant at E = 0 vs reference electrode, cm/sec. j_0 = exchange current density in A/cm^2.

transfer coefficient at 10°C were determined to be 0.6 ± 0.1 A/cm^2 and 0.15 ± 0.05, respectively, in 10.8 M KOH for the electrode reaction

$$MnO_4^{2-} + e = MnO_4^{3-} \tag{3.2}$$

The permanganate-manganate redox system was studied by Plieth [96] in NaOH solutions at both platinum and gold electrodes at 25°C. In order to eliminate diffusion polarization, potentiostatic pulse measurements were carried out in this study. Initial current densities obtained by extrapolating the current-time curves at controlled potentials to zero time were used in the construction of the linear current density-overpotential plots at potentials near the equilibrium potential and the current density-potential plots in the Tafel region. Accordingly, the following results were obtained:

1. At electrodes pretreated in nitric acid, the value of exchange current density j_0 obtained from the charge transfer resistance

$$R_D = \left(\frac{\partial \eta}{\partial j}\right)_{E \to E^\circ}$$

(η = overpotential; j = initial current density) is different from that obtained from the extrapolation of the Tafel plot to the equilibrium potential. Furthermore, the exchange current density from the anodic branch of the Tafel plot j_{0+} is different from that of the cathodic branch j_{0-}. The sum of the apparent charge transfer coefficient for the cathodic process and the anodic process from the Tafel plots is not unity.

2. At platinum electrodes which had not been cleaned in nitric acid, the permanganate-manganate behaved in a manner conforming to the simple charge transfer mechanism

$$MnO_4^{-} + e = MnO_4^{2-} \tag{3.3}$$

The exchange current density obtained from R_D is identical with that obtained from the Tafel plot. The transfer coefficients add up to unity with the cathodic transfer coefficient equal to 0.62. In a 1 M NaOH solution containing 10^{-2} M MnO_4^- and 10^{-2} M MnO_4^{2-}, the exchange current density at the platinum electrode without the nitric acid pretreatment was found to be 0.11 mA/cm^2, about two orders of magnitude lower than that at the nitric acid treated platinum electrode which was 20 mA/cm^2.

These results led Plieth to the conclusion that on a "clean" electrode an adsorption/desorption process is involved in the MnO_4^-/MnO_4^{2-} electrode reaction as an intermediate step. The mechanism can be expressed as

$$MnO_4^{-} \rightleftharpoons (MnO_4)^*_{ad} + \lambda e \tag{3.4a}$$

$$(MnO_4)^*_{ad} + (1 - \lambda)e \rightleftharpoons MnO_4^{2-} \tag{3.4b}$$

TABLE 3.2. Exchange Current Density of the MnO_4^-/MnO_4^{2-} Electrode [96]

Concentration (M)		$j_0\left(=\frac{RT}{FR_D}\right)$ (mA/cm^2)		j_{0^+} (anodic Tafel plot) (mA/cm^2)		j_{0^-} (cathodic Tafel plot) (mA/cm^2)		$j_0\left(=\frac{j_{0^+} \times j_{0^-}}{j_{0^+} + j_{0^-}}\right)$ (mA/cm^2)	
MnO_4^-	MnO_4^{2-}	Pt electrode	Au electrode	Pt electrode	Au electrode	Pt electrode	Au electrode	Pt electrode	Au electrode
1×10^{-3}	5×10^{-3}	11.7	7.6	27	18	20	14.5	11.5	8.0
5×10^{-3}	1×10^{-3}	9.5	5.3	22.5	9.4	18.5	15	9.9	5.5
5×10^{-3}	5×10^{-3}	11.7-12.8	10.3	30	20-21	21-22	20-21	12.4-12.7	10-10.5
5×10^{-3}	10×10^{-3}	14.6		38		22.5		14.1	
5×10^{-3}	50×10^{-3}	18.3	23.3	44	52	28.5	32	17.3	20
10×10^{-3}	5×10^{-3}	15.1	13.9	33	23	27.5	31	15	13.2
50×10^{-3}	5×10^{-3}	18.4		40		35		18.7	

where $(MnO_4)^*_{ad}$ represents the adsorbed species on the electrode and λ the partial charge transfer coefficient. According to Plieth and Vetter [97, 98], the charge transfer resistance R_D in this case can be related to the exchange current densities from the Tafel plots, j_{0+} and j_{0-} by

$$R_D = \frac{RT}{F}\left(\frac{1}{j_{0+}} + \frac{1}{j_{0-}}\right) \tag{3.5}$$

and the nominal exchange current density j_0 can be represented by:

$$j_0 = \frac{j_{0+}j_{0-}}{j_{0+} + j_{0-}} \tag{3.6}$$

The values of the exchange current density calculated from j_{0+} and j_{0-} according to Eq. (3.6) agree very well with those calculated from the charge transfer resistance as shown in Table 3.2.

It is noted that if the electrode were not cleaned by nitric acid, very few active sites would be available for the adsorption/desorption process. Therefore, the permanganate/manganate reaction would be forced to proceed via the more difficult route of the simple charge transfer process as shown by Eq. (3.3). Accordingly, the exchange current density is very much lower than the values shown in Table 3.2, as discussed previously.

4. ELECTROCHEMICAL REACTIONS

4.1. ELECTROCHEMICAL BEHAVIOR OF MANGANESE — DEPOSITION, HYDROGEN EVOLUTION, DISSOLUTION, AND PASSIVATION

Manganese is the least noble metal which can be electrodeposited from aqueous solutions on a technical scale. The current efficiency and the crystal structure of the electrodeposited manganese are influenced by factors such as the substrate, the impurities in the electrolyte, the current density, and the temperature. In carefully prepared and purified NH_4Cl-$MnCl_2$ [99, 100], or $(NH_4)_2SO_4$-$MnSO_4$ solution [101], coarse grained γ-manganese was deposited on platinum, cobalt, nickel, and brass electrodes. The current efficiency was found to be as high as 80 to 90% at current densities of 40 to 100 mA/cm^2. However, in the presence of small amounts such impurities as SO_3^{2-} and SeO_3^{2-}, α-manganese was deposited with reduced current efficiency. Despite the decrease in the current efficiency for the deposition of manganese, an increase in hydrogen evolution was not observed. Presumably, the current was consumed by the reduction of the impurities. Furthermore, an increase in hydrogen overpotential was observed in the presence of SO_3^{2-} or SeO_3^{2-} [101].

The kinetics of the cathodic codeposition of manganese and hydrogen in various chloride or sulfate solutions has been investigated by various authors [92, 94, 95, 102, 103].

Galushko and Loshkarev [102, 103] found that the slopes of the Tafel lines for the discharge of H^+ and Mn^{2+} are, respectively, 33 and 25 mV per decade of current density. The low value for the slope of the Tafel line for hydrogen evolution was attributed to a catalytic mechanism such as

$$Mn^{2+} + ne \rightarrow Mn^{(2-n)+} \tag{4.1.1a}$$

$$Mn^{(2-n)+} + nH^+ \rightarrow Mn^{2+} + \frac{n}{2} H_2 \tag{4.1.1b}$$

where n = 1 or 2. Furthermore, the similarity of the Tafel slopes for the deposition of manganese and the evolution of hydrogen suggested that the rate of the catalytic evolution of hydrogen is determined by the stage of the electron addition to manganese or by the formation of an active complex of the type $H_2O\text{-}Mn^{2+}\text{-}e$.

Recently, Heusler and Bergmann [95] investigated the deposition of manganese in the $MnSO_4$-$(NH_4)_2SO_4$ and $MnCl_2$-NH_4Cl solutions. Contrary to the findings of Galushko and Loshkarev, these investigators found that the slope of the Tafel line for hydrogen evolution is 120 mV per decade of current density at 25°C and the cathodic steady-state polarization curve is pH independent, which agrees with the result obtained by Hurlen and Valand [92] in the investigation of the electrochemical behavior of manganese in $MnSO_4$-K_2SO_4 and $MnCl_2$-KCl solutions. These results indicate a direct hydrogen evolution rather than a catalytic process proposed by Galushko and Loshkarev [102, 103].

It is conceivable that both processes can take place, depending on the experimental conditions. As Gamali and Stender [100] pointed out, the most essential requirement in the study of hydrogen overpotential on manganese is the purity of the electrolyte. The contradictory results obtained by different investigators may be attributed to differences in impurities in their electrolyte solutions.

According to Heusler and Bergmann [95], the slope of the steady-state Tafel line for the deposition of manganese is 29 mV per decade of current density. Furthermore, a reaction order of 1.5 with respect to the manganese ions was determined. This rather unusual reaction order can be explained by assuming that the charge transfer reaction takes place only at specific sites at the solid manganese surface (e.g., semicrystal sites), the surface activity of which depends upon the potential and the concentration of the manganese ion.

It is expected from the thermodynamic properties that corrosion or dissolution of manganese will take place in aqueous solution. Spontaneous corrosion and anodic dissolution of manganese does indeed occur in acid sulfate, chloride, perchlorate, and acetate solutions [104-106]. Evidently, this can be attributed to the fact that the manganese salts of these anions are extremely soluble. In sulfate or chloride solution with pH values between 0 and 3 the corrosion rate of manganese is determined by the diffusion rate of H^+ to the surface of the metal [104].

In acid phosphate and arsenate solutions, manganese can be passivated by applying a high anodic current as shown by Melendez and Brenet [107] and Heusler [106]. As expected, the corrosion rate of the passive manganese is extremely low. Heusler [106] found that in 0.5 M Na_3PO_4-H_3PO_4 solutions the corrosion current of the passive manganese electrode is 1 μA/cm^2 at pH 1 and 4 μA/cm^2 at pH 4, respectively, at 22°C after the electrode has been maintained at a potential of about 1.3 V vs the saturated calomel electrode for 4 days. Similar to the behavior of passive iron electrodes, the passive manganese electrodes can be self-activated. When the passivation potential is interrupted and the passive manganese electrode left at open circuit, the potential of the electrode decreases slowly at first. Then the rate of potential decay begins to increase sharply and the electrode becomes active. The electrode potential where the inflection point occurs in the potential decay curve is known as the Flade potential. According to Heusler, the pH dependence of the Flade potential of manganese, E_F, is

$$E_F = 1.50 - 0.118 \text{ pH} \qquad (4.1.2)$$

The Flade potential of the manganese electrode is determined by the kinetics of the dissolution of the passive layer rather than the thermodynamics of the manganese electrode. Furthermore, the passive layer, by and large, is a thin film of manganese oxides with the oxidation state of manganese being determined by the electrolyte and other experimental conditions.

4.2. ELECTROCHEMISTRY OF MANGANESE(III)

It has been discussed previously that manganese(III) is not stable in aqueous solutions with respect to its reduction by water and to the precipitation of its oxide. Therefore, the oxidation and reduction of the manganese(III) ions in aqueous solution has been studied in the presence of complexing agents for manganese(III) ions or in highly concentrated acid solutions. Gorbachev and Belyaeva [108, 109] studied the oxidation and reduction of the manganese(III)-manganese(II) system at platinum electrodes in phosphate solution. They concluded that the rate of the cathodic process was determined by the discharge process of Mn^{3+} from a complex anion such as $[Mn(Na_2P_2O_7)_3]^{3-}$. In the anodic process, on the other hand, the rate was diffusion limited.

The manganese(III)-manganese(II) and the manganese(IV)-manganese(III) redox systems at platinum electrodes were investigated in 15 N H_2SO_4 by Vetter and Manecke [12, 13, 110]. They found that the kinetics of the Mn^{3+}/Mn^{2+} electrodes are simple. The charge transfer reaction is also the overall electrode reaction, and the reaction rate is diffusion limited. However, a more complicated reaction mechanism was proposed for the Mn^{4+}/Mn^{3+} electrode:

$$2Mn^{3+} \rightleftharpoons Mn^{2+} + Mn^{4+} \qquad (4.2.1a)$$

$$Mn^{2+} \rightleftharpoons Mn^{3+} + e \qquad (4.2.1b)$$

$$Mn^{3+} \rightleftharpoons Mn^{4+} + e \text{ . . . overall electrode reaction} \qquad (4.2.1)$$

Further studies of the manganese(IV)-manganese(III) and the manganese(III)-manganese(II) systems are found in more recent publications. Guidelli and Piccardi [89, 111] determined the diffusion coefficients of Mn^{2+}, Mn^{3+}, and Mn^{4+} in 15 N H_2SO_4 to be 0.98×10^{-6}, 1.10×10^{-6}, and 0.90×10^{-6} cm^2/sec, respectively.

Selim and Lingane [112] investigated the stability of manganese(III) in sulfuric acid media. The equilibrium constant for the disproportionation reaction determined by these authors ranges between 10^{-3} in 4 M H_2SO_4 and 10^{-4} in 7 M H_2SO_4 at 25°C. These values are one to two orders of magnitude smaller than those determined by Grube and Huberich [113] who evaluated the constant from emf measurements of the Mn^{3+}/Mn^{2+} and Mn^{4+}/Mn^{3+} couples. However, as Selim and Lingane noted, it is extremely difficult to attain a steady electrode potential for the Mn^{4+}/Mn^{3+} couple and the thermodynamic validity of potentials obtained under such conditions is questionable. Therefore, it is reasonable to assume that the values of the disproportionation constants obtained by Selim and Lingane are more reliable and manganese(III) is rather stable in sulfuric acid, presumably due to its complexation with sulfate or bisulfate anions.

4.3. ELECTROCHEMISTRY OF MANGANESE DIOXIDE

Manganese dioxide is undoubtedly one of the most studied compounds in the field of electrochemistry. Since 1866, when Leclanché invented the Zn-MnO_2 cell, manganese dioxide has been an important battery reactant.

In a report entitled "Literature Search on Dry Cell Technology with Special Reference to Manganese Dioxide and Methods for its Synthesis," Bolen and Weil [114] compiled abstracts for no less than 3654 publications for the period from 1872 through 1947. In 1952 Heise and Cahoon [115] published a review paper on the Leclanché cells. In 1959 Vosburgh [35] reviewed the structures and properties of manganese dioxide and the discharge mechanisms of manganese dioxide electrodes. More recent developments in the mechanistic studies of the manganese dioxide electrode will be reviewed here. Furthermore, an attempt will be made to formulate a unified theory compatible with the various findings from previous investigations of the electrochemical behavior of the manganese dioxide electrode.

Manganese dioxide is usually classified according to crystal structure into α, β, γ, δ, etc. varieties. They are more or less different substances of the approximate composition MnO_2 with or without other elements. In a recent review of the solid state properties of manganese dioxides, Malati [116] discussed the different crystal structures, the nonstoichiometry, the electrical and magnetic properties, and the general compositions of various types of manganese dioxide. The electrochemical properties also vary with the different types of manganese dioxide. As described in the thermodynamic discussion, a stable standard potential of the MnO_2/Mn^{2+} electrode can be measured with β-MnO_2, but it is difficult to do this with γ-MnO_2 because of its ability to form solid solutions with lower manganese oxides over a wide range of compositions. In view of the fact that γ-MnO_2 can be discharged more efficiently than other types of MnO_2, it is widely used in batteries as the cathode material and most of the following discussion will be concerned with γ-MnO_2.

4.3.1. Cathodic Reactions of Manganese Dioxide in Neutral or Acidic Solutions

The determination of the mechanism of the manganese dioxide electrode has been approached from 1) analyses of the reaction products (by both physical and chemical methods), 2) analyses of the pH-potential relationship, and 3) analyses of the discharge characteristics.

Cahoon and co-workers [117-119] investigated the manganese dioxide electrode in NH_4Cl or $NH_4Cl + ZnCl_2$ solutions with various pH values. The findings were 1) the slope of the potential-pH plot was 60 mV/pH, 2) Mn^{2+} was found in the electrolyte, and 3) MnO(OH) and, in the presence of zinc, $ZnO.Mn_2O_3$ were identified by x-ray diffraction measurements as solid products at the cathode. From these findings, Cahoon, Johnson, and Korver [118] concluded that two types of reactions take place at the cathode: manganese dioxide is reduced to Mn^{2+} as the primary cathodic reaction, and the product Mn^{2+} is then oxidized by the remaining MnO_2 to form MnO(OH) and $ZnO.Mn_2O_3$ at the cathode. These reactions can be summarized as

$$MnO_2 + 4H^+ + 2e \longrightarrow Mn^{2+} + 2H_2O \qquad (4.3.1)$$

$$Mn^{2+} + MnO_2 + 2OH^- \longrightarrow 2MnO(OH) \qquad (4.3.2)$$

$$Mn^{2+} + Zn^{2+} + MnO_2 + 4OH^- \longrightarrow ZnO.Mn_2O_3 + 2H_2O \qquad (4.3.3)$$

The formation of Mn^{2+} and the lower oxidation state manganese oxides which can be found in the Leclanché type of electrolytes and in the electrodes from the cathodic reduction of manganese dioxide has been observed by other investigators. Vosburgh and co-workers [35-39, 120-126] thoroughly investigated the characteristics of manganese dioxide electrodes. The presence of manganeous ion in the electrolyte as a reduction product of manganese dioxide electrodes was observed. However, the analyses of the electrolyte at different stages of discharge revealed that the formation of Mn^{2+} ions occurred only during the later stages of discharge. In fact, the initial product of the cathodic reaction at the manganese dioxide was MnO(OH) [35]. Very little manganese(II) entered the solution during the first third of the discharge in ammonium chloride solution at pH 7. However, it appeared in increasing amounts thereafter [35, 38, 121, 124]. Mn^{2+} ions were detected much earlier in solutions with pH values of less than 5 [35, 38, 122-124]. These observations led Vosburgh and co-workers to conclude:

1. The initial product of the cathodic reaction is MnO(OH):

$$MnO_2 + H + e \rightarrow MnO(OH) \qquad (4.3.4)$$

2. In acidic electrolytes MnO(OH) can react with the electrolyte to form Mn^{2+} and MnO_2:

$$2MnO(OH) + 2H^+ \rightarrow MnO_2 + Mn^{2+} + 2H_2O \qquad (4.3.5)$$

3. In neutral electrolyte the MnO(OH) accumulated on the surface of the electrode can be further reduced to $Mn(OH)_2$ which dissolves in the electrolyte as Mn^{2+}:

$$MnO(OH) + H^+ + e \rightarrow Mn(OH)_2 \qquad (4.3.6)$$

$$Mn(OH)_2 + 2H^+ \rightarrow Mn^{2+} + 2H_2O \qquad (4.3.7)$$

Furthermore, the diffusion of protons and electrons in the MnO_2 lattice is considered to be involved in the formation of MnO(OH) [35]. When current is applied, electrons from the electrode and protons from the electrolyte meet at the surface of MnO_2 exposed to the electrolyte and also beneath the surface as protons penetrate the MnO_2 crystal lattice. The electrons are retained by Mn^{4+} to form Mn^{3+} and protons are attached to O^{2-} ions to form OH^- ions.

The manganese dioxide crystal structure has a profound effect on the discharge behavior of the electrode. Under similar discharge conditions in an electrolyte of pH 7.5, Vosburgh and Lau [39] found that the overpotential of a γ-manganese dioxide electrode is substantially less than that of an α-manganese dioxide electrode with identical capacity and physical dimensions. From their analytical results these authors also found that the α-manganese dioxide electrode produces substantially larger amounts of Mn^{2+} in the electrolyte than the γ-manganese dioxide electrode during discharge. Furthermore, the oxide composition at the end of discharge was found to be $MnO_{1.57}$ for the γ-electrode and $MnO_{1.66}$ for the α-electrode. These differences between the two electrodes are consistent with the theory that the diffusion of protons in the lattice is involved in the discharge process and that α-MnO_2 is less penetrable for protons than γ-MnO_2.

The mechanism proposed by Vosburgh's school on the cathodic process of manganese dioxide has been generally confirmed by recent studies. Vetter and Jaeger [24, 32] have shown through thermodynamic studies the wide range of homogeneity of γ-MnO_2 and the ability of γ-MnO_2 to form solid solutions with lower manganese oxides. Bode and Schmier [34, 127-129] found that the reduction of γ-MnO_2 proceeds in a homogeneous phase to $MnO_{1.5}$, α-MnO_2 in a homogeneous phase to $MnO_{1.86\pm0.01}$. On the other hand, the reduction of β-MnO_2 takes place heterogeneously as early as the composition of manganese dioxide reaches $MnO_{1.97}$. These results again point out the fact that the capability of γ-MnO_2 in forming solid solution with lower manganese oxides and that the diffusion of protons takes place more readily in the lattice of γ-MnO_2 than that of α- or β-MnO_2.

The diffusion of protons and electrons in the lattice of manganese dioxide was proposed by Brenet years ago [130]. Ghosh and Brenet [131, 132] studied the reaction products at various discharge stages of γ-MnO_2 by means of x-ray and electron diffraction, electron microscopy, differential thermal analysis, thermogravimetry, and magnetic susceptibility. They concluded that during the initial stages of discharge the manganese dioxide lattice is dilated by proton addition. This conclusion is in agreement with that reached by Vosburgh. However, there is a difference between the Brenet theory and the Vosburgh theory. Brenet considered that the initial discharge process involved the random reduction of Mn^{4+} to Mn^{2+} in the lattice whereas Vosburgh considered the electrochemical process to produce MnO(OH). In the case of γ-MnO_2 it is immaterial whether the reduction of MnO_2 proceeds through manganese(III) or manganese(II) oxides since it is recognized [24, 32] that γ-MnO_2 possesses a wide range of homogeneity and the formation of the manganese oxides proceeds homogeneously in the lattice, resulting in a continuous change of the stoichiometry of "manganese dioxide." In fact, Gabano, Morignot, and Laurent [133, 134], using the theory of homogeneity of Vetter and Jaeger [24, 32], proposed a general equation for the reduction of γ-manganese dioxide:

$$MnO_{2n-2}(OH)_{4-2n} + 2dnH^+ + 2dne \rightarrow MnO_{2n-2+2dn}(OH)_{4-2n+2dn} \qquad (4.3.8)$$

where n is the degree of oxidation having any value between 1.5 and 2. This equation presents a good concept of the continuity existing between the two infinitesimally close states of reduction. Nevertheless, since MnO(OH) has been identified analytically as the electrochemical reduction product from MnO_2 [35, 131, 132], the Vosburgh equation

$$MnO_2 + H^+ + e \rightarrow MnO(OH) \qquad (4.3.9)$$

is more acceptable as a special case for the general expression shown by Eq. (4.3.8).

From the above discussion, one can conclude that the reduction of manganese dioxide (at least for the gamma variety) proceeds through the diffusion of protons and electrons in the lattice.

Johnson and Vosburgh [37] found that the potential of a MnO_2/MnO(OH) electrode is a function of the mole ratio R of MnO_2 and MnO(OH) in the solid solution. At 25°C it can be expressed as

$$E = E_0 + 0.073 \log R \qquad (4.3.10)$$

where E is the potential of the MnO_2/MnO(OH) electrode; E_0 has an average value of 0.416 V vs the saturated calomel electrode at pH 7.5. The difference between the value of the constant 0.073 and the expected value of 0.059 is attributed to the fact that R is the mole ratio rather than the activity ratio of MnO_2 and MnO(OH) in the electrode. Benson, Price, and Tye [135] examined the equilibria involved in a manganese dioxide electrode and concluded that in the absence of Mn^{2+} ion in the electrolyte the potential is determined by the surface condition of the electrode in the event the surface is not in equilibrium with the bulk of the electrode.

Based on these findings and conclusions, an examination can be made for the detailed mechanisms of the polarization and the open circuit potential recuperation processes. Upon discharge of the MnO_2 electrode, MnO(OH) is formed on the surface of the electrode, causing a decrease in the electrode potential. At the same time the removal of MnO(OH) from the electrode surface will also take place either by diffusing toward the interior of the electrode

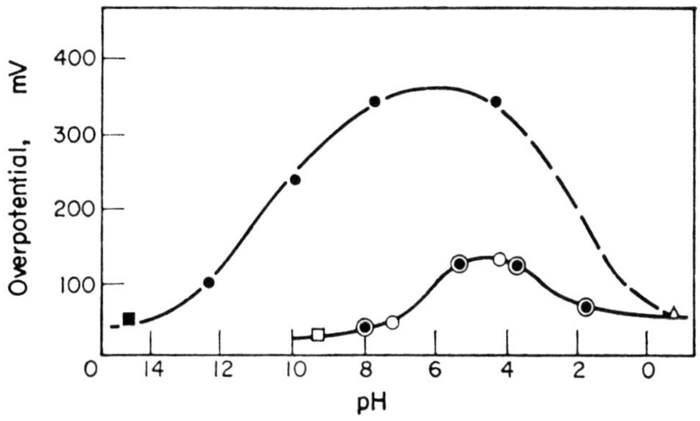

FIG. 4.3.1. Effect of pH and $[NH_4^+]$ on the overpotential of the γ-manganese dioxide electrode. Electrode: 17 mg MnO_2 electrodeposited on a 0.6-cm diameter graphite rod (geometric area = 10 cm^2). Discharge current: 1 mA. Temperature: 25°C. Electrolytes: (■) 3 M KOH; (●) 3 M KCl; (□) 3 M NH_3; (◉) 1.5 M $(NH_4)_2SO_4$; (○) 3 M NH_4Cl; (△) 1.5 M H_2SO_4 [136].

(involving protons and electrons) or by disproportionation. In acidic electrolytes the removal of the reduction product is primarily through the reaction [123, 136]

$$2MnO(OH) + 2H^+ \rightarrow MnO_2 + Mn^{2+} + 2H_2O \qquad (4.3.11)$$

The rate of this reaction is higher than that of the diffusion of protons. As a result, the overpotential of the manganese dioxide electrode in an acidic electrolyte is considerably lower than that in a neutral electrolyte where the removal of MnO(OH) is increasingly dependent on the slower process of proton and electron diffusion. Figure 4.3.1 shows the discharge overpotential* of a γ-MnO_2 electrode as a function of the pH value of the electrolyte as reported by Era, Takehara, and Yoshizawa [136]. It is noted from Fig. 4.3.1 that under a similar discharge current of 0.1 mA/cm^2, the overpotential of the γ-MnO_2 electrode is less than 100 mV in 3 N H_2SO_4 but more than 300 mV in 3 M KCl. It is also noted that in electrolytes of similar pH value, the overpotential is much lower in the presence of NH_4^+ ions. Indeed, the NH_4^+ ion is assumed to act as a proton-donor to MnO(OH):

$$2MnO(OH) + 2NH_4^+ \rightarrow MnO_2 + Mn^{2+} + 2NH_3 + 2H_2O \qquad (4.3.12)$$

Furthermore, in the electrolytes containing NH_4^+ ions, NH_3 can complex metal ions such as Zn^{2+}, Ni^{2+}, or Mn^{2+} and lower the overpotential as predicted by Eq. (4.3.12). The effect of NH_4^+ on the discharge overpotential of the manganese dioxide electrode was also shown

*Defined by Era, Takehara, and Yoshizawa as the difference between the open circuit potential in the steady state attained after a discharge and the closed circuit potential at the end of discharge.

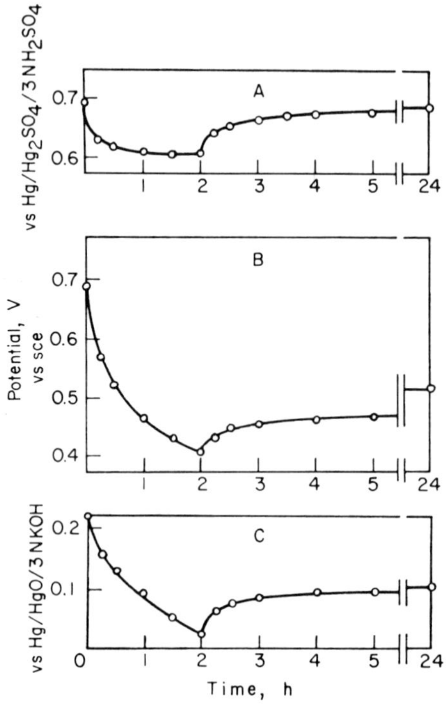

FIG. 4.3.2. Comparative growth of polarization and recovery of open circuit potential of γ-manganese dioxide electrode in various electrolytes. Electrode: disk type; 20 mm diameter, 3 mm thick (0.5 g electrolyte MnO_2, 0.2 g acetylene black). Discharge current: 10 mA. Temperature: 25°C. Electrolytes: (A) 1.5 M H_2SO_4; (B) 3 M NH_4Cl; (C) 3 M KOH [136].

by Chreitzberg and Vosburgh [122]. Figure 4.3.2, also from the work by Era, Takehara, and Yoshizawa [136], shows the discharge and open circuit recuperation curves of a γ-MnO_2 electrode in 3 N H_2SO_4, 3 N NH_4Cl, and 3 N KOH solutions, respectively. In the 3 N H_2SO_4 electrolyte the electrode reaches a practically constant potential within 1 hr of discharge, indicating a steady-state condition with respect to the formation and removal of MnO(OH) on the electrode surface. On the other hand, the electrode potential in the 3 N NH_4Cl electrolyte continues to decrease even after 2 hr. Figure 4.3.2 also shows a profound effect of the pH value of the electrolytes on the open circuit potential recuperation. In 3 N H_2SO_4 the rate of recuperation is relatively high. Furthermore, the open circuit potential of the electrode approximates the original value, indicating that the MnO(OH) is removed completely by the disproportionation. Indeed, the products of the disproportionation are Mn^{2+} and MnO_2. Mn^{2+} dissolves in the electrolytes and only MnO_2 remains in the electrode, and the original open circuit potential of the electrode is restored. On the other hand, in 3 N NH_4Cl the rate of recuperation is lower than that in 3 N H_2SO_4 and the final open circuit potential is lower than that of the original electrode. In NH_4Cl the disproportionate reaction of MnO(OH) takes place at a lower rate and to a lesser extent than in H_2SO_4. The diffusion of protons and electrons becomes more important. As the electrode approaches the equilibrium state, the electrode potential is determined by the Johnson-Vosburgh equation. The presence of MnO(OH) in the solid solution causes the lowering of the open circuit potential after discharge. Evidently the diffusion resistance of protons in the manganese dioxide lattice is substantially lower in strong alkaline solution than in neutral solutions as shown by the low discharge overpotential of the manganese dioxide electrode in KOH solutions (Figs. 4.3.1 and 4.3.2). Era, Takehara, and Yoshizawa [136] proposed the following to explain this phenomenon: When manganese dioxide is in contact with a solution of high pH value, the OH group present in the lattice [137, 138] polarizes and weakens the bond between the O and H. As a result, protons can move with relative ease in the solid phase.

The diffusion of protons and electrons in manganese dioxide has been examined quantitatively [40, 41, 139-146]. Scott [40] considered a one dimensional semi-infinite solid manganese dioxide electrode and obtained the following equations for the diffusion of protons in the manganese dioxide electrode:

$$C(t, 0) = \frac{2F_0}{(D\pi)^{1/2}}[t^{1/2} - (t - T)^{1/2}] \qquad (4.3.13)$$

$$C(t, x) = \frac{2F_0}{D^{1/2}}\left[t^{1/2}\left(i\,\mathrm{erfc}\,\frac{x}{2(Dt)^{1/2}}\right) - (t - T)^{1/2}\left(i\,\mathrm{erfc}\,\frac{x}{2D^{1/2}(t - T)^{1/2}}\right)\right] \qquad (4.3.14)$$

where $C(t, 0)$ = concentration of MnO(OH) on the surface of the electrode ($x = 0$), $C(t, x)$ = concentration of MnO(OH) in the interior, ($x > 0$), F_0 = equivalents of MnO(OH) produced per unit area per unit time during discharge at a constant current density, D = diffusion coefficient of proton (accompanied by electron) in the solid, t = time following the start of discharge, and T = time at which discharge stops and recuperation begins. For $t < T$ the term involving $(t - T)$ is to be taken as zero. By substituting the concentration-time relation on the surface of the electrode into the Johnson-Vosburgh equation and taking account of the fact that the sum of the concentration of MnO_2 and that of MnO(OH) is constant, the general potential-time relation can be computed:

$$E = E_0 + k \log\left[\frac{(\pi D)^{1/2}}{2F_0 t^{1/2} - (t - T)^{1/2}} - 1\right] \qquad (4.3.14)$$

The polarization and recuperation behavior of the manganese dioxide electrode in the NH_4Cl-NH_4OH electrolyte (pH = 7.5) was investigated experimentally by Kornfeil [141] and compared with the results calculated from Eq. (4.3.14). The agreement was good. However, deviations occurred at the start of the polarization and after relatively extended periods of potential recuperation. The deviation at the beginning of the discharge can be attributed to the assumption of zero MnO(OH) concentration making $E \to \infty$. The deviation during the latter part of the recuperation is due to the fact that the conditions for semi-infinite diffusion are no longer realized. Although the potential-time relation is derived from the surface conditions of the electrode, it is important to realize that as equilibrium approaches during recuperation, the concentration ratio of manganese dioxide and the lower manganese oxides in the entire electrode determines the electrode potential.

The water in the manganese dioxide lattice plays an important role in the electrochemical process. It has been suggested [138] that the presence of water in the lattice promotes the diffusion of protons and that the water molecule in the manganese dioxide lattice can be present as H_2O, H_3^+O, and OH^-. Johnson and Vosburgh [27] used the formula $H_2MnO \cdot xMnO_2$ to designate the hydrated manganese dioxide. Vetter and Jaeger [24, 32] used the general formula of $MnO_{2n-2}(OH)_{4-2n}$ or $MnO_n \cdot (2 - n)H_2O$ to show the homogeneity and the presence of H_2O in the lattice of γ-manganese dioxide. Brenet and co-workers [142, 143] proposed the formula $MnO_x(OH)_{4-2x}$ and later [31] another formula $MnO_{n-z}(OH)_{2z} \cdot mH_2O$ to show the active acidic OH group. The presence of water in manganese dioxide has been well established. Tvarusko [138] reviewed briefly the findings in this respect. Furthermore, he also found from analyses of a wide variety of manganese dioxide samples that the total water content (both in the lattice and adsorbed) varied from 1% for a highly crystalline β-MnO_2 to 20% for synthetic, hydrous MnO_2 when equilibrated in an atmosphere of 75% relative humidity. The effect of the water in the lattice on the diffusion of proton during discharge was clearly shown by Era, Takehara, and Yoshizawa [136]. Manganese dioxide of the gamma variety was heated at 200°C for 3 hr to remove part of the crystalline water while maintaining the gamma structure. The polarization behavior of the heat-treated γ-MnO_2 was compared with that of the original γ-MnO_2. In 3 N H_2SO_4 no significant change was observed in the overpotential (85.5 vs 88.2 mV) under the following discharge conditions:

Electrode:	disk type (0.5 g MnO_2, 0.2 g acetylene black)
Geometric area:	2 cm^2
Discharge current:	10 mA
Temperature:	25°C
Discharge duration:	2 hr

On the other hand, in 3 M NH_4Cl under the same discharge condition the overpotential increased from 201.5 mV for the original γ-MnO_2 to 285.0 mV for the heat-treated γ-MnO_2. The overpotential increased from 78.7 to 145 mV after the heat treatment when the electrodes were discharged in 3 N KOH under the same conditions. In acid solution the removal of the discharge product does not depend upon proton diffusion and the overpotential is not affected by the water content in the lattice. As the pH of the electrolyte increases, the proton diffusion becomes more important for the removal of MnO(OH). Therefore the amount of water in the lattice which facilitates the process has a profound effect on the overpotential, especially in alkaline solutions where the removal of MnO(OH) depends solely on the proton diffusion process.

4.3.2. Manganese Dioxide in Alkaline Solutions

The cathodic reduction of electrolytic manganese dioxide was studied by Cahoon and Korver [144] in 7.6 N KOH using Zn(Hg) as the counter electrode. Based on analyses of the products at various stages of discharge, the authors postulated the formation of $Mn(OH)_2$ from the initial stage of the discharge and also suggested the formation of a hypothetical oxide Mn_4O_7. However, the formation of $Mn(OH)_2$ at the beginning of the discharge has not been confirmed by later investigators. It is conceivable that the $Mn(OH)_2$ formation observed by Cahoon and Korver was due to the high current density (approximately 15 mA/cm^2 or 60 mA/g) used in the reduction process, and under these conditions reduction to $Mn(OH)_2$ may have taken place on the surface of the electrode [145]. Furthermore, Cahoon and Korver used 500 cc of NH_4Cl solution to extract the manganous ion from the discharged electrode. Although the pH value of the suspension adjusted to 5.4 wherein Mn_2O_3 or MnO(OH) may be stable, the presence of $NH_4{}^+$ and Zn^{2+} ions may cause the disproportionation of Mn_2O_3 according to

$$2MnO(OH) + 2NH_4{}^+ \rightarrow Mn^{2+} + MnO_2 + 2H_2O + 2NH_3 \qquad (4.3.15)$$

$$Zn^{2+} + nNH_3 \rightarrow Zn(NH_3)_n{}^{2+} \qquad (4.3.16)$$

Bell and Huber [145] investigated the cathodic reduction of both γ- and β-manganese dioxide in electrolytes containing 40 wt% KOH and 7.5 wt% ZnO. At low rates (0.5 mA/g) and intermittent discharge (100 hr/week), the γ-MnO_2 electrode exhibits a three-step discharge curve. The electrode potential decreases gradually until the manganese oxide reaches the composition $MnO_{1.7}$. Between $MnO_{1.7}$ and $MnO_{1.625}$ the electrode potential shows a rather rapid decrease. From $MnO_{1.625}$ to $MnO_{1.47}$ the potential is approximately constant. Between $MnO_{1.47}$ and $MnO_{1.43}$ the electrode potential shows another rapid decrease and remains constant beyond $MnO_{1.43}$. Furthermore the rest potentials of the electrode at various stages of discharge exhibit similar behavior as the potential under load, viz., an initial continuous decrease as the oxidation state of the manganese decreases and remains constant after the composition reaches $MnO_{1.625}$ and $MnO_{1.47}$, respectively. Based on these results, Huber and Bell concluded that for the γ-MnO_2 electrode the reduction proceeds in three steps:

1. MnO_2-$MnO_{1.7}$ homogeneous phase reduction.
2. $MnO_{1.7}$-$MnO_{1.47}$ heterogeneous system.
3. Below $MnO_{1.47}$ heterogeneous system.

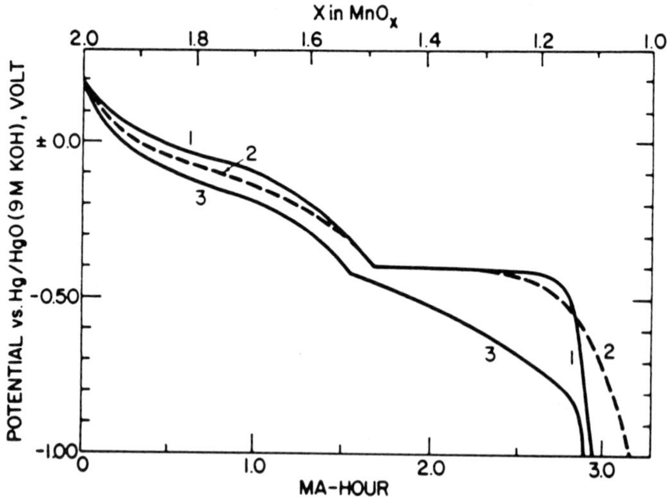

FIG. 4.3.3. Discharge curves of the γ-manganese dioxide electrode in 9 M KOH solution. Electrode: 10.6 mg MnO_2 electrodeposited on a 7.7-cm long, 0.45-cm diameter graphite rod. Discharge current: (1) 0.11 mA; (2) 0.33 mA; (3) 3.0 mA. Temperature: 23°C [14].

During the reduction in the homogeneous phase, lattice dilation has been observed by x-ray diffraction. It should be pointed out that under the low current density and intermittent discharge condition, recrystallization of the lower manganese oxides may occur. Indeed, when the discharge stage reached $MnO_{1.625}$, Bell and Huber [145] observed cementation of the electrode and γ-Mn_2O_3 in the electrode from the x-ray diffraction pattern.

The behavior of the β-MnO_2 electrode is distinctly different from that of the γ-MnO_2 electrode. According to Bell and Huber [145], the following steps are involved in the reduction of β-MnO_2 electrode:

1. MnO_2-$MnO_{1.96}$ homogeneous phase reduction.
2. $MnO_{1.96}$-$MnO_{1.8}$ heterogeneous system.
3. $MnO_{1.8}$-$MnO_{1.6}$ homogeneous phase reduction.
4. $MnO_{1.6}$-$MnO_{1.48}$ heterogeneous system.
5. Below $MnO_{1.48}$ heterogeneous system.

Recently Kozawa, Yeager, and Powers [14, 15, 47, 146, 147] conducted detailed investigations on the γ-MnO_2 electrode in alkaline electrolytes. In order to eliminate any possible complications in interpreting the data, the authors avoided the use of zinc anodes and zinc oxide in the electrolyte. The discharge behavior of the γ-MnO_2 electrode depends upon the OH^- ion concentration of the electrolyte. In concentrated alkaline solutions such as 9 M

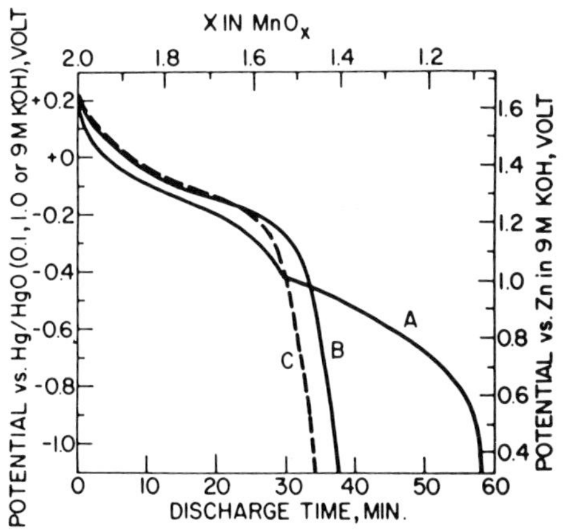

FIG. 4.3.4. Effect of KOH concentration on the discharge behavior of the γ-manganese dioxide electrode: (A) 9 M KOH; (B) 1 M KOH; (C) 0.1 M KOH. Electrode: 10.6 mg MnO_2 electrodeposited on a 7.7-cm long, 0.45-cm diameter graphite rod. Discharge current: 3 mA. Temperature: 23°C [14].

KOH, two distinct discharge steps are shown in the discharge curves (Fig. 4.3.3). In the first step MnO_2 is reduced to $MnO_{1.5}$ and the potential decreases continuously to about -0.4 V vs the Hg/HgO (9 M KOH) reference electrode. In the second step $MnO_{1.5}$ is reduced to $MnO_{1.0}$ and the potential remains practically constant especially under low current densities. On the other hand, in dilute KOH solutions such as 0.1 or 1.0 M KOH, the second step discharge is absent and the potential drops rapidly to a value lower than -1.0 V vs Hg/HgO as the electrode reaches $MnO_{1.5}$ as shown by Curves B and C in Fig. 4.3.4. Furthermore, in the dilute KOH solutions, the addition of triethanolamine, which is a complexing agent for Mn^{2+} and Mn^{3+}, can restore the second discharge step as shown by Fig. 4.3.5. The open circuit equilibrium potential of the electrode also decreases with the oxidation state of the electrode in the first step of discharge; however, it remains unchanged in the second step of discharge as shown by Fig. 4.3.6. These observations led Kozawa and coworkers to the following conclusions:

1. In concentrated alkali solutions the reduction of the MnO_2 electrode involves two distinct steps.

2. The first step, reduction of MnO_2 to $MnO_{1.5}$, involves a homogeneous process. As shown by the schematics in Fig. 4.3.7, electrons are introduced into the MnO_2 lattice and Mn^{4+} is reduced to Mn^{3+}. As a result of electron exchange between Mn^{4+} and Mn^{3+} in the lattice, the position of Mn^{3+} moves around in the entire lattice. At the same time, H_2O decomposes at the electrolyte-electrode interface and protons are introduced into the lattice to form OH^-. OH^- also moves around in the entire lattice by means of proton jumping from

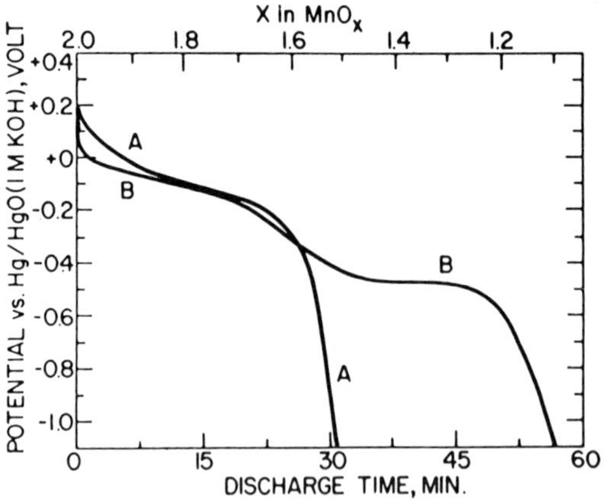

FIG. 4.3.5. Effect of triethanolamine on the discharge behavior of the γ-manganese dioxide electrode in 1 M KOH: (A) 1 M KOH; (B) 1 M KOH + 20% by volume triethanolamine. Electrode: 10.6 mg MnO_2 electrodeposited on a 7.7-cm long, 0.45-cm diameter graphite rod. Discharge current: 3 mA. Temperature: 23°C [14].

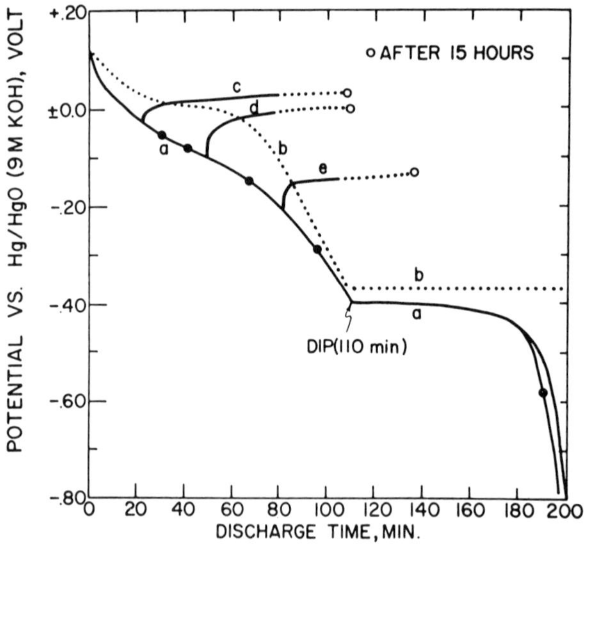

FIG. 4.3.6. Discharge curve and open circuit potential recovery of the MnO_2 electrode in 9 M KOH: (a) discharge curve; (b) open circuit equilibrium potential; (c, d, e) recovery curve. Electrode: 3 mg MnO_2 electrodeposited on a 7.7-cm long, 0.45-cm diameter graphite rod. Discharge current: 0.25 mA. Temperature: 23°C [47].

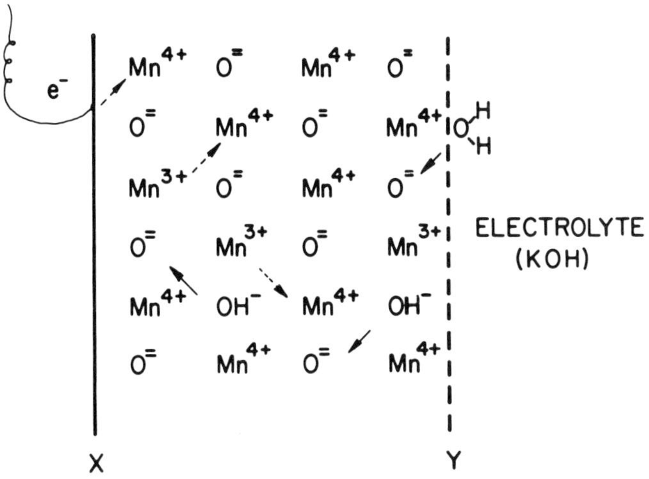

FIG. 4.3.7. Diffusion of protons and electrons in the manganese dioxide lattice during discharge. - - →: Electron movement. —►: Proton movement. X: Interface of MnO_2 and electronic conductor. Y: Interface of MnO_2 and electrolyte [47].

one O^{2-} site to another. The cathodic reduction of the first step may be expressed as:

$$MnO_2 + H_2O + e \rightarrow MnO(OH) + OH^- \qquad (4.3.17)$$

[It must be pointed out, however, that this is a simplified picture of the first discharge step. In fact, Kozawa and Powers [47], in discussing the details of the discharge mechanism, suggested that while the reduction of MnO_2 to $MnO_{1.5}$ proceeds in a homogeneous phase, a thin layer of Mn_2O_3 or MnO(OH) may begin to form in a separate phase at the later stage of discharge. A detailed discussion will be given later in this section.]

3. The second step, reduction of $MnO_{1.5}$ to $MnO_{1.0}$, occurs heterogeneously. Due to the formation of complex ions $Mn(OH)_4^-$ and $Mn(OH)_4^{2-}$, manganese(III) and manganese(II) oxides are quite soluble in concentrated KOH solutions. The electrochemical reduction of $MnO_{1.5}$ to $MnO_{1.0}$ takes place in the solution phase. These processes can be expressed as

$$MnO(OH)(s) \rightleftharpoons Mn(III)\ (\text{in solution}) \qquad (4.3.18a)$$

$$Mn(III) + e \rightarrow Mn(II)\ \ldots\ \text{electrochemical reduction} \qquad (4.3.18b)$$

$$Mn(II)\ (\text{in solution}) \rightleftharpoons Mn(OH)_2(s) \qquad (4.3.18c)$$

Our discussion of the manganese dioxide electrode indicates repeatedly that the electrochemical reduction of manganese dioxide proceeds homogeneously both in neutral and alkaline

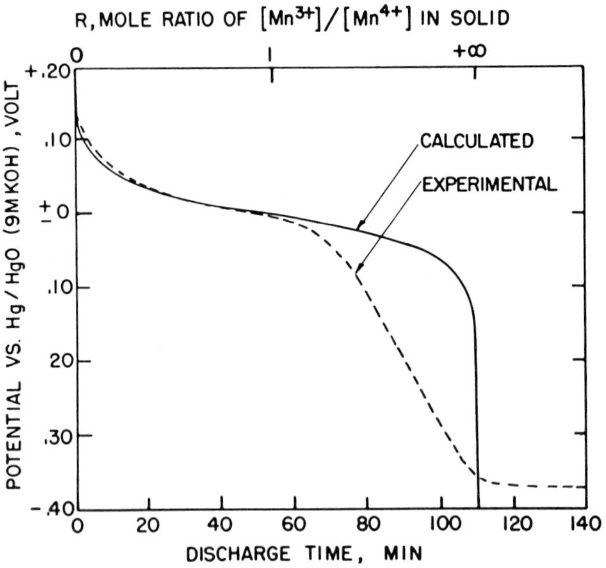

FIG. 4.3.8. Equilibrium potential of the manganese dioxide electrode in 9 M KOH as a function of $[Mn^{3+}]/[Mn^{4+}]$ in the electrode [47].

electrolytes via the proton and electron diffusion until the lattice becomes so strained it recrystallizes to another structure. Kozawa and Powers [47] considered the potential-generating mechanism of the homogeneous Mn^{3+}-Mn^{4+}-O^{2-}-OH^- phase in contact with the KOH electrolyte analogous to an aqueous redox system. Therefore the open circuit potential of the electrode at any stage of discharge between MnO_2 and $MnO_{1.5}$ may be expressed by Eq. (4.3.19) with the concentration of OH^- and H_2O assumed constant, and $[Mn^{3+}]$ and $[Mn^{4+}]$ being the concentration of Mn^{3+} and Mn^{4+}, respectively, in the solid electrode:

$$E = E^\circ - \frac{RT}{F} \ln \frac{[Mn^{3+}]}{[Mn^{4+}]} \tag{4.3.19}$$

Figure 4.3.8 shows a comparison between the calculated potential vs log $[Mn^{3+}]/[Mn^{4+}]$ relation from Eq. (4.3.19) and the experimental results. A good agreement is noted in the first part of the curve. However, disagreement between the calculated and the experimental curve occurs at the later stage. According to Kozawa and Powers [47], this is due to the formation of MnO(OH) or Mn_2O_3 in a new phase on the surface of the Mn^{4+}-Mn^{3+}-O^{2-}-OH^- system at the later discharge stage, and the electrode potential is controlled by the lower manganese oxides on the surface.

General agreement exists among various investigators [14, 47, 145, 148] in regard to the initial homogeneous process involved in the cathodic reduction of γ-MnO_2. Nonetheless,

Kozawa and co-workers [14, 15, 146] were the first to study the effect of the OH^- ion concentration on the cathodic reduction of manganese dioxide. They found that the further reduction of manganese(III) oxides depends on the solubility of manganese(III) in the electrolyte. In low concentration KOH solutions where the solubilities of both manganese(III) and manganese(II) oxides are extremely low, no further reduction of manganese(III) oxides can take place unless a complexing agent such as triethanolamine is added to increase the solubility of manganese(III) ions. On the other hand, in highly concentrated KOH solutions where the solubilities of manganese(III) and manganese(II) oxides are relatively large due to the complex ions $Mn(OH)_4^-$ and $Mn(OH)_4^{2-}$ (e.g., in 9 M KOH the solubilities of Mn^{3+} and Mn^{2+} are 4.4×10^{-3} and 0.4×10^{-3} M, respectively), the reduction of manganese(III) oxides to $Mn(OH)_2$ takes place and the electrode potential is independent of the depth of discharge, which characterizes the heterogeneous system. Furthermore these authors found that the limiting current for the reduction at this stage depends upon the apparent surface area rather than the true surface area of the manganese dioxide particles. In other words, the rate of dissolution of manganese(III) ions into the KOH solution controls the rate of reduction. Consequently the charge transfer step in the reduction of manganese(III) oxides to $Mn(OH)_2$ must be taking place in the solution phase.

The electrochemical characteristics of γ-MnO_2 in alkaline electrolytes were investigated by Liang and co-workers, and the results of the cathodic reduction of the γ-MnO_2 electrode are in excellent agreement with those obtained by Kozawa and co-workers. Furthermore, investigations were conducted on the anodic oxidation of the discharged manganese dioxide electrodes in alkaline electrolytes [149, 150]. The results of these investigations indicate that the state of discharge determines the rechargeability of the electrode as seen from the following discussion.

When the γ-MnO_2 electrode is reduced to -0.4 V vs Hg/HgO in both 10 and 1 M KOH solutions or to -1.0 V vs Hg/HgO in 1 M KOH solutions, the product is manganese(III) oxide. This reduced electrode can be re-oxidized efficiently as shown by the similarity of the discharge behavior of the reoxidized electrode and that of the original manganese dioxide electrode shown in Figs. 4.3.9(b) and 4.3.10. However, when the γ-MnO_2 electrode is discharged to $Mn(OH)_2$, which can be accomplished by reducing the electrode to -1.0 V vs Hg/HgO in high concentration KOH solutions, the electrode cannot be reoxidized efficiently and the reoxidized electrode behaves differently from the original manganese dioxide electrode [Fig. 4.3.9(a)]. In view of the fact that the reduction of manganese(III) oxide to $Mn(OH)_2$ occurs primarily in the solution phase and that the solubility of manganese(III) ions increases with the concentration of KOH, it is evident that the efficiency of manganese(III) oxide reduction decreases with the KOH concentration when the manganese dioxide is cathodized to -1.0 V vs Hg/HgO. Accordingly, the rechargeability of the manganese dioxide electrode which has been cathodized to -1.0 V vs Hg/HgO increases as the KOH concentration decreases as shown in Table 4.3.1. From the analytical results it is obvious that $Mn(OH)_2$ can be oxidized anodically to manganese(III) oxide. However, further oxidation of this manganese(III) oxide to

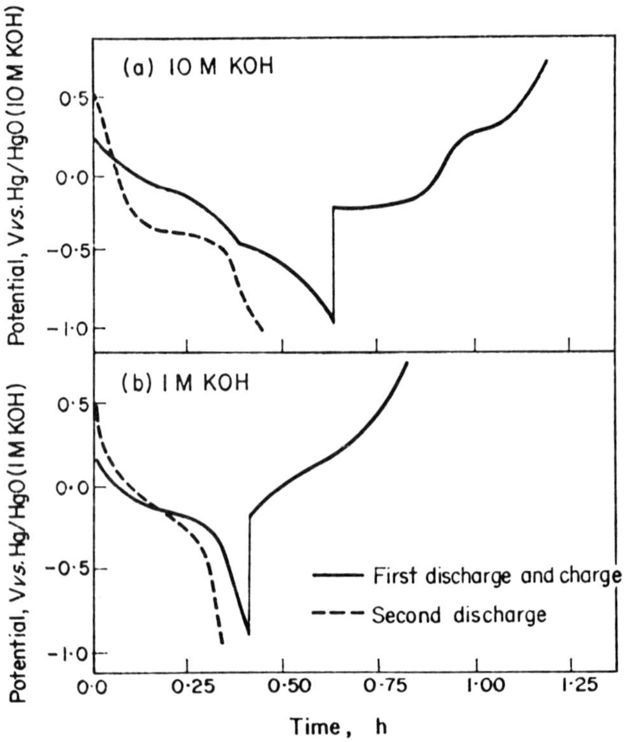

FIG. 4.3.9. Discharge-recharge curves of the manganese dioxide electrode. Electrode: 6 mg MnO_2 electrodeposited on a 5-cm^2 graphite rod. Current density: 1 mA/cm^2. Temperature: 25°C [150].

manganese(IV) oxide does not occur. It has been suggested that the manganese(III) oxide produced from the cathodic reduction of the MnO_2 electrode is different from that produced from the anodic oxidation of $Mn(OH)_2$. The former is believed to be α-MnO(OH) which is electrochemically active, while the latter is electrochemically inactive γ-Mn_2O_3.

Consequently, the electrochemical redox cycle of the γ-MnO_2 electrode in highly concentrated KOH solutions is postulated to be

$$\gamma\text{-}MnO_2$$
$$\downarrow\uparrow$$
$$\alpha\text{-}MnO(OH) \rightarrow Mn(III)\ (\text{in solution}) \rightleftharpoons \gamma\text{-}Mn_2O_3$$
$$\downarrow\uparrow$$
$$Mn(II)\ (\text{in solution}) \rightleftharpoons Mn(OH)_2$$

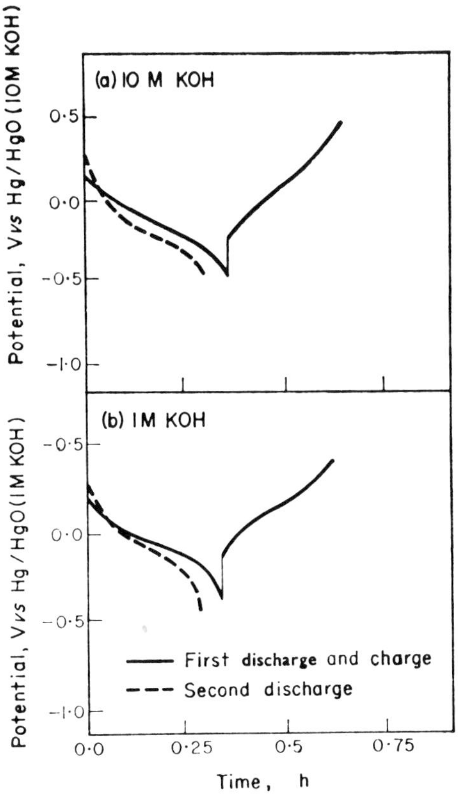

FIG. 4.3.10. Discharge-recharge curves of the manganese dioxide electrode. Electrode: 6 mg MnO_2 electrodeposited on a 5-cm^2 graphite rod. Current density: 1 mA/cm^2. Temperature: 25°C [150].

The first step, oxidation of $Mn(OH)_2$ to manganese(III) oxide, may involve the dissolution of manganese(II), its solution oxidation to manganese(III), followed by the precipitation of manganese(III). The precipitated form of manganese(III) is thought to be γ-Mn_2O_3 which is inactive to further oxidation. When the discharge process of γ-MnO_2 is stopped at the end of the first step, the reoxidation of the product manganese(III) oxide, which is α-MnO(OH), is efficient. Also, once manganese(III) is dissolved in KOH solution, reprecipitation to α-MnO(OH) seems to be inhibited.

The effect of the discharge depth on the rechargeability of the manganese dioxide electrode was also observed by Boden, Venuto, Wisler, and Wylie [151] who studied the discharge and charge behavior in 7 M KOH by x-ray diffraction and charge and discharge curves. However, these authors concluded that the formation of Mn_3O_4 upon deep discharge

TABLE 4.3.1. Effect of KOH Concentration on the Recharge Efficiency of the Discharged Manganese Dioxide Electrode at an Apparent Current Density of 1 mA/cm^2

KOH concentration (M)	Anodization time (sec)	Potential at the end of charge volt (vs Hg/HgO)	x in MnO_x in electrode		Percent[b] MnO_2 recovery efficiency	Percent[c] Mn in electrolyte
			Analytical	Calculated[a]		
1	0	Initial[d]	1.51	1.52	-	0
	850	0.2	1.74	1.78	50.0	0
	1100	0.4	1.85	1.89	73.9	0
	1300	0.6	1.93	1.96	91.3	0
2	0	Initial	1.45	1.43	-	0
	900	0.2	1.70	1.73	48.1	0
	1300	0.4	1.81	1.88	69.2	0
	1500	0.6	1.89	1.95	84.6	0
4	0	Initial	1.35	1.32	-	0
	1100	0.2	1.64	1.75	46.8	0
	1500	0.4	1.76	1.84	66.1	0
	1800	0.6	1.78	1.91	69.4	0
6	0	Initial	1.24	1.26	-	0
	1300	0.2	1.56	1.70	42.1	0
	1700	0.4	1.71	1.79	61.8	0
	1900	0.6	1.72	1.94	63.2	5
10	0	Initial	1.14	1.15	-	4
	650	-0.3	1.24	1.24	12.0	4
	1300	0.2	1.52	1.59	41.9	5
	1600	0.2	1.62	1.69	57.8	10
	1800	0.4	1.68	1.79	65.1	17
	1900	0.55	1.65	1.87	61.4	20

[a]Calculated on the basis of the total charge (current X time) involved in discharge and recharge, and the total amount of Mn in the electrode.

[b]Fraction of oxidized manganese to the total amount of manganese oxidizable to MnO_2.

[c]Fraction of manganese in the electrolyte at the end of each electrolysis to the total amount of manganese initially present.

[d]Initial state of the electrode is that obtained by cathodically discharging the original MnO_2 electrode to -1.0 V (vs Hg/HgO) in each KOH solution.

hinders the rechargeability whereas Kang and Liang [150] concluded that γ-Mn_2O_3 cannot be anodized to MnO_2 efficiently. It should be pointed out that the distinction between Mn_3O_4 and γ-Mn_2O_3 is difficult based on the x-ray diffraction patterns. Therefore, Boden, Venuto, Wisler, and Wylie did not rule out the existence of γ-Mn_2O_3 in the products of discharge although they gave more weight to Mn_3O_4 based on the observation of a Mn_3O_4 line of d = 1.57 Å in the x-ray diffraction pattern.

Ambrose and Briggs [152] studied the anodic oxidation of $Mn(OH)_2$ and the subsequent reduction of the anodized electrode in 1 M KOH solutions. The electrode was a cathodically formed $Mn(OH)_2$ layer on a platinum electrode in a manganese nitrate solution. The electrochemical behavior of the thin film $Mn(OH)_2$ electrode observed by Ambrose and Briggs is similar to that of the completely discharged manganese dioxide electrode observed by Kang and Liang [150]. During the anodic process, the electrode potential exhibits two arrests at about -0.2 and +0.2 V vs the Hg/HgO reference electrode before the oxygen evolution potential. During the cathodic process, the potential of the oxidized thin film electrode drops rapidly to a relatively constant value of about -0.35 V vs Hg/HgO where the mangaese(III) to manganese(II) process takes place.

It is interesting to note, however, that the reduction process was carried out in 1 M KOH solution by Ambrose and Briggs where the solubilities of manganese(III) and manganese(II) are small (estimated at 10^{-5} M [15]) and cannot sustain relatively high rate of discharge for the manganese(III) to manganese(II) process according to the mechanism proposed by Kozawa and Yeager [14, 47, 146]. Indeed, Ambrose and Briggs have found that during the cathodic process of the oxidized $Mn(OH)_2$ electrode, the potential only shows a long arrest at about -0.35 V vs Hg/HgO when the current density is as low as 30 $\mu A/cm^2$. When the current density is 0.2 mA/cm^2, the electrode polarizes to -1.0 V vs Hg/HgO rather rapidly. Although the electrochemical behavior of the $Mn(OH)_2$ thin film electrode observed by Ambrose and Briggs is in good agreement with that observed by Liang and Kang, they concluded that the oxygen to manganese ratio of the oxidation product is 1.85 ± 0.05 whereas the analytical results by Kang and Liang [150] have shown that $MnO_{1.65\pm0.05}$ is the product when $Mn(OH)_2$ is oxidized to 0.55 to 0.60 V vs Hg/HgO. These differences may be attributed to the fact that Ambrose and Briggs anodized the $Mn(OH)_2$ thin film electrode to and held at 0.65 V vs Hg/HgO for several minutes. That the adsorbed oxygen on the electrode caused a higher ratio of O/Mn is indeed a possibility.

5. APPLIED ELECTROCHEMISTRY

5.1. ELECTROPLATING AND ELECTROWINNING OF MANGANESE

Ungiadze and Agladze [153] investigated the electroplating of manganese in acid sulfate solution and found that high purity manganese (≥99.92%) could be deposited with a current efficiency of 40 to 45% under the following conditions:

Plating solution:	Mn^{2+} (25 g/l) and $(NH_4)_2SO_4$ (180 g/l), pH 2
Current density:	100 mA/cm^2
Temperature:	18 to 20°C

In the presence of a small amount of H_2SeO_3 (0.01-0.1 g/l), Banerjee and Dhananjayan [154] obtained a current efficiency of 90% for the manganese deposition in $MnSO_4$ solutions.

It was established by various investigators [101, 155-157] that the presence of impurities such as SO_3^{2-}, SeO_3^{2-}, TeO_3^{2-}, and CS_2 in the plating solution results in the deposition of α-manganese while manganese of the gamma variety can be deposited in the absence of these impurities. Shvab and Zosimovich [101, 155] observed that the presence of impurities such as CS_2, SeO_3^{2-}, and SO_3^{2-} reduces the current efficiency for the manganese deposition while Polukarov and Shul'gina [158-161] found that impurities such as H_2SeO_3, H_2TeO_3, and H_6TeO_6 increase the current efficiency. Both groups agree that these impurities cause an increase in the hydrogen overpotential at the cathode [101, 155, 161].

The electrodeposition of manganese from amide solutions of $MnCl_2$ was studied by Qazi and Leja [162] in a diaphragm cell. It was found that manganese with a purity of 98.6% can be deposited with a current efficiency of 76% under the following condition:

Catholyte:	1-1.5% Mn as $MnCl_2$ in $HCONH_2$, pH 7.5
Anolyte:	90-100 g/l NH_4Cl at about pH 2.5
Current density:	50 mA/cm^2
Temperature:	35°C

The electrowinning process of manganese metal beginning with the manganese ore was reviewed by Bacon [163]. The principal procedures may be summarized as follows:

1. The ore is crushed and roasted to MnO.

2. The roasted ore is leached with a solution of $(NH_4)_2SO_4$ (concentration maintained at 135-140 g/l) and H_2SO_4 (pH maintained at 2.5). The leaching process requires a period of 1.25 to 1.5 hr. Insoluble sulfates such as $CaSO_4$ are precipitated.

3. The solution is then neutralized to a pH of 6.5 to precipitate aluminum and iron as hydroxides; and molybdenum, arsenic, and silica are removed by coprecipitations.

4. Copper, zinc, nickel, and cobalt are quantitively removed by the addition of S^{2-}.

5. The electroplating is carried out in a diaphragm cell. The cathode is lead metal containing 1% silver while the anode may be Hastelloy, type 316 stainless steel, or titanium. The catholyte contains 30-40 g/l Mn as $MnSO_4$, 125-150 g/l $(NH_4)_2SO_4$, 0.1 g/l SO_2 and 0.008-0.016 g/l glue and is adjusted to a pH value of 6-7.2. The anolyte contains 10-20 g/l Mn as $MnSO_4$, 25-40 g/l H_2SO_4, and 125-150 g/l $(NH_4)_2SO_4$. Under a current density of 40-65 mA/cm^2, a current efficiency of 60-65% can be obtained for the deposition of manganese.

5.2. ELECTROPLATING OF MANGANESE DIOXIDE

Gamma or "battery grade" manganese dioxide can be anodically deposited from a hot acidic manganese sulfate solution. Typical operating conditions for the electroplating of γ-MnO_2 are [164]:

Electrolyte:	50-100 g/l $MnSO_4$, 50-75 g/l H_2SO_4
Cathode:	Graphite
Anode:	Graphite
Anode current density:	5-10 mA/cm^2
Temperature:	90-94°C

The current efficiency may be as high as 90 to 95%. In industrial processes, deposition of manganese dioxide is continued until the deposit is approximately 1 in. thick before it is stripped from the electrode. Typical battery grade manganese dioxide contains approximately 90% MnO_2.

Since the graphite anode is slowly consumed during the electrolysis, other anode materials such as titanium [164-168], platinized titanium [164, 166], lead dioxide [169], and lead [164] have been used in the manganese dioxide electroplating.

It was also noted that both manganese chloride and acid manganese nitrate can be used in place of manganese sulfate in plating solutions [170-172]. Indeed, Amano et al. [170] found that 100% current efficiency for the deposition of manganese dioxide is attainable in a plating bath containing 1.5 M $MnCl_2$ and 0.3-1 M HCl at a current density of 25 mA/cm^2 and 95°C.

On December 1, 1971, the Electrochemical Society announced the establishment of the Distribution Center for the International Common Sample of Manganese Dioxide [173]. Battery grade manganese dioxide samples from a variety of suppliers are available for distribution at a nominal price. When an appropriate new type of manganese dioxide is introduced to the market, the supplier may offer it to the center for sample distribution. The international common samples may serve as a control or a reference in order to study, evaluate, and compare various battery grade manganese dioxides. As of September 1, 1971, ten common samples were available in the distribution center.

5.3. BATTERY APPLICATIONS

The electrochemistry of manganese dioxide in chloride and alkaline solutions was discussed previously. In LeClanché and alkaline manganese batteries, manganese dioxide is used as the active cathode material with zinc being the active anode material. In addition to ammonium chloride, other chlorides such as zinc chloride and calcium chloride are used in electrolytes of the LeClanché batteries. The electrochemical system of the alkaline manganese battery is Zn/40% KOH/MnO_2. The use of the alkaline electrolytes in the Zn/MnO_2 system results in a substantial improvement in energy densities (watt-hours/unit volume and watt-hours/unit weight), shelf life, and rate capability.

5.4. MANGANESE DIOXIDE IN SOLID ELECTROLYTE CAPACITORS

Manganese dioxide of the beta variety is used as the "solid electrolyte" in tantalum capacitors. The solid electrolyte tantalum capacitors are based on the system Ta-Ta_2O_5-MnO_2 [174, 175] and are prepared by pyrolyzing manganese nitrate to form manganese dioxide on the anodized tantalum. The manganese dioxide serves to "heal" the flaws in the tantalum oxide phase by oxidizing the bare tantalum to tantalum oxide and to maintain the highly insulating properties of the dielectric and thus replace the electrolytic solution which acts as the cathode plate in conventional electrolytic capacitors.

ACKNOWLEDGMENT

The author wishes to express his appreciation to Drs. N. I. Jaeger, P. Bro, and R. G. Selim for their invaluable assistance during the preparation of this chapter.

REFERENCES

1. W. M. Latimer, Oxidation Potentials, 2nd ed., Prentice-Hall, Englewood Cliffs, New Jersey, 1952, pp. 234-241.
2. A. D. Mah, Selected Values of Chemical Thermodynamic Properties, U.S. Bureau of Mines Report of Investigations No. 5600, U.S. Govt. Printing Office, Washington, D.C., 1960.
3. J. J. Morgan, in Principles and Applications of Water Chemistry (S. D. Faust and J. V. Hunter, eds.), Wiley, New York, 1967, pp. 561-624
4. T. A. Zordan and L. G. Hepler, Chem. Rev., 68, 737 (1968).
5. D. D. Wagman, W. H. Evans, V. B. Parker, I. Harlow, S. M. Bailey, and R. H. Schuman, Selected Values of Chemical Thermodynamics, National Bureau of Standards Technical Note 270-4, U.S. Govt. Printing Office, Washington, D.C., May 1969.

6. F. D. Rossini, D. D. Wagman, W. H. Evans, and I. Jaff, Selected Values of Chemical Thermodynamic Properties, National Bureau of Standards Circular 500, U. S. Govt. Printing Office, Washington, D.C., 1952.

7. P. Pascal, Nouveau Traité de Chimie Minérale, Vol. 16, Masson, Paris, 1960, pp. 1028-1030.

8. R. Thiele and R. Landsberg, Z. Phys. Chem. (Leipzig), 236, 95 (1967).

9. I. Ya Garkavi, Sb. Nauch. Tr. Kemerov. Gorn. Inst., 1954, 225; Ref. Zh. Khim., 1956, Abstr. No. 6482; Chem. Abstr., 52, 130e (1958).

10. A. K. Covington, T. Cressey, B. G. Lever, and H. R. Thirsk, Trans. Faraday Soc., 58, 1975 (1962).

11. G. Jangg, Monatsh. Chem., 95, 1103 (1964).

12. K. J. Vetter and G. Manecke, Z. Phys. Chem., 195, 270 (1950).

13. K. J. Vetter, Electrochemical Kinetics, Theoretical and Experimental Aspects, Academic, New York, 1967, pp. 460, 461.

14. A. Kozawa and J. F. Yeager, J. Electrochem. Soc., 112, 959 (1965).

15. A. Kozawa, T. Kalnoki-Kis, and J. F. Yeager, ibid., 113, 405 (1966).

16. D. J. Brown and H. A. Liebhafsky, J. Amer. Chem. Soc., 52, 2595 (1930).

17. A. W. Hutchinson, ibid., 69, 3051 (1947).

18. A. D. Wadsley and A. Walkey, J. Electrochem. Soc., 95, 11 (1949).

19. J. Brenet and A. M. Moussard, in CITCE, 6th Meeting, Poitiers, 1954, Butterworths, London, 1955, p. 415.

20. K. H. Maxwell and H. R. Thirsk, J. Chem. Soc., 1955, 4054.

21. K. H. Maxwell and H. R. Thirsk, ibid., 1955, 4057.

22. J. P. Gabano and J. P. Brenet, Z. Elektrochem., 62, 497 (1958).

23. J. P. Gabano and J. P. Brenet, Electrochim. Acta, 1, 242 (1959).

24. K. J. Vetter and N. Jaeger, ibid., 11, 401 (1966).

25. A. K. Covington, P. K. Talukdar, and H. R. Thirsk, Electrochem. Technol., 5, 523 (1967).

26. B. N. Ghosh, J. Chem. Soc., 1926, 2605.

27. R. S. Johnson and W. C. Vosburgh, J. Electrochem. Soc., 99, 317 (1952).

28. A. Kozawa, ibid., 106, 552 (1959).

29. K. J. Vetter, Z. Elektrochem., 66, 577 (1962).

30. K. J. Vetter, J. Electrochem. Soc., 110, 597 (1963).

31. J. Brenet, in Power Sources (D. H. Collins, ed.), Pergamon, London, 1967, p. 39.

32. N. Jaeger, Dissertation, Freie Univ., Berlin, 1968.

33. H. Bode, Angew. Chem., 73, 553 (1961).

34. H. Bode, A. Schmier, and D. Berndt, Z. Elektrochem., 66, 586 (1962).

35. W. C. Vosburgh, J. Electrochem. Soc., 106, 839 (1959).

36. D. T. Ferrell, Jr. and W. C. Vosburgh, ibid., 98, 334 (1951).

37. R. S. Johnson and W. C. Vosburgh, ibid., 98, 471 (1951).

38. A. M. Chreitzberg, D. R. Allenson, and W. C. Vosburgh, ibid., 102, 557 (1955).

39. W. C. Vosburgh and P. S. Lau, ibid., 108, 485 (1961).

40. A. B. Scott, ibid., 107, 941 (1960).

41. J. J. Coleman, Trans. Electrochem. Soc., 90, 545 (1946).

42. K. Neumann and W. Fink, Z. Elektrochem., 62, 114 (1958).

43. A. M. Moussard, J. Brenet, F. Jolas, M. Pourbaix, and J. Van Muylder, in CITCE, 6th Meeting, Poitiers, 1954, Butterworths, London, 1955, p. 190.

44. M. V. Ionin and G. I. Zvorykina, Tr. Po. Khim. i Khim. Tekhnol., 1962, 24; Chem. Abstr., 58, 12179b (1963).

45. K. Neumann and E. von Roda, Ber. Bunsenges. Phys. Chem., 69, 349 (1965).

46. H. Bode and A. Schmier, Naturwissenschaften, 49, 465 (1962).

47. A. Kozawa and R. A. Powers, J. Electrochem. Soc., 113, 870 (1966).

48. L. V. Andrews and D. J. Brown, J. Amer. Chem. Soc., 57, 254 (1935).

49. A. Carrington and M. C. R. Symons, J. Chem. Soc., 1956, 3373.

50. A. Moussard, J. Brenet, F. Jolas, M. Pourbaix, and J. Van Muylder, "Section 11.1, Manganese," in Atlas D'Equilibres Electrochemiques (M. Pourbaix, ed.), Gauthier-Villars, Paris, 1963, pp. 286-293.

51. J. Brenet, J. P. Gabano, and J. Reynaud, Electrochim. Acta, 8, 207 (1963).

52. W. J. Hamer, M. S. Malmberg, and B. Rubin, J. Electrochem. Soc., 103, 8 (1956).

53. W. J. Hamer, M. S. Malmberg, and B. Rubin, ibid., 112, 750 (1965).

54. W. J. Hamer, J. Electroanal. Chem., 10, 140 (1965).

55. S. N. Flengas and T. R. Ingraham, J. Electrochem. Soc., 106, 714 (1959).

56. H. A. Laitimer and C. H. Lin, J. Amer. Chem. Soc., 80, 1015 (1958).

57. H. C. Gaus and H. C. Jindal, Electrochim. Acta, 15, 1113 (1970).

58. H. C. Gaus and W. K. Behl, in Proceedings of the 1st Australian Conference on Electrochemistry (A. Friend and F. Gutman, eds.), Pergamon, London, 1964, p. 543.

59. Iu. K. Delimarsky and B. F. Markov, Electrochemistry of Fused Salts (translated by R. E. Wood), Sigma Press, Washington, D.C., 1961, pp. 159-197.

60. A. Eluard and B. Tremillon, J. Electroanal. Chem., 26, 259 (1970).

61. I. M. Kolthoff and J. J. Lingane, Polarography, Vol. 2, 2nd ed., Interscience, New York, 1952, pp. 468-471.

62. E. T. Veklier, Collect. Czech. Chem. Commun., 11, 216 (1939).

63. G. F. Reynolds, H. I. Shalzosky, and T. J. Webber, Anal. Chim. Acta, 9, 91 (1953).

64. L. Meites, Handbook of Analytical Chemistry, McGraw-Hill, New York, 1963, Section 5.

65. J. N. Gaur and N. K. Goswami, Electrochim. Acta, 12, 1483 (1967).

66. S. A. Moros, Ph.D. Dissertation, Polytech Institute of Brooklyn, Brooklyn, New York, 1961; University Microfilms (Ann Arbor, Michigan), Order No. 61-3756.

67. D. S. Jain, N. K. Goswami, and J. N. Gaur, Electrochim. Acta, 13, 1757 (1968).

68. J. W. Grenier and L. Meites, Anal. Chim. Acta, 14, 482 (1956).

69. G. S. Deshmukh, A. L. J. Rao, and S. V. S. S. Murty, Z. Anal. Chem., 196, 183 (1963).

70. S. Baumgarten, R. E. Cover, H. Hopass, S. Karp, P. B. Pinches, and L. Meites, Anal. Chim. Acta, 20, 397 (1959).

71. S. A. Moros and L. Meites, J. Electroanal. Chem., 5, 90 (1963).

72. P. Zuman, Collect. Czech. Chem. Commun., 15, 1107 (1950).

73. E. J. Breda, L. Meites, T. B. Reddy, and P. W. West, Anal. Chim. Acta, 14, 390 (1956).

74. H. A. Catherino and L. Meites, ibid., 23, 57 (1960).

75. I. M. Issa, A. A. El Sammahy, and M. M. Ghoneim, Electrochim. Acta, 16, 1655 (1971).

76. I. M. Kolthoff and J. F. Coetzee, J. Amer. Chem. Soc., 79, 1852 (1957).

77. R. C. Larson and R. T. Iwamoto, ibid., 82, 3239 (1960).

78. R. C. Larson and R. T. Iwamoto, ibid., 82, 3526 (1960).

79. G. H. Brown ahd H-S. Hsiung, J. Electrochem. Soc., 107, 56 (1960).

80. J. B. Headridge, M. Ashraf, and H. L. H. Dodds, J. Electroanal. Chem., 16, 114 (1968).

81. J. N. Gaur and N. K. Goswami, Electrochim. Acta, 12, 1489 (1967).

82. G. den Boef and B. C. Poeder, Rec. Trav. Chim. Pays-Bas, 77, 1071 (1958).

83. D. Cozzi and P. G. Desideri, J. Electroanal. Chem., 1, 301 (1959/1960).

84. P. G. Desideri, ibid., 4, 359 (1962).

85. P. G. Desideri, ibid., 6, 344 (1963).

86. H. Schurig and K. E. Heusler, Fresenius' Z. Anal. Chem., 224, 45 (1967).

87. R. Landsberg, J. Hendel, and W. Muller, J. Electroanal. Chem., 2, 484 (1961).

88. F. J. Miller and H. E. Zittel, ibid., 7, 116 (1963).

89. R. Guidelli and G. Piccardi, Electrochim. Acta, 13, 99 (1968).

90. T. Amemiya and H. Hirata, Nippon Kagaku Zasshi, 89, 58 (1968).

91. T. Fujinaga and K. Hagiwara, Bunseki Kagaku, 11, 442 (1962).

92. T. Hurlen and T. Valand, Electrochim. Acta, 9, 1077 (1964).

93. T. Hurlen, ibid., 7, 653 (1962).

94. T. Hurlen and T. Valand, ibid., 9, 1087 (1964).

95. K. E. Heusler and M. Bergmann, ibid., 15, 1887 (1970).

96. W. J. Plieth, Ber. Bunsenges. Phys. Chem., 74, 1042 (1970).

97. W. J. Plieth and K. J. Vetter, Z. Phys. Chem., [NF] 61, 282 (1968).

98. W. J. Plieth, ibid., [NF] 67, 178 (1969).

99. G. N. Znamenskii, I. V. Gamali, and V. V. Stender, Doklady Akad. Nauk SSSR, 137, 335 (1961); Chem. Abstr., 56, 268f (1962).

100. I. V. Gamali and V. V. Stender, Zh. Pribl. Khim., 35, 127 (1962); Chem. Abstr., 56, 13952g (1962).

101. N. A. Shrab and D. P. Zosimovich, Ukr. Khim. Zh., 34, 569 (1968).

102. V. P. Galushko and Yu. M. Loshkarev, Tr. Dnepropetr. Khim.-Technnol. Inst., 16, 59 (1962).

103. Yu. M. Loshkarev, Ukr. Khim. Zh., 29, 918 (1963).

104. T. Hurlen and T. Valand, Corros. Sci., 4, 253 (1964).

105. Ya. M. Kolotyrkin and T. R. Agladze, Zashchita Metallov., 3, 413 (1967) (English Translation).

106. K. E. Heusler, J. Electrochem. Soc., 110, 703 (1963).

107. M. Melendez and J. Brenet, Electrochim. Acta, 16, 61 (1971).

108. S. V. Gorbachev and V. A. Belyaeva, Russ. J. Phys. Chem., 36, 114 (1962).

109. S. V. Gorbachev and V. A. Belyaeva, ibid., 37, 97 (1963).

110. K. J. Vetter and G. Manecke, Z. Phys. Chem., 195, 337 (1950).

111. G. Piccardi and R. Guidelli, Sci. Chim., 38, 46 (1968).

112. R. G. Selim and J. J. Lingane, Anal. Chim. Acta, 21, 536 (1959).

113. G. Grube and H. Huberich, Z. Elektrochem., 29, 8 (1929).

114. D. Bolen and B. H. Weil, Literature Search on Dry Cell Technology with Special Reference to MnO_2 and Methods for Its Synthesis, Georgia Inst. Technol., State Eng. Expt. Sta., Special Rept. No. 27, 1948.

115. G. W. Heise and N. C. Cahoon, J. Electrochem. Soc., 99, 179C (1952).

116. M. A. Malati, Chem. Ind. (London), 17, 446 (1971).

117. N. C. Cahoon, J. Electrochem. Soc., 99, 343 (1952).

118. N. C. Cahoon, R. S. Johnson, and M. P. Korver, ibid., 105, 296 (1958).

119. M. P. Korver, R. S. Johnson, and N. C. Cahoon, ibid., 107, 587 (1960).

120. C. W. Jennings and W. C. Vosburgh, ibid., 99, 309 (1952).

121. W. C. Vosburgh, R. S. Johnson, J. S. Reiser, and D. R. Allenson, ibid., 102, 151 (1955).

122. A. M. Chreitzberg and W. C. Vosburgh, ibid., 104, 1 (1957).

123. S. Yashizawa and W. C. Vosburgh, ibid., 104, 399 (1957).

124. W. C. Vosburgh, M. J. Pribble, A. Kozawa, and A. Sam, ibid., 105, 1 (1958).

125. W. C. Vosburgh and J. H. Delap, ibid., 107, 255 (1960).

126. H. B. Mark and W. C. Vosburgh, ibid., 108, 615 (1961).

127. H. Bode and A. Schmier, in Batteries (D. H. Collins, ed.), Macmillan, New York, 1963, p. 329.

128. H. Bode and A. Schmier, Ber. Bunsenges. Phys. Chem., 68, 954 (1964).

129. H. Bode and A. Schmier, Chem.-Ing.-Tech., 38, 651 (1966).

130. J. Brenet, in CITCE, 8th Meeting, Madrid, 1956, Butterworths, London, 1958, p. 394.

131. S. Ghosh and J. P. Brenet, Electrochim. Acta, 7, 449 (1962).

132. S. Ghosh and J. Brenet, Ber. Bunsenges. Phys. Chem., 67, 723 (1963).

133. J. P. Gabano, B. Morignot, and J. F. Laurent, Electrochem. Technol., 5, 531 (1967).

134. J. P. Gabano, J. Seguret, and J. F. Laurent, J. Electrochem. Soc., 117, 147 (1970).

135. P. Benson, W. B. Price, and F. L. Tye, Electrochem. Technol., 5, 517 (1967).

136. A. Era, Z. Takehara, and S. Yoshizawa, Electrochim. Acta, 12, 1199 (1967).

137. O. Glemser, G. Gattow, and H. Meisek, Z. Anorg. Allgem. Chem., 309, 121 (1961).

138. A. Tvarusko, J. Electrochem. Soc., 111, 125 (1964).

139. P. Brouillet, A. Grund, F. Jolas, and P. Mellet, C.R. Acad. Sci., Paris, 257, 3390 (1965).

140. P. Brouillet, A. Grund, F. Jolas, and P. Mellet, ibid., 261, 3392 (1967).

141. F. Kornfeil, J. Electrochem. Soc., 109, 349 (1962).

142. J. P. Brenet, G. Coeffier, and J. P. Gabano, Electrochim. Acta, 8, 273 (1963).

143. J. Brenet, in Batteries (D. H. Collins, ed.), Macmillan, New York, 1963, p. 357.

144. N. C. Cahoon and M. P. Korver, J. Electrochem. Soc., 106, 745 (1959).

145. G. S. Bell and R. Huber, ibid., 111, 1 (1964).

146. A. Kozawa and J. F. Yeager, ibid., 115, 1003 (1968).

147. A. Kozawa and R. A. Powers, ibid., 115, 122 (1968).

148. D. Boden, C. J. Venuto, D. Wisler, and R. B. Wylie, ibid., 114, 415 (1967).

149. H. Y. Kang and C. C. Liang, Electrochim. Acta, 13, 277 (1968).

150. H. Y. Kang and C. C. Liang, J. Electrochem. Soc., 115, 6 (1968).

151. D. Boden, C. J. Venuto, D. Wisler, and R. B. Wylie, ibid., 115, 333 (1968).

152. J. Ambrose and G. W. D. Briggs, Electrochim. Acta, 16, 111 (1971).

153. E. M. Ungiadze and R. I. Agladze, Sb. Inst. Neorg. Khim. Elektrokhim., Akad. Nauk Gruz. SSR, 7, 52 (1967); Chem. Abstr., 68, 89174j (1968).

154. T. Banerjee and N. Dhananjayan, Indian Patent 81,402 (May 23, 1964).

155. N. A. Shvab and D. P. Zosimovich, Elektrokhim. Protsessy Electroosazhdenii Anodnom Rastvorenii Metal, 1969, 47-51; Chem. Abstr., 73, 126369c (1970).

156. N. Dhananjayan, J. Electrochem. Soc., 117, 1006 (1970).

157. M. Kurachi and T. Kudo, Nippon Kinzoku Gakkaishi, 35, 313 (1971); Chem. Abstr., 75, 11454a (1971).

158. M. N. Polukarov and N. P. Shul'gina, Elektrokhim. Protsessy Elektroosazhdenii Anodnom Rastvorenii Metal, 1969, 66-73; Chem. Abstr., 73, 83092n (1970).

159. N. P. Shul'gina and M. N. Polukarov, Uch. Zap., Permsk. Gos. Univ., 141, 86 (1966); Chem. Abstr., 68, 83643q (1968).

160. N. P. Shul'gina and M. N. Polukarov, ibid., 159, 95 (1966); Chem. Abstr., 68, 83644r (1968).

161. N. P. Shul'gina and M. N. Polukarov, ibid., 159, 112 (1966); Chem. Abstr., 68, 83645s (1968).

162. M. A. Qazi and J. Leja, J. Electrochem. Soc., 118, 548 (1971).

163. F. E. Bacon, "Manganese Electrowinning," In The Encyclopedia of Electrochemistry (C. A. Hampel, ed.), Reinhold, New York, 1964, pp. 792-795.

164. T. W. Clapper, "Electrolytic Manganese Dioxide," in The Encyclopedia of Electrochemistry (C. A. Hampel, ed.), Reinhold, New York, 1964, pp. 789-792.

165. K. Shimizu and I. Shirobata, Furukawa Denko Jiho, 43, 91 (1967).

166. K. Ohzawa, K. Shimizu, and T. Takasae, Japanese Patent 70-36,473 (November 20, 1970); Chem. Abstr., 72, 104561y (1970).

167. T. Ishimo, H. Tanura, and M. Yanokava, Osaka Univ. Tech. Report, 6, 359 (1956).

168. R. I. Agladze, L. A. Zautashvili, and K. Sh. Vanidze, Soobshch, Akad. Nauk Gruz. SSR, 56, 581 (1969); Chem. Abstr., 72, 106543f (1970).

169. M. G. Potdar and H. V. K. Udupa, Bull. Acad. Pol. Sci., Ser. Sci. Chem., 16, 39 (1968).

170. Y. Amano, H. Kumano, A. Nishino, and Y. Noguchi, South African Patent 69 08,746 (August 24, 1970); Chem. Abstr., 74, 106618a (1971).

171. A. Nishino, H. Kumano, Y. Noguchi, and Y. Amano, German Offen. 2,063,906 (July 29, 1971); Chem. Abstr., 75, 94091d (1971).

172. I. Tanabe, German Offen. 2,051,917 (April 29, 1971); Chem. Abstr., 74, 23342a (1971).

173. "Bulletin of International Common Sample of Manganese Dioxide," Electrochemical Society, December 1, 1971.

174. D. M. Smyth, J. Electrochem. Soc., 113, 19 (1966).

175. M. Berteleau, Rev. Gen. Sci. Pures Appl. Bull. Assoc. Fr. Avan. Sci., 74, 7 (1967).

Chapter I-7

CALCIUM, STRONTIUM, BARIUM, and RADIUM

SHINOBU TOSHIMA

Department of Applied Chemistry
Faculty of Engineering
Tohoku University
Sendai, Miyagi, Japan

1. STANDARD AND FORMAL POTENTIALS

1.1. AQUEOUS SOLUTION

Since the alkaline earth metals, Ca, Sr, Ba, and Ra, are violently attacked by all aqueous solutions, it is very difficult to obtain the precise values of the standard potential of the electrochemical reaction in aqueous solutions. As with the alkali metals, two different approaches have been taken to evaluate the values. One is by calculation from thermodynamic data, and the other is by an indirect method of Lewis [1] who determined the potential difference between a reference electrode and a dilute alkali metal amalgam in water, and the potential difference between the same amalgam and the alkali metal electrode in a dry nonaqueous solvent. Devoto [2] proposed the theoretical potentials of alkaline earth metals calculated from free energy changes. In the case of calcium, he obtained $E^\circ = -2.888$ V from the energy change of the reaction

$$Ca + \frac{1}{2} O_2 + H_2O(aq) = Ca(OH)_2 \qquad (1.1.1)$$

and $E^\circ = -2.903$ V from that of the reaction

$$C(graphite) + \frac{3}{2} O_2 + Ca = CaCO_3(calcite) \qquad (1.1.2)$$

The average of these potentials, -2.895 V [3], was formulated to be $E^\circ = -2.90 \pm 0.02$ V after taking errors into consideration. This value agrees well with that of Drucker and Luft [4]. In the case of strontium and barium, Devoto stated that $E^\circ(Sr/Sr^{2+}) = -2.945 \pm 0.05$ V and $E^\circ(Ba/Ba^{2+}) = -2.96 \pm 0.05$ V. As the more noble standard potentials, Latimer [5] computed that $E^\circ(Ca/Ca^{2+}) = -2.87$ V, 1 mole/1000 g H_2O; $E^\circ(Sr/Sr^{2+}) = -2.89$ V, 1 mole/1000 g H_2O; $E^\circ(Ba/Ba^{2+}) = -2.90$ V, 1 mole/1000 g H_2O; and $E^\circ(Ra/Ra^{2+}) = -2.92$ V, 1 mole/1000 g H_2O. It is believed, however, that the most reliable calculated values [3, 6, 7] are those of Bockris and Herringshaw [8] at ion activity = 1, i.e., $E^\circ(Ca/Ca^{2+}) = -2.84$, $E^\circ(Sr/Sr^{2+}) = -2.89$, and $E^\circ(Ba/Ba^{2+}) = -2.92$ V. These values are cited in the textbooks of Kortüm and Bockris [9], Bockris and Reddy [10], Koryta, Dvorak, and Bohackova [11], Conway [12], and the tables of Landolt-Börnstein [13].

Tamele [14] applied the indirect method of Lewis for determining the potential of calcium. Pyridine was used as a nonaqueous solvent. The potential of calcium was evaluated to be $E^\circ = -2.758 \pm 0.004$ V. A metallic calcium electrode was scraped with a glass scraper during the measurement. After prolonged, very intensive scratching, the emf fluctuated toward a still greater value, which was most probably caused by the warming of the rubbed surface. If the electrode was not scratched, the value was about 0.6 V and indistinct. Drucker and Luft [4] introduced stirring of the solution and vibrating of the metal rod against the glass wall to prevent passivity, and produced the highest voltage after renewal of an amalgam surface. Shibata [15] pointed out two difficulties in these experiments: one is the decomposition of calcium amalgam in contact with water, which lowers the concentration of calcium, and the other is the oxidation of the calcium surface, even in a nonaqueous solvent. He recommended continuous rubbing of the surface of the calcium electrode on a rough surface inside the vessel. Butler [16] attributed these phenomena to atmospheric contamination with oxygen or water. It is clear from these experiments that passivation of the calcium surfaces is an important problem. Butler said that it might be possible to obtain a more nearly thermodynamic value by being extremely careful to eliminate oxygen and water from the nonaqueous solvent, and by introducing freshly prepared amalgam and freshly polished solid calcium simultaneously into the electrolyte. If the potential is read as a function of time during the first few minutes of cell operation, a better value may be obtained by extrapolating to zero time.

The activity coefficient of calcium in the amalgam may be evaluated as unity. This can be confirmed by the fact that the amalgam concentrations used in both the aqueous and nonaqueous studies are roughly the same [16]. The calcium ion concentration in the electrolyte does not enter into the calculation, but the nonaqueous solution is, in most cases, saturated with calcium salt.

Standard and formal potentials of alkaline earth metals in aqueous solutions are listed in Table 1.1.1 (see page 437). Isothermal and thermal temperature gradients of standard potentials are listed in Table 1.1.2 (see page 440). The electrochemical series of calcium, strontium, and barium in aqueous solution at 25°C are described as follows:

- − Li, Rb, Cs, K, Ba, Sr, Ca, Na, La, Mg + [3],[6],[9],[19]
- − Li, Ba, Sr, K, Ca, Na, Mg + [3],[6],[27],[29]

Potential-pH diagrams of alkaline earth metals in aqueous solution are shown in Figs. 1.1.1 (Ca), 1.1.2 (Sr), 1.1.3 (Ba), and 1.1.4 (Ra) [24].

The potential of the amalgam electrode in aqueous solution is closely related to that of polarography, which is described in Section 2.1. However, the calcium amalgam electrode has been investigated because of the interesting fact that its standard potential is considerably more negative than that of sodium amalgam, in spite of their reverse order in pure metals.

Fosbinder [30] determined the potential of the calcium amalgam electrode, but his work was not compiled in thermodynamic data, because calcium electrode behaved with considerable irreversibility and because corrections for activity coefficients, temperature, liquid junction potentials, or ion pairing were not sufficient. Butler [16] recalculated the standard potential of the calcium amalgam electrode in aqueous solution, beginning with the actual experimental measurements, and by applying the most recent data on activity coefficients, liquid junction potentials, reference electrode potentials, and ion-pairing equilibria. Figure 1.1.5 illustrates the standard potential of the calcium amalgam electrode vs the concentration of the salt.

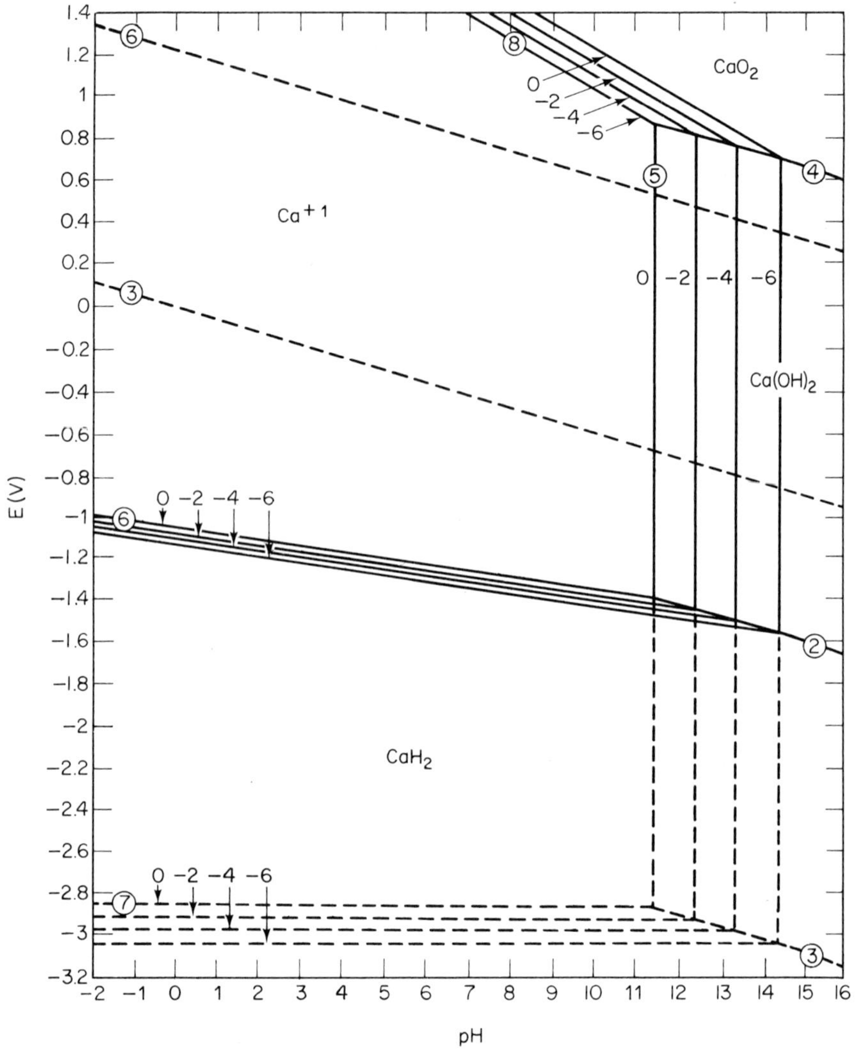

FIG. 1.1.1. Potential-pH equilibrium diagram for the system calcium-water at 25°C [24]. (Permission of Pergamon Press.)

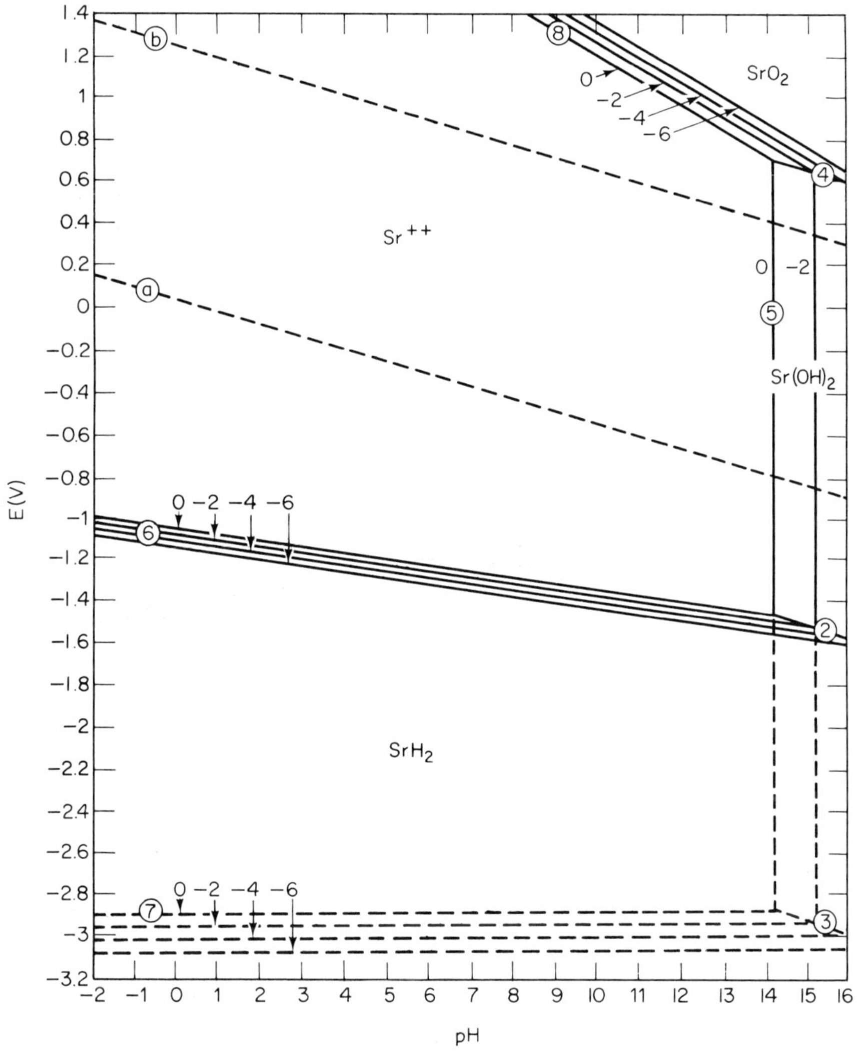

FIG. 1.1.2. Potential-pH equilibrium diagram for the system strontium-water at 25°C [24]. (Permission of Pergamon Press.)

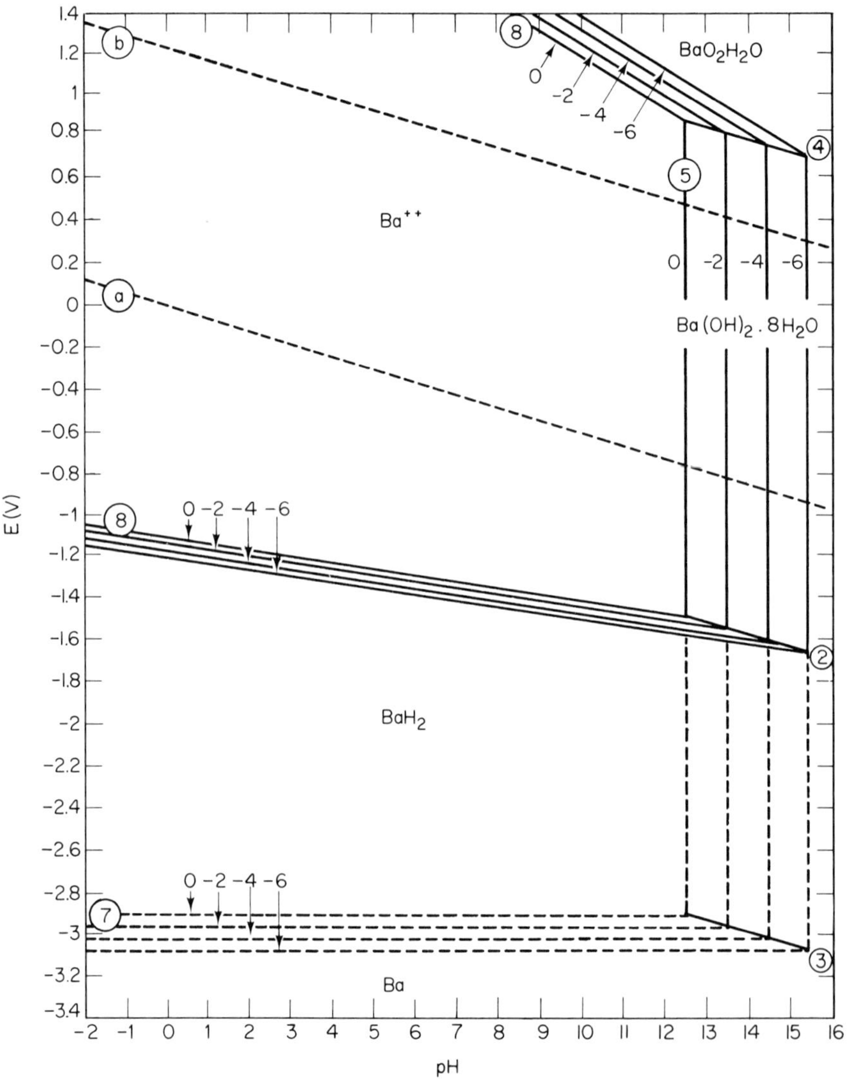

FIG. 1.1.3. Potential-pH equilibrium diagram for the system barium-water at 25° C [24]. (Permission of Pergamon Press.)

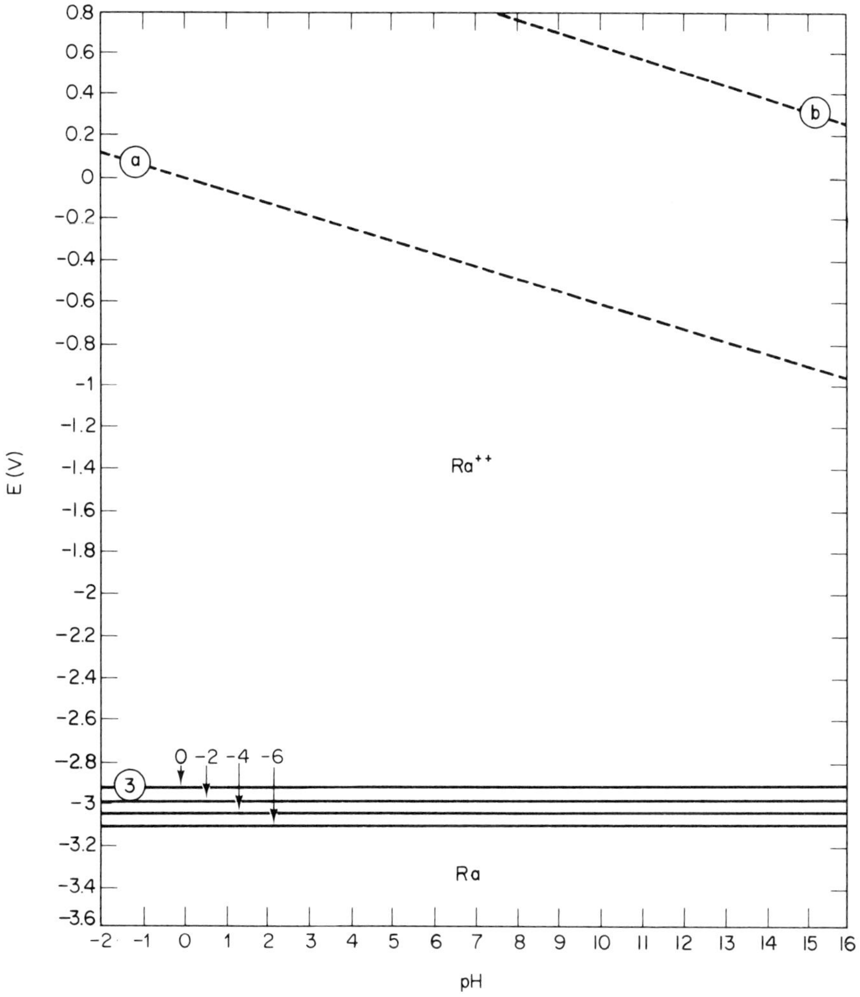

FIG. 1.1.4. Potential-pH equilibrium diagram for the system radium-water at 25°C [24]. (Permission of Pergamon Press.)

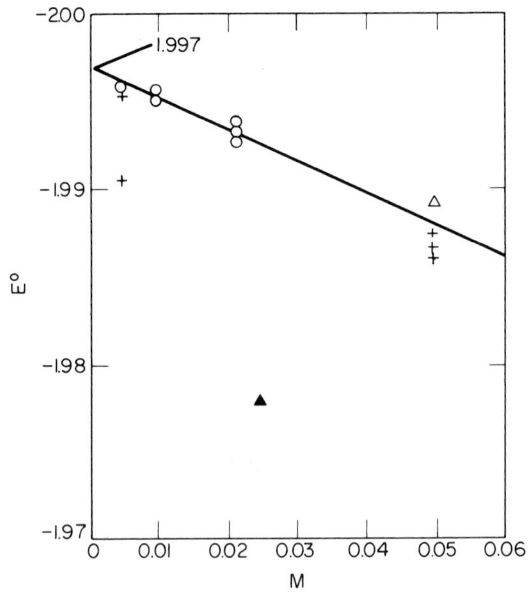

FIG. 1.1.5. Standard potential of Ca(Hg) electrode. (⊙) Shibata ($Ca(OH)_2$); (△) Shibata ($CaCl_2$); (+) Tamele ($CaCl_2$) [16]. Tamele's cell [14]:

$Ca(Hg)|CaCl_2|Hg_2Cl_2(s)|Hg$

Shibata's cell [15]:

$Ca(Hg)|CaCl_2|Hg_2Cl_2(s)|Hg$

$Ca(Hg)|Ca(OH)_2|HgO(s)|Hg$

(Permission of Elsevier Publishing Co.)

1.2. NONAQUEOUS SOLUTION

In order to determine a standard potential of metal ion in nonaqueous solution, we must take into account the free energy of solvation, which is a step in Born-Haber cycle for calculating the standard free energy change. Therefore, we may compare the potential values using the standard hydrogen electrode as a reference if the same solvent is employed. But it is very difficult to compare the values of standard potential when different kinds of solvent are used because the values include the change of free energy due to the variation of solvation energy and because elimination of liquid junction potential is very difficult. In principle the static part of the free energy change between two solvents, 1 and 2, may be represented by Born's equation:

$$\Delta G = \frac{NZ^2e^2}{2r}\left(\frac{1}{\epsilon_2} - \frac{1}{\epsilon_1}\right) \qquad (1.2.1)$$

where ϵ is dielectric constant and r is the ionic radius. This estimation, however, may be inaccurate when the value of r changes with the kinds of Solvents 1 and 2. These phenomena suggest the difficulty in selecting a suitable reference electrode in nonaqueous solutions.

For a reference electrode, it is necessary for the solvation energy to differ little among the different kinds of solvent. Pleskov [31] found that the rubidium ion had almost the same, small energy of solvation in different solvents. He proposed using the rubidium ion as the most favorable reference electrode in nonaqueous solvents, and he determined the standard potential of the other elements vs rubidium. Strehlow [32] gave a correction for the solvation energy of Rb^+ in various kinds of solvents, assuming that the standard potential of rubidium in aqueous solution equals -2.92 V. The potentials of the standard rubidium electrode, E_{Rb}, are shown in Table 1.2.1 (see page 440) relative to the normal hydrogen electrode potential E°_H. Khomutov [25] recently presented equations relating the standard electrode potentials in nonaqueous solvents to those in aqueous solutions and to the free energies of solvation and hydration of ions. For cations with the electronic configuration of noble gases, an empirical relation between the free energy of solvation of ions in nonaqueous solvents and the free energy of hydration was found. He proposed two kinds of standard potentials, E°_I and E°_{II}, based on different assumptions. In Table 1.2.2 (see page 441) the values of standard potentials in nonaqueous solutions are shown. In liquid ammonia the electrochemical series is [3, 34]:

- Li, K, Rb, Na, Ca +

The potential difference between calcium metal and its amalgam electrode in a dry nonaqueous solvent has been measured by several authors in order to obtain a standard potential as described in Section 1.1. MeOH, pyridine, EtOH, DMSO and propylene carbonate were used as nonaqueous solvents. Butler [16] summarized these data (Table 1.2.3, page 442), where

$$\Delta E^\circ = E^\circ_{Ca(Hg)} - E^\circ_{Ca} = E + \frac{RT}{2F} \ln X_{Ca}\gamma_{Ca} \qquad (1.2.2)$$

X_{Ca} = mole fraction of calcium in amalgam

γ_{Ca} = activity coefficient

1.3. FUSED SALTS

1.3.1. Single Electrode Potential

In analogy with nonaqueous solutions, there is a problem in selecting a reference electrode for fused salts. High temperature, corrosiveness of the electrolyte, and difficulties of thermostating make the selection more tedious under the conditions prevailing in work with various fused salts. However, there exists the possibility of working with a reversible reference electrode because chemical equilibrium can be rapidly realized at high temperature. It is necessary for the reference electrode reaction in fused salts to be simple and not affected by the presence of air or water vapor. Thus the number of reference electrodes proposed for fused salts is rather large. They include the hydrogen electrode, the oxygen electrode, the halogen electrode, the metal electrode, the silver chloride electrode, and the glass electrode [37, 38].

Jellinek [21] proposed the standard electrode potential of hydrogen be set equal to zero for fused salts as well as for aqueous solutions. The emf of a chemical cell, $H_2|HX|X_2$, where X is halogen, is represented by

$$E = E_{X_2} - E_{H_2} = E_{X_2} = -\frac{RT}{nF} \ln K \qquad (1.3.1)$$

where E_{X_2}, E_{H_2}, and K are the electrode potential of halogen, hydrogen (equal zero), and the equilibrium constant, respectively, of the reaction, $H_2 + X_2 \rightleftharpoons 2HX$. He proposed calculating the value of E_{X_2} from that of K, employing approximate forms of the Nernst equation:

$$\begin{aligned} \log K_{HCl} &= -\frac{9554}{T} + 0.553 \log T - 2.42 \\ \log K_{HBr} &= -\frac{5223}{T} + 0.553 \log T - 2.72 \qquad (1.3.2) \\ \log K_{HI} &= -\frac{5404}{T} + 0.503 \log T - 2.35 \end{aligned}$$

If it is assumed that the electrode potential of the halogen is a constant value in all fused electrolytes, it would be possible to obtain the value of the single electrode potential of metal relative to hydrogen by calculating the electrode potential of the halogen from the decomposition potential or the emf of the cell of the fused halide. In practice, the values of E_{Cl_2}, E_{Br_2}, and E_{I_2} were assumed to be 1.02, 0.63, and 0.146 V, respectively. Results of these calculations are listed in Table 1.3.1 (see page 442) as $E_{H_2} = 0$.

Solvation is either completely absent or plays an insignificant part when we deal with fused salts. Much greater significance is assumed by the interaction of ions, which often leads to their mutual polarization and deformation. The rubidium electrode, which is a good reference electrode in nonaqueous solvents, has such a large polarizability that it is not an adequate reference in fused salts. Metals whose ions are least polarizable include lithium, sodium, beryllium, magnesium, and aluminum. However, the last three metals easily form complex ions in fused electrolytes and may not be chosen as zero electrodes. Since the sodium electrode (in glass-sodium electrodes) has been studied in detail, sodium has been adopted as a standard electrode for fused salts in equilibrium with a single fused sodium salt, which is a good electric conductor. Therefore, if we know the values of decomposition potentials of pure fused salts and of their solutions in other fused electrolytes, and if we also use single electrode potentials determined with the glass-sodium reference electrode, it is possible to compute the values of single electrode potentials of metals relative to sodium, arbitrarily set at zero. Results of these procedures are also listed in Table 1.3.1 with $E_{Na} = 0$.

Values of the electrode potentials of metals in fused chlorides, calculated on the basis of theoretical values of the decomposition potentials given by Hamer, Malmberg, and Rubin [39], are shown in Table 1.3.2 (see page 443).

Experimental methods for determining the single electrode potentials in fused salts may be classified in the following three ways: 1) from measurement of the equilibrium potential between a reliable reference electrode and a test electrode, 2) from measurement of the half-wave potential by polarography, and 3) from the values of decomposition voltage, assuming that the overvoltage of the electrode reaction is negligibly small. Method 1 is, of course, the most accurate, but we cannot make these measurements on alkaline earth metals. Method 3 includes the conventional measurement using two electrodes in fused salt and a measurement of the polarization emf obtained just after interruption of the current. The single electrode potentials of alkaline earth metals in fused salt determined only by Method 3 are listed in Table 1.3.3 (see page 443) for simple salts and in Table 1.3.4 (see page 444) for mixed salts.

Johnson [46] recently presented a short communication on the theoretical decomposition potentials for binary calcium compounds in which, on the basis of Hamer et al.'s work, he predicted the standard potentials of the nonmetal/anion couples as well as the Ca^{2+}/Ca couple in melts of $CaCl_2$ at 1200° K and CaF_2 at 1800° K. The values of standard potential of Ca^{2+}/Ca given by him are -3.25 and -4.77 V, respectively.

Electrochemical series of alkaline earth metals in fused salt solvents are shown in Table 1.3.5 (see page 444).

1.3.2. Decomposition Voltage

The decomposition voltage is the minimum applied voltage required for prolonged electrolysis in fused salt, and it is an important factor in the electrochemistry of fused salts. Generally, if we pass a direct current through an electrolyte, there arises an emf of polarization whose direction is opposite to that of the applied voltage. The origin of the emf of polarization consists of the emf of the chemical cell, concentration polarization, and the electrode overvoltages. In single fused salts the decomposition voltage may be equal to the emf of the corresponding chemical cell only when overvoltage and depolarization are absent, because concentration polarization may be negligibly small in the single melt. When only one electrode reaction occurs at each electrode, the decomposition voltage is larger than the emf of the reversible cell in the presence of overvoltage on the electrodes. The overvoltage on electrodes in fused salts is very small in the electrolysis of most single salts because of the high temperature. Therefore, if polarization may be neglected, the decomposition voltage agrees with the emf of the cell.

To calculate a decomposition voltage at a given temperature, we must know the free energy change at that temperature from thermochemical data. Hamer, Malmberg, and Rubin [39] computed the decomposition voltages of a large number of halides over a wide range of temperatures by using the relations

$$E^\circ = -\Delta G_f^\circ/nF \tag{1.3.3}$$

$$\Delta G_f^\circ = H_0^\circ - (\Delta a)\, T \ln T - \frac{(\Delta b)T^2}{2} - \frac{\Delta c}{2T} + IT \tag{1.3.4}$$

where ΔG_f° is the standard free energy of formation of the electrolyte; ΔH_0° is the heat of formation at 298.16°K; T is the absolute temperature; and Δa, Δb, and Δc are known numerical values appearing in an expression of heat capacity change as a function of T:

$$\Delta C_p = \Delta a + (\Delta b)T + \frac{\Delta c}{T^2} \tag{1.3.5}$$

and I is the integration constant. The results obtained are shown in Table 1.3.6 (see page 445) in which the solid and aqueous states at 25°C are also written for reference. The references show that the relative position of the compound is nearly the same whether the electrolyte phase is solid or molten. If the solid phase is anhydrous, the values for the solid electrolyte and aqueous systems differ in the free energy associated with the process of dilution of the saturated solution of the electrolyte to a solution of unit mean activity $a_\pm$. The difference is described by

$$E_{aq}^\circ - E_s^\circ = \frac{2.303RT}{nF} \log a_\pm{}^r \tag{1.3.6}$$

where r is the sum of the number of positive ions, r^+, and the number of negative ions, r^-, dissociating in the electrolyte. E_{aq}° is higher than E_s° for highly soluble salts.

Johnson [46] plotted the relation between decomposition potentials and temperature on the basis of the works of Hamer et al. [39], of Hamer [49], for sulfides [50], and for carbide and nitride [51]. This is shown in Fig. 1.3.1.

Experimental methods for determining the decomposition voltage are:

1. Determining the emf of chemical cells.
2. Determining from an inflection point in current-voltage relation.
3. Determining the polarization emf by employing an interruptor.
4. Determining with suitable auxiliary electrodes.

The most reliable values are considered to be those obtained by Method 3 [37]. Method 2 has been employed most frequently in the case of alkaline earth metals. In all the experimental methods, account should be taken of the peculiar difficulties encountered in the

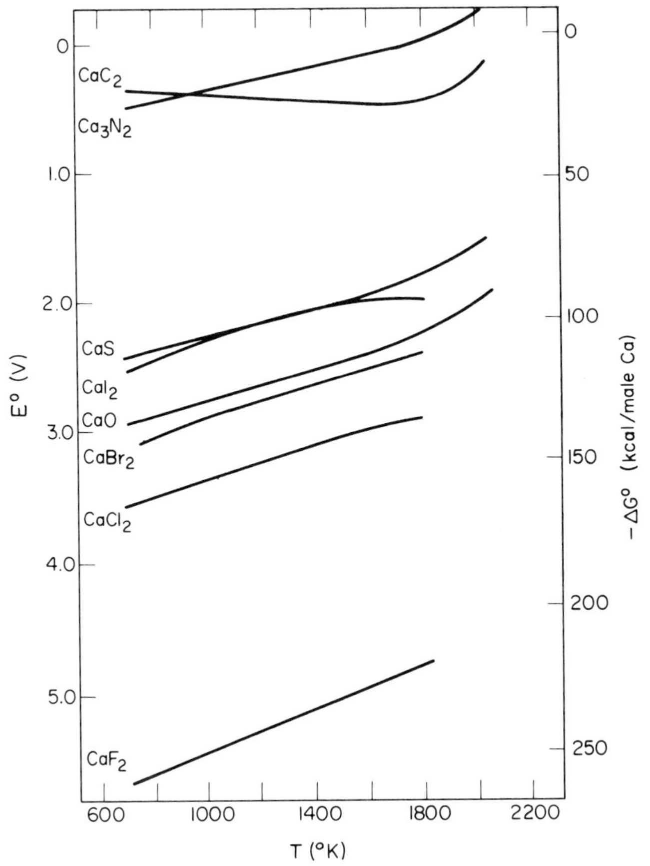

FIG. 1.3.1. Decomposition potentials of binary calcium compounds as functions of temperature [46]. (Permission of Pergamon Press.)

handling of fused salts: pyrosol phenomena, reaction occurring between the electrolysis products and the electrode materials, interaction of the salt with the oxygen and moisture of the atmosphere, corrosiveness, etc.

The values of decomposition voltages of compounds of the alkaline earth metals are listed in Table 1.3.7 (see page 446), in which "most reliable" means the most reliable values for the decomposition potentials of pure fused halides as presented in the text of Delimarskii [37]. The values for salts at their melting points are shown in Table 1.3.8 (see page 449).

Johnson [46] inspected the decomposition voltage of calcium, lithium, and potassium halides at 450°C [39], and potentials measured in the LiCl-KCl eutectic at 450°C [52]. These are shown in Table 1.3.9 (see page 449). The experimental value for the Br_2/Br^- potential lies between the theoretical values for the constituent salts, whereas the value for I_2/I potential is less than either theoretical value. Thus the values are very sensitive to the cation present. Johnson considered that such sensitivity is to be expected because the cations are the nearest neighbors of the potential-determining ions.

2. VOLTAMMETRIC CHARACTERISTICS

2.1. POLAROGRAPHIC CHARACTERISTICS

Polarographic behavior employing the dropping mercury electrode as a cathode shows a distinct difference between calcium and the other alkaline earth metals. Zlotowski and Kolthoff [53] systematically studied its characteristics. In aqueous tetramethylammonium iodide as a supporting electrolyte, the polarographic wave of calcium is distorted by a large and sharp maximum, whereas that of strontium is very well defined and that of barium is an excellently developed S shaped wave [54]. The wave of calcium has the maximum even in a mixed solution of up to 80 vol % ethanol in the aqueous solution. The height of the maximum decreases with an increasing content of ethanol, and the maximum potential is shifted to more positive values. This maximum can be suppressed by the addition of lanthanum ion in aqueous solution or of a barium ion in ethanolic aqueous solution. The necessary concentrations of ion to be added to the solution to eliminate the maximum are [3, 53]:

	H_2O	25% ethanol	50% ethanol	80% ethanol
Mole Ba^{2+}/l	$\sim 10^{-4}$	5×10^{-5}	3×10^{-5}	1×10^{-5}
Mole La^{3+}/l	5×10^{-5}	3×10^{-5}	-	-

The lanthanum ion serves better than the barium ion to suppress the maximum in aqueous solution. The magnesium ion interferes seriously with the calcium wave when it is present in concentrations greater than 3×10^{-5} M.

Strontium and barium reductions proceed reversibly both in aqueous medium and in media containing up to 80% ethanol. The half-wave potentials vs SCE in ethanol-water mixtures decrease linearly with increasing volume percent of ethanol, P, according to [54]

$$\text{Sr: } E_{1/2} = -2.110 + 0.0010P$$
$$\text{Ba: } E_{1/2} = -1.940 + 0.0015P \qquad (2.1.1)$$

Nyman studied [55] the polarographic behavior of calcium, strontium, and barium in anhydrous liquid ammonia at -36°C, and he observed well-developed waves in the presence of tetraethylammonium iodide.

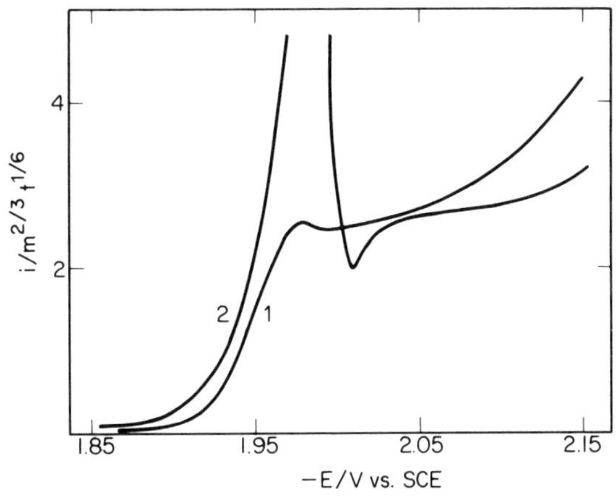

FIG. 2.1.1. Dc polarograms for 0.5 m$\underline{M}$ Ba^{2+} in 1 $\underline{M}$ LiCl (1) and 0.1 $\underline{M}$ LiCl (2). Currents are given at maximum drop life, normalized for $m^{2/3}t^{1/6}$ [56]. (Permission of Elsevier Publishing Co.)

The polarographic characteristics appearing in references or texts are listed in Table 2.1.1 (see page 450); $E_{1/2}$ represents the half-wave potential; I, the value of diffusion current constant $i_d/Cm^{2/3}t^{1/6}$; and $E_{3/4} - E_{1/4}$, the steepness of the polarogram.

A linear relation exists between the diffusion current and the concentration of calcium ion when the maximum is eliminated by the addition of lanthanum ion in aqueous solution. To obtain proportionality within a 1-2% error, the concentration of calcium ion should be from 0.00025 to 0.001 $\underline{M}$ in aqueous solution, and be less than 0.0025 $\underline{M}$ in alcoholic aqueous solution [53]. The values of diffusion current I_D (μA) of strontium and barium in the mixed solution of water and ethanol are shown in Tables 2.1.2 and 2.1.3 (see page 452) at 25°C, where C denotes the concentration of strontium and barium in m$\underline{M}$, and x is the volume % of ethanol. The diffusion current obeys the Ilkovic equation in liquid ammonia at -36°C [54, 55].

In Fig. 2.1.1 the dc polarogram of a 0.5-m$\underline{M}$ Ba^{2+} solution in 1 $\underline{M}$ LiCl is presented; the y-axis is the diffusion current constant. A small maximum is present which increases with Ba^{2+} concentration. Moreover, the lithium and barium wave are reasonably well separated although at some intermediate potentials there will be some overlap of the two ac waves.

When 0.1 $\underline{M}$ LiCl is used as a supporting electrolyte, the barium wave has a rounded maximum which may not be suppressed by the addition of gelatin, thymol, Methyl Red, or Methylene Blue. But when 0.05 $\underline{M}$ $CaCl_2$ is used, the barium wave is usually well developed [54, 57]. As with alkalies, the reactions of the heavier elements tend to be more nearly reversible than those of the lighter elements. The behavior in dimethylacetamide is unusual in that no reactions are observed unless a small amount of water is introduced, in which case the expected two electron reductions occur [59, 60]. This is attributed to the formation of solvated ions in the anhydrous solvent which show a very pronounced overvoltage at mercury. On addition of water, hydrates are formed which show sufficiently small overvoltage to permit observation of the reaction.

The half-wave potentials of alkaline earth metals in nonaqueous solvents have a close relation to the standard potential, solubility of the metal in mercury, and the free energy change on amalgamation. The latter two are indifferent to the kind of solvent. If the electrode reaction proceeds reversibly, the observed value of the half-wave potential indicates the amount of interaction between the metal ion and the molecule of the solvent. If the electrode reaction proceeds irreversibly, the value of the potential depends not only on the standard potential but also on the polarographic overvoltage, which may be considered to be a function of solvation. Mann and Barnes [59] list a large number of polarographic characteristics in nonaqueous solutions in their text, from which the characteristics concerning compounds of alkaline earth metals were selected and are shown in Table 2.1.4 (see page 453).

To compare the values of the half-wave potential of metal ions in nonaqueous solvents, measurements of the potential should be made against a well-defined reference electrode, and the liquid junction potential should be eliminated. Of course, the formation of complex compounds with the supporting electrolyte must be avoided. It is usually convenient to employ a solvent with a high dielectric constant ($\epsilon > 20$). On comparing the half-wave potentials in nonaqueous solvents, the activity coefficient and the magnitude of the electron donating ability of the solute are more important factors than the dielectric constant. The values of half-wave potential of alkaline earth metals are shown in Table 2.1.5 (see page 456) [72], both vs SCE and vs rubidium ion, which was proposed as a reference ion by Pleskov [31]. From Table 2.1.5 it may be seen that the half-wave potentials in propanediolecarbonate (PDC) are almost equal to that in CH_3CN, despite the large difference in dielectric constant [$\epsilon = 69$ (PDC) and $\epsilon = 38.8$ (CH_3CN)]. On the other hand, although the dielectric constant of CH_3CN is almost equal to that of dimethylacetamide (DMA) ($\epsilon = 37.8$), the half-wave potentials in CH_3CN are more positive than those in DMA. This suggests that electron donating ability plays an important role in electrochemistry in nonaqueous solvents.

Polarography in fused salts is described in detail in Delimarskii and Markov's text [37], but we could not find the polarographic characteristics of alkaline earth metals in these media.

2.2. VOLTAMMETRIC CHARACTERISTICS

No studies on the voltammetric characteristics of alkaline earth metals could be found except for a few investigations in nonaqueous solvents with calcium as an active metal anode in a galvanic cell. In the galvanic cell the activity of calcium lies between lithium and magnesium. Elliot, Huff, Adler, and Towle [73] investigated the current density capabilities of the calcium anode in various nonaqueous electrolytes; the results are shown in Table 2.2.1

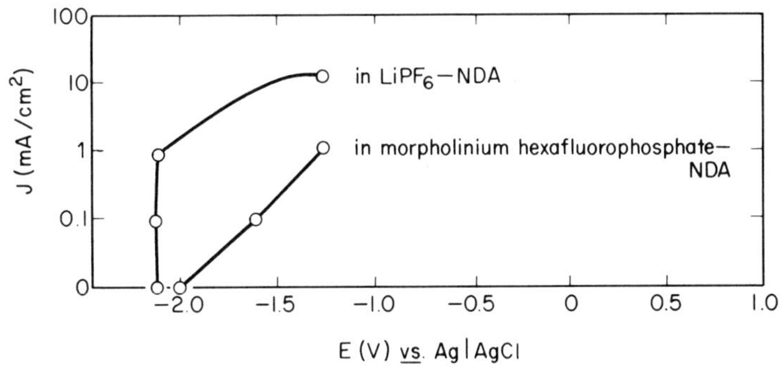

FIG. 2.2.1. Polarization curves of a Ca anode in nonaqueous electrolytes [73]. (Permission of PSC Publication Committee.)

(see page 456). For example, the polarization curves of a calcium anode in both $LiPF_6$-N-nitrosodimethylamine (NDA) and morpholinium hexafluorophosphate-NDA are shown in Fig. 2.2.1. Kurygina and Iofa [74] studied the anodic characteristics of calcium in a galvanic cell containing methanolic solution, and they measured the quantity of calcium consumed. The results are shown in Table 2.2.2 (see page 456). Jasinski [75] has reviewed electrochemistry in propylene carbonate; data from this review on a cell containing calcium are shown in Table 2.2.3 (see page 457).

As another example of an electroanalytical technique for Ca and Ba, reduction at the amperometric titration of calcium and barium ion involving a dropping mercury electrode can be mentioned. Typical conditions for the titration are shown in Table 2.2.4 (see page 457) [22].

3. KINETIC PARAMETERS AND DOUBLE-LAYER PROPERTIES

3.1. KINETIC PARAMETERS

Kinetic parameters of the electrode reaction of alkaline earth metals are not found in the literature except for the work of Timmer, Sluyters-Rehback, and Sluyters [56], who studied the influence of nonspecific adsorption of reactants on the electrode impedance. In a theoretical study of the coupling of double-layer charging and faradaic processes, initiated by Delahay, they presented some necessary conditions for finding experimentally the effects caused by nonspecific adsorption. They tested these theoretical predictions on the barium ion/barium amalgam electrode reaction in LiCl solutions.

From the experimental charge transfer resistance values obtained from the impedance analysis, the standard heterogeneous rate constant of the electrode reaction can be computed. The results are shown in Table 3.1.1 (see page 457) for studies in both 1 M LiCl and 0.1 M LiCl. To obtain the cathodic transfer coefficient, α, two values of rate constant in 1 and 0.1 M LiCl are used by applying the Frumkin correction

$$k = k_{true} u_2^{z-\alpha n} \tag{3.1.1}$$

where k_{true} is the rate constant corrected for double-layer effects and $u_2 = \exp(-F\varphi_2/RT)$ (φ_2 is a potential of the plane of closest approach). When z (charge of ion) = n (number of electrons involved in the electrode reaction) = 2 is introduced, together with the u_2 values, an α of 0.83 is obtained. This value is in good agreement with the potential dependence of the charge transfer resistance calculated from ac measurements. They state that the barium electrode reaction appears to be too irreversible for detecting nonspecific adsorption effects in the ac peak potential region, although the barium adsorption is rather strong.

3.2. DOUBLE-LAYER PROPERTIES

Differential capacities, which are an important factor in discussing double-layer properties, have been measured on mercury in 0.1 N aqueous solutions by Grahame [76]. Table 3.2.1 (see page 457) shows the values in $CaCl_2$, $SrCl_2$, and $BaCl_2$.

Timmer et al. [56] measured the double-layer capacity values of the barium ion/DME interface at 25°C with an ac bridge in order to detect an influence of nonspecific adsorption on the electrode impedance. The results in 1 M LiCl and 0.1 M LiCl are shown in Tables 3.2.2 and 3.2.3 (see page 458), respectively. The point of zero charge, obtained from droptime-potential curves, is -0.52 V (vs SCE) in 1 M LiCl and is -0.47 V in 0.1 M LiCl, which are independent on the concentration of barium ion up to 1.5 mM, both in 1 and 0.1 M LiCl solution.

Timmer et al. [56] also carried out some experiments for 1 mM barium ion in 0.1 M tetraethylammonium iodide, which has a very negative reduction potential. The double-layer capacity vs potential curve, shown in Fig. 3.2.1, shows a maximum. In this peak potential region an increase of differential capacity, C_d, would be expected when barium ion is adsorbed at the interface. A change in C_d with barium ion concentration was found to be caused by nonspecific adsorption, in quantitative agreement with double-layer theory [56].

The potentials of the electrocapillary maximum, E_{ecm}, of mercury in aqueous solutions of alkaline earth chlorides at 25°C are listed in Table 3.2.4 (see page 459). The relations between E_{ecm} of mercury in $CaCl_2$ solution and its concentration are listed in Table 3.2.5 (see page 459).

The surface charge densities of the electrical double-layer at the mercury electrode in 0.1 N aqueous solutions of alkaline earth chlorides have been obtained by Grahame [76] and

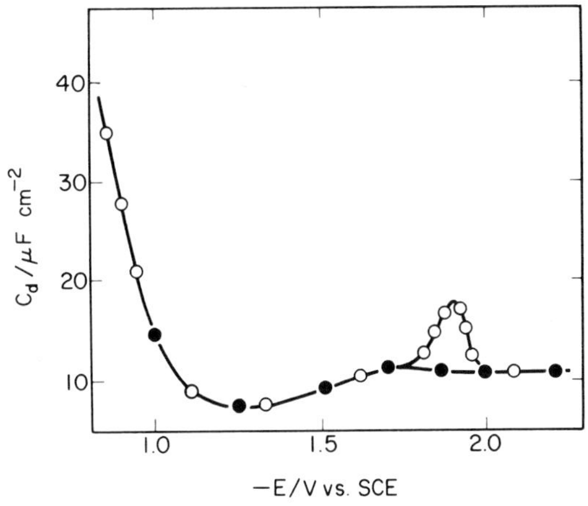

FIG. 3.2.1. Double-layer capacity curves for 0.1 M ethylammonium iodide with O (●) and 0.5 mM Ba^{2+} (O). (Permission of Elsevier Publishing Co.)

are shown in Table 3.2.6 (see page 460). In this table, E is the electrode potential of mercury electrode with respect to a calomel electrode in the same solution, and σ is the surface charge density in μC/cm^2. Timmer et al. [56] obtained the surface charge density of barium ion by integration of the double-layer capacity given in Tables 3.2.2 and 3.2.3. On the basis of theoretical models, they discussed the relevant contributions to the charge density. Integration of the C_d curves of Table 3.2.2 from the point of zero charge, -0.52 V, to -1.95 V yields q = -27.4 C/cm^2 at -1.95 V. In the absence of specific adsorption, the following expression from diffuse double-layer theory holds for mixed electrolytes:

$$q = -2\left\{\frac{RT\varepsilon}{8\pi}\left[\Sigma C_j(u_2^{zj} - 1)\right]\right\}^{1/2} \tag{3.2.1}$$

From this equation u_2 = 23.4 when C_0 = 0.5 mM (which corresponds to a 1 mM Ba^{2+} solution at the half-wave potential). The resulting partial derivatives, starting from u_2 and C_d, are shown in Table 3.2.7 (see page 461), Column A, where C_0 is the concentration of oxidant (barium ion) at the electrode surface, E is the electrode potential, C_{LF} is the low frequency capacitance, and Γ_0 is the surface excess of oxidant. It was concluded from these quantities that the net effect of adding barium ion is the replacement of lithium ion in the diffuse double-layer with barium ion, whereas the charge density q remains practically constant. A similar argument holds for a 0.1-M solution as is evident from the other columns in Table 3.2.7.

When the C_d curve for 0.5 mM barium ion is integrated from -0.47 to -1.95 V, a q = -26.5 μC/cm^2 is obtained, and u_2 = 126 if C_0 = 0.5 mM. The parameters are shown in Column B. Rather large values result because the barium adsorption is strong (7.7 μC/cm^2). There exists a small (0.9 μF/cm^2) difference between $(\partial q/\partial E)_{C_0}$ and the C_d. Therefore,

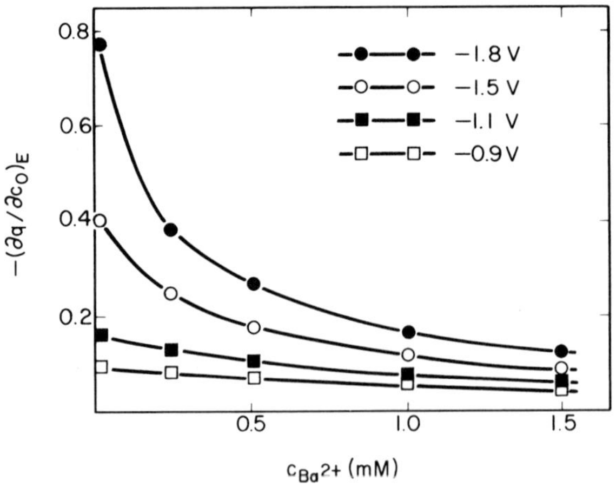

FIG. 3.2.2. Values of $(\partial q/\partial C_0)_E$ vs $C_{Ba^{2+}}$, solutions in 0.1 M LiCl at different electrode potentials [56]. (Permission of Elsevier Publishing Co.)

in Column C a new calibrated set of parameters is given with $(\partial q/\partial E)_{C0} = C_d + 0.9 = 24.2$ $\mu F/cm^2$. In Columns D and E, sets of parameters for 0.5 and 1.5 mM barium ion in 0.1 M LiCl are given. From the values of $nF\Gamma_0$ it may be seen that no linear adsorption isotherm holds. A Langmuir isotherm is obeyed for Γ_0 of barium ion between 0.25 and 1.5 mM ($C_{Ba^{2+}}$ in 1 mM)

$$2F\Gamma_{Ba^{2+}} = \frac{17.4}{1 + 0.63/C_{Ba^{2+}}} \quad (\mu C/cm^2) \tag{3.2.2}$$

Experimental values of the charge on the electrode for 0.1 M LiCl solutions in the presence of barium ion are given as a function of electrode potential in Table 3.2.8 (see page 462). The values of $(\partial q/\partial C_0)_E$ vs the concentration of barium ion in 0.1 M LiCl are given at some potentials in Fig. 3.2.2.

4. ELECTROCHEMICAL STUDIES

4.1. ACTIVITY COEFFICIENT

Electrochemical properties, including standard potential and voltammetry, are given in Gmelin's texts [3, 6, 7], which contain the values of ionic mobility in aqueous, nonaqueous,

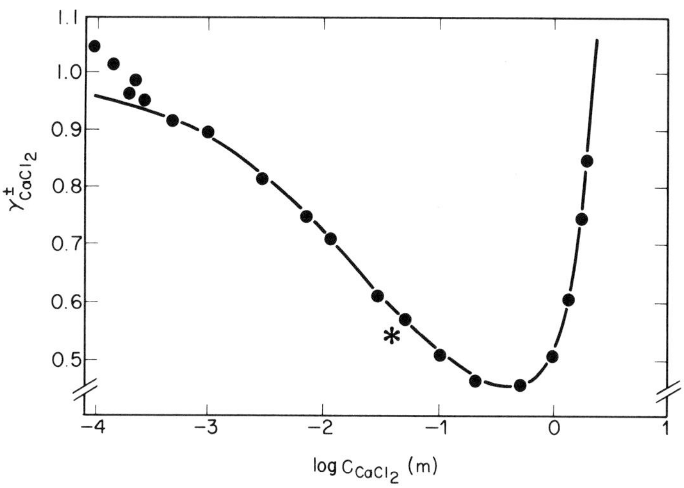

FIG. 4.1.1. Mean activity coefficient of $CaCl_2$ [81]. (*) Reference value 0.569 [83]; (•) calculated from experimental values. (Permission of Pergamon Press.)

and other solvents, and absolute ionic mobility. In addition, the cell voltage values of many cells involving calcium are found in the text on calcium [3], and alkaline earth metals are discussed in the text of Landolt-Börnstein [13].

Shatkay [81] obtained the individual activities of calcium ion and chloride ion in pure aqueous solutions of calcium chloride by measuring the emf's using a calcium ion specific electrode and a chloride ion specific electrode, each opposed by a calomel electrode with a KCl or a KNO_3 bridge. The liquid junction potentials were evaluated for the two bridges using Henderson's equation. Activity coefficients were calculated from the emf's obtained, both for calcium chloride and for the ionic species, using various assumptions concerning ion activities. The calcium ion specific electrode is an Orion model 92-17 liquid membrane ion-exchange calcium electrode with an internal solution of 0.101 or 0.0117 M $CaCl_2$. The mean activity coefficients of calcium chloride are shown in Fig. 4.1.1 coupled with the curve interpolated from literature values [82, 83]. The consistency of the experimental results of the activity coefficient for calcium ion was checked against the three assumptions

1. $\gamma_{Ca^{2+}} = \gamma_{\pm CaCl_2}$
2. $\gamma_{Ca^{2+}} = (\gamma_{\pm CaCl_2})^2$
3. $\gamma_{Ca^{2+}} = (\gamma_{\pm CaCl_2})^3/(\gamma_{\pm KCl})^2$ (the so-called McInnes convention)

The experimental results, combined with liquid junction potentials, are consistent with the McInnes convention and are shown in Fig. 4.1.2. The other assumptions considered did not

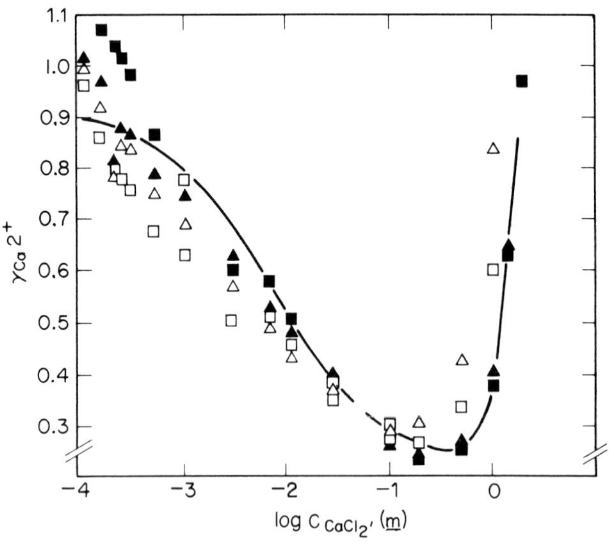

FIG. 4.1.2. Activity coefficient of calcium ion. The curve was calculated by employing the McInnes convention, the points were calculated from experimental values. The asterisks are reference values, the triangles are for the KNO_3 bridge, the squares are for the KCl bridge, the open figures are without correction for liquid junction potentials, and the filled figures are after correction for liquid junction potentials. (Permission of Pergamon Press.)

give consistent results. He also obtained the activity coefficients of calcium ions in calcium chloride solutions (approximately 10^{-3} M in the presence of NaCl and $MgCl_2$) in the range of foreign salt concentrations from 0 to 1 M [84]. Although electrodes have been developed to measure calcium ion activity, there seems to be little reliable data for the calculation of activity coefficients in concentrated solutions [85]. Specific ion glass electrodes useful for this purpose are described below.

4.2. SPECIFIC ION GLASS ELECTRODES

Calcium is an element essential to life, and it is also important in geochemistry and the hardness of water. In these cases the activity of the ion plays a more important role than does the concentration. Moreover, in general, the magnesium ion coexists with the calcium ion to form complexes, and the solutions are often saturated with calcite or dolomite. The activity coefficient of calcium ion or magnesium ion in these solutions cannot be calculated by the Debye-Hückel theory except for very dilute solutions so that a direct method for their measurement is a real necessity. Truesdell and Christ [86] reviewed work on the divalent cation-sensitive glass electrode, one of the most promising tools for these measurements.

When two cations are present at a glass electrode, an ion-exchange equilibrium is attained. If the two cations are calcium and magnesium ion and the anion is monovalent X, the exchange reaction is

$$CaX_{2gl} + Mg_{aq}^{2+} = MgX_{2gl} + Ca_{aq}^{2+} \tag{4.2.1}$$

where gl denotes the glass phase. From the law of mass action, one can write the ion-exchange constant K as

$$K_{Ca}^{Mg} = \frac{[Ca^{2+}][MgX_2]}{[Mg^{2+}][CaX_2]} \tag{4.2.2}$$

If we introduce λ, the rational activity coefficients of the adsorbed ions, which correspond to their mole fractions, the bi-ionic potential is given by

$$E = C_{Ca} + \frac{RT}{2F} \ln\left(\frac{[Ca^{2+}]}{\lambda CaX_2} + K_{Ca}^{Mg} \frac{[Mg^{2+}]}{\lambda MgX_2}\right) \tag{4.2.3}$$

In the same way, if the cations are calcium and monovalent sodium ions, the bi-ionic potential is

$$E = C_{Ca} + \frac{RT}{2F} \ln\left(\frac{[Ca^{2+}]}{\lambda CaX_2} + K_{Ca}^{Na} \frac{[Na^{+}]^2}{\lambda Na_2X_2}\right) \tag{4.2.4}$$

where C_{Ca} is a constant dependent upon the glass, the ion, and the construction of the electrode.

The constant K in the electrode equation is the selectivity constant and is a direct measure of the specificity of an electrode toward a given ion in a bi-ionic system. The larger the value of K_{Ca}^{Mg}, the more specific the electrode is toward Mg^{2+}. The larger the value of K_{Ca}^{Na}, the more specific the electrode is toward Na^+. Such electrodes are less useful as calcium specific ion electrodes. The value of K can be calculated from appropriate emf measurements in solutions containing a single ion. If a given glass electrode is sensitive to A^{2+} and B^{2+}, K_A^B can be expressed as

$$\frac{RT}{2F} \ln K_A^B = C_B - C_A \tag{4.2.5}$$

where C_B and C_A are constants readily obtained from separate measurements of E_A and E_B for known values of $[A^{2+}]$ and $[B^{2+}]$. The component and electrode constants K_A^B of some divalent cation sensitive glasses are shown in Table 4.2.1 (see page 462). NG-2 and NG-6 are natural glasses whose details are shown in the table footnotes; 916-P is a synthetic glass. The

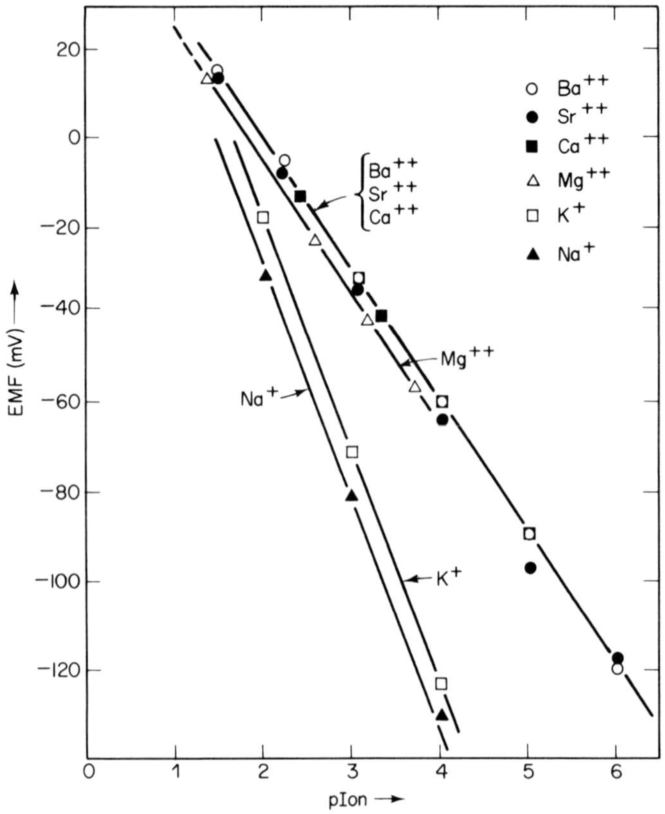

FIG. 4.2.1. Emf vs pIon plots for various ions with glass electrode NG-6 [86].

relations between emf and pIon for various ions, obtained with a ground membrane electrode made from a taktite, NG-6, and from 916-P are shown in Figs. 4.2.1 and 4.2.2, respectively. The emf's are referred to a saturated calomel electrode and the temperature is 25°C. In single-cation solutions these glasses have the theoretical slope for alkaline earth ions over a range of activities of 10^{-1} to 10^{-6}. In solutions containing two different cations the electrodes made from these glasses obey the bi-ionic potential. For example, when the activity of calcium is 4.7×10^{-4} and that of sodium is 0.12, K_{Ca}^{Na} should be no larger than 0.0033 for the electrode to respond selectively to Ca^{2+} without the need for a Na^{+} correction.

A phosphate glass seems to have significant specificity toward alkaline-earth ions. This glass is more specific toward divalent ions than the silicate glasses and shows strong selectivity among the divalent ions. In single cation solutions this glass has the distinct characteristics shown in Fig. 4.2.3. In general, as the divalent ion-sensitive electrode glass

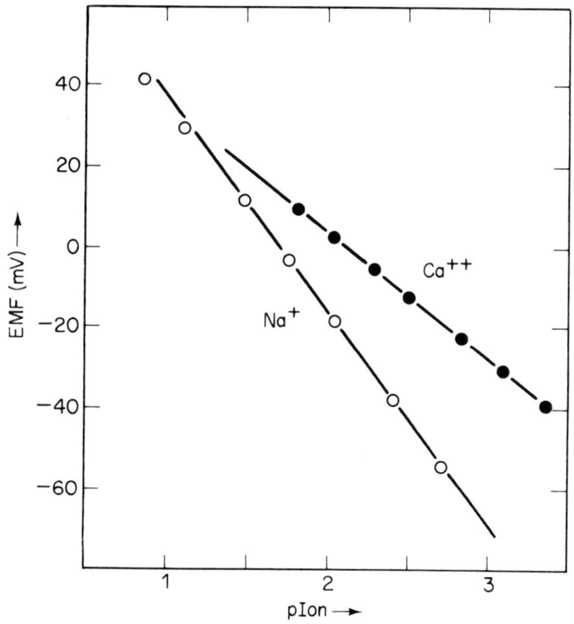

FIG. 4.2.2. Emf vs pIon plots for Ca^{2+} and Na^{+} with glass electrode 916-P [86].

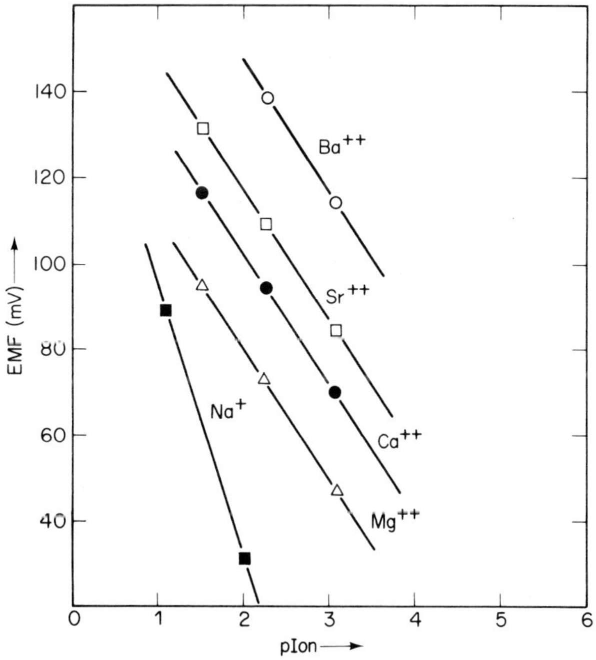

FIG. 4.2.3. Emf vs pIon plots for various ions with phosphate glass electrode [86].

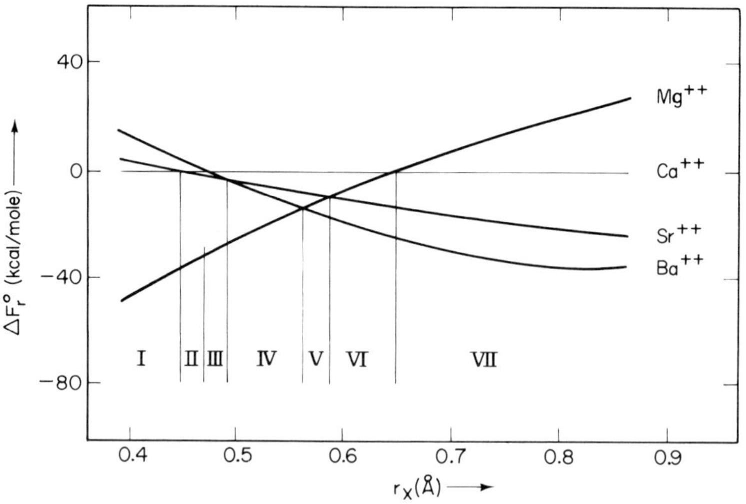

FIG. 4.2.4. Standard free energy of the exchange reaction vs anion radius at 25°C [86].

has a high specific resistance, the electrode membrane must be very thin in order for the resistance to be less than $10^9\,\Omega$.

The selectivity between different divalent ions depends primarily on the equivalent anion radius and secondarily on equivalent site separation. The standard free energies of the exchange reaction

$$CaX_2 + M^{2+} = MX_2 + Ca^{2+}$$

are calculated as a function of the equivalent anion radius r_X^- for the fixed site separation of x = 4.5 Å, shown in Fig. 4.2.4. The ion-selectivity order for each region in Fig. 4.2.4 is listed in Table 4.2.2 (see page 463) together with actual ion-exchange materials.

4.3. BARIUM SULFATE MEMBRANE

The barium sulfate-cellophane membrane is a precipitate-impregnated membrane which is semipermeable to ions. Hirsch-Ayalon [87] obtained some characteristic current-potential curves with different combinations of solutions in a study of permeability. The membrane is impermeable for Ba^{2+}, SO_4^{2-}, and HSO_4^- ions, while it has good permeability to other ions and water. A platinum electrode is attached to the membrane and the counterelectrode is a reference calomel electrode. Current-voltage curves obtained with a $BaSO_4$ membrane and different solution compositions are shown in Fig. 4.3.1, where the concentration was

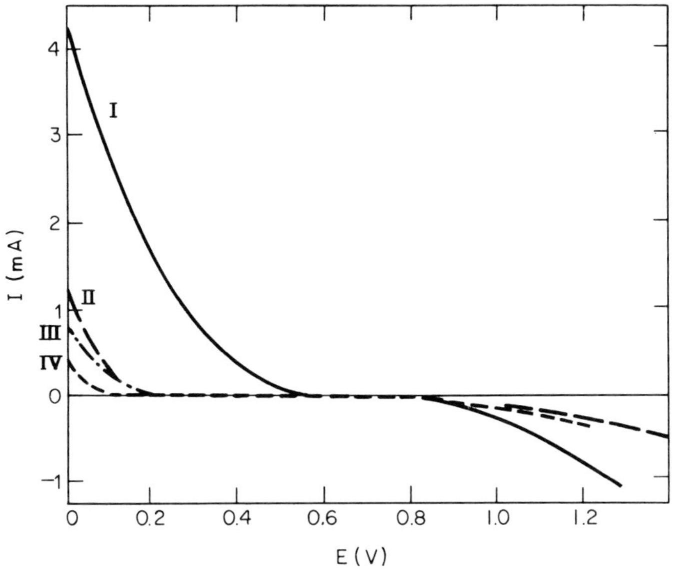

FIG. 4.3.1. Current-voltage curves of $BaSO_4$ membrane.

I $Ba(OH)_2$ || H_2SO_4

II $Ba(OH)_2$ || K_2SO_4

III $BaCl_2$ || H_2SO_4

IV $BaCl_2$ || K_2SO_4

(Permission of Pergamon Press.)

0.05 N and the exposed area of the membrane was 0.79 cm^2. It is found from Fig. 4.3.1 that H^+ and OH^- can transfer through the membrane, and this membrane can transform the concentration energy of H^+ and OH^- ions directly and reversibly into electrical energy.

The permeability of the membrane to one ion species (e.g., K^+) can be modified within a wide range, and it differs widely for different ions. Typical behavior is shown in Fig. 4.3.2, representing the permeation of K^+, and in Fig. 4.3.3, representing that of other solutions. The current-voltage curves of a $BaSO_4$ membrane between 0.1 N $Ba(OH)_2$ and 0.1 N H_2SO_4 with the addition of KCl are shown in Fig. 4.3.4. From these, the current-voltage curves of $BaSO_4$ membranes seem to be similar to those of an anion- and a cation-exchange membrane in series.

Recently Honig and Hengst [88] performed similar studies on $BaSO_4$ membranes to obtain more information about the mechanism of electrical transport through the membranes.

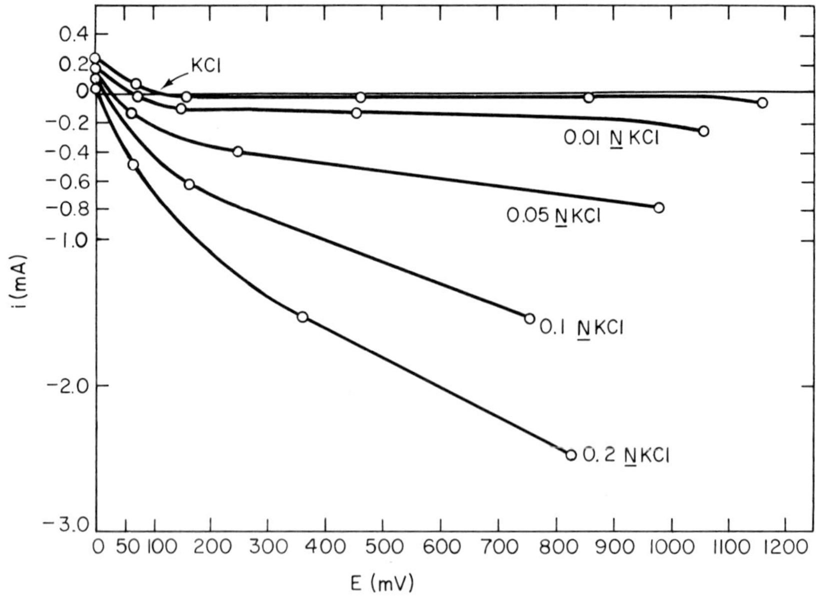

FIG. 4.3.2. Permeation of K^+ through $BaSO_4$ membrane. 0.1 N $BaCl_2$ ‖ 0.1 N K_2SO_4 with addition of different amounts of KCl to Ba side. (Permission of Pergamon Press.)

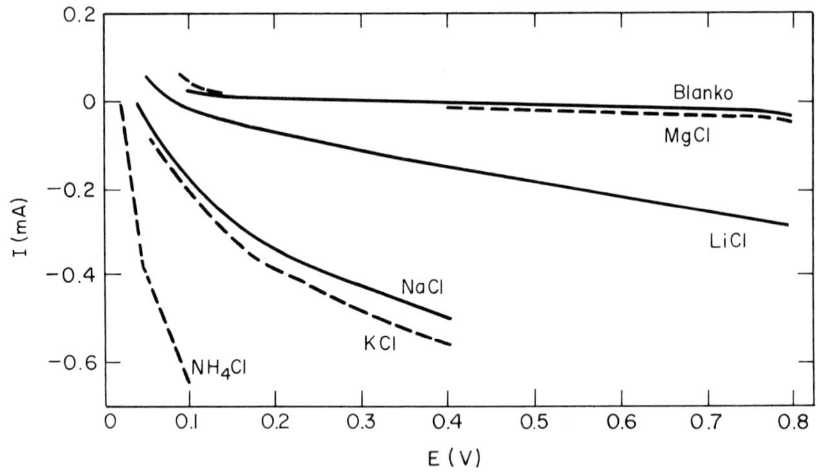

FIG. 4.3.3. Current voltage curves of $BaSO_4$ membrane

0.020 M $BaCl_2$ + 0.005 M Ba-borate ‖ 0.025 M K_2SO_4 + 0.003 M citric acid
(pH = 8.40) (pH = 5.50 by adding NaOH)

with addition of different chloride solutions of 0.03 N to Ba side. (Permission of Pergamon Press.)

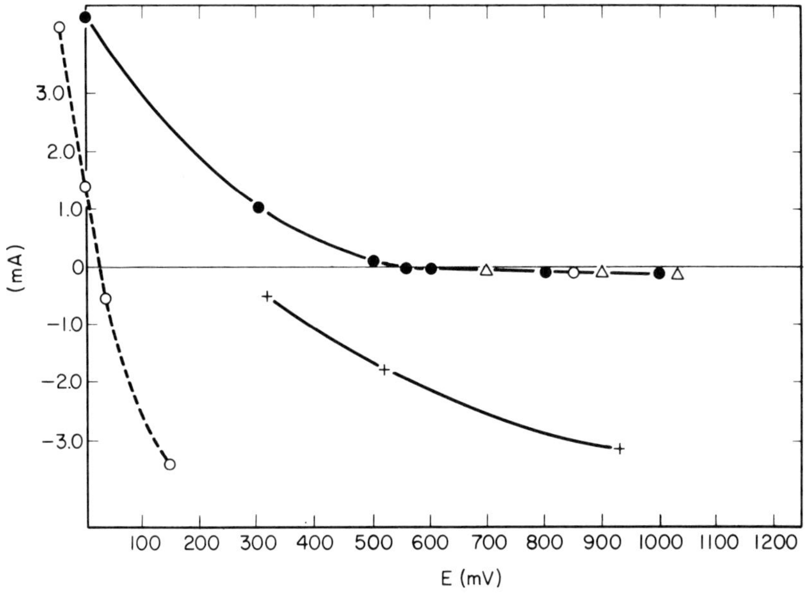

FIG. 4.3.4. Current voltage curves of $BaSO_4$ membrane. 0.1 N $Ba(OH)_2$ ‖ 0.1 N H_2SO_4 with addition of KCl. (●) Without KCl; (x) 0.01 N KCl in the $Ba(OH)_2$ side; (△) 0.1 N KCl in the $Ba(OH)_2$ side; (o) 0.2 N KCl in the $Ba(OH)_2$ side; (+) 0.5 N in the $Ba(OH)_2$ side; and (⊕) 1.0 N KCl in both sides. (Permission of Pergamon Press.)

They found that reproducible results are obtained if one uses the same membrane for all experiments and if one waits a few days for each new set of experimental conditions, which were not fully regulated by Hisch-Ayalon. Measured current-voltage curves of $BaSO_4$ membranes are very similar to those of bipolar ion-exchange membranes, but no satisfactory theory has been given. The small currents at potentials above the rest membrane potential difference (mpd) of the four combinations $Ba(OH)_2$ or $Ba(NO_3)_2$ ‖ H_2SO_4 or Na_2SO_4 are almost identical at a given temperature. This suggests the same mechanism, presumably decomposition of water into H^+ and OH^-. Below the mpd the current first increases nearly exponentially with decreasing voltage and then becomes linear with the voltage; the slope is only slightly dependent on temperature, but strongly dependent on the system. Typical current-voltage curves of $BaSO_4$ membranes, shown in Fig. 4.3.5, illustrate a combination of 0.1 N $Ba(OH)_2$ ‖ 0.1 N Na_2SO_4.

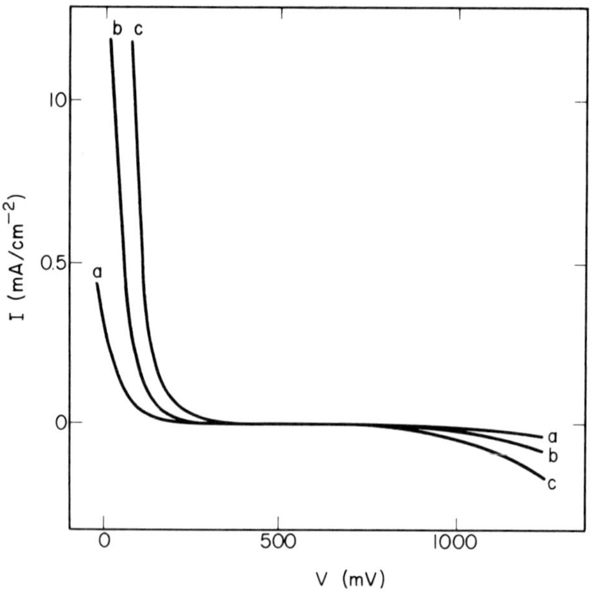

FIG. 4.3.5. Current-voltage curves of $BaSO_4$ membranes. Curves have been shifted horizontally so that the membrane potential difference (mpd) coincides with the reference potential. 0.1 N $Ba(OH)_2$–0.1 N Na_2SO_4; (a) 2.5°C mpd, 498 mV; (b) 25.6°C mpd, 492 mV; (c) 49.5°C mpd, 470 mV. Reference potential, 490 mV. (Permission of Pergamon Press.)

5. APPLIED ELECTROCHEMISTRY

5.1. ELECTROWINNING

Electrolytic separation of calcium from aqueous solutions is possible only on the mercury cathode by the formation of the calcium amalgam. Several examples, found in Gmelin's text [3], are from aqueous solutions of calcium hydroxide, calcium chloride, calcium acetate, and other calcium salts. Electrowinning of calcium on the mercury from aqueous $Ca(OH)_2$ solutions at 17°C was carried out by Eilert [89] who investigated the conditions of winning. The current efficiency is as high as 90% and the amalgam contains calcium up to 2.4 wt %. Electrowinning of calcium from nonaqueous solvents is possible; for example, from calcium nitrate in pyridine and from calcium chloride in acetone [3], but the procedure is so tedious that this process is not practical.

Practical electrowinning of calcium metal, therefore, is performed by the electrolysis of a fused salt of calcium, because calcium halide cannot be reduced chemically to metal, even by alkali metal, at high temperature. In a fused salt, calcium is more basic than alkali metal as shown in Section 1. By the electrolysis of water-free calcium chloride in a closed vessel at a temperature above its melting point, the porous matter at an iron cathode can be found. The anode is the graphite vessel which contains the fused salt. The porous material is separated from the melt and contains metallic calcium. Some difficulties occur from the use of calcium chloride due to its deliquescence and its high melting point. When hydrated calcium chloride is heated to expel the water, calcium hydroxide is formed to some extent and this often covers the calcium metal in the melt. If the separated calcium metal makes contact with the air at high temperature, it tends to burn, forming the oxide. Therefore the temperature range for electrolysis is too narrow to allow a stable winning method. Rathenau and Suter discovered a contact electrode (Berührungselektrod) for overcoming these difficulties [90]. It utilizes the fact that the melting point of calcium metal (845°C) is slightly higher than that of calcium chloride (772°C). When an iron cathode is drawn upward after contact with the melt, the calcium deposited on the cathode is covered with the melt, free from the atmosphere. This device has been employed in the industrial production of calcium.

Some attempts have been made to lower the temperature of the melt by the addition of other compounds to calcium chloride. A mixed melt of $CaCl_2$ (100) + CaF_2 (15) has a melting point of 660°C, while mixtures of $CaCl_2$ (75-85%) and KCl (15-25%) have a melting point range from 630 to 690°C. These mixtures, as well as a pure calcium chloride melt, are used in practice. The cell of Bitterfeld shows the following characteristics [90]:

capacity, 5.000 A; current efficiency, 65-82%
cathodic current density, 10-11 A/cm^2
anodic current density, 0.3 A/cm^2
cell voltage, 25 V
energy consumed, 40-50 kWh/kg Ca

The decomposition voltages of a $CaCl_2$ melt at various temperatures are 3.534 V (500°C), 3.462 V (600°C), and 3.323 V (800°C), and the specific conductivities are 2.02-2.13 at 800°C and 2.17-2.34 Ω^{-1} cm^{-1} at 850°C. The yield of calcium ion from the process is 60-70%, and the product contains a small amount of impurities; for example, 0.31% Fe, 0.03% Al, 0.1-0.2% $CaCl_2$, and 1% others. If the crude calcium is remelted in vacuum or in argon gas, the purity of calcium can be as high as 98.8%. Purification of $CaCl_2$ is essential for obtaining a pure deposit [91].

If a molten lead cathode is used for an electrolysis in a mixed melt of calcium chloride and sodium chloride, a comparatively good current efficiency can be obtained at low current densities. The calcium deposited forms an alloy with lead. The relation between current efficiencies and cathodic current densities is shown in Fig. 5.1.1 [37].

The drop of the current efficiency is caused by the decrease of the surface concentration of calcium, which becomes such that part of the metal goes into the electrolyte and is oxidized by chlorine.

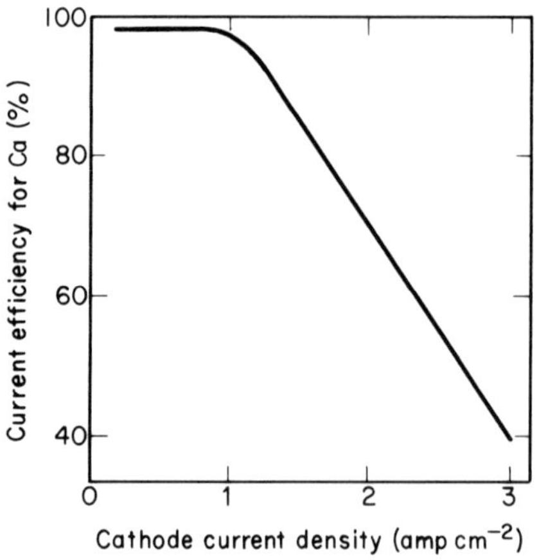

FIG. 5.1.1. Current efficiency vs cathodic current density for electrolysis of $CaCl_2$-NaCl melt at a Pb cathode [37]. (Permission of Andromeda Books.)

5.2. USES

Although the compounds of alkaline earth metals with carbonate, sulfate, and halide are important in inorganic chemistry (including the ceramic industry), the metals themselves are useless except for a few cases where calcium metal is used for purification of rare gas, or the elimination of oxygen, carbon, and nitrogen from iron in metallurgy, based on calcium's high affinity for oxygen, nitrogen, and hydrogen at high temperatures. Calcium hydride (CaH_2) or hydrides of calcium alloys (Ca-Cu-H_2, Ca-Zn-H_2) have been used as a convenient source of hydrogen gas outdoors because 1065 cc of hydrogen gas can be evolved from 1 g of CaH_2 reacting with water. In lead acid batteries an alloy of calcium and lead is suitable for the base plate of the electrode. The reason why an industry based on the alkaline earth metals has not developed may be due to their lack of characteristic or particular properties.

Metallic calcium has recently been used as an anode for thermal batteries. These are primary batteries containing a solid and virtually nonconductive electrolyte at ambient temperatures, which becomes highly conductive when heated to temperatures above its melting point. The electrolyte often used is a LiCl-KCl eutectic because of its relatively low melting point, good thermal stability, and high electrical conductivity. Calcium is chosen as an anode because of its large negative potential, low equivalent weight, high melting point, low polarization, and ready availability. Unfortunately, calcium displaces lithium from the molten electrolyte by the reaction

$$Ca(s) + 2Li^+ \xrightarrow{352^\circ} Ca^{2+} + 2Li \qquad (5.2.1)$$

and the molten lithium in turn reacts with additional calcium to form a calcium-lithium alloy, $CaLi_2$, with a melting point of 230°C. This is obviously an undesirable situation because of the possibility of shorting [92]. Clark and Grothaus [93] presented a technique for chemically treating the calcium anode so that $CaLi_2$ alloy formation is no longer a significant problem. The process involves the production of a high concentration of calcium ions at the anode-electrolyte interface, thereby greatly reducing the formation of Li. This is accomplished by depositing a layer of $3Ca(CH_3COO)_2 \cdot 2CH_3COOH$ on the Ca anode through the reaction of calcium with acetic acid. This double salt decomposes to $CaCO_3$ either prior to or at the beginning of the cell discharge. The resulting high concentration of calcium ion at the anode-electrolyte interface retards the displacement by Ca of Li from the lithium chloride in the electrolyte. Cell and battery test results were described which demonstrate the improvement in performance.

TABLES

TABLE 1.1.1. Standard and Formal Potentials in Aqueous Solutions

Half-reaction	Standard or formal potential (V)	Conditions	Refs.
$Ca(OH)_2(s) + 2e = Ca(s) + 2OH^-$	-3.03		[3], [5], [17], [18]
	-3.02		[19], [20]
$Ca^{2+} + 2e = Ca$	-2.895		[2]
	-2.87	1 $\underline{M}$/1000 g H_2O	[3], [17], [19], [21-23]
	-2.866		[18], [20], [24], [25]
	-2.84	a = 1	[3], [5], [9], [10], [12], [13]
	-2.835	calcd	[26]
	-2.810	a = 1, 15°C	[15]
	-2.809		[16]
	-2.76		[4], [27]
	-2.758		[14]
$Ca^{2+} + 2e = Ca(Hg)$	-2.76	calcd	[5]
	-1.996	obsd	[16]
$Ca^{2+} + PbCO_3 + 2e = Pb + CaCO_3$	-2.868		[16], [28]

(continued)

TABLE 1.1.1 (continued)

Half-reaction	Standard or formal potential (V)	Conditions	Refs.
$CaO + 2H^+ + 2e = Ca + H_2O$	-2.189 -1.902	CaO hyd CaO anhyd	[24]
$Ca^{2+} + 2H^+ + 4e = CaH_2$	-1.045		[24]
$CaO + 4H^+ + 4e = CaH_2 + H_2O$	-0.706 -0.563	CaO hyd CaO anhyd	[24]
$Ca + 2H^+ + 2e = CaH_2$	0.776		[24]
$CaO_2 + 2H^+ + 2e = CaO + H_2O$	1.547 1.260	CaO hyd CaO anhyd	[24]
$CaO_2 + 4H^+ + 2e = Ca^{2+} + 2H_2O$	2.224		[24]
$Sr(OH)_2 \cdot 8H_2O + 2e = Sr(s) + 2OH^- + 8H_2O$	-2.99	calcd	[5], [6], [19]
$Sr^{2+} + 2e = Sr$	-2.89		[5], [6], [10], [13], [17], [19], [23]
	-2.888		[18], [20], [24], [25]
$Sr^{2+} + PbCO_3 + 2e = Pb + SrCO_3$	-2.886		[28]
$Sr(OH)_2 + 2e = Sr + 2OH^-$	-2.88		[20]
$SrO + 2H^+ + 2e = Sr + H_2O$	-2.047 -1.672	SrO hyd SrO anhyd	[24]
$Sr^{2+} + 2H^+ + 4e = SrH_2$	-1.085		[24]
$SrO + 4H^+ + 4e = SrH_2 + H_2O$	-0.665 -0.477	SrO hyd SrO anhyd	[24]
$Sr + 2H^+ + 2e = SrH_2$	0.718		[24]
$SrO_2 + 2H^+ + 2e = SrO + H_2O$	1.492 1.116	SrO hyd SrO anhyd	[24]
$SrO_2 + 4H^+ + 2e = Sr^{2+} + 2H_2O$	2.333		[24]

(continued)

TABLE 1.1.1 (continued)

Half-reaction	Standard or formal potential (V)	Conditions	Refs.
$Ba(OH)_2 \cdot 8H_2O + 2e = Ba(s) + 2OH^- + 8H_2O$	-2.99		[20]
	-2.97		[5], [19]
$Ba^{2+} + 2e = Ba$	-2.92		[7], [9], [10], [12], [13], [19]
	-2.917		[7], [26]
	-2.906		[20], [25]
	-2.905		[24]
	-2.90		[5], [17], [19], [22], [23]
	-2.89		[5], [7], [21]
$Ba^{2+} + PbCO_3 + 2e = BaCO_3 + Pb$	-2.912		[28]
$Ba(OH)_2 + 2e = Ba + 2OH^-$	-2.81		[20]
$BaO + 2H^+ + 2e = Ba + H_2O$	-2.166	BaO hyd	[24]
	-1.509	BaO anhyd	
$Ba^{2+} + 2H^+ + 4e = BaH_2$	-1.110		[24]
$BaO + 4H^+ + 4e = BaH_2 + H_2O$	-0.741	BaO hyd	[24]
	-0.412	BaO anhyd	
$Ba + 2H^+ + 2e = BaH_2$	-0.685		[24]
$BaO_2 + 2H^+ + 2e = BaO + H_2O$	1.626	BaO hyd, BaO_2 hyd	[24]
	1.047	BaO anhyd, BaO_2 hyd	
	1.679	BaO hyd, BaO_2 anhyd	
	1.023	BaO anhyd, BaO_2 anhyd	
$BaO_2 + 4H^+ + 2e = Ba^{2+} + 2H_2O$	2.365	BaO_2 hyd	[24]
	2.419	BaO_2 anhyd	
$Ra^{2+} + 2e = Ra$	-2.92		[5], [17], [19], [24]
	-2.916		[20]
$RaO + 2H^+ + 2e = Ra + H_2O$	-1.319		[24]

TABLE 1.1.2. Isothermal and Thermal Temperature Gradients of Standard Potentials in Aqueous Solutions

Half-reaction	Standard potential, E (V)	(dE/dT) isoth (mV/°C)	(dE/dT) therm (mV/°C)	Refs.
$Ca^{2+} + 2e = Ca$	-2.866	-0.175	0.696	[18], [20]
$Sr^{2} + 2e = Sr$	-2.888	-0.191	0.680	[18], [20]
$Ba^{2+} + 2e = Ba$	-2.906	-0.395	0.48	[18], [20]
$Ra^{2+} + 2e = Ra$	-2.916	-0.59	0.28	[18], [20]
$Ca(OH)_2 + 2e = Ca + 2OH^-$	-3.02	-0.965	-0.094	[18], [20]
$Ba(OH)_2 \cdot 8H_2O + 2e = Ba + 2OH^- + 8H_2O$	-2.99	0.38	1.25	[20]
$Sr(OH)_2 + 2e = Sr + 2OH^-$	-2.88	-0.96	-0.09	[20]
$Ba(OH)_2 + 2e = Ba + 2OH^-$	-2.81	-0.03	-0.06	[20]

TABLE 1.2.1. Values of E°_{Rb} and E°_H in Various Media [32]

Solvent	E°_{Rb} (V)	E°_H (V)
H_2O	-2.92	0.00
CH_3OH	-2.91	-0.03
CH_3CN	-3.03	+0.14
HCOOH	-2.98	+0.47
$HCONH_2$	-2.94	-0.09

TABLE 1.2.2. Standard Potentials in Nonaqueous Solutions

Solvent	Electrode reaction	E°	Potential (V) E_I°	E_H°	Reference electrode	Conditions	Refs.
MeOH	$Ca^2 + 2e = Ca$		-2.897	-2.934			[25]
	$Ca^{2+} + PbCO_3 + 2e = Pb + CaCO_3$	-2.929			SCE		[28]
	$Sr^{2+} + 2e = Sr$		-3.104	-2.924			[25]
	$Sr^{2+} + PbCO_3 + 2e = Pb + SrCO_3$	-2.938			SCE		[28]
	$Ba^{2+} + 2e = Ba$		-3.034	-3.040			[25]
	$Ba^{2+} + PbCO_3 + 2e = Pb + BaCO_3$	-2.943			SCE		[28]
EtOH	$Ca^{2+} + 2e = Ca$		-3.040	-2.870			[25]
	$Sr^{2+} + 2e = Sr$		-3.062	-2.892			[25]
	$Ba^{2+} + 2e = Ba$		-3.035	-3.040			[25]
BuOH	$Ca^{2+} + 2e = Ca$			-2.869			[25]
	$Sr^{2+} + 2e = Sr$			-2.991			[25]
	$Ba^{2+} + 2e = Ba$			-2.930			[25]
AmOH	$Ca^{2+} + 2e = Ca$			-2.808			[25]
	$Sr^{2+} + 2e = Sr$			-2.830			[25]
	$Ba^{2+} + 2e = Ba$			-2.912			[25]
$(CH_3)_2CO$	$Ca^{2+} + 2e = Ca$			-2.895			[25]
	$Sr^{2+} + 2e = Sr$			-2.902			[25]
	$Ba^{2+} + 2e = Ba$			-2.870			[25]
HCOOH	$Ca^{2+} + 2e = Ca$	-2.67			$E_{H(H_2O)}^\circ = 0$		[32]
	$Ca^{2+} + 2e = Ca$	-3.20			Ag|Ag picrate, 0.01 N		[17], [33]
	$Ca^{2+} + 2e = Ca$	0.25			$E_{Rb} = 0$		[12], [33]
CH_3CN	$Ca^{2+} + 2e = Ca$	-2.50			$E_{H(H_2O)}^\circ = 0$		[32]
	$Ca^{2+} + 2e = Ca$	-2.75					[17]
	$Ca^{2+} + 2e = Ca$	0.42			$E_{Rb} = 0$		[3], [33]
NH_3	$Ca^{2+} + 2e = Ca$	-2.63			$E_{H(H_2O)}^\circ = 0$		[32]
	$Ca^{2+} + 2e = Ca$	-2.39				-50°C	[3], [34]
	$Ca^{2+} + 2e = Ca$	0.29			$E_{Rb} = 0$	-50°C	[12], [33]
	$Ca^{2+} + 2e = Ca$	-2.185			Pb|$PbNO_3$		[35]
N_2H_4	$Ca^{2+} + 2e = Ca$	-2.82			$E_{H(H_2O)}^\circ = 0$		[32]
	$Ca^{2+} + 2e = Ca$	-1.91					[17]
	$Ca^{2+} + 2e = Ca$	0.10			$E_{Rb} = 0$		[3], [33]

TABLE 1.2.3. Potential between Calcium Amalgam and Solid Calcium Ca(s) | Ca^{2+} in Nonaqueous Solution | Ca(Hg)

Salt	Concn (M)	Solvent	X_{Ca} in amalgam	T (°C)	E	ΔE°	Refs.
$CaCl_2$	0.25	MeOH	0.0148 (satd)	-80	0.502	0.467	[16]
$CaCl_2$	0.25	MeOH	0.01	-80	0.811	0.773	[16]
$CaCl_2$	0.35	MeOH	0.01	-80	0.829	0.791	[16]
CaI_2	0.0093	Pyridine	0.01	25	0.149	0.089	[16]
CaI_2	satd?	Pyridine	0.000385	17.5	0.88	0.782	[14], [16]
					0.90	0.802	[14], [16]
CaI_2	satd	Pyridine	0.00126	17	0.843	0.759	[4], [16]
CaI_2	satd	Pyridine	0.00133	15	0.895	0.813	[15], [16]
CaI_2	satd?	EtOH	0.00133	16	0.885	0.803	[15], [16]
$CaCl_2$	0.0288	DMSO	0.0051	25	0.310	0.242	[16], [36]
			0.0551	25	0.316	0.248	[16], [36]
$CaCl_2$	0.010	Propylene carbonate	0.000148	25	0.802	0.789	[16]
		Propylene carbonate	0.000136		0.876	0.762	[16]

TABLE 1.3.1a. Electrode Potentials of Metals in Single Fused Halides at 700°C (V)

Electrode	Anion				Condition	Refs.
	F^-	Cl^-	Br^-	I^-		
Ca | Ca^{2+}		-2.36	-2.25	-2.09		[37]
Sr | Sr^{2+}		-2.52	-2.41	-2.40	$E_{H_2} = 0$	[37]
Ba | Ba^{2+}		-2.60	-2.62	-		[37]
Ca | Ca^{2+}	-0.29	+0.01	+0.10	+0.18		[37]
Sr | Sr^{2+}	-0.40	-0.15	-0.06	-0.13	$E_{Na} = 0$	[37]
Ba | Ba^{2+}	-0.47	-0.23	-0.27	-0.01		[37]

TABLE 1.3.1b. Electrode Potentials of Metals in Single Fused Halides at the Melting Points of the Salts (V)

Electrode	Anion				Condition	Refs.
	F^-	Cl^-	Br^-	I^-		
Ca | Ca^{2+}	0.32	-0.03	-0.11	-0.01	$E_{Na} = 0$	[37]
Sr | Sr^{2+}	0.27	-0.16	-0.15	-0.19	$E_{Na} = 0$	[37]
Ba | Ba^{2+}	-0.04	-0.19	-0.12	-	$E_{Na} = 0$	[37]

TABLE 1.3.2. Electrode Potentials of Metals in Fused Chlorides (V) [37]

Electrode	Temperature (°C)			Condition
	700	800	1000	
Ba|Ba^{2+}	(-0.314)		-0.393	$E_{Na} = 0$
Sr|Sr^{2+}	(-0.208)		-0.314	$E_{Na} = 0$
Ca|Ca^{2+}	(-0.060)	-0.083	-0.189	$E_{Na} = 0$

TABLE 1.3.3. Single Electrode Potentials in Simple Fused Salts

Fused Salt	Temp (°C)	Potential (E)	dE/dT	Conditions	Refs.
CaF_2	783	0.74	4.045×10^{-3}	Two electrodes	[38], [40]
	827	0.56			
	872	0.38			
SrF_2	783	0.91	4.045×10^{-3}	Two electrodes	[38], [40]
	827	0.73			
	872	0.55			
BaF_2	783	0.99	4.045×10^{-3}	Two electrodes	[38], [40]
	827	0.81			
	872	0.63			
CaF_2	1400	2.40			[38], [41]
SrF_2	1400	2.43			[38], [41]
BaF_2	1400	2.58			[38], [41]
$CaCl_2$	585	2.85	0.685×10^{-3}	Two electrodes	[38], [42]
	660	2.8			
$BaCl_2$	615	3.0		Two electrodes	[38], [42]
	685	2.95			
$SrCl_2$	615	3.0	0.715×10^{-3}	Two electrodes	[38], [42]
$CaCl_2$	900	3.045 = 3.254 -0.00164 (t - 772)		Polarization emf	[38], [43]
$SrCl_2$	900	3.24		Polarization emf	[38], [43]
$BaCl_2$	900	3.38		Polarization emf	[38], [43]
$CaCl_2$	638	2.82	0.704×10^{-3}		[38], [40]
	675	2.79			
	737	2.75			
	773	2.725			
$SrCl_2$	615	3.0	0.714×10^{-3}	Two electrodes	[38], [40]
	633	2.99			
	675	2.96			
	685	2.95			
$BaCl_2$	647	3.06	0.714×10^{-3}	Two electrodes	[38], [40]
	689	3.03			
	745	2.99			
CaI_2	700	2.19			[38], [44]
SrI_2	700	2.41			[38], [44]
BaI_2	700	2.29			[38], [44]

TABLE 1.3.4. Single Electrode Potentials in Fused Salts Solvents

Solvents	Solute	Temp (°C)	Potential	dE/dT	Conditions	Refs.
NaBr-KBr(1:1)	$CaBr_2$	700	2.86	1.4 10^{-3}	Two electrodes	[38],[44]
Na_3AlF_6	Al_2O_3, BeO MgO, CaO	1020	1.07 ± 0.01		Two electrodes	[38],[45]
Na_3AlF_6	SrO, BaO Cr_2O_3	1020	0.75 ± 0.05		Two electrodes	[38],[45]

TABLE 1.3.5. Electrochemical Series in Fused Salts Solvent

Solvents	Temp (°C)	Electrochemical series	Refs.
Individual fluorides	1000	Ba, Sr, Ca, Na, K, Mg, Li	[37]
Individual chlorides	700	Ba, Sr, K, Li, Na, Ca, Mg, Be	[37]
Individual bromides	700	Ba, K, Sr, Li, Na, Ca, Mg, Mn	[37]
NaBr-KBr	700	Na, Ca, Mn, Be, Tl(I), Zn	[3],[37],[38],[47]

TABLE 1.3.6. Theoretical Values of Decomposition Voltages (V) [38, 39]

Salt	Aqueous (25°C)	Solid (25°C)	Uncertainty (25°C)	Conditions (°C) 100	200	300	350	400	450	500	550	600	800	1000	1500
CaF_2	5.736	3.021	±0.01	5.953	5.864	5.776	5.732	5.689	5.646	5.603	5.560	5.517	5.350	5.182	4.785
SrF_2	5.756	6.012	±0.002	5.943	5.854	5.768	5.726	5.684	5.643	5.602	5.562	5.222	5.364	5.203	4.768
BaF_2	5.766	5.952	±0.02	5.885	5.797	5.712	5.670	5.629	5.588	5.547	5.507	5.468	5.310	5.154	4.803
RaF_2	5.786	5.941	±0.07	5.871	5.781	5.694	5.651	5.608	5.563	5.524	5.483	5.442	5.277	5.111	4.566
$CaCl_2$	4.230	3.888	±0.02	3.830	3.754	3.680	3.643	3.607	3.570	3.534	3.498	3.462	3.323	3.208	2.926
$SrCl_2$	4.25	4.048	±0.01	3.987	3.909	3.832	3.794	3.757	3.720	3.684	3.648	3.612	3.469	3.333	2.977
$BaCl_2$	4.26	4.202	±0.01	4.139	4.056	3.975	3.935	3.888	3.848	3.808	3.768	3.728	3.568	3.412	3.079
$RaCl_2$	4.28	4.336	±0.07	4.272	4.189	4.108	4.068	4.029	3.990	3.952	3.913	3.876	3.723	3.569	3.098
$CaBr_2$	3.951	3.416	±0.02	3.357	3.280	3.206	3.170	3.134	3.099	3.664	3.028	2.994	2.861	2.740	2.447
$SrBr_2$	3.971	3.621	±0.01	3.559	3.479	3.402	3.363	3.326	3.288	3.251	3.214	3.178	3.053	2.930	2.595
$BaBr_2$	3.981	3.814	±0.01	3.750	3.669	3.590	3.552	3.514	3.476	3.438	3.401	3.364	3.216	3.084	2.791
$RaBr_2$	4.001	3.957	±0.07	3.890	3.803	3.718	3.676	3.635	3.594	3.553	3.513	3.473	3.310	3.154	2.667
CaI_2	3.506	2.845	±0.01	2.784	2.705	2.628	2.591	2.554	2.517	2.480	2.444	2.407	2.272	2.150	2.015
SrI_2	3.526	3.014	±0.01	2.953	2.874	2.797	2.759	2.722	2.685	2.648	2.617	2.589	2.480	2.370	2.063
BaI_2	3.536	3.187	±0.01	3.124	3.042	2.963	2.924	2.886	2.848	2.811	2.773	2.737	2.598	2.479	2.205
RaI_2	3.556	3.350	±0.07	3.283	3.195	3.110	3.068	3.027	2.986	2.945	2.905	2.865	2.713	2.578	2.111

TABLE 1.3.7. Decomposition Voltage of Compounds of Ca, Sr, and Ba

Substances	Temp (°C)	Decomposition voltage (V)		Conditions	Refs.
CaF_2	782	0.74		Method 2, graphite electrode, Acheson crucible, CaF_2 + KCl	[38], [40]
	827	0.56		Method 2, graphite electrode, Acheson crucible, CaF_2 + KCl	[38], [40]
	872	0.38		Method 2, graphite electrode, Acheson crucible, CaF_2 + KCl	[38], [40]
	1400	2.40		Temp coeff 0.0020 V/°C	[37]
$CaCl_2$	587	2.85		Method 2, graphite electrode, graphite crucible, $CaCl_2$ + KCl	[38], [42]
	660	2.8		Method 2, graphite electrode, graphite crucible, $CaCl_2$ + CKl	[38], [42]
	638	2.82		Method 2, graphite electrode, Tammann tube cell	[38], [40]
	675	2.79		Method 2, graphite electrode, Tammann tube cell	[38], [40]
	737	2.75		Method 2, graphite electrode, Tammann tube cell	[38], [40]
	773	2.725		Method 2, graphite electrode, Tammann tube cell	[38], [40]
	700	3.38		"Most reliable"	[37]
$CaBr_2$	700	2.88		"Most reliable"	[37]
CaI_2	700	2.24		"Most reliable"	[37]
		(Pt anode)	(C anode)		
CaO	900	2.70	1.50	Method 2, graphite crucible, mixed bath of CaF_2 (150 g) + NaF (75 g) saturated with CaO	[38], [48]
	950	2.50	1.32	Method 2, graphite crucible, mixed bath of CaF_2 (150 g) + NaF (75 g) saturated with CaO	[38], [48]
	1000	2.25	1.19	Method 2, graphite crucible, mixed bath of CaF_2 (150 g) + NaF (75 g) saturated with CaO	[38], [48]
	1050	1.98	0.98	Method 2, graphite crucible, mixed bath of CaF_2 (150 g) + NaF (75 g) saturated with CaO	[38], [48]
	1100	1.75	0.81	Method 2, graphite crucible, mixed bath of CaF_2 (150 g) + NaF (75 g) saturated with CaO	[38], [48]
	1000	2.25		In fused fluoride	[37]
SrF_2	783	0.91		Method 2, graphite electrode, Acheson crucible, SrF_2 + KCl	[38], [40]
	827	0.73		Method 2, graphite electrode, Acheson crucible, SrF_2 + KCl	[38], [40]
	872	0.55		Method 2, graphite electrode, Acheson crucible, SrF_2 + KCl	[38], [40]
	1400	2.43		Temp coeff 0.0020 V/°C	[37]

$SrCl_2$	615	3.0		Method 2, graphite electrode, graphite crucible, $SrCl_2$ + KCl + CaF_2	[38], [42]
	685	2.95		Method 2, graphite electrode, graphite crucible, $SrCl_2$ + KCl + CaF_2	[38], [42]
	633	2.99		Method 2, graphite electrode, Tammann tube cell	[38], [40]
	675	2.96		Method 2, graphite electrode, Tammann tube cell	[38], [40]
	700	3.54		"Most reliable"	[37]
$SrBr_2$	700	3.04		"Most reliable"	[37]
SrI_2	700	2.55		"Most reliable"	[37]
		(Pt anode)	(C anode)		
SrO	900	1.06	0.20	2SrO = Sr_2O + $1/2O_2$, Method 2, graphite crucible, mixed bath of SrF_2 (150 g) + NaF (75 g) saturated with SrO	[38], [48]
	950	1.02	0.14	2SrO = Sr_2O + $1/2O_2$, Method 2, graphite crucible, mixed bath of SrF_2 (150 g) + NaF (75 g) saturated with SrO	[38], [48]
	1000	0.90	0.17	2SrO = Sr_2O + $1/2O_2$, Method 2, graphite crucible, mixed bath of SrF_2 (150 g) + NaF (75 g) saturated with SrO	[38], [48]
	1050	0.87	0.16	2SrO = Sr_2O + $1/2O_2$, Method 2, graphite crucible, mixed bath of SrF_2 (150 g) + NaF (75 g) saturated with SrO	[38], [48]
	1100	0.83	0.10	2SrO = Sr_2O + $1/2O_2$, Method 2, graphite crucible, mixed bath of SrF_2 (150 g) + NaF (75 g) saturated with SrO	[38], [48]
	900	1.56	1.28	Values after electrolysis with 50 A/hr, F_2 evolved, Method 2, graphite crucible, mixed bath of SrF_2 (150 g) + NaF (75 g) saturated with SrO	[38], [48]
	950	1.42	1.10	Values after electrolysis with 50 A/hr, F_2 evolved, Method 2, graphite crucible, mixed bath of SrF_2 (150 g) + NaF (75 g) saturated with SrO	[38], [48]
	1000	1.17	0.98	Values after electrolysis with 50 A/hr, F_2 evolved, Method 2, graphite crucible, mixed bath of SrF_2 (150 g) + NaF (75 g) saturated with SrO	[38], [48]
	1050	0.92	0.74	Values after electrolysis with 50 A/hr, F_2 evolved, Method 2, graphite crucible, mixed bath of SrF_2 (150 g) + NaF (75 g) saturated with SrO	[38], [48]
	1100	0.78	0.64	Values after electrolysis with 50 A/hr, F_2 evolved, Method 2, graphite crucible, mixed bath of SrF_2 (150 g) + NaF (75 g) saturated with SrO	[38], [48]
	1000	1.17		In fused fluoride	[37]
BaF_2	783	0.99		Method 2, graphite electrode, Acheson crucible, BaF_2 + KCl	[38], [40]
	827	0.81		Method 2, graphite electrode, Acheson crucible, BaF_2 + KCl	[38], [40]
	872	0.63		Method 2, graphite electrode, Acheson crucible, BaF_2 + KCl	[38], [40]
	1400	2.58		Temp coeff 0.0020 V/°C	[37]

(continued)

TABLE 1.3.7 (continued)

Substances	Temp (°C)	Decomposition voltage (V) (Pt anode)	(C anode)	Conditions	Refs.
$BaCl_2$	750	1.45		Method 2, graphite electrode, graphite crucible, $BaCl_2 + CaCl_2 + LiCl$	[38], [42]
	647	3.06		Method 2, graphite electrode, Tammann tube cell	[38], [40]
	689	3.03		Method 2, graphite electrode, Tammann tube cell	[38], [40]
	745	2.99		Method 2, graphite electrode, Tammann tube cell	[38], [40]
	700	3.62		"Most reliable"	[37]
$BaBr_2$	700	3.25		"Most reliable"	[37]
BaO	900	0.72	0.34	$2BaO = Ba_2O + 1/2O_2$, Method 2, graphite crucible, mixed bath of BaF_2 (250 g) + NaF (75 g) saturated with BaO	[38], [48]
	950	0.82	0.12	$2BaO = Ba_2O + 1/2O_2$, Method 2, graphite crucible, mixed bath of BaF_2 (250 g) + NaF (75 g) saturated with BaO	[38], [48]
	1000	0.62	0.16	$2BaO = Ba_2O + 1/2O_2$, Method 2, graphite crucible, mixed bath of BaF_2 (250 g) + NaF (75 g) saturated with BaO	[38], [48]
	1050	0.63	0.08	$2BaO = Ba_2O + 1/2O_2$, Method 2, graphite crucible, mixed bath of BaF_2 (250 g) + NaF (75 g) saturated with BaO	[38], [48]
	1100	0.42	0.10	$2BaO = Ba_2O + 1/2O_2$, Method 2, graphite crucible, mixed bath of BaF_2 (250 g) + NaF (75 g) saturated with BaO	[38], [48]
	900	1.44	1.21	Values after electrolysis with 50 A/hr, F_2 evolved, Method 2, graphite crucible, mixed bath of BaF_2 (250 g) + NaF (75 g) saturated with BaO	[38], [48]
	950	1.10	1.08	Values after electrolysis with 50 A/hr, F_2 evolved, Method 2, graphite crucible, mixed bath of BaF_2 (250 g) + NaF (75 g) saturated with BaO	[38], [48]
	1000	1.09	0.78	Values after electrolysis with 50 A/hr, F_2 evolved, Method 2, graphite crucible, mixed bath of BaF_2 (250 g) + NaF (75 g) saturated with BaO	[38], [48]
	1050	0.69	0.68	Values after electrolysis with 50 A/hr, F_2 evolved, Method 2, graphite crucible, mixed bath of BaF_2 (250 g) + NaF (75 g) saturated with BaO	[38], [48]
	1100	0.55	0.49	Values after electrolysis with 50 A/hr, F_2 evolved, Method 2, graphite crucible, mixed bath of BaF_2 (250 g) + NaF (75 g) saturated with BaO	[38], [48]
	1000	1.09		In fused fluoride	[37]
$BaCO_3$	600	-2.90		emf series relative to E, arbitrary zero for the CO_2/O_2 electrode with $CO_2:O_2 = 2:1$ at a total pressure of 1 atm	[38]

TABLE 1.3.8. Decomposition Voltage of Salts at their Melting Points (V) [37]

Cation	Anion			
	F^-	Cl^-	Br^-	I^-
Ca^{2+}	2.46	3.28	2.82	2.50
Sr^{2+}	2.51	3.41	3.08	2.68
Ba^{2+}	2.82	3.44	3.05	2.36

TABLE 1.3.9. Potentials Measured in LiCl-KCl Eutectic at 450°C

Cation	Br_2/Br^- potential	I_2/I^- potential	F_2/F^- potential
Ca^{2+}	-0.471	-1.069	2.076
Li^+	-0.435	-1.081	1.927
K^+	-0.273	-0.779	1.265
$(Li\text{-}K)^+$	-0.356	-0.740	

TABLE 2.1.1. Polarographic Characteristics

Substance	Product	Conditions	$E_{1/2}$ (V)	I	$E_{3/4} - E_{1/4}$	Remarks	Refs.
Ca(II)	Ca(0)	0.1 N $NaClO_4$ or 0.1 N KCl	-2.23fw				[3], [9], [12], [17], [19]
		1 N alkaline soln	-2.23fw				[12]
		Purely aqueous	-2.220	3.39		La^{3+} added	[3], [53], [54]
		25% ethanol	-2.180			Ba^{2+} added	[3], [53]
		50%	-2.155			Ba^{2+} added	[3], [53]
		80%	-2.125	2.47		Ba^{2+} added	[3], [53]
		90%	-2.115			Ba^{2+} added	[3], [53]
		100% ethanol	-2.10	2.6		Extrapolated	[3]
		Liquid NH_3	-1.96			-36°C, Hg pool	[3], [55]
		Liquid NH_3	-1.64	5.27	-50	-36°C, $Pb(NO_3)_2$	[22]
		Acetonitrile	-1.82	4.61	-43		[22]
		Ethylenediamine	-1.70		irr		[22]
Sr(II)	Sr(0)	0.1 N $NaClO_4$ or 0.1 N KCl	-2.13fw				[6], [12], [17]
		1 N alkaline soln	-2.13fw				[12]
		Purely aqueous	-2.110w	3.44			[6], [53], [54]
		Purely aqueous	-2.110	3.46			[22]
		25% C_2H_5OH	-2.095				[6]
		50%	-2.060				[6]
		80%	-2.030	2.50			[6], [53], [54]
		90%	-2.020				[6]
		100%	-2.010fi	2.67		Extrapolated	[6], [53], [54]
		Liquid NH_3	-1.69fw	2.39		-36°C, Hg pool, $Sr(NO_3)_2$ up to 0.37 mM	[6], [55]

			-1.68fw	3.27		Up to 0.52 mM	[6], [55]
			-1.67fw			Up to 0.78 mM	[6], [55]
			-1.36w	5.48	-80	-36°C, Pb | $Pb(NO_3)_2$	[22]
		Acetonitrile	-1.76	5.43	-69		[22]
		Ethylenediamine	-1.64			23°C	[22]
Ba(II)	Ba(0)	1 M LiCl	-1.95w				[56]
		0.1 N $NaClO_4$ or 0.1 N KCl	-1.90fw				[17]
		Neutral or acid	-1.94fw				[12], [17]
		1 N alkaline soln	-1.96fw				[17]
		1 N alkaline soln	-1.94fw				[12]
		Purely aqueous	-1.940w	3.57			[7], [9], [19], [53], [54], [57]
		Purely aqueous	-1.92	3.58			[22]
		25% C_2H_5OH	-1.915				[7]
		50%	-1.875				[7]
		80%	-1.835	2.67			[7], [53], [54]
		90%	-1.820				[7]
		100%	-1.8fi	2.82		Extrapolated	[7], [53], [54]
		Liquid NH_3	-1.54	2.65	[0.43 mM $Ba(NO_3)_2$]	-36°C, Hg pool, Ba^+ 0.58-0.63 mM	[7], [55]
				3.90	[0.63 mM $Ba(NO_3)_2$]		[7], [55]
			-1.22	4.94	-34	Pb | $Pb(NO_3)_2$ - 36°C	[22]
		Acetonitrile	-1.63	5.37	-42		[22]
		Ethylenediamine	-1.76			23°C	[22]
Ra(II)	Ra(0)	0.1 N $NaClO_4$	-1.89fw				[12], [17]
		0.1 N KCl	-1.89fw				[12], [17]
		Acid soln	-1.89fw				[12]
		1 N alkaline soln	-1.89fw				[12]
		KCl	-1.84fw				[22], [54], [58]

TABLE 2.1.2. Diffusion Current of Strontium [6, 53]

H_2O		x = 50% vol EtOH		x = 80% vol EtOH	
C	I_D	C	I_D	C	I_D
0.36	2.12	0.33	1.80	0.18	0.75
0.83	5.00	0.625	3.47	0.305	1.30
		0.85	4.72	1.07	4.58
1.38	8.35	1.65	9.13	2.05	8.74
3.15	19.10	3.55	19.60		

TABLE 2.1.3. Diffusion Current of Barium [7, 53]

H_2O			25% vol EtOH			50% vol EtOH			80% vol EtOH		
C	I_D	I_D/C	C	I_D	I_D/C	C	I_D	I_D/C	C	I_D	I_D/C
0.335	2.06	6.16	0.256	1.48	5.59	0.335	1.78	5.01	0.355	1.56	4.66
0.74	4.65	6.28	0.76	4.35	5.72	0.63	3.21	5.10	0.575	2.59	4.51
1.13	7.17	6.35	1.56	8.85	5.69	1.37	6.94	5.05	1.00	4.55	4.55
						2.07	10.52	5.09			
2.74	17.22	6.28	2.45	14.12	5.77	2.54	12.91	5.09	1.83	8.27	4.51
4.5	28.53	6.32				4.08	20.86	5.11			

TABLE 2.1.4. Polarographic Characteristics in Nonaqueous Solvent[a]

Compound	Solvent	Supporting electrolyte	Reference electrode	Temp (°C)	Half-wave potential (V)	Slope of log $i/(i_d - i)$ vs E (V)	Refs.
CaI_2	MeCN	0.1 M TBAI	Hg pool	25	-1.24	0.045	[59], [61]
$Ca(ClO_4)_2 \cdot 6aq$	MeCN	0.1 M TEAP	SCE	25	-1.82	0.045	[59], [62]
$Ca(ClO_4)_2$	MeCN	0.1 M TBAP	Ag| 0.01 M $AgNO_3$	25	-2.14	-	[59], [63]
	PhCN	0.1 M TEAP	SCE	25	-1.73	0.058	[59], [64]
	$PhCH_2CN$	0.1 M TEAP	SCE	25	-1.79	0.040	[59], [64]
	EtCN	0.1 M TEAP	SCE	25	-1.83	0.045	[59], [64]
$CaCl_2$	DMF	0.1 M TBAI	Hg pool	25	-1.78	-	[59], [65]
$Ca(ClO_4)_2$	DMF	0.1 M TEAP	Ag| AgCl(s)	25	-1.95	0.045	[59], [66]
	CH_3CONMe_2	0.1 M TEAP	SCE	25	-2.37	0.056	[59], [60]
$Ca(NO_3)_2$	DMSO	0.1 M TEAN	SCE	21	-2.30	-	[59], [67]
$Ca(ClO_4)_2$	PC	0.1 M TEAP	SCE	25	-1.92	0.077	[59], [68]
CaI_2	NH_3	0.021 M TEAI	Hg pool	-36	-1.96	0.044	[55], [59]
CaI_2	MOR	0.1 M TBAI	NCE	20	-1.9	-	[59], [69]
$Ca(NO_3)_2$	EDA	0.1 M $NaNO_3$	NCE	23	-1.74	-	[59], [67]
$Sr(ClO_4)_2$	MeCN	0.1 M TEAP	SCE	25	-1.76	0.072	[59], [62]
	MeCN	0.1 M TBAP	Ag| 0.01 M $AgNO_3$	25	-2.08	-	[59], [63]
	C_3H_5CN	0.1 M TEAP	SCE	25	-1.72	0.074	[59], [64]

(continued)

TABLE 2.1.4 (continued)

Compound	Solvent	Supporting electrolyte	Reference electrode	Temp (°C)	Half-wave potential (V)	Slope of log $i/(i_d - i)$ vs E (V)	Refs.
$Sr(ClO_4)_2$	$PhCH_2CN$	0.1 M TEAP	SCE	25	-1.73	-	[59], [64]
	EtCN	0.1 M TEAP	SCE	25	-1.78	0.031	[59], [64]
	DMF	0.1 M TBAI	Hg pool	25	-1.68	0.059	[59], [65]
	DMF	0.1 M TEAP	Ag| AgCl(s)	25	-1.80	0.044	[59], [66]
	CH_3CONMe_2	0.1 M TEAP	SCE	25	-2.23	0.050	[59], [60]
$Sr(NO_3)_2$	DMSO	0.1 M TEAN	SCE	21	-2.10	-	[59], [67]
$Sr(ClO_4)_2$	DMSO	0.1 M TEAP	Zn-Hg(s)| $Zn(ClO_4)_2$(s)	25	-1.229	0.047	[59], [70]
	PC	0.1 M TEAP	SCE	25	-1.83	0.030	[59], [68]
$Sr(NO_3)_2$	NH_3	0.021 M TEAI	Hg pool	-36	-1.67	0.117	[55], [59]
	EDA	0.1 M $NaNO_3$	NCE	23	-1.68	-	[59], [67]
$Ba(ClO_4)_2$	MeCN	0.1 M TBAP	Ag| 0.01 M $AgNO_3$	25	-1.92	(I = 4.62)	[59], [63]
	MeCN	0.1 M TEAP	SCE	25	-1.63	(I = 5.37)	[59], [62]
	PhCN	0.1 M TEAP	SCE	25	-1.58	0.032	[59], [64]
	PhCN	0.1 M $LiClO_4$	SCE	25	-1.69	0.034	[59], [64]

	$PhCH_2CN$	0.1 M TEAP	SCE	25	-1.58	0.033	[59], [64]
	EtCN	0.1 M TEAP	SCE	25	-1.63	0.037	[59], [64]
	DMF	0.1 M TBAI	Hg pool	25	-1.49	(I = 2.36)	[59], [65]
	DMF	0.1 M TEAP	Ag|AgCl(s)	25	-1.62	0.033	[59], [66]
	CH_3CONMe_2	0.1 M TEAP	SCE	25	-2.02	0.042	[59], [60]
$BaCl_2$	DMSO	0.1 M TEAN	SCE	21	-2.09	-	[59], [67]
$Ba(ClO_4)_2$	DMSO	0.1 M TEAP	Zn-Hg(s)| $Zn(ClO_4)_2$(s)	25	-0.992	0.034	[59], [70]
	PC	0.1 M TEAP	SCE	25	-1.67	0.028	[59], [68]
$Ba(NO_3)_2$	NH_3	0.021 M TEAI	Hg pool	-36	-1.54	4.95	[55], [59]
BaI_2	MOR	0.1 M TBAI	NCE	20	-1.70	-	[59], [69]
$BaCl_2$	EDA	0.1 M $NaNO_3$	NCE	23	-1.80	-	[59], [67]
$Ba(ClO_4)_2$	SUL	0.1 M TEAP	Ag|AgCl TEAC(s)	40	-1.33	0.041	[59], [71]

[a]MeCN, acetonitrile; PhCN, benzonitrile; EtCN, propionitrile; DMF, N,N'-dimethylformamide; DMSO, dimethyl sulfoxide; PC, propylene carbonate; EDA, ethylenediamine; SUL, sulfolane; MOR, morpholine; TEAP, tetraethylammonium perchlorate; TBAI, tetrabuthylammonium iodide; TEAN, tetraethylammonium nitrate; TEAI, tetraethylammonium iodide; and TEAC, tetraethylammonium chloride.

TABLE 2.1.5. Half-Wave Potentials in Various Media [72]

Ion	Solvent							
	DMSO	DMA	DMF	H_2O	PDC	CH_3CN	PhCN	Ref. elec.
Ca^{2+}	-2.30	-2.37	-1.84	-2.20	-1.92	-1.82	-1.73	SCE
Sr^{2+}	-2.10	-2.23	-1.64	-2.11	-1.83	-1.76	-1.72	SCE
Ba^{2+}	-2.09	-2.02	-1.49	-1.92	-1.67	-1.63	-1.58	SCE
Ca^{2+}	-0.24	-0.33	-0.31	-0.07	0.05	0.16	0.15	$Rb/Rb^+ = 0$
Sr^{2+}	-0.04	-0.19	-0.12	0.02	0.14	0.22	0.16	$Rb/Rb^+ = 0$
Ba^{2+}	-0.03	0.02	0.03	0.21	0.30	0.35	0.30	$Rb/Rb^+ = 0$

TABLE 2.2.1. Current Density Capabilities of the Ca Anode in Various Nonaqueous Electrolytes, j, in mA/cm^2 [73]

Solute	Solvent			
	MeCN	DMF	NDA	EC/PC (ethylene carbonate 80 wt% propylene carbonate 20 wt%)
$AlCl_3$	10[a]	-	10	10
KPF_6	1	10	10	10
$LiPF_6$	10	-	10	-
$NaPF_6$	0.1	10	-	-
$KAsF_6$	-	10	-	-
NH_4PF_6	100[a]	10	-	-
$(n\text{-}C_3H_7)_4NPF_6$	100	10	10	0.1
$(C_6H_5)(CH_3)_3NPF_6$	10	50	10	10
ONH_2PF_6	100[a]	10	1	10
$(n\text{-}C_4H_9)_2NH_2AsF_6$	10	1	0.1	1
$(n\text{-}C_3H_7)_3NHAsF_6$	100	1	0.1	1
$(n\text{-}C_3H_7)_4NAsF_6$	1	0.1	<0.1	-
$(n\text{-}C_4H_9)_4NCl$	1	10	1	10
$(C_6H_5)_4PCl$	-	10	0.1	-
$(CH_3)_4NPF_6$	10	10	10	10
$(n\text{-}C_3H_7)_4NSbF_6$	<0.1	0.1	-	-

[a]System pretreated with anode metal is also capable of 100 mA/cm^2.

TABLE 2.2.2. Quantity of Ca Anode Consumed in a Galvanic Cell in CH_3OH Solution [74]

Electrolyte	Concn (g/l)	Temp (°C)	Current density (mA/cm^2)	Quantity consumed (g/cm^2 hr)
$CaBr_2$	100	-40	0	7.6×10^{-4}
$CaBr_2$	100	-40	2.5	3.1×10^{-3}
$CaBr_2$	200	20	0	2.7×10^{-4}
$CaBr_2$	200	20	5	5.2×10^{-3}

TABLE 2.2.3. Galvanic Cell with Ca in Propylene Carbonate

Cell reaction	Cell potential (V)	Watt-hr/lb
$Ca + CuF_2 \rightarrow CaF_2 + Cu$	3.51	604
$Ca + NiF_2 \rightarrow CaF_2 + Ni$	2.82	501

TABLE 2.2.4. Amperometric Titration using DME [22]

Substance	Reagent	Solvent	Supporting electrolyte	Indicator electrode potential (V)	Reference electrode
Ba^{2+}	0.01 $\underline{M}$ Li_2SO_4	H_2O	0.05 $\underline{M}$ Et_4NBr in EtOH	-2.00	SCE
Ca^{2+}	0.05 $\underline{M}$ EDTA	H_2O	0.16-0.23 $\underline{M}$ KOH + 0.08 $\underline{M}$ KCl + 2 m$\underline{M}$ K_2ZnO_2	-1.70	SCE

TABLE 3.1.1. Kinetic Parameters [56]

Reaction	Conditions	Electrode	Technique	Rate constant (cm/sec)	α
Ba(II) + Hg + 2e = Ba(Hg)	1 $\underline{M}$ LiCl	DME	farad. imp	$(9 \pm 1) \times 10^{-3}$	0.83
	0.1 $\underline{M}$ LiCl	DME	farad. imp	$(1.6 \pm 0.3) \times 10^{-2}$	

TABLE 3.2.1. Differential Capacities of Mercury in 0.1 $\underline{N}$ Aqueous Solutions at 25°C ($\mu F/cm^2$) [12, 76]

E (V vs SCE)	$CaCl_2$	$SrCl_2$	$BaCl_2$
0.063	99.2	96.4	103.6
0.08	75.9	77.5	78.7
0.12	54.3	54.9	55.3
0.16	45.82	46.09	46.36
0.24	39.16	39.27	39.34
0.32	38.65	38.71	38.73
0.41	39.82	40.09	40.16
0.46	38.51	38.92	38.98
0.54	32.10	32.50	32.59
0.62	26.12	26.49	26.53
0.70	22.85	23.06	23.11
0.86	18.81	18.89	18.89
1.00	16.92	16.98	16.97
1.16	16.12	16.25	16.25
1.32	16.40	16.59	16.62
1.48	17.49	17.63	17.64
1.64	19.13	19.11	19.11
1.80	20.91	20.63	20.96

TABLE 3.2.2. Double-Layer Capacity Values for Ba^{2+} Solutions in 1 M LiCl at 25°C [56]

Electrode potential (V)	C_d (μF/cm²) 0 mM Ba^{2+}	0.5 mM Ba^{2+}	1.3 mM Ba^{2+}
-0.5	39.1	38.4	40.0
-0.7	22.1	22.0	22.7
-0.9	16.4	16.3	16.6
-1.1	15.3	15.5	15.6
-1.3	16.0	16.0	15.8
-1.5	17.0	17.0	17.1
-1.7	18.9	18.6	19.0
-1.8	20.2	20.0	20.1
-1.9	21.4	21.1	21.7
-1.93	21.7	21.6 ± 0.3	22.4 ± 0.5
-1.96	22.1	22.3 ± 0.3	22.5 ± 1
-2.0	22.4 ± 0.2	22.3 ± 0.2	22.7 ± 0.5
-2.05	22.4 ± 0.3	22.3 ± 0.3	22.7 ± 0.4

TABLE 3.2.3. Double-Layer Capacity Values for Ba^{2+} Solutions in 0.1 M LiCl [56]

Electrode potential (V)	C_d (μF/cm²) 0 mM Ba^{2+}	0.5 mM Ba^{2+}	1 mM Ba^{2+}	1.5 mM Ba^{2+}
-0.5	25.4	25.6	25.6	25.7
-0.7	18.5	18.6	18.6	18.7
-0.9	16.0	16.1	16.2	16.2
-1.0	15.4	15.5	15.5	15.6
-1.1	15.2	15.3	15.4	15.5
-1.2	15.1	15.3	15.5	15.6
-1.3	15.5	15.7	15.8	16.0
-1.4	15.9	16.2	16.5	16.5
-1.5	16.6	17.0	17.2	17.3
-1.6	17.3	17.9	18.2	18.2
-1.7	18.1	18.8	19.2	19.3
-1.8	19.2	20.1	20.5	20.7
-1.9	20.4	21.0 ± 0.4	22.6 ± 0.2	22.6 ± 0.2
-1.93	20.6	22.2 ± 0.3	22.9 ± 0.4	23 ± 1
-1.96	20.8	22.0 ± 0.5	23.5 ± 1	
-2.0	22.1	21.9 ± 0.5	23 ± 1.5	22 ± 2
-2.05	22.6	22.9 ± 0.3	23.5 ± 0.2	23.7 ± 0.2

TABLE 3.2.4. Potential of the Electrocapillary Maximum (ecm) of Mercury in Aqueous Solutions at 25°C

Substance	Concn (N)	Method[a]	E_{ecm} (V vs SCE)	Before correction for liquid junction potential, E_{ecm} (vs N/10 CE)	E_{ecm} (NCE)	Refs.
$CaCl_2$	0.1	I	-0.5589	-0.5491	-0.5060	[12]
		II	-0.559	-0.549	-0.506	[12]
		III	-0.5586	-0.5488	-0.5057	[12]
$SrCl_2$	0.1	I	-0.5582	-0.5482	-0.5053	[12]
		II	-0.559	-0.549	-0.506	[12]
		III	-0.5588	-0.5488	-0.5069	[12]
$BaCl_2$	0.1	I	-0.5581	-0.5483	-0.5052	[12]
		II	-0.560	-0.550	-0.507	[12]
		III	-0.5587	-0.5489	-0.5058	[12]
			-0.52 (in 1 M LiCl)			[56]
			-0.47 (in 0.1 M LiCl)			[56]

[a]I, measurement of isotention potentials [77]; II, determination of isotension potentials close to E_{ecm}; and III, use of one reference electrode in the streaming method together with suitable correction for liquid junction potential.

TABLE 3.2.5. The influence of Concentration on E_{ecm} of Hg in $CaCl_2$ Solution

Substance	Concn (N)	E_{ecm}	Refs.
$CaCl_2$	1.0	-0.552	[12], [78]
		-0.55	[12], [79]
$CaCl_2$	0.1	-0.495	[12], [78]
		-0.496	[12], [80]
$CaCl_2$	0.01	-0.481	[12], [78]

TABLE 3.2.6. Surface Charge Density of the Electrical Double Layer at the Hg Electrode in 0.1 N Aqueous Solutions at 25°C [12, 76]

	σ (μc/cm²)		
-E (V)[a]	$CaCl_2$	$SrCl_2$	$BaCl_2$
0.08	19.87	19.97	20.04
0.10	18.51	18.60	18.64
0.12	17.36	17.43	17.46
0.14	16.32	16.39	16.42
0.16	15.38	15.44	15.46
0.18	14.49	14.55	14.57
0.20	13.64	13.70	13.72
0.25	11.64	11.70	11.71
0.30	9.72	9.77	9.78
0.35	7.78	7.83	7.84
0.40	5.81	5.84	5.85
0.45	3.83	3.85	3.85
0.50	1.949	1.949	1.950
0.55	0.274	0.256	0.253
0.60	-1.180	-1.216	-1.217
0.65	-2.474	-2.525	-2.529
0.70	-3.66	-3.72	-3.73
0.80	-5.80	-5.87	-5.88
0.90	-7.70	-7.78	-7.79
1.00	-9.44	-9.53	-9.54
1.10	-11.09	-11.19	-11.20
1.20	-12.71	-12.81	-12.82
1.30	-14.33	-14.45	-14.46
1.40	-15.98	-16.12	-16.14
1.50	-17.71	-17.86	-17.88
1.60	-19.52	-19.68	-19.70
1.70	-21.45	-21.61	-21.63
1.80	-23.48	-23.62	-23.66
Electrocapillary max V	0.5589	0.5582	0.5581

[a]Versus calomel electrode in the same solution.

TABLE 3.2.7. Double-Layer Parameters [56]

Solution	A: 1 m$\underline{M}$ Ba^{2+} + 1 $\underline{M}$ LiCl	B: 1 m$\underline{M}$ Ba^{2+} + 0.1 $\underline{M}$ LiCl	C: 1 m$\underline{M}$ Ba^{2+} + 0.1 $\underline{M}$ LiCl	D: 0.5 m$\underline{M}$ Ba^{2+} + 0.1 $\underline{M}$ LiCl	E: 1.5 m$\underline{M}$ Ba^{2+} + 0.1 $\underline{M}$ LiCl	F: 1 m$\underline{M}$ Ba^{2+} + 0.1 $\underline{M}$ LiCl
q ($\mu C/cm^2$)	-27.4	-26.5	-26.5	-26.3	-26.7	-37.0
C_d ($\mu F/cm^2$)	22.0	23.3	23.3	22.0	23.4	21.4
$(\partial q/\partial E)_{co}$ ($\mu F/cm^2$)	22.0	23.3	24.2	22.7	24.7	21.4
u_2	23.4	126	126	147	113	199
$(\partial \varphi_2/\partial E)_{co}$	3.73×10^{-2}	3.20×10^{-2}	3.34×10^{-2}	3.45×10^{-2}	3.20×10^{-2}	1.96×10^{-2}
$(\partial q/\partial C_0)_E$ (cm/Cmole)	-0.013	-0.333	-0.347	-0.493	-0.266	-0.366
$(\partial \Gamma_0/\partial C_0)_E$ (cm)	2.46×10^{-6}	4.00×10^{-5}	4.01×10^{-5}	6.63×10^{-5}	2.84×10^{-5}	5.83×10^{-5}
$(\partial \Gamma_0/\partial E)_{co}$ (mole V^{-1} cm^{-2})	-2.60×10^{-12}	-6.73×10^{-11}	-7.02×10^{-11}	-4.98×10^{-11}	-8.04×10^{-11}	-7.38×10^{-11}
C_{LF} ($\mu F/cm^2$)	26.1	85.7	86.2	75.5	89.3	117
$2F\Gamma_0$ ($\mu C/cm^2$)	0.24	7.7	7.7	5.0	9.4	14.1

[a]All calculations are for $E_{1/2} = -1.95$ V, where the surface concentrations of oxidant and reductant are the same and equal to half the bulk concentration of oxidant. Column F has been calculated according to Delahay, hence u_2 and $(\partial \varphi_2/\partial E)_{co}$ were taken from the supporting electrolyte; the other parameters are calculated starting from these two values.

TABLE 3.2.8. Experimental Values of the Charge on the Electrode for 0.1 M LiCl Solutions in the Presence of Ba^{2+} [56]

Electrode potential, $C_{Ba^{2+}}$ (mM)	-0.9 V		-1.1 V		-1.5 V		-1.8 V	
	$-q_{obs}$	$nF\Gamma_{Ba^{2+}}$	$-q_{obs}$	$nF\Gamma_{Ba^{2+}}$	$-q_{obs}$	$nF\Gamma_{Ba^{2+}}$	$-q_{obs}$	$nF\Gamma_{Ba^{2+}}$
0	8.47		11.57		17.81		23.14	
0.5	8.53	0.6	11.65	1.3	17.99	3.5	23.51	6.1
1.0	8.54	1.1	11.67	2.1	18.08	5.2	23.70	8.5
1.5	8.55	1.5	11.70	2.8	18.15	6.3	23.80	10.1
	±0.05		±0.07		±0.10		±0.13	

TABLE 4.2.1. Compositions (wt%) and Electrode Constants (K_A^B) of Some Divalent Cation-Sensitive Glasses [86]

Component or K_A^B	Glass			
	NG-2[a]	NG-6[a]	916-P[b]	Phosphate[c]
SiO_2	76.4	71.6	50	3.1
Al_2O_3	12.7	13.5	25	5.6
Fe_2O_3	0.56	0.70		16.0
FeO	0.58	4.3		
MgO	0.4	2.5		
CaO	0.30	2.80		0.2
Na_2O	4.1	1.4	6	6.1
K_2O	4.57	2.45		0.09
H_2O	0.67	0.20		0.19
P_2O_5				66.7
Cs_2O			6	
BaO			13	
K_{Ca}^{Na}	24	11	13	0.56
K_{Ca}^{K}	87	43	50	0.83
K_{Ca}^{Mg}	1.0	0.77		0.19

[a]Glass NG-2 is Puerto de Abrigo Obsidian, Jamez Caldera, New Mexico. Glass NG-6 is Tektite, Marulas, Philippine Islands.

[b]Synthetic glass, batch composition listed.

[c]Synthetic glass, glass analysis listed. Total iron given as Fe_2O_3.

TABLE 4.2.2. Calculated Selectivity Orders and Observed Selectivities of Ion Exchangers [86]

I.	Mg> Ca> Sr> Ba	Zeolon carboxylic and phosphate resins
		kaolinite, beidellite-2nd site, Ag-ultramarine
II.	Ma> Sr> Ca> Ba	
III.	Mg> Sr> Ba> Ca	
		bentonite-2nd site
IV.	Mg> Ba> Sr> Ca	NH_4-kaolinite
V.	Ba> Mg> Sr> Ca	K-biotite
		bentonite-1st site, K-feldspar, aluminoborosilicate glasses
VI.	Ba> Sr> Mg> Ca	NH_4-kaolinite, K-muscovite
		montmorillonite
VII.	Ba> Sr> Ca> Mg	NH_4-feldspar, NH_4-biotite, NH_4-muscovite, bentonite, phosphate glass, Dowex 50, beidellite-1st site, Zeolite 13X, mordenite, clinoptilotile, chabazite, erionite, Na-faujasite, Zr-phosphate, illite, sulfonated resin, Zeolite A

REFERENCES

1. G. N. Lewis, J. Amer. Chem. Soc., 32, 1459 (1910); 34, 119 (1912); 35, 340 (1913); 37, 1983 (1915).
2. G. Devoto, Z. Elektrochem., 34, 19, 326 (1928).
3. Gmelin's Handbuch der Anorganischen Chemie, "Calcium," 8th ed., Verlag Chemie, Weinheim/Bergstrasse, 1957.
4. C. Drucker and F. Luft, Z. Phys. Chem., 121, 307, 318 (1926).
5. W. M. Latimer, Oxidation Potentials, 2nd ed., Prentice-Hall, Englewood Cliffs, New Jersey, 1952.
6. Gmelin's Handbuch der Anorganischen Chemie, "Strontium," 8th ed., Verlag Chemie, Weinheim/Bergstrasse, 1960.
7. Gmelin's Handbuch der Anorganischen Chemie, "Barium," 8th ed., Verlag Chemie, Weinheim/Bergstrasse, 1960.
8. J. O'M. Bockris and J. F. Herringshaw, Discussions Faraday Soc., 1, 328 (1947).
9. G. Kortum and J. O'M. Bockris, Textbook of Electrochemistry, 2nd ed., Elsevier, Amsterdam, 1951.
10. J. O'M. Bockris and A. K. N. Reddy, Modern Electrochemistry, Vol. 2, Plenum Press, New York, 1970.
11. J. Koryta, J. Dvorak, and V. Bohackova, Electrochemistry, Methuen, London, 1966.
12. B. E. Conway, Electrochemical Data, Elsevier, Amsterdam, 1952.

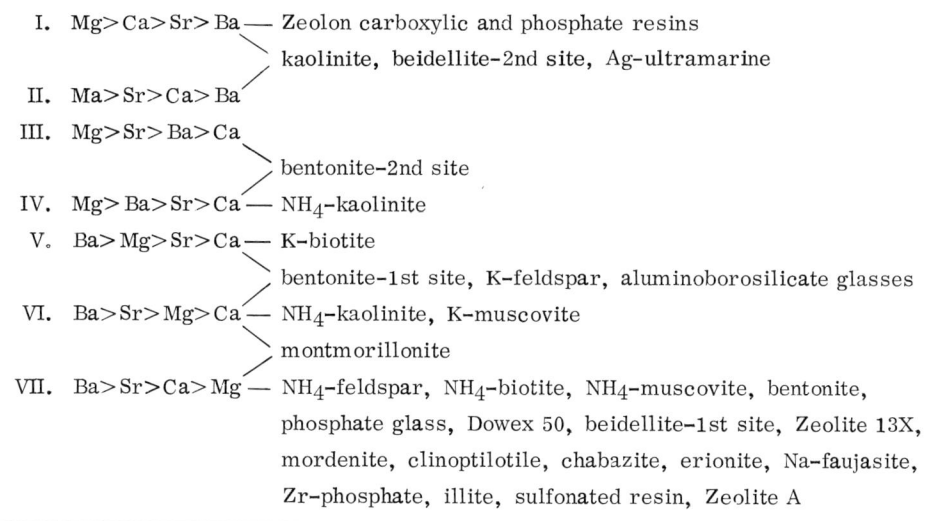

13. Landolt-Börnstein, Zahlenwerte und Funktion, Vol. 2, Part 7, Springer, Berlin, 1960.
14. M. Tamele, J. Phys. Chem., 23, 502 (1924).
15. F. L. E. Shibata, J. Sci. Hiroshima Univ., Ser. A, 1 147, 155 (1931).
16. J. N. Butler, J. Electroanal. Chem., 17, 309 (1968).
17. R. Parsons, Handbook of Electrochemical Constants, Butterworths, London, 1959.
18. S. Toshima, Fundamental Electrochemistry, Asakura, Tokyo, 1965 (in Japanese).
19. G. Milazzo, Electrochemistry, Elsevier, Amsterdam, 1963.
20. A. J. de Bethune, J. S. Licht, and N. S. Swendemann, J. Electrochem. Soc., 106, 616 (1959).
21. K. Jellinek, Lehrbuch der physikalische Chemie, 2nd ed., Enke, Stuttgart, Vol. 3, 1930; Vol. 4, 1933.
22. L. Meites, Handbook of Analytical Chemistry, McGraw-Hill, New York, 1963.
23. K. Vetter, Elektrochemische Kinetik, Springer, Berlin (1961).
24. M. Pourbaix, Atlas of Electrochemical Equilibria in Aqueous Solutions, Pergamon, Oxford, 1966.
25. N. E. Khomutov, Zh. Fiz. Khim., 42, 2223 (1968); Chem. Abstr., 70, 61823n (1969).
26. S. Makishima, Z. Elektrochem., 41, 697 (1935).
27. Iu. K. Delimarskii and A. A. Kolotti, Zh. Fiz. Khim., 23, 339 (1949); Chem. Abstr., 43, 6089 (1949).
28. B. Jakuszewski and S. Taniewska-Osenka, Rocz. Chem., 36, 329 (1962), Chem. Abstr., 57, 9568i (1962).
29. G. Tammann and H. O. v. Samson-Himmelstjerna, Z. Anorg. Allg. Chem., 216, 288 (1934).
30. R. J. Fosbinder, J. Amer. Chem. Soc., 51, 1345 (1929).
31. V. A. Pleskov, Usp. Khim., 16, 254 (1947).
32. H. Strehlow, Z. Elektrochem., 56, 827 (1952).
33. V. A. Pleskov, Acta Physicochim. URSS, 21, 41 (1946); J. Phys. Chem. (URSS), 20, 153 (1946).
34. V. A. Pleskov, Acta Physicochim. URSS, 6, 1 (1937); Zh. Fiz. Khim., 9, 12 (1937).
35. F. E. Rosztoozy and C. W. Tobias, Electrochim. Acta, 11, 857 (1966).
36. W. H. Smyrl, Thesis, University of California, 1966.
37. Iu. K. Delimarskii and B. F. Markov, Electrochemistry of Fused Salts (English Translation from Russian), Andromeda Books, Washington, D.C., 1961.
38. The Electrochemical Society of Japan, Physico-Chemical Constants of Fused Salts, Kagaku Dojin, Tokyo, 1963 (in Japanese).
39. W. J. Hamer, N. S. Malmberg, and B. Rubin, J. Electrochem. Soc., 103, 8 (1956); 112, 750 (1965).
40. B. Neuman and H. Richter, Z. Elektrochem., 31, 296, 481 (1925).
41. Iu. K. Delimarskii and F. F. Grigorenko, Ukr. Khim. Zh., 21, 561 (1955); Chem. Abstr., 50, 9176h (1956).
42. B. Neuman and E. Bergve, Z. Elektrochem., 21, 152 (1915).
43. L. Cambi and G. Devoto, Giorn. Chim. Applicata, 8, 303 (1927); Chem. Abstr., 21, 2837 (1927).
44. Iu. K. Delimarskii and A. A. Kolotti, Zh. Fiz. Khim., 23, 90 (1949); Chem. Abstr., 43, 4581 (1949).

45. P. Mergault, C. R. Acad. Sci., Paris, 240, 864 (1955).

46. K. E. Johnson, Electrochim. Acta, 13, 1715 (1968).

47. Iu. K. Delimarskii, Zh. Fiz. Khim., 29, 28 (1955).

48. M. D. Thompson and A. L. Kaye, Trans. Electrochem. Soc., 67, 169 (1935).

49. W. J. Hamer, J. Electroanal. Chem., 10, 140 (1965).

50. F. D. Richardson and J. H. E. Jeffes, J. Iron Steel Inst. (London), 160, 261 (1948).

51. J. F. Elliott, M. Gleiser, and V. Ramakrishna, Thermochemistry for Steelmaking, Addison-Wesley, Reading, Massachusetts, 1963.

52. H. A. Laitinen and C. H. Liv, J. Amer. Chem. Soc., 80, 1015 (1958).

53. I. Zlotowski and I. M. Kolthoff, ibid., 66, 1431 (1944); J. Phys. Chem., 49, 386 (1945).

54. I. M. Kolthoff and J. J. Lingane, Polarography, Vol. 2, Interscience, New York, 1952.

55. C. J. Nyman, J. Amer. Chem. Soc., 71, 3914 (1949).

56. B. Timmer, M. Sluyters-Rehback, and J. H. Sluyters, J. Electroanal. Chem., 24, 287 (1970).

57. I. M. Kolthoff and H. P. Gregov, Anal. Chem., 20, 541 (1948).

58. J. Heyrovsky and S. Berezicky, Collect. Czech. Chem. Commun., 1, 19 (1920).

59. C. K. Mann and K. K. Barnes, Electrochemical Reactions in Nonaqueous Systems, Dekker, New York, 1970.

60. V. Gutmann, M. Michlmayr, and G. Peychal Heiling, J. Electroanal. Chem., 17, 153 (1968).

61. S. Wawzonek and M. E. Runner, J. Electrochem. Soc., 99, 457 (1952).

62. I. M. Kolthoff and J. F. Coetzee, J. Amer. Chem. Soc., 79, 870 (1957).

63. A. I. Popov and D. H. Geske, ibid., 79, 2074 (1957).

64. R. C. Larson and R. T. Iwamoto, ibid., 82, 3239, 3526 (1960).

65. G. H. Brown and R. Al-Urfali, ibid., 80, 2113 (1958).

66. J. B. Headridge, M. Ashraf, and H. L. H. Dodds, J. Electroanal. Chem., 16, 114 (1968).

67. V. Gutmann and G. Schober, Z. Anal. Chem., 171, 339 (1959).

68. V. Gutmann, M. Kogelnig, and M. Michlmayr, Monatsh., 99, 693 (1968).

69. V. Gutmann and E. Nedbalek, ibid., 88, 320 (1957).

70. D. L. McMasters, R. B. Dunlop, J. R. Knempel, L. W. Kreider, and T. R. Shearer, Anal. Chem., 39, 103 (1967).

71. J. B. Headridge, D. Pletcher, and M. Callingham, J. Chem. Soc., A, 1967, 684.

72. V. Gutmann, G. Peychal-Heiling, and M. Michlmayr, Inorg. Nucl. Chem. Lett., 3, 501 (1967).

73. W. E. Elliot, J. R. Huff, R. W. Adler, and W. L. Towle, Proc. Ann. Power Sources Conf., 20, 67 (1966).

74. D. V. Kurygina and Z. A. Iofa, Zashchitu Metal i Oksiduge Pokrytiya, Korroziya Metal, i Issled, v Obl. Elektrokhim. Akad. Nauk SSSR, Otd. Obshch. i Tekhu Khim. Sb. Statei, 1965, 270; Chem. Abstr., 65, 5013b (1966).

75. R. Jasinski, "Electrochemistry and Application of Propylene Carbonate," in Advances in Electrochemistry and Electrochemical Engineering, Vol. 8 (C. W. Tobias, ed.), Wiley (Interscience), New York, 1971.

76. D. C. Grahame, J. Amer. Chem. Soc., 71, 2975 (1949); J. Electrochem. Soc., 98, 343 (1951).

77. D. C. Grahame, R. P. Larsen, and M. A. Poth, J. Amer. Chem. Soc., 71, 2978 (1949).

78. T. Erdey-Gruz and P. Szarvas, Z. Phys. Chem., A177, 277 (1936).

79. G. Gouy, Ann. Chim. Phys., 29, 145 (1903).

80. D. C. Grahame, E. M. Coffin, and J. I. Cummius, Office of Naval Research Tech. Report No. 2. August 11, 1950, Research Contract N8-onv-66903.

81. A. Shatkay, Electrochim. Acta, 15, 1759 (1970).

82. R. Huston and J. N. Butler, Anal. Chem., 41, 200 (1969).

83. R. A. Robinson and R. H. Stokes, Electrolyte Solutions, Butterworths, London, 1965.

84. A. Shatkay, J. Phys. Chem., 71, 3858 (1967).

85. R. M. Garrels, M. E. Thompson, and A. H. Truesdell, Science, 135, 1045 (1962).

86. A. H. Truesdell and C. L. Christ, "Glass Electrodes for Calcium and Other Divalent Cations," in Glass Electrodes for Hydrogen and Other Cations (G. Eisenman, ed.), Dekker, New York, 1967.

87. P. Hirsch-Ayalon, Electrochim. Acta, 10, 773 (1965).

88. E. P. Honig and J. H. Th. Hengst, Electrochim. Acta, 15, 491 (1970).

89. A. Eilert, Z. Elektrochem., 31, 176 (1925).

90. N. Kameyama, "Theory and Application of Electrochemistry," in Electrolysis of Melt, Maruzen, Tokyo, 1955 (in Japanese).

91. The Electrochemical Society of Japan, Denki Kagaku Benran, Maruzen, Tokyo, 1964, p. 1046.

92. S. M. Selis and L. P. McGinnis, J. Electrochem. Soc., 108, 191 (1961).

93. R. P. Clark and K. R. Grothaus, ibid., 118, 1680 (1971).

Chapter I-8

INERT GASES

BRUNO JASELSKIS and R. H. KRUEGER*

Department of Chemistry
Loyola University
Chicago, Illinois

1. INTRODUCTION

Noble gases enter into chemical combinations to a rather limited extent and because of their inertness the preparation of these compounds was not achieved until 1962! Well characterized compounds of krypton, xenon, and radon have been prepared to date. The ease of combination increases with the decreasing ionization potential of the noble gases. The ionization potentials for noble gases are summarized in Table 1.1.

Helium and argon form no stable compounds. Krypton forms only krypton difluoride, while xenon forms compounds having the oxidation states +2, +4, +6, and +8. Radon

*Present address: Borg-Warner Corporation, Des Plaines, Illinois.

TABLE 1.1. Ionization Potentials of Noble Gases

Element and configuration of outer electrons		Atomic number	1st IP (eV)	ΔH_{vap} (kcal/mole)	Promotion energy (eV), $ns^2np^6 \rightarrow ns^2np^5(n+1)s$
Helium	$1s^2$	2	24.58	0.022	-
Neon	$2s^22p^6$	10	21.56	0.44	16.6
Argon	$3s^23p^6$	18	15.76	1.50	11.5
Krypton	$4s^24p^6$	36	14.00	2.31	9.9
Xenon	$5s^25p^6$	54	12.13	3.27	8.3
Radon	$6s^26p^6$	86	10.75	4.3	6.8

apparently forms divalent radon compounds comparable to krypton. Furthermore, radon is not only highly radio-active but also quite scarce since it is obtained from the nuclear decay of ^{238}U and ^{232}Th.

Inasmuch as reactions of noble gas compounds are highly irreversible, the oxidizing power can only be measured thermodynamically by using the following methods primarily: 1) studying xenon-fluorine equilibria as a function of temperature and relating the change of enthalpy or heat of reaction by the van't Hoff equation

$$d(\ln K)/dT = \Delta H^\circ_{react}/(RT^2)$$

as described by Weinstock et al. [1]; 2) direct calorimetric measurements of the reaction of the noble gas compound with hydrogen as illustrated by Stein and Plurien [2]:

$$XeF_4(g) + 2H_2(g) \rightarrow Xe(g) + 4HF(g)$$

$$\Delta H_{react} = 4\Delta H_f(HF) - \Delta H_f(XeF_4)$$

3) measuring enthalpies of the reactions between a noble gas compound and HI as described by Gunn and Williamson [3] and O'Hare et al. [4]; and 4) calorimetric measurement of heats of combustion or decomposition (explosive decomposition) as determined by Russian workers [5]. The free energy is calculated using experimentally determined enthalpies. Enthalpy and entropy values are tabulated in National Bureau of Standards Circular 500 and Technical Note 270-3. The electrochemical potentials then can be found from the free energy and appropriate half-cell potentials.

2. KRYPTON COMPOUNDS

Krypton forms only krypton difluoride, KrF_2 [6, 7], which is apparently thermodynamically unstable and slowly decomposes spontaneously. In aqueous solution it decomposes to krypton gas and hydrofluoric acid. Higher oxidation states of krypton fluorides were not confirmed [7].

3. XENON COMPOUNDS

The principal oxidation states of xenon compounds are +2, +4, +6, and +8. Xenon compounds containing fluorine are XeF_2, XeF_4, XeF_6, and some oxyfluorides [8]. All of these compounds are strong oxidizing and fluorinating agents, and they decompose in water according to the stoichiometric reactions [9]

$$XeF_2 + H_2O \rightarrow Xe(g) + 1/2O_2(g) + 2HF$$

$$3XeF_4 + 6H_2O \rightarrow XeO_3 + 2Xe(g) + 12HF + 3/2O_2(g)$$

$$XeF_6 + 3H_2O \rightarrow XeO_3 + 6HF$$

Hydrolysis of XeF_4 and XeF_6 yields xenon trioxide, XeO_3, which is quite stable in water and forms numerous salts. In strongly alkaline solution xenon trioxide yields insoluble perxenate, Na_4XeO_6. Barium perxenate in anhydrous sulfuric acid yields xenon tetroxide, XeO_4 [10].

Thermodynamic studies are limited to xenon fluorides, xenon trioxide and perxenate. Available information is summarized according to the compound.

3.1. XENON DIFLUORIDE

Xenon difluoride is readily prepared from xenon and fluorine mixtures using rather mild conditions. It dissolves in pure water to the extent of 2.5 g/100 ml at 0°C, and it decomposes on standing according to

$$XeF_2 + H_2O \rightarrow Xe(g) + 1/2O_2(g) + 2HF$$

Heats of formation, sublimation, and decomposition for xenon difluoride, XeF_2, are summarized in Table 3.1.1. The heat of sublimation of XeF_2 was determined by Jortner,

TABLE 3.1.1. Summary of Thermodynamic Quantities for XeF_2

	kcal/mole	Refs.
Heat of sublimation	12.5 ± 0.2	[11]
Heat of formation	-25.9	[1]
	-37 ± 10	[12]
	-25 to -45	[13]
	-41.5 ± 0.6	[5]
Heat of combustion	42.1	[5]

Wilson, and Rice using UV absorption measurements as a function of pressure and temperature. The heat of formation was estimated by various methods. Weinstock, Weaver, and Knop obtained the heat of formation from the equilibrium constant measurements at various temperatures, while Svec and Flesch used mass spectrometric methods for measuring the appearance potential at a source temperature of 150°C. Pitzer calculated the heat of formation on a purely theoretical basis. Russian workers determined the heat of combustion calorimetrically.

A saturated aqueous XeF_2 solution at 0°C was reported by Appelman [14] to have a specific conductance of 4×10^{-4} ohm^{-1} cm^{-1}. The molar conductance of XeF_2 solution in the concentration range 5×10^{-3} to 2×10^{-2} M is < 0.4 ohm^{-1} cm^2 $mole^{-1}$, indicating that the electrolytic dissociation of XeF_2 cannot be detected by conductance measurements [15].

3.2. XENON TETRAFLUORIDE

Xenon tetrafluoride may be prepared by the reaction of xenon (1.7 atm) and fluorine (8 atm) at 400°C. It reacts with water to form xenon trioxide and xenon. No evidence has been found for any Xe(IV) that is stable in water.

The heats of sublimation and formation (298.15°K) for xenon tetrafluoride are summarized in Table 3.2.1. Gunn and Williamson obtained the heat of formation from calorimetric measurements, while Stein and Plurien obtained it from the reduction of XeF_4 with hydrogen using isothermal calorimetry at 120–130°C.

The heat capacity of solid XeF_4 was measured by Johnston, Pilipovich, and Sheehan [16] and found to range from 1.933 cal/mole deg at 20°K to 28.457 cal/mole deg at 300°K. From this data they obtained the entropy of solid XeF_4 at 298.16°K to be 35.0 cal deg^{-1} $mole^{-1}$. Weinstock et al. [1] estimated S° for XeF_4 to be 75.3 cal deg^{-1} $mole^{-1}$ at 298.16°K. This value is in good agreement with their value of 75.6 cal deg^{-1} $mole^{-1}$ calculated from molecular properties.

TABLE 3.2.1. Summary of Thermodynamic Quantities for XeF_4

	kcal/mole	Refs.
Heat of sublimation	15.3 ± 0.2	[11]
Heat of formation	-48.0	[3], [1]
	-51.5	[1]
	-53 ± 5	[12]
	-57.6	[2], [1]

3.3. XENON HEXAFLUORIDE

Xenon hexafluoride is prepared by heating a 1 to 20 mixture of xenon and fluorine at a pressure of 60 to 100 atm at 300°C for several hours. It dissolves in water to form xenon trioxide:

$$XeF_6 + 3H_2O \rightarrow XeO_3 + 6HF$$

The heats of sublimation and formation (298.15°K) for xenon hexafluoride are summarized in Table 3.3.1. The heat of sublimation of xenon hexafluoride was determined by Weinstock, Weaver, and Knop by determining the vapor pressure of XeF_6 over a temperature range of 0 to 23°C. Claassen determined the heat of sublimation from the IR absorption as a function of pressure and temperature. The heat of formation of XeF_6 was determined by methods given in the previous sections.

Heat capacity and other thermodynamic functions of xenon hexafluoride from 5 to 350°K were determined by Schreiner, Osborne, Malm, and McDonald [18]. At 298.15°K the values of C_p°, S°, $H^\circ - H_0^\circ$, and $(G - H_0^\circ)/T$ for solid XeF_6 are 41.00 cal K^{-1} $mole^{-1}$, 50.27 cal K^{-1} $mole^{-1}$, 7407 cal/mole, and -25.42 cal K^{-1} $mole^{-1}$, respectively. At 335°K the standard entropy S° is 61.05 cal K^{-1} $mole^{-1}$ for liquid XeF_6 and 96.20 cal K^{-1} $mole^{-1}$ for the gas.

TABLE 3.3.1. Summary of Thermodynamic Quantities for XeF_6

	kcal/mole	Refs.
Heat of sublimation	15.3	[1]
	14.9	[17]
Heat of formation	-70.4	[1]
	-82.9	[2], [1]

TABLE 3.3.2. Ideal Gas Thermodynamic Properties at 298.15°K for the Xenon Compounds[a]

	Compounds				
Property	XeF_2	XeF_4	XeF_6	XeO_3	XeO_4
C_p° (cal K^{-1} $mole^{-1}$)	12.94	21.45	30.15	14.89	18.37
$(H^\circ - H_0^\circ)/T$ (cal K^{-1} $mole^{-1}$)	10.07	15.13	19.34	11.07	12.80
$-(G^\circ - H_0^\circ)/T$ (cal K^{-1} $mole^{-1}$)	51.94	61.92	70.05	58.02	57.62
S° (cal K^{-1} $mole^{-1}$)	62.01	77.05	89.38	69.09	70.42

[a]Taken in part by permission from Ref. 18a.

Recently the ideal gas thermodynamics functions from 100 and 1500°K have been computed by Kudchadker and Kudchadker [18a] for xenon fluorides, xenon oxyfluorides and xenon oxides, using published spectroscopic data as shown in Table 3.3.2. These values are within a reasonable agreement with those published by Weinstock et al. [1].

Bond energies varying from 29.7 to 39 kcal have been calculated for XeF_2, XeF_4, and XeF_6. The average bond energies derived by Weinstock et al. [1] were calculated using ΔH_0° for the several equilibria and the value of 36.71 kcal/mole for ΔH_0° of dissociation of F_2. They found that the average bond energies for XeF_2 and XeF_4 were 31.0 and 30.9 kcal, respectively, and the average energy for the formation of the last two bonds in XeF_6 was 27.3 kcal.

3.4. XENON TRIOXIDE

The hydrolysis of XeF_4 and XeF_6 yields solutions from which colorless crystals of XeO_3 can be obtained by evaporation. This compound is thermodynamically unstable and a powerful explosive. However, solutions of XeO_3 in dilute aqueous acid are very strong oxidizers and show no evidence of spontaneous decomposition.

Solid XeO_3 has a heat of formation of 96 ± 2 kcal/mole [19]. The heat of formation of aqueous XeO_3 at 298.15°K was found by O'Hare, Johnson and Appelman [4] to be 99.94 ± 0.24 kcal/mole. This value was used in calculations to determine the thermodynamic oxidizing power of aqueous XeO_3. Since reactions of aqueous XeO_3 are irreversible, the oxidizing power can only be measured thermodynamically. The electrode potentials of the Xe-XeO_3 couple in acidic solution and of the Xe-$HXeO_4^-$ couple in basic solution were experimentally determined by Johnson et al. [4] as 2.10 ± 0.01 and 1.24 ± 0.01 V, respectively, in the following manner:

1. The heat of formation of aqueous XeO_3 was determined from calorimetric measurements of the enthalpics of the reaction between XeO_3(aq) and HI(aq).

$$\Delta H_f^\circ\ (XeO_3 \cdot \infty H_2O) = 99.94 \pm 0.24\ \text{kcal/mole}$$

2. The heat of solution for the reaction

$$XeO_3 + \infty H_2O(l) \rightarrow XeO_3 \cdot \infty H_2O(aq)$$

was calculated based on the literature value [19] for ΔH_f° ($XeO_{3(c)}$): $\Delta H_{soln} = 3.9 \pm 2.0$ kcal/mole

3. The standard partial molar entropy was estimated by the expression [20]

$$\bar{S}^\circ\ (XeO_3|aq|) = S^{int} + \frac{3}{2} R \ln \underline{M} + 10 - 0.22V_m$$

where

$$S^{int} = S^\circ\ (XeO_3, g) - 6.86 \log \underline{M} - 11.44 \log T + 2.31$$

$$S^\circ\ (XeO_3, g) = 68.9\ \text{cal deg}^{-1}\ \text{mole}^{-1}\ [21]$$

$$M = 179.298$$

$$T = 298.15$$

$$S^{int} = 27\ \text{cal deg}^{-1}\ \text{mole}^{-1}$$

$$V_M = 39.4\ \text{cm}^3/\text{mole}\ [22]$$

$$\bar{S}^\circ\ (XeO_3, aq) = 44 \pm 4\ \text{cal deg}^{-1}\ \text{mole}^{-1}$$

4. The standard entropy of formation, ΔS_f°, may be calculated for the reaction

$$Xe(g) + \frac{3}{2} O_2(g) \rightarrow XeO_3(aq)$$

Since S° values of 40.53 and 49.00 cal deg^{-1} $mole^{-1}$ for Xe(g) and O_2(g) are known [23]·

$$\Delta S_f^\circ = (44 \pm 4) - 40.53 - \frac{3}{2}(49) = -(70 \pm 4)\ \text{cal deg}^{-1}\ \text{mole}^{-1}$$

5. The standard free energy of formation, ΔG_f°, may then be calculated:

$$\Delta G_f^\circ = \Delta H_f^\circ - T\Delta S_f^\circ = 120.8 \pm 1.2\ \text{kcal/mole}$$

6. The combination of ΔG_f° (XeO_3, aq) with ΔG_f° (H_2O, l) equal to -56.69 ± 0.01 kcal/mole [23] is then used to calculate E° for the half reaction:

$$Xe(g) + 3H_2O(l) \rightarrow XeO_3(aq) + 6H^+(aq) + 6e^-$$

$$\Delta G^\circ = \Delta G_f^\circ (XeO_3, aq) - 3\,\Delta G_f^\circ (H_2O, l)$$

$$= 120.8 \pm 1.2 - 3(-56.69)$$

$$= 290 \text{ kcal/mole}$$

$$E^\circ = \Delta G^\circ/(6 \times 23.060) = 2.10$$

7. The potential E_B° is calculated by the same method for the corresponding reaction in alkaline solution:

$$Xe(g) + 7(OH^-)aq \rightarrow HXeO_4^-(aq) + 3H_2O(l) + 6e^-$$

$$\Delta G^\circ = \Delta G_f^\circ (HXeO_4^-, aq) + 3\Delta G_f^\circ (H_2O, l) - 7\Delta G_f^\circ (OH^-, aq)$$

using 6.7 ± 0.5 X 10^{-14} as the equilibrium constant [24] for the reaction

$$HXeO_4^-(aq) \rightarrow XeO_3(aq) + OH^-(aq)$$

and

$$\Delta G_f^\circ (OH^-, aq) = -37.59 \text{ kcal/mole}$$

yields E_B° = 1.24 ± 0.01 V.

Xenon trioxide is a powerful oxidant. Aqueous xenon trioxide readily oxidizes ammonia, presumably to nitrogen [24, 25], Fe(II) to Fe(III) [25], Hg^0 to Hg(II) [25], I^- to I_3^- [24, 26, 27], Br^- to Br_3^- [25], Cl^- to Cl_2 [28], Mn(II) to MnO_2 [28], Mn(II) to MnO_4^- [28], I_2 to IO_3^- [28], Br_2 to BrO_3^- [28], Pu(III) to Pu(IV) [29, 30], Np(V) to Np(VI) [31], and numerous organic compounds to yield carbon dioxide and other by-products [32-38].

3.5. PERXENATE

The electrode potential of 2.18 V for the perxenate-xenate couple:

$$H_4XeO_6 + 8H^+ + 8e^- \rightarrow Xe^0 + 6H_2O$$

has been calculated using the following: 1) the heat of formation, ΔH_f (XeO_4, g) = 153.5 kcal/mole [39]; 2) the estimated entropy of 66 e.u. for XeO_4(g); and 3) the free energy, $\Delta G \sim 0$ for the reaction

$$XeO_4(g) + 2H_2O \rightleftarrows H_4XeO_6(aq)$$

The potential of 1.18 V for the perxenate-xenon couple in alkaline solution:

$$HXeO_6^{-3} + 5H_2O + 8e^- \rightarrow Xe + 11OH^-$$

has been calculated using ionization constants of 10^{-2} and 10^{-6}, respectively, for the reactions:

$$H_4XeO_6 \rightarrow H^+ + H_3XeO_6^- \quad \text{and} \quad H_3XeO_6^- \rightarrow H^+ + H_2XeO_6^{-2}$$

as reported by Appelman and Malm [40].

Combination of these results with the Xe(VI)-Xe couple yields potentials of approximately 2.36 and 0.94 V, respectively, for the acid and alkaline perxenate reductions

$$H_4XeO_6 + 2H^+ + 2e^- \rightarrow XeO_3 + 3H_2O$$

$$HXeO_6^{-3} + 2H_2O + 2e^- \rightarrow HXeO_4^- + 4OH^-$$

These calculated potentials are consistent with perxenate oxidation reactions: perxenate oxidizes Ag(I) to Ag(II) in acid, IO_3^- to IO_4^- in base, and XeO_3 is oxidized to XeO_6^{-4} by ozone in base.

3.6. SUMMARY OF ELECTRODE POTENTIALS FOR AQUEOUS XENON COMPOUNDS

Xenon trioxide and perxenate are powerful oxidants in aqueous solutions. The potentials for xenon compounds in aqueous solutions have been calculated from the available thermodynamic data and are summarized in Table 3.6.1.

TABLE 3.6.1. Summary of Electrode Potentials

Half-cell reaction	Potential[a]
Acid media	
$XeO_3(aq) + 6H^+ + 6e \rightarrow Xe(g) + 3H_2O(l)$	2.10
$H_4XeO_6(aq) + 8H^+ + 8e \rightarrow Xe(g) + 6H_2O(l)$	2.18
$H_4XeO_6(aq) + 2H^+ + 2e \rightarrow XeO_3(l) + 3H_2O(l)$	2.42
Basic media	
$HXeO_6^{-3}(aq) + 5H_2O + 8e \rightarrow Xe(g) + 11OH^-$	1.18
$HXeO_4^-(aq) + 3H_2O + 6e \rightarrow Xe(g) + 7OH^-$	1.24
$HXeO_6^{-3}(aq) + 2H_2O + 2e \rightarrow HXeO_4^-(aq) + 4OH^-$	0.94

[a]All half-cell reactions are highly irreversible and the calculated potential values are referred to the standard hydrogen electrode.

3.7. ELECTRODE POTENTIAL DIAGRAM OF AQUEOUS XENON COMPOUNDS

Xenon trioxide and perxenate are powerful oxidants as indicated in the electrode potential diagram for aqueous solutions:

<u>Acid</u>

$$Xe \xrightarrow{2.10V} XeO_3(aq) \xrightarrow{2.42\ V} H_4XeO_6$$

$$Xe \xrightarrow{2.64V} XeF_2(aq)$$

<u>Base</u>

$$Xe \xrightarrow{1.24\ V} HXeO_4^- \xrightarrow{0.94\ V} HXeO_6^{-3}$$

3.8. POLAROGRAPHIC CHARACTERISTICS OF AQUEOUS XENON TRIOXIDE, SODIUM PERXENATE, AND XENON DIFLUORIDE

At the dropping mercury electrode, aqueous xenon trioxide, sodium perxenate, and xenon difluoride are reduced in a single step involving 6, 8, and 2 electrons, respectively [41, 42]. Against an Hg_2SO_4-Hg reference electrode the half-wave potentials are -0.32 V for xenon trioxide at pH 6.4; 0 V for xenon difluoride at pH 4.5; and -0.21 to -0.31 V for xenon perxenate in the pH range 10.1 to 11.3. For XeF_2 the reduction wave is followed by a broad maximum which can be effectively suppressed by fluoride ion. The diffusion current varies in linear fashion with concentration.

4. RADON COMPOUNDS

Radon forms only radon difluoride, RaF_2 [43]. The chemistry of radon has been studied primarily in tracer experiments because of the absence of stable isotopes.

Aqueous chemistry of radon difluoride is not comparable to that of xenon. Ionic species of radon difluoride have been observed in several nonaqueous solutions [6]. The upper and the lower limits for the free energy for radon difluoride have been deduced from the free energies of formation of halogen fluorides and arsenic fluorides [44, 45], which are used to oxidize radon. Since ClF_5, ClF, ClF_3, IF_7, BrF_5, and BrF_3 oxidize radon but not AsF_5 or IF_5, and since BrF_3 is last in the series to oxidize the radon, the free energy for the formation of the radon difluoride in solution must be between the ΔG's for the reduction of BrF_3 to BrF and AsF_5 to AsF_3. Thus the standard free energy of formation of the dissolved radon compound must be between -29 and -51 kcal/mole.

5. ANHYDROUS SOLUTIONS

Most of the noble gas fluorides are soluble in anhydrous hydrogen fluoride, HF. Although xenon difluoride is more soluble than xenon tetrafluoride, the two solutions are colorless and nonconducting [9]. However, xenon hexafluoride is ionized and a 0.02 M solution has a molar conductivity of 150 cm^2 ohm^{-1} $mole^{-1}$ at 0°C, corresponding to a degree of ionization possibly as high as 50%. Anhydrous hydrogen fluoride as a solvent has not been used in electrochemical studies, and yet it appears as the most promising solvent to study the stepwise reductions of noble gas fluorides.

Xenon difluoride dissolves in acetonitrile and can be kept at -10°C without decomposition [46]. The solutions are nonconducting.

Other solvents such as halogen fluorides (IF_5, BrF_3, IF_5-BrF_3, etc. [47-52]), fluorosulfuric acid, and fluorotelluric acid [53] have been used in preparative or solution studies. However, the electrochemical behavior of noble gas fluorides has not been investigated.

Tertiary butyl alcohol has been used as a solvent for the solid xenon trioxide [54]. In this solvent, xenon trioxide behaves as a weak acid comparable to acetic acid.

REFERENCES

1. B. Weinstock, E. E. Weaver, and C. P. Knop, Inorg. Chem., 5, 2189 (1966).
2. G. Stein and P. Plurien, in Noble Gas Compounds (H. H. Hyman, ed.), Univ. Chicago Press, Chicago, 1963, p. 147.
3. S. R. Gunn and S. M. Williamson, Science, 140, 177 (1963).
4. P. A. G. O'Hare, G. K. Johnson, and E. H. Appelman, Inorg. Chem., 9, 332 (1970).
5. V. I. Pepekin, Yu. A. Lebedev, and A. Ya. Apin, Zh. Fiz. Khim., 43, 1564 (1969).
6. A. V. Grosse, A. D. Kirshenbaum, A. G. Streng, and L. V. Streng, Science, 139, 1047 (1963).
7. F. Schreiner, J. G. Malm, and J. C. Hindman, J. Amer. Chem. Soc., 87, 25 (1965).
8. H. H. Hyman, ed., Noble Gas Compounds, Univ. Chicago Press, Chicago, 1963, pp. 106, 47.
9. H. H. Claassen, in The Noble Gases, Heath, Boston, 1966, pp. 61-63.
10. H. Selig, H. H. Claassen, C.L. Chernick, J. G. Malm, and J. L. Huston, Science, 143, 1322 (1964).
11. J. Jortner, E. G. Wilson, and S. A. Rice, J. Amer. Chem. Soc., 85, 814 (1963).
12. H. J. Svec and G. D. Flesch, Science, 142, 954 (1963).
13. K. S. Pitzer, ibid., 139, 414 (1963).
14. E. H. Appelman, Inorg. Chem., 6, 1268 (1967).
15. I. Feher and M. Semptey, Magy. Kem. Foly., 76, 954 (1963).

16. W. V. Johnston, D. Pilipovich, and D. E. Sheehan, in Noble Gas Compounds (H. H. Hyman, ed.), Univ. Chicago Press, Chicago, 1963, pp. 139-143.

17. H. H. Claassen, in The Noble Gases, Heath, Boston, 1966, p. 51.

18. F. Schreiner, D. W. Osborne, J. G. Malm, and G. N. McDonald, J. Chem. Phys., 51, 4838 (1969).

18a. S. A. Kudchadker, and A. P. Kudchadker, Proc. Indian Acad. Sci., Sect. A, 73(5), 261 (1971).

19. S. R. Gunn, in Noble Gas Compounds (H. H. Hyman, ed.), Univ. Chicago Press, Chicago, 1963, p. 149.

20. R. E. Powell and W. M. Latimer, J. Chem. Phys., 19, 1139 (1951).

21. G. Narajan, Bull. Soc. Chim. Belges, 73, 665 (1964).

22. D. H. Templeton, A. Zalkin, J. D. Forrester, and S. M. Williamson, in Noble Gas Compounds (H. H. Hyman, ed.), Univ. Chicago Press, Chicago, 1963, p. 229.

23. D. D. Wagman, W. H. Evans, V. B. Parker, I. Halow, S. M. Bailey, and R. H. Schumm, National Bureau of Standards Technical Note 270-3, U.S. Govt. Printing Office, Washington, D.C., 1968

24. D. F. Smith, J. Amer. Chem. Soc., 85, 816 (1963).

25. S. M. Williamson and C. W. Koch, Science, 139, 1046 (1963).

26. C. W. Koch and S. M. Williamson, in Noble Gas Compounds (H. H. Hyman, ed.), Univ. Chicago Press, Chicago, 1963, p. 181.

27. F. B. Dudley, G. Gard, and G. H. Cady, Inorg. Chem., 2, 228 (1963).

28. E. H. Appelman and J. G. Malm, J. Amer. Chem. Soc., 86, 2141 (1964).

29. J. M. Cleveland, Inorg. Chem., 6, 1302 (1967).

30. J. M. Cleveland, J. Amer. Chem. Soc., 87, 1816 (1965).

31. J. M. Cleveland and G. J. Werkema, Nature, 215, 732 (1967).

32. R. H. Krueger, J. P. Warriner, and B. Jaselskis, Talanta, 15, 741 (1968).

33. R. H. Krueger, Ph.D. Thesis, Loyola University, Chicago, 1967.

34. B. Jaselskis and S. Vas, J. Amer. Chem. Soc., 86, 2078 (1964).

35. B. Jaselskis and J. P. Warriner, Anal. Chem., 38, 563 (1966).

36. H. J. Rhodes and M. I. Blake, J. Pharm. Sci., 56, 1352 (1967).

37. H. J. Rhodes, R. P. Shian, and M. I. Blake, ibid., 57, 1706 (1968).

38. B. Jaselskis and R. H. Krueger, Talanta, 13, 945 (1966).

39. S. R. Gunn, J. Amer. Chem. Soc., 87, 2290 (1965).

40. E. H. Appelman and J. G. Malm, ibid., 86, 2141 (1964).

41. B. Jaselskis, Science, 146, 263 (1964).

42. B. Jaselskis, ibid., 143, 1324 (1964).

43. P. R. Fields, L. Stein, and M. H. Zirin, J. Amer. Chem. Soc., 84, 4164 (1962).

44. L. Stein, in Halogen Chemistry, Vol. 1 (V. Gutman, ed.), Academic, London, 1967, pp. 133-224.

45. P. A. G. O'Hare and W. N. Hubbard, J. Phys. Chem., 69, 4358 (1965).

46. H. Meinert and S. Ruediger, Z. Chem., 7, 239 (1967).

47. H. Meinert, G. Kauschka, and S. Ruediger, Z. Chem., 7, 111 (1967).

48. A. V. Nikolaev, A. A. Opalskii, A. S. Nazarov, and G. V. Tretyakov, Dokl. Akad. Nauk SSSR, 191(3), 629 (1970).

49. A. V. Nikolaev, A. A. Opalskii, A. S. Nazarov, and G. V. Tretyakov, ibid., 189, 1025 (1969).

50. L. Stein, J. Amer. Chem. Soc., 91, 5396 (1969).

51. D. Martin, C. R. Acad. Sci. Paris, Ser. C, 268(12), 1145 (1969).

52. F. O. Sladky, P. A. Bulliner, and N. Bartlett, J. Chem. Soc., A14, 1969, 2179.

53. F. O. Sladky, Angew. Chem. Int. Ed. Eng., 8(5), 373 (1969).

54. B. Jaselskis and J. P. Warriner, J. Amer. Chem. Soc., 91, 210 (1968).

SUBJECT INDEX

C

E

F

G

H

K

L

R